YOUR KNOWLEDGE I

- We will publish your bachelor's and master's thesis, essays and papers

- Your own eBook and book - sold worldwide in all relevant shops

- Earn money with each sale

Upload your text at www.GRIN.com and publish for free

GRIN

Bibliographic information published by the German National Library:

The German National Library lists this publication in the National Bibliography; detailed bibliographic data are available on the Internet at http://dnb.dnb.de .

Imprint:

Print and binding: Books on Demand GmbH, Norderstedt Germany
ISBN: 9783668999053

This book at GRIN:

https://www.grin.com/document/492165

Alauddin Khan

Quantum Mechanics. Basic Concepts, Mathematical Structure and Applications

GRIN Verlag

QUANTUM MECHANICS

Muhammad Alauddin Khan
M.Sc., PhD

- **Brief description of the book**
- **PREFACE**
- **Acknowledgements**
- ***Dedication***
- **Contents**

Brief description of the book

This book has been written with the notion that a wave is associated with a material particle i.e. waves and particles coexist. Heisenberg's uncertainty principle has been described taking this into account. The book consists of a total of 17 chapters. The two initial chapters discuss the development and basic concepts of Quantum Mechanics. The third chapter is dedicated to the mathematical structure of Quantum Mechanics. The fourth deals with Matrix formulation after which the fifth chapter discusses applications to one-dimensional [1D] problems. The sixth chapter is on Quantum Mechanics of Linear Harmonic Oscillator. Discussion on Atomic Orbitals of a hydrogen atom and a hydrogen atom of Quantum Mechanics are treated in two separate chapters namely the seventh and eighth chapters. Orbital and general angular momentums are treated in two separate chapters. Among numerous other topics Matrix formulation of Quantum Mechanics, Quantum theory of scattering, Quantum dynamics,[Three pictures of time development]Dirac's Relativistic Quantum Mechanics, Born approximation, Time dependent and Time independent perturbation theory[both generate and non-degenerate case],Variational method, and WKB approximation method have also been discussed.

PREFACE

In bygone days, Physics was often referred to as "Natural Philosophy". Physics has been the product of the overarching desire of the human mind to understand and explain the diverse phenomenon of the natural world. Quantum Mechanics is a young and vigorous subject and only through restless enquiry can the depth of understanding be enriched. This volume has a threefold purpose: [i] to explain the physical concepts of quantum mechanics, [ii] to describe the mathematical formalism, and [iii] to present illustrative examples of both the ideas and the methods. The book is intended as a text book on Quantum Mechanics for undergraduate levels and Masters Levels and also as a reference book for anyone who is interested in this field of enquiry. This book is based on my lecture plans for courses on Quantum Mechanics that I have been conducting for the last 25 years. For the lack of a book on Quantum Mechanics, students were facing serious obstacles in their learning process. This book has been written in the form of lecture plans and hence teachers of courses on Quantum Mechanics can use the book as their own lecture plans without any modification. It is to be noted that the purpose of this book is to cover the basic principles and methods of Quantum Mechanics which are usually included in the course of teaching Physics at the undergraduate levels and Masters levels. I hope that this book will be useful to the students and teachers in the different universities.

Muhammad Alauddin Khan
M.Sc., PhD

Acknowledgements

I firstly express all of my admiration and devotion to the almighty Allah, the most beneficial and merciful who has enabled me to write this book. I am obliged to my respected teacher Professor Dr. M. Shamsher Ali, Department of Physics, University of Dhaka, Bangladesh, who taught me Quantum Mechanics at the Graduate and Masters Level. I am deeply grateful to my Ph.D Supervisor Professor Dr. M. Habibul Ahsan, Department of Physics, Shahjalal University of Science & Technology for his kind suggestions. I remain deeply indebted to my Ph.D Co-Supervisor Professor Dr. H. Suzuki Department of Physics, Kanazawa University, Kakuma-machi, Kanazawa, 920-1192, Japan, for his constant suggestions, guidance and patience at every step of my work. I am extremely grateful for the moral support and constant inspiration of my dear wife Professor Dr. Shamsun Naher Begum, Head, Department of Physics, Shahjalal University of Science and Technology. She continuously encouraged me in every step throughout the work. The dissertation would have never been possible without her sincere help. I like to thank Professor Ahmed Hossain who encouraged me to teach this course, when I joined MC (MurariChand) College [Under National University] in August 1996. Discussions with my friend Professor Dr. M. Arshad Momen, Professor in Theoretical Physics Department, Dhaka University, Bangladesh have enriched my thoughts.

I am also thankful to my colleague Professor Shafiul Alam, Department of English, Murari Chand (M.C.) College, who helped me in clarifying my own ideas. I have completed this dissertation for my beloved son ***ASIF SHADMAN KHAN,*** Grade Eight student, in Cambridge curriculum who has influenced me and given much more patience in every step of my work. I also like to thank my former student Borhanul Alom, Senior Scientific Officer [S.S.O.], Atomic Energy Commission, Dhaka Center, Dhaka, Bangladesh who helped me to type the book. I would like to thank Jahirul Islam, Masters level student Department of English, Shahjalal University of Science and Technology, and my dear wife Professor Dr. Shamsun Naher Begum, Head, Department of Physics, Shahjalal University of Science and Technology. They checked my whole manuscripts and helped me in the proof reading. I would also like to thank my Undergraduate level student Mir Saifur Rahman, Department of Mathematics, MurariChand Collge [M.C.], Sylhet, and S. Afroz Tuhin Department of Mathematics, P.C.[Profullo Chandra] College, Bagerhat, Bangladesh who helped me to type the book and in the proof reading.

Dedicated

To

My wife

Prof. Dr. Shamsun Naher Begum

&

My beloved son

Asif Shadman Khan

QUANTUM MECHANICS

483 pages and 105 figures

Muhammad Alauddin Khan M.Sc., PhD

Department of Physics,

M.C. [MurariChand] College, [Under National University, Bangladesh]

Sylhet-3114

Bangladesh

[This book to be published as ISBN]

FOR SALE IN INDIA, PAKISTAN, BANGLADESH, SRILANKA, BHUTAN, NEPAL MALAYSIA ONLY

Chapter 1 Development of Quantum Mechanics

1.1 Quantum Mechanics: an important intellectual achievement of the 20th century

Quantum mechanics is an important intellectual achievement of the 20th century. It is one of the most sophisticated fields in physics that has influenced our understanding of nano-meter length scale systems important for chemistry, materials, optics, electronics, and quantum information. The existence of orbitals and energy levels in atoms can only be explained by quantum mechanics. Quantum mechanics can explain the behaviors of insulators, conductors, semi-conductors, and giant magneto-resistance. It can explain the quantization of light and its particle nature in addition to its wave nature (known as particle-wave duality). Quantum mechanics can also explain the radiation of hot body or black body, and its change of color with respect to temperature. It explains the presence of holes and the transport of holes and electrons in electronic devices.

Quantum mechanics is important in order to illustrate photonics, quantum electronics, nano and micro-electronics, nano- and quantum optics, quantum computing, quantum communication and crytography, solar and thermo-electricity, nano-electromechacnical systems, etc. Many emerging technologies require the application of quantum mechanics; and hence, it is essential that scientists and engineers have to have the solid understanding of quantum mechanics better. Due to the recent advent of nano-fabrication techniques; nano-meter size systems become more and more common in the field of nano-technology. In electronics, as transistor devices become smaller, the behavior of electronics in the device is quite different than that of larger device: nano-electronic transport is quite different from micro-electronic transport.

The quantization of electromagnetic field is important in the area of nano-optics and quantum optics. It explains how photons interact with atomic systems or materials. It also allows the use of electromagnetic or optical field to carry quantum information. Quantum mechanics is certainly giving rise to interest in quantum information, quantum communication, quantum cryptography, and quantum computing. Moreover, quantum mechanics is also needed to understand the interaction of photons with materials in solar cells, as well as many topics in material science.

When two objects are placed close together, they experience a force called the Casimir force that can only be explained by quantum mechanics. Moreover, the understanding of spins is important in spintronics, another emerging technology where giant magneto-resistance, tunneling magneto-resistance, and spin transfer torque are being used. It is obvious that the richness of quantum physics will greatly affect the future generation technologies in many aspects.

1.2 Quantum Mechanics is Insubstantial

The development of quantum mechanics is a great intellectual achievement, but at the same time, it is bizarre too. The reason is that quantum mechanics is quite different from classical physics. The development of quantum mechanics is likened to watching two players having a game of chess, but the observers have not a clue as to what the rules of the game are. By observations, and conjectures, finally the rules of the game are outlined. Often, equations are conjectured like conjurors pulling tricks out of a hat to match experimental observations. It is the interpretations of these equations that can be quite bizarre.

Quantum mechanics equations were postulated to explain experimental observations, but the deeper meanings of the equations often confused even the most gifted. Even though Einstein received the Nobel Prize for his work on the photo-electric effect that confirmed that light energy is quantized, he himself was not totally at ease with the development of quantum mechanics as charted by the younger physicists. He was never comfortable with the probabilistic interpretation of quantum mechanics by Born and the Heisenberg uncertainty principle. "God doesn't play dice," was his statement assailing the probabilistic interpretation. He proposed "hidden variables" to explain the random nature of many experimental observations. He was thought of as the "old fool" by the younger physicists during his time. Schrödinger came up with the bizarre "Schrödinger cat paradox" that showed the struggle that physicists had with quantum mechanics interpretation. But with today's understanding of quantum mechanics, the paradox is a thing of yester-year.
The latest twist to the interpretation in quantum mechanics is the parallel universe view that explains the multitude outcomes of the prediction of quantum mechanics. All outcomes are possible, but with each outcome occurring in different universes that exist in parallel with respect to each other.

1.3 Classical Mechanics and Some Mathematical Preliminaries

Quantum mechanics cannot be derived from classical mechanics, but classical mechanics can inspire quantum mechanics. Quantum mechanics is richer and more sophisticated than classical mechanics. *Quantum mechanics was developed during the period when physicists had rich knowledge of classical mechanics.* The fundamental concepts of classical mechanics are prerequisite in terms of understanding the development of Quantum Mechanics. *Classical mechanics can be considered as a special case of quantum mechanics.* We will review some classical mechanics concepts here[18]. In classical mechanics, a particle moving in the presence of potential $V(q)$ will experience a force given by

$$F(q) = -\frac{dV(q)}{dq} \quad [1]$$

Where, q represents the coordinate or the position of the particle. Hence, the particle can be described by the equations of motion

$$\frac{dp}{dt} = F(q) = -\frac{dV(q)}{dq}, \frac{dq}{dt} = \frac{p}{m} \quad [2]$$

For example, when a particle is attached to a spring and moves along a frictionless surface, the force of the particle experiences is $F(q) = kq$ where k is the spring constant. Then the equations of motion of this particle are

$$\frac{dp}{dt} = \dot{p} == kq, \frac{dq}{dt} = \dot{q} = \frac{p}{m} \quad [3]$$

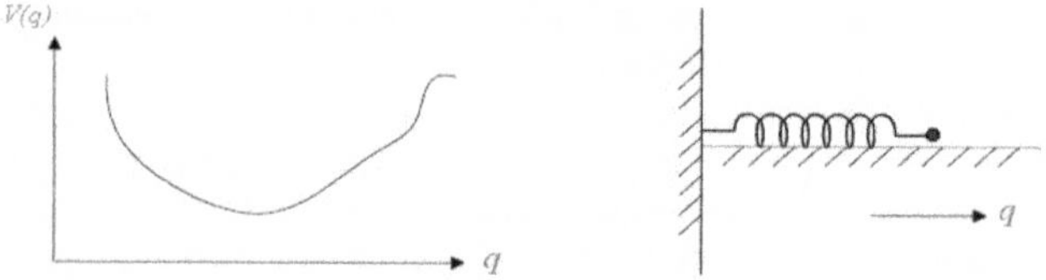

Fig.[1.3]: The left side shows a potential well in which a particle can be trapped. The right side shows a particle attached to a spring. The particle is subject to the force due to the spring, but it can also be described by the force due to a potential well.

Given p and q at some initial time t_0, one can integrate [2] or [3] to obtain p and q for all later time. A numerical analyst can think of that [2] or [3] can be solved by the finite difference method, where time-stepping can be used to find p and q for all later times. For instance, we can write the equations of motion more compactly as

$$\frac{du}{dt} = f(u) \quad [4]$$

Where, $u = [p, q]^t$ and f is a general vector function of u. It can be non-linear or linear; in the event, if it is linear, then $f(u) = \bar{A}.u$

Using finite difference approximation, we can rewrite the above as

$$u(t + \Delta t) - u(t) = \Delta t f\big(u(t)\big) \quad or, u(t + \Delta t) = \Delta t f\big(u(t)\big) + u(t)$$

The above can be used for time marching to derive the future values of u from past values. *The above equations of motion are essentially derived using Newton's law.* However, there exist other methods of deriving these equations of motion. Notice that only two variables p and q are sufficient to describe the state of a particle.

❖ Lagrangian Formulation[18]

Another way to derive the equations of motion for classical mechanics is via the use of the Lagrangian and the principle of least action[18]. A Lagrangian is usually defined as the difference between the kinetic energy and the potential energy, i.e.,

$$L(\dot{q}, q) = T - V \qquad [1]$$

Where, $\dot{q}$ is the velocity. For a fixed *t*, q and $\dot{q}$ are independent variables, since $\dot{q}$ cannot be derived from q if it is only known at one given *t*. The equations of motion are derived from the principle of least action which says that $q(t)$ that satisfies the equations of motion between two times t_1 and t_2 should minimize the action integral

$$S = \int_{t_1}^{t_2} L(\dot{q}(t)q(t))dt \qquad [2]$$

Assuming that $q(t_1)$ and $q(t_2)$ are fixed then the function $q(t)$ between t_1 and t_2 should minimize *S*, the action. In other words, a first order perturbation in q from the optimal answer that minimizes S should give rise to second order error in *S*. Hence, taking the first variation of [2], we have

$$\delta S = \delta \int_{t_1}^{t_2} L(\dot{q}, q)dt = \int_{t_1}^{t_2} L(\dot{q} + \delta\dot{q}\,,\ q + \delta q)dt - \int_{t_1}^{t_2} L(\dot{q}, q)dt$$

$$= \int_{t_1}^{t_2} \delta L(\dot{q}, q)dt = \int_{t_1}^{t_2} \left(\delta\dot{q}\frac{\partial L}{\partial \dot{q}} + \delta q\frac{\partial L}{\partial q}\right)dt = 0 \qquad [3]$$

In order to take the variation into the integrand, we have to assume that $\delta L(\dot{q}, q)$is taken with constant time. At constant time, q and $\dot{q}$ are independent variables; hence the partial derivatives in the next equality above follow. Using integration by parts on the rest term, we have

$$\delta S = \delta q\left[\frac{\partial L}{\partial \dot{q}}\right]_{t_1}^{t_2} - \int_{t_1}^{t_2} \delta q\frac{d}{dt}\left(\frac{\partial L}{\partial \dot{q}}\right)dt + \int_{t_1}^{t_2} \delta q\frac{\partial L}{\partial q}dt$$

$$= \int_{t_1}^{t_2} \delta q\left[-\frac{d}{dt}\left(\frac{\partial L}{\partial \dot{q}}\right)dt + \frac{\partial L}{\partial q}\right]dt = 0 \qquad [4]$$

The first term vanishes because $\delta q(t_1) = \delta q(t_2) = 0$ because $q(t_1)$ and $q(t_2)$ are fixed. Since $\delta q(t)$ is arbitrary between t_1 and t_2, we must have[18]

$$\boxed{\frac{d}{dt}\left(\frac{\partial L}{\partial \dot{q}}\right) dt + \frac{\partial L}{\partial q} = 0} \qquad [5]$$

The above is called the Lagrange equation[18], from which the equation of motion of a particle can be derived. The derivative of the Lagrangian with respect to the velocity $\dot{q}$ is the momentum

$$p = \frac{\partial L}{\partial \dot{q}} \qquad [6]$$

The derivative of the Lagrangian with respect to the coordinate q is the force

$$F = \frac{\partial L}{\partial q} \qquad [7]$$

The above equation of motion is then

$$\dot{p} = F \qquad [8]$$

Equation [6] can be inverted to express $\dot{q}$ as a function of p *and* q namely

$$\dot{q} = f(p, q) \qquad [9]$$

Equations [8] and [9] can be solved in tandem to find the time evolution of p *and* q.
For example, the kinetic energy T of a particle is given by

$$T = \frac{1}{2} m\dot{q}^2 \qquad [10]$$

Then from equation [1], and the fact that V is independent of $\dot{q}$,

$$p = \frac{\partial L}{\partial \dot{q}} = \frac{\partial T}{\partial \dot{q}} = m\,\dot{q} \qquad [11]$$

$$\dot{q} = \frac{p}{m} \qquad [12]$$

Also from equation [1], [7] and [8]

$$\dot{p} = F = -\frac{dV}{dq} \qquad [13]$$

The above equations [12] and [13], form *the equations of motion* for this problem. The above can be generalized to multi-dimensional problems. For example, for a one particle system in three dimensions, q_i has three degrees of freedom, and $i = 1, 2, 3$.[The q_i can represent x, y, z in Cartesian coordinates, but r, θ, ϕ in spherical coordinates] But for N particles in three dimensions, there are 3N degrees of freedom, and $i = 1, 2, 3, \ldots \ldots \ldots \ldots, 3N$. The formulation can also be applied to particles constraint in motion. For instance, for N particles in three dimensions, q_i may run from $i = 1, 2, 3, \ldots \ldots \ldots \ldots, 3N - k$, representing k constraints on the motion of the particles. This can happen, for example, if the particles are constraint to move in a mainfold [surface], or a line [ring] embedded in a three dimensional space. Going through similar derivation, we arrive at the equation of motion

$$\frac{d}{dt}\left(\frac{\partial L}{\partial \dot{q}_i}\right) dt + \frac{\partial L}{\partial q_i} = 0 \qquad [14]$$

In general, q_i may not have a dimension of length and it is called the generalized coordinate [Also called conjugate coordinate]. Also, $\dot{q}_i$ may not have a dimension of velocity and it is generalized velocity. The derivative of the Lagrangian with respect to the generalized velocity is the generalized momentum [Also called conjugate momentum], namely[18]

$$p_i = \frac{\partial L}{\partial \dot{q}_i} \qquad [15]$$

The generalized momentum may not have a dimension of momentum. Hence, the equation of motion [14] can be written as

$$\dot{p}_i = \frac{\partial L}{\partial q_i} \qquad [16]$$

Equation [15] can be inverted to yield an equation for $\dot{q}$ as a function of the other variables. This equation can be used in tandem [16] as time-marching equations of motion.

❖ Hamiltonian Formulation[18]

For multi-dimensional system, or a many particle system in multi-dimensions, the total time derivative of L is

$$\frac{dL}{dt} = \sum_i \left(\frac{\partial L}{\partial q_i}\dot{q}_i + \frac{\partial L}{\partial \dot{q}_i}\ddot{q}_i\right) \qquad [1]$$

Since $\frac{\partial L}{\partial q_i} = \frac{d}{dt}\left(\frac{\partial L}{\partial \dot{q}_i}\right)$ from the Lagrangian equation, we have

$$\frac{dL}{dt} = \sum_i \left(\frac{d}{dt}\left(\frac{\partial L}{\partial \dot{q}_i}\right)\dot{q}_i + \frac{\partial L}{\partial \dot{q}_i}\ddot{q}_i\right) = \frac{d}{dt}\sum_i \frac{d}{dt}\left(\frac{\partial L}{\partial \dot{q}_i}\right)\dot{q}_i \qquad [2]$$

$$\frac{d}{dt}\sum_i \left(\frac{\partial L}{\partial \dot{q}_i}\dot{q}_i - \mathrm{L}\right) = 0 \qquad [3]$$

The quantity $H = \sum_i \left(\frac{\partial L}{\partial \dot{q}_i}\dot{q}_i - \mathrm{L}\right)$ [is known as the Hamiltonian of the system] [4]

And is a constant of motion, namely $\frac{dH}{dt} = 0$

As shall be shown, the Hamiltonian represents the total energy of the system. It is a constant of motion because of the conversation of energy. The Hamiltonian of the system can also be written, $H = \sum_i p_i \dot{q}_i - \mathrm{L}$ [5]

Where $p_i = \frac{\partial L}{\partial \dot{q}_i}$ is the generalized momentum[18]. The first term has a dimension of energy, and in Cartesian coordinates, for a simple particle motion, it is easily seen that it is twice the kinetic energy. Hence, the above indicates that the Hamiltonian[18]

$$H = T + V \qquad [6]$$

The total variation of the Hamiltonian is

$$\delta H = \delta(\textstyle\sum_i p_i \dot{q}_i) - \delta \mathrm{L}$$

$$\delta H = (\textstyle\sum_i \dot{q}_i \delta p_i + p_i \delta \dot{q}_i) - \sum_i \left(\frac{\partial L}{\partial q_i} \delta q_i + \frac{\partial L}{\partial \dot{q}_i} \delta \dot{q}_i\right) \qquad [7]$$

From previous section, using [15] and [16], we have

$$\delta H = \textstyle\sum_i (\dot{q}_i \delta p_i + p_i \delta \dot{q}_i) - \sum_i (\dot{p}_i \delta q_i + p_i \delta \dot{q}_i)$$

$$\delta H = \textstyle\sum_i (\dot{q}_i \delta p_i - \dot{p}_i \delta q_i) \qquad [8]$$

From the above, since the first variation of the Hamiltonian depends only on δp_i and δq_i, we gather that the Hamiltonian is a function of p_i and q_i. Taking the first variation of the Hamiltonian with respect to these variables, we arrive at another expression for its first variation, namely,

$$\delta H = \textstyle\sum_i \left(\frac{\partial H}{\partial p_i} \delta p_i + \frac{\partial H}{\partial q_i} \delta q_i\right) \qquad [9]$$

Comparing the above with [8], we gather that

$$\boxed{\dot{q}_i = \frac{\partial H}{\partial p_i}} \qquad [10]$$

$$\boxed{\dot{p}_i = -\frac{\partial H}{\partial q_i}} \qquad [11]$$

These are the equations of motion known as the Hamiltonian equations[18].

The [4] is also known as the Legendre transformation. The original function L is a function of $\dot{q}_i, q_i$. Hence, δL depends on both $\delta \dot{q}_i$ and δq_i. After the Legendre transformation, δH depends on the differential δp_i and δq_i as indicated by [8]. This implies that H [Hamiltonian] is a function of p_i and q_i. The equation of motion then can be written as in [10] and [11].

❖ More on Hamiltonian[18]

The Hamiltonian of a particle in Classical Mechanics is given by $H = T + V$, and it is a function of p_i and q_i. For a non- relativistic particle in three dimensions, the kinetic energy

$$T = \frac{\bar{p}.\bar{p}}{2m} \qquad [1]$$

And the potential energy is a function of $\boldsymbol{q}$. Hence the Hamiltonian in three dimensions can be expressed as

$$H = \frac{\bar{p}.\bar{p}}{2m} + V(\boldsymbol{q}) \qquad [2]$$

When the electromagnetic field is present, the Hamiltonian for an electron can be derived by letting the generalized momentum

$$p_i = m\dot{q}_i + eA_i \qquad [3]$$

Where $e = -|e|$ is the electron charge and A_i is the component of the vector potential $\bar{A}$. Consequently, the Hamiltonian of an electron in the presence of an electromagnetic field is

$$H = \frac{(\bar{p}-e\bar{A}).(\bar{p}-e\bar{A})}{2m} + e\varphi(\boldsymbol{q}) \qquad [4]$$

The equation of motion of an electron is an electromagnetic is governed by the Lorentz force law, which can be derived from the above Hamiltonian using the equations of motion provided by

$$\boxed{\dot{q}_i = \frac{\partial H}{\partial p_i} \text{ and } \dot{p}_i = -\frac{\partial H}{\partial q_i}}$$

❖ Poisson Bracket[18]

Yet another way of expressing equations of motions in Classical Mechanics is via the use of Poisson bracket. *This is interesting because Poisson bracket has a close Quantum Mechanics analogue.* A Poisson bracket of two scalar variables *u and v* that are functions of *q and p* is defined as

$$\{u, v\} = \frac{\partial u}{\partial q}\frac{\partial v}{\partial p} - \frac{\partial v}{\partial q}\frac{\partial u}{\partial p} \qquad [1]$$

In this notation, using $\dot{q}_i = \frac{\partial H}{\partial p_i}$ and $\dot{p}_i = -\frac{\partial H}{\partial q_i}$

$$\frac{du}{dt} = \frac{\partial u}{\partial q}\frac{\partial q}{dt} - \frac{\partial u}{\partial p}\frac{dp}{dt} = \frac{\partial u}{\partial q}\frac{\partial H}{\partial p} - \frac{\partial u}{\partial p}\frac{\partial H}{\partial q} \qquad [2]$$

which is valid for any variable u that is a function of *p and q*. Hence, we have the equations of motion as in the Poisson bracket notation.

$$\dot{q} = \{q, H\}, \;\; \dot{p} = \{p, H\} \qquad [3]$$

As we shall see later, similar equations will appear in Quantum Mechanics. The algebraic properties of Poisson bracket are

Or, $$\{u, v\} = -\{v, u\} \qquad [4]$$

Or, $\{u + v, \omega\} = \{u, \omega\} + \{v, \omega\}$ [5]

Or, $\{uv, \omega\} = \{u, \omega\}v + u\{v + \omega\}$ [6]

Or, $\{u, v\omega\} = \{u, v\}\omega + v\{u, \omega\}$ [7]

Or, $\{\{u, v\}, \omega\} + \{\{v, \omega\}, u\} + \{\{\omega, u\}, v\}$ [8]

These properties are anti-symmetry, distributive, associative and Jacobi's identity. If we define a commutator operation between two non commuting operator $\hat{u}$ *and* $\hat{v}$ as

$$[\hat{u}\,, \hat{v}] = (\hat{u}\,\hat{v} - \hat{v}\hat{u}) \quad [9]$$

It can be shown that the above commutators have the same algebraic properties as the Poisson bracket. *An operator in Quantum Mechanics can be a matrix operator or a differential operator. In general, operators do not commute unless under very special circumstances.*

❖ Some useful knowledge of Matrix Algebra[20,21]

Matrix algebra (or linear algebra) forms the backbone of many quantum mechanical concepts. Hence, it is prudent to review some useful knowledge of matrix algebra. Many of the mathematical manipulations in quantum mechanics can be better understood if we understand matrix algebra. A matrix is a mathematical linear operator that when operate (also called "act") on a vector produces another vector, or

$$\boldsymbol{b} = \bar{A}.\boldsymbol{a} \quad [1]$$

Where, **a** and **b** are distinct vectors, and $\bar{A}$ is a matrix operator other than the identity operator. The inner product between two vectors can be of the form of reaction inner product

$$\boldsymbol{v}^{transpose[t]}.\boldsymbol{\omega} \quad [2]$$

Or, energy inner product

$$\boldsymbol{v}^{daggar[\text{f}]}.\boldsymbol{\omega} \quad [3]$$

Where $daggar[\text{f}]$ implies conjugation transpose or that $\boldsymbol{v}^{daggar[\text{f}]} = (v^{\times})^{daggar[\text{f}]}$. For a finite dimensional matrix and vectors, the above can be written as

$$b_j = \sum_{i=1}^{N} A_{ji} a_i \quad [4]$$

$$\boldsymbol{v}^{transpose[t]}.\boldsymbol{\omega} = \sum_{i=1}^{N} v_i \omega_i \quad [5]$$

$$\boldsymbol{v}^{daggar[\text{f}]}.\boldsymbol{\omega} = \sum_{i=1}^{N} v_i^{\times} \omega_i \quad [6]$$

The above are sometimes written with the summation sign removed, namely, as

$$b_j = A_{ji} a_i \quad [7]$$

$$\boldsymbol{v}^{transpose[t]}.\boldsymbol{\omega} = v_i\omega_i \qquad [8]$$

$$\boldsymbol{v}^{daggar[f]}.\boldsymbol{\omega} = v_i^{\times}\omega_i \qquad [9]$$

The summation is implied whenever repeated indices occur. *This is known varyingly as the index notation, indicial notation, or Einstein notation.* The above concepts can be extended to infinite dimensional system by letting $N \rightarrow \infty$. It is quite clear that

$$\boldsymbol{v}^{daggar[f]}.\boldsymbol{v} = \sum_{i=1}^{N} v_i^{\times}v_i = \sum_{i=1}^{N}|v_i|^2 > 0 \qquad [10]$$

Or, $\boldsymbol{v}^{daggar[f]}.\boldsymbol{v}$ is positive definite.

Furthermore, matrix operators satisfy associativity but not commutativity, namely

$$(\bar{A}.\bar{B}).\bar{C} = \bar{A}.(\bar{B}.\bar{C}) \qquad [11]$$

$$\bar{A}.\bar{B} \neq \bar{B}.\bar{A} \qquad [12]$$

❖ Identity, Hermitian, Symmetric, Inverse and Unitary Matrices

For discrete, countable systems, the definition of the above is quite straightforward. The identity operator $\bar{I}$ is defined such that

$$\bar{I}.a = a \qquad [1]$$

or the ij element of the matrix is

$$[\bar{I}]_{ij} = \delta_{ij} \qquad [2]$$

or it is diagonal matrix with one on the diagonal. A Hermitian matrix A_{ij} has the property that

$$A_{ij} = A_{ij}^{\times} \qquad [3]$$

or,

$$\bar{A}^{daggar[f]} = \bar{A} \qquad [4]$$

A symmetric matrix A_{ij} is such that

$$A_{ij} = A_{ji} \qquad or,\ A^{transpose[t]} = \bar{A} \qquad [5]$$

The inverse of the matrix $\bar{A}$, denoted as $\bar{A}^{-1}$ and has the property

$$\bar{A}^{-1}.\bar{A} = \bar{I} \qquad [6]$$

So given the equation

$$\bar{A}.x = b \quad [7]$$

x can be found once $\bar{A}^{-1}$ is known. Multiplying the above by $\bar{A}^{-1}$, we have

$$x = \bar{A}^{-1}.b \quad [8]$$

A unitary matrix $\bar{U}$ has the property that

$$\bar{U}^{\,daggar[f]}\bar{U} = \bar{I} \quad [9]$$

In other words

$$\bar{U}^{\,daggar[f]} = \bar{U}^{-1} \quad [10]$$

❖ Formulation of Planck's radiation law: Planck's hypothesis[25]

According to Planck, the Quantum theory is applicable only to the process of emission and absorption of energy. In his theory he made the following two radical assumptions about the atomic oscillators:

- A simple harmonic oscillator cannot have any value of total energy. It can have only those values of the total energy E which satisfy the relation

$$E = nh\upsilon \quad [1]$$

Where, $n = 0, 1, 2, 3, \ldots\ldots\ldots\ldots\ldots\ldots\ldots\ldots\ldots\ldots\ldots\ldots\ldots\ldots$; n is called the quantum number. υ is the frequency of oscillation, h is the universal constant called Planck's constant $[h = 6.625 \times 10^{-34} Joule - Second]$. In equation [1], $h\upsilon$ is the basic unit of energy and is called the Quantum. Its value depends only on the frequency of radiation.

- The oscillator does not emit or absorb energy continuously. The emission or absorption of energy occurs only when the oscillator jumps from one energy level to another. The loss or gain of energy is in the form of discrete quanta each of energy $h\upsilon$.

So, each Planck oscillator does not hold any other energy levels but only can hold $0, \varepsilon, 2\varepsilon, 3\varepsilon, \ldots\ldots\ldots\ldots\ldots$; let, at absolute temperature T, for any system, the energy levels $0, \varepsilon, 2\varepsilon, 3\varepsilon, \ldots\ldots\ldots\ldots\ldots$ etc. of Planck oscillator have the number $n_0, n_1, n_2, \ldots\ldots\ldots\ldots$ etc. respectively. Therefore, the total number of the Planck oscillator

$$N = n_0 + n_1 + n_2 + \cdots \ldots\ldots\ldots\ldots\ldots\ldots\ldots\ldots\ldots\ldots.. = \sum_{j=0}^{\infty} n_j \quad [2]$$

And total energy $E = (0 \times n_0) + (\varepsilon \times n_1) + (2\varepsilon \times n_2) + \cdots \ldots\ldots\ldots\ldots\ldots\ldots\ldots\ldots\ldots\ldots..$

$$E = \varepsilon(0 + n_1 + 2n_2 + \cdots \ldots\ldots\ldots\ldots\ldots\ldots\ldots\ldots\ldots\ldots\ldots.. = \varepsilon \sum_{j=0}^{\infty} j n_j \qquad [3]$$

According to Maxwell's Boltzmann distribution law $n_j \propto e^{-\frac{j\varepsilon}{kT}}$

$$\therefore \qquad n_j = n_0\, e^{-\frac{j\varepsilon}{kT}} = n_0\, e^{-jx} \qquad [4]$$

Here $x = \frac{\varepsilon}{kT}$; k is the Boltzmann constant and n_0 = constant = Number of particles with zero energy. Then the average energy for each Planck oscillator[25]

$$\bar{E} = \frac{E}{N} = \frac{\varepsilon \sum_{j=0}^{\infty} j n_j}{\sum_{j=0}^{\infty} n_j} = \frac{\varepsilon \sum_{j=0}^{\infty} j\, e^{-jx}}{\sum_{j=0}^{\infty} e^{-jx}} = -\varepsilon \frac{d}{dx}\left(\log \sum_{j=0}^{\infty} e^{-jx}\right)$$

$$= -\varepsilon \frac{d}{dx}\{\log(1 + e^{-x} + e^{-2x} + \cdots \ldots\ldots\ldots\ldots\ldots\ldots\ldots\ldots..)\}$$

$$= -\varepsilon \frac{d}{dx}\{\log(1 - e^{-x})^{-1}\} = \varepsilon \frac{e^{-x}}{1-e^{-x}} = \varepsilon \frac{1}{e^{x}-1} = \frac{\varepsilon}{e^{\varepsilon/kT}-1} = \frac{h\upsilon}{e^{h\upsilon/kT}-1} \qquad [5]$$

The number of oscillator per unit volume between the ranges υ *and* $\upsilon + d\upsilon$

$$= \frac{8\pi\upsilon^2}{c^3} d\upsilon . \frac{h\upsilon}{e^{h\upsilon/kT}-1} \qquad [6]$$

The energy density between the ranges υ *and* $\upsilon + d\upsilon$

$$u_\upsilon d\upsilon = \frac{8\pi\upsilon^2}{c^3} d\upsilon . \bar{E} = \frac{\mathbf{8\pi h\upsilon^3}}{\mathbf{c^3}} \frac{\mathbf{d\upsilon}}{\mathbf{e^{h\upsilon/kT}-1}} \qquad [7]$$

Equation [7] represents Planck's radiation law.
Again, if the wave length of radiation is λ then $\upsilon = \frac{c}{\lambda}$ $\quad \therefore |d\upsilon| = c d\lambda/\lambda^2$

From equation [7] $\quad u_\lambda d\lambda = \frac{\mathbf{8\pi hc}}{\mathbf{\lambda^5}} \frac{\mathbf{d\lambda}}{\mathbf{e^{hc/\lambda kT}-1}}$ [8]

Equation [7] represents another form of Planck's radiation law.

❖ Formulation of another radiation laws from Planck's radiation law[26]

- **Rayleigh- Jeans law:** For small values of $e^{hc/\lambda kT}$ I.e., in the region of long wave lengths

$$e^{hc/\lambda kT} \ll 1, \ so \quad e^{hc/\lambda kT} - 1 \approx hc/\lambda kT$$

In the region of long wave lengths equation [8] becomes $\quad u_\lambda d\lambda = \frac{8\pi kT}{\lambda^4} d\lambda$ [9]

- **Wien's law:** In the region of low wave lengths, $e^{hc/\lambda kT}$ becomes large. I.e.,

$$e^{hc/\lambda kT} \gg 1, \ so \quad e^{hc/\lambda kT} - 1 \approx \quad e^{hc/\lambda kT}$$

In the region of low wave lengths equation [8] becomes $u_\lambda d\lambda = \frac{8\pi hc}{\lambda^5} e^{-hc/\lambda kT} d\lambda$ [10]

- **Stefan- Boltzmann law:** The energy density of the total radiation in a black body enclosure is given by $u = \int_0^\infty u_\upsilon \, d\upsilon$

$$u = \frac{8\pi h}{c^3}\int_0^\infty \frac{\upsilon^3 d\upsilon}{e^{h\upsilon/kT}-1} = \frac{8\pi^5 k^4}{15c^3h^3}T^4 = \sigma T^4 \quad [11]$$

Where Stefan- Boltzmann constant $\sigma = \frac{8\pi^5 k^4}{15c^3h^3}$ [12]

1.4 Introduction to Quantum Mechanics: Preliminary remarks

In classical mechanics, Newton's second law (**F** = m**a**) has been used to make a mathematical prediction as to what path a given physical system will take over time following a set of known initial conditions. Solving this equation gives the position, and the momentum of the physical system as a function of the external force F on the system. Those two parameters are sufficient to describe its state at each time instant. Newton's laws of motion are the basis of the most elementary principles of Classical mechanics.

1. Classical Mechanics: (a) Newtonian: $F = m\vec{a} = m\frac{d^2\vec{r}}{dt^2}$

(b) Lagrangian
(c) Hamiltonian } Equivalent to Newtonian

Equations based on these laws are the simplest and they are suitable for the solution of simple dynamical problems such as the motion of macroscopic bodies. Lagrangian equations, Hamiltonian equations and Hamilton's principle are also the fundamental principles of Classical Mechanics, because they are consistent with each other and with Newton's law of motion. That's why both the Lagrangian and Hamiltonian are equivalent to Newtonian[24].

2. Classical electrodynamics (Maxwell's theory of Electrodynamics)
3. Einstein's theory of relativity

These three constitute what is called classical; otherwise we try to use quantum mechanics. The properties of bulk matter must be deducible from the properties of electrons and atomic nuclei of which they are composed. However, it is found that the observed properties of matter cannot be explained on the assumption that the particles obey the laws of Classical Mechanics. Classical Mechanics explained successfully the motion of the objects which are directly observable or observable with the help of instrument like microscope. When the objects such as electron in the atom are not observable even with the help of instruments, classical concepts cannot be applied. Therefore, classical concepts do not hold in the region of atomic dimensions. So, on the basis of above explanation, a new concept had to be developed. This concept led to a new mechanics called Quantum Mechanics. It was developed by Schrödinger, Heisenberg, Dirac and other scientists. At the present stage of human knowledge, Quantum Mechanics can be regarded as the fundamental theory of atomic phenomena.

The development of Quantum Mechanics took place in two stages[26]. The first stage began with Max Planck's presentation at a meeting of the German Physical Society on December 14, 1900. His hypothesis that Radiation is emitted or absorbed by matter in discrete packets or quanta each of energy $h\upsilon$, where υ is the frequency of radiation and h is the Planck's constant and $h = 6.623\times10^{-34}$ joule-second. The theory based on the hypothesis consisted of a mixture of classical and non-classical concepts.

The second stage is that Quantum Mechanics began from two different viewpoints in 1925. A particular form of this mechanics is called Matrix Mechanics which was introduced by Warner Heisenberg in 1925. In this mechanics, unobserved quantities such as positions and velocities etc. in electronic orbits are omitted. But only observed quantities such as frequencies and intensities of spectral lines are taken into account.

Another kind of Quantum Mechanics is called wave mechanics. The mathematical theory of this mechanics was developed by Erwin Schrödinger in 1926. In this mechanics, earlier ideas of classical wave theory are combined with Louis de-Broglie's Wave-Particle relationship. Finally, it has to be concluded that the mathematical theories of Wave Mechanics and Matrix Mechanics appear to be different but in fact, they are completely equivalent.

1.5 Brief Description of Quantum Mechanics: Preliminary remarks uniquely

Quantum mechanics is the extension of classical mechanics into the microscopic world, the world of atoms and molecules and of atomic nuclei and elementary particles. The need for a revision of the foundations of mechanics arises as a result of the wave-particle duality of matter, which manifests itself in systems of atomic dimensions. Wave-particle duality means that particles, such as electrons or protons, have properties characteristic of both waves and particles. Wave-particle duality was first found for electromagnetic radiation and later understood to be a general property of all matter. It all started with Max Planck's work on black body radiation.

Black Body Radiation:

In 1900 Max Planck found the formula that describes all data on black body radiation. To do so he had to demand that radiation can be absorbed and emitted only in *"Portions of Energy"* or quanta. *This novel concept was the birth of Quantum Mechanics*[22,2]

- In 1900 Max Planck found the formula that describes all data on black body radiation
- To do so he had to demand that radiation can be absorbed and emitted only in *"Portions of Energy"* or quanta. This novel concept was the birth of Quantum Mechanics
- This is Planck's formula

$$\rho(\upsilon,\mathrm{T})=\frac{8\pi h\upsilon^3}{c^3}\frac{1}{e^{h\upsilon/k\mathrm{T}}-1}$$

$$u(\upsilon)\,d\upsilon=\frac{8\pi h}{c^3}\frac{\upsilon^3 d\upsilon}{e^{h\upsilon/k\mathrm{T}}-1}$$

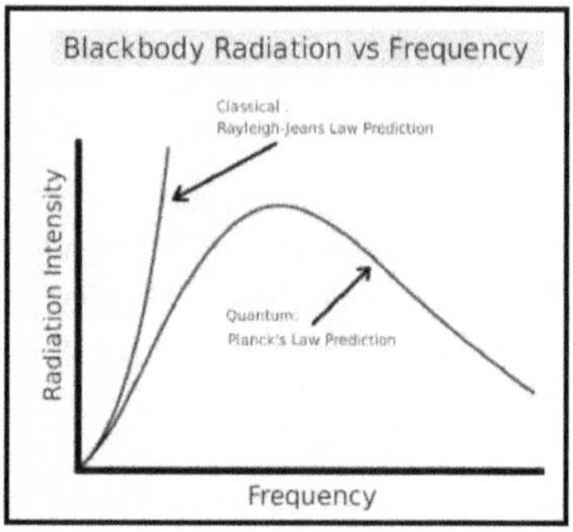

Fig.[1.5]: Shows Max Planck's black body radiation

This is Planck's formula

$$\rho(\upsilon,\mathrm{T})=\frac{8\pi h\upsilon^3}{c^3}\frac{1}{e^{h\upsilon/k\mathrm{T}}-1}$$

Or.

$$u(\upsilon)\,d\upsilon=\frac{8\pi h}{c^3}\frac{\upsilon^3 d\upsilon}{e^{h\upsilon/k\mathrm{T}}-1}$$

Planck's idea was further developed by Einstein. First, in 1905, he applied it to explain the photoelectric effect: he showed that the kinetic energy of the electrons emitted from a metal surface illuminated with light of frequency v is equal to

$$\mathrm{E}=h\upsilon-W$$

Where h is the Planck constant and W is a constant specific for the metal (this constant is called the work function of the metal). Later Einstein repeatedly returned to the question of the interaction of radiation with matter. He himself was finally convinced of the reality of light quanta only in 1918. The word photon was coined only in 1926 by the American physical chemist Gilbert Lewis. In today's language the observation of wave properties and particle properties in one and the same object is called *wave-particle duality* (or particle-wave duality, as you like)

- ## Particle-Wave Duality of the Electron

The wave nature of electrons is demonstrated for instance in the Davisson-Germer experiment. In this experiment a beam of electrons is incident on a crystal; the reflected electrons show a diffraction pattern similar to that observation when x-rays are made to be reflected from a crystal. Independently Louis de Broglie applied the particle-wave duality of radiation to the electron. This enabled him to deduce the quantum condition introduced by Niels Bohr in an *ad hoc* fashion in order to derive the Balmer formula of the hydrogen spectrum. For this work de Broglie was awarded the Nobel Prize. The basic idea of quantum theory is, of course, the impossibility of considering an isolated fragment of energy without assigning a certain frequency to it, which I will call the quantum relation: Energy $E = h\upsilon$; where υ frequency h is Planck's constant. de-Broglie's idea remains basic for quantum mechanics; it is alive in the terminology we use the quantity $\lambda = h/p$, where h is the Planck constant and p is the momentum of a particle, is called the de Broglie wavelength of that particle. As a consequence of wave-particle duality, the position x and momentum p of a particle cannot be measured simultaneously with absolute precision; the uncertainties Δx and Δp in these measurements satisfy the relation *(Heisenberg's uncertainty relation)* $\Delta x \Delta p_x \geq h$; Where h [= 6.6023×10^{-34} joule – second] is the Planck's constant and h $\left[\hbar = \frac{h}{2\pi}\right]$ is the reduced Planck's constant

The uncertainty relation destroys the basis of Newton's equation of motion which has solutions only if the position and momentum of a particle are simultaneously specified with absolute precision. However, this does not mean that Classical Mechanics must be completely abandoned.

- Newton's 1st law of mechanics, the relativity principle, remains in force
- Newton's 3rd law of mechanics, "action equals reaction", also remains in force

The first one of these was later extended to electromagnetism; it is one of the postulates of Einstein's (special) theory of relativity. Newton's second law, the equation of motion, remains valid in an average sense. This is the statement of Ehrenfest's theorem. A consistent theory, based on the wave nature of matter, was formulatedby Schrödinger. For a single particle of mass m in a field of potential energy V, the Schrödinger equation is in the form

$$ih\frac{\partial}{\partial t}\psi(\vec{r},t) = -\frac{\hbar^2}{2m}\nabla^2\psi(\vec{r},t) + V(\vec{r},t)\psi(\vec{r},t)$$

Where $\psi(\vec{r},t)$ is the wave function and $\nabla^2 = \frac{\partial^2}{\partial x^2} + \frac{\partial^2}{\partial y^2} + \frac{\partial^2}{\partial z^2}$; is the Laplacian operator.

1.6 Rigid descriptions: Quantum mechanics versus Classical mechanics

In classical mechanics, Newton's second law (**F** = m**a**) has been used to make a mathematical prediction as to what path a given physical system will take over time following a set of known initial conditions. Solving this equation gives the position, and the momentum of the physical system as a function of the external force F on the system. Those two parameters are sufficient to describe its state at each time instant.

In quantum mechanics, the analogue of Newton's law is Schrödinger's equation for a quantum system (usually atoms, molecules, and subatomic particles whether free, bound, or localized).

The equation is mathematically described as a linear partial differential equation, which describes the time-evolution of the system's wave function (also called a "state function").

Quantum mechanics versus Classical mechanics

- If any physical quantity describing the system and having the dimensions of an action, has a value of the order of Planck's constant h(=6.6023×10^{-34} joule-second), then the behavior of the system is to be governed by the laws of Quantum Mechanics[17]. i.e. Quantum Mechanics is something which deals with the laws of small
- If any physical quantity describing the system and having the dimensions of an action, has a value of very large compared to the Planck's constant h (=6.6023×10^{-34} joule-second), then the behavior of the system is to be governed by the laws of Classical mechanics. i.e. Classical Mechanics is something which deals with the laws of large
- In Quantum physics, an object can have only certain amounts of energy. The word that came to be known for this is *quantized*
- In classical physics, an object can have any amount of energy
- In Quantum Mechanics, physical quantities are discrete
- In Classical Mechanics, physical quantities are continuous
- Quantum Mechanics can explain the stability of atom

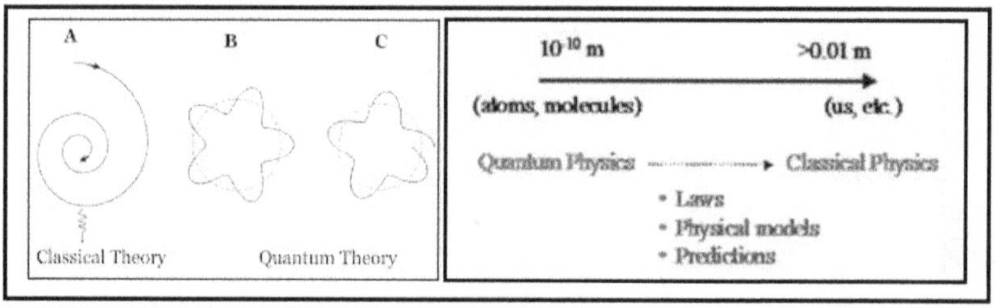

Fig.[1.6(a)]: Shows Quantum mechanics versus Classical mechanics

- Classical Mechanics can't explain the stability of atom
- In Quantum Mechanics, matter has both particle nature and wave nature
- In Classical Mechanics, matter has only particle nature
- In Classical Mechanics, the minimum energy is zero
- In Quantum Mechanics, the minimum energy $\frac{1}{2}\hbar\omega$ for linear harmonic oscillator
- Quantum Mechanics explains the spectrum of hydrogen atom
- Classical Mechanics can't explain the spectrum of hydrogen atom

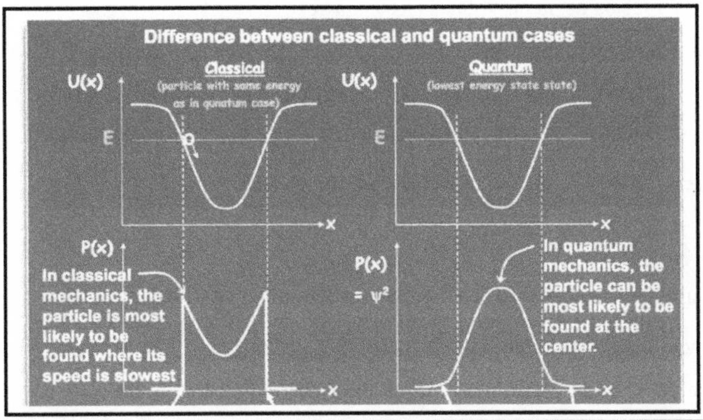

Fig.[1.6(b)]: Graphical presentation of Quantum mechanics versus Classical Mechanics

1.7 Drawbacks of the Old Quantum Theory

Planck's quantum hypothesis with application and extension to explain the black body radiation, the photo electric effect, the Compton Effect, variation of specific heat of solids with temperature and the spectrum of hydrogen is now *called the Old Quantum Theory.* Though these phenomena are successfully explained by the theory, there are numerous drawbacks of the theory[27]. A few of them are as follows:

- Bohr's quantization rules are arbitrary. The theory does not provide physical explanation for the assumptions
- The old quantum theory cannot be applied to explain the spectra of helium and a more complex atom
- It cannot explain the dispersion of light
- It can provide only a qualitative and incomplete explanation of the intensities of the spectral lines
- The theory cannot be applied to explain non-harmonics vibrations of systems

1.8 Basic Postulates of Quantum Mechanics

- Quantum mechanically, Operators are assigned to the physically measureable quantities of classical mechanics
- The state of a system is described by the wave function $\psi(x,t)$; And $|\psi(x,t)|^2 d^3x$ expresses the probability of finding the particle at any instant, somewhere on the X-axis

in a certain region i.e. in the volume element d^3x

- The average or expectation value of any dynamical observable is given by

  $\langle\widehat{\Omega}\rangle = \int \psi^{\times}(x,t)\widehat{\Omega}\,\psi(x,t)d^3x$; If the wave function has been normalized
- The time evolution of the state vectors are described by Schrödinger equation $\widehat{\mathrm{H}}\psi(x,t) = i\hbar\frac{\partial}{\partial t}\psi(x,t)$; with the Hamiltonian operator

  $$\widehat{H} = -\frac{\hbar}{2m}^{2}\nabla^2 + V$$
- The wave function $\psi(\bar{r},t)$ and its partial derivatives $\frac{\partial\psi}{\partial x},\frac{\partial\psi}{\partial y}$ $and\,\frac{\partial\psi}{\partial z}$ must be finite, continuous and singled-valued for all values of x, y, z and t
- A linear vector space H describes ensembles of physical systems of the same type
- If two systems A and A' are elements of ensembles described by the kets $|\psi >$ and $|\psi' >$ respectively, then the ensemble of the compound system is described by the ket $|\psi >$ and $|\psi' >$ which is an element of the product space $H \times H'$; where H and H' are the vector spaces corresponding to A and A' respectively.

With every observable dynamical quantity there is an operator. The operators corresponding to the pertinent dynamical quantities are[26]:

Dynamical variable	Symbol	Quantum Mechanical Operator
Positon	x y z	x y z
Momentum	P_x, P_y, P_z	$\frac{\hbar}{i}\frac{\partial}{\partial x}$, $\frac{\hbar}{i}\frac{\partial}{\partial y}$, $\frac{\hbar}{i}\frac{\partial}{\partial z}$
Total energy	E	$i\hbar\frac{\partial}{\partial t}$
Total energy	E	$-\frac{\hbar^2}{2m}\nabla^2 + V$
Kinetic energy	K	$-\frac{\hbar^2}{2m}\nabla^2$
Potential energy	V(x,y,z)	V(x,y,z)

1.9 Some Phenomena that could not be explained classically

1. The nature of energy distribution in the spectrum of black body radiation could not be explained classically[24].

2. Electrical conductivity of solids has an enormous range. The conductivity of silver is more than 10^{24} times larger than that of fused quartz. This could not be explained classically. Band theory of solids could explain this using quantum mechanics. Max Planck used an adhoc hypothesis that emission and absorption of radiation by matter take place not as continuous process but in discrete amount or quanta of energy hυ. Where h is called Planck's constant

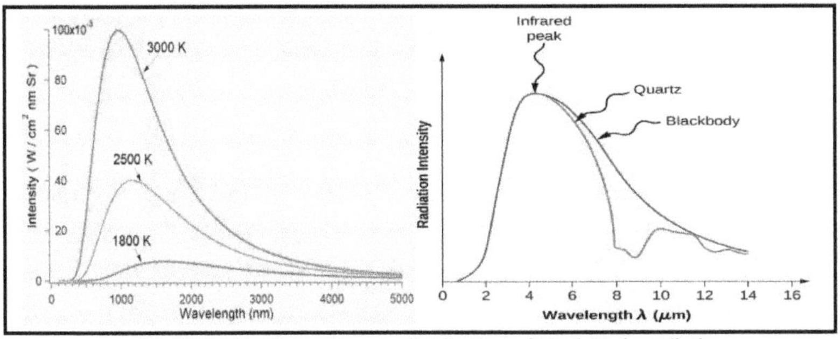

Fig.[1.9(a)]: Shows Energy distribution of black body radiation

h = 6.627×10^{-34} joule-second and υ is the frequency of radiation. With this hypothesis, Planck could arrive at an equation which reproduces the experimental energy distribution of black body radiation.

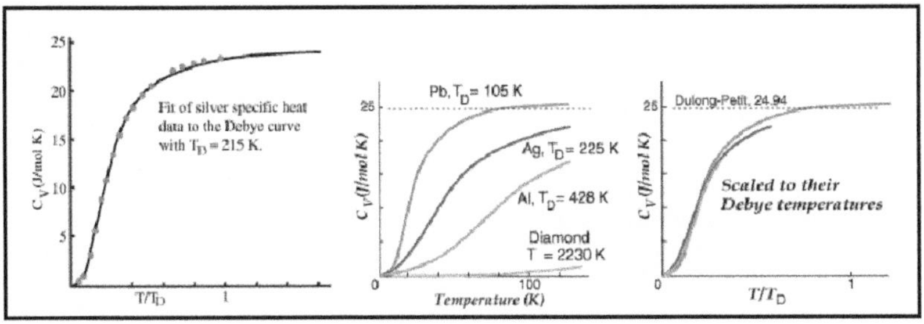

Fig.[1.9(b)]: Shows Temperature Dependence Specific heat of Solids

3. The decrease of specific heat of solids with decreasing temperature could not be explained classically. Einstein could explain it assuming that the energy of oscillator of any atom in a solid can take only a discrete set of values E = n hυ ; Where n = 0, 1, 2, 3 ....

4. The emission spectrum of atoms consisting of discrete frequencies or lines could not be

explained classically. Bohr could explain the spectrum of hydrogen assuming that orbital angular momentum of the electron can be

$$L=n\frac{h}{2\pi}=n\hbar \quad ; \text{ Where n} = 1, 2, 3 \ldots\ldots\ldots\ldots\ldots\ldots\ldots \text{ and } \hbar=\frac{h}{2\pi}$$

5. Frank-Hertz experiment could not be explained classically. The way of this experiment could be explained supports that energy of an electron in an atom can have only some discrete allowed values.

1.10 Mass less Particles:

In Classical Mechanics, a particle must have rest mass in order to have energy & momentum. But in relativistic mechanics this requirement does not hold. We can learn from the relativistic formulas for total energy & linear momentum

$$E = \frac{m_0c^2}{\sqrt{1-\frac{v^2}{c^2}}} \qquad [\because E = mc^2 \ \& \ m = \frac{m_0}{\sqrt{1-\frac{v^2}{c^2}}}] \qquad [1]$$

And $$p = mv = \frac{m_0 v}{\sqrt{1-\frac{v^2}{c^2}}} \qquad [2]$$

- When $m_0 = 0$ and $m_0 \; is < c$; It is clear that $E = P = 0$ i.e. "A mass less particle with a speed less than that of light can have neither energy nor momentum
- When $m_0 = 0$ and $m_0 = c$; It is clear that $E = P = \frac{0}{0}$; which is in determined. E & P can have any values

Thus equation [1] and [2] are consistent with the existence of mass less particles that possess energy and momentum provided that they travel with the speed of light.

From equation [1] $$E^2 = \frac{{m_0}^2c^4}{1-\frac{v^2}{c^2}} \qquad [3]$$

And from equation [2] $$P^2 = \frac{{m_0}^2v^2}{1-\frac{v^2}{c^2}} \quad \text{or, } P^2c^2 = \frac{{m_0}^2v^2c^2}{1-\frac{v^2}{c^2}} \qquad [4]$$

Now subtracting [4] from [3]

$$E^2 - P^2c^2 = {m_0}^2c^4$$

Or, $$E^2 = {m_0}^2c^4 + P^2c^2 \qquad [5]$$

Equation [5] is valid for all Particles. If a Particle exists with $m_0 = 0$ then equation [5] becomes

$$E = pc \qquad [6]$$

Equation [6] is valid for mass less particles. In fact, mass less particles of two different kinds: the photon & the neutrino.

1.11 Quanta or Photon/Quantum:

Max Planck in 1900, when he was able to explain the black body spectrum in terms of the assumed emission & absorption of electromagnetic radiation in discrete quanta, each of which contains an amount of energy E that is equal to the frequency of the radiation υ muttiplied by a universal constant h i.e.

$$E = h\upsilon \qquad [1]$$

This amounts to a particle theory of light such a particle of electromagnetic energy we call a photon or quantum. This quantum idea was later used by Albert Einstein in accordance for some of the experimental observations on the photo electric effect. *In this way the dual character of electromagnetic radiation became established.* It sometimes behaves like a wave motion & sometimes like a stream of corpuscular quanta.

1.12 Radiation possesses dual character: Wave-Particle duality

In order to explain these phenomena, radiant energy is considered as a stream of small packets of energy. These packets of energy are known as light quanta or photons.

The amount of energy assigned to each photon is given by

$$E = h\upsilon \qquad [1]$$

The frequency υ is determined from the measurement of the wavelength λ of radiation

$$\upsilon = \frac{c}{\lambda} \qquad [2]$$

From the above equation the frequency or wave length is a concept relevant to a wave & quantum having the isolated energy $h\upsilon$ is the concept of a particle. So it has been concluded that-*radiation possesses dual character.*de-Broglie proposed that the wave particle duality may not be a monopoly of light but a universal characteristic of nature which becomes evident where the magnitude of 'h' cannot be neglected. From de-Broglie relation

$$\lambda = \frac{h}{p} \qquad [3]$$

This relation establishes the contact between the wave & the particle pictures. The finiteness of Planck's constant is the basic point here. For if h were Zero, then no matter what momentum a particle had, the associated wave would always corresponds to $\lambda = 0$ and would follow the laws of classical mechanics, which can be regarded as the short wavelength limit of wave mechanics. Indeed a free particle would then not be diffracted but go on a straight rectilinear path, just as we expect classically.

1.13 Significance of Wave-Particle duality

De-Broglie proposed that the wave particle duality may not be a monopoly of light but a universal characteristic of nature which becomes evident where the magnitude of 'h' cannot be neglected. From de-Broglie relation

$$\lambda = \frac{h}{p} \qquad [1]$$

This relation establishes the connection between the wave & the particle pictures. The finiteness of Planck's constant is the basic point here[22,23]. For if h were Zero, then no matter what momentum a particle had, the associated wave would always corresponds to $\lambda = 0$ and would follow the laws of classical mechanics, which can be regarded as the short wavelength limit of wave mechanics. Indeed a free particle would then not be diffracted but go on a straight rectilinear path, just as we expect classically.

1.14 de-Broglie Hypothesis

In 1900, Max Planck's was able to explain the black body spectrum in terms of the assumed emission & absorption of electromagnetic radiation in discrete quanta. Each of which contains an amount of energy E that is equal to the frequency of the radiation υ, multiplied by a universal constant h.

$$E = h\upsilon \qquad [1]$$

From Dirac relativistic equation

$$E^2 = mo^2C^4 + p^2c^2 \qquad [2]$$

Since Particle has no rest mass. So

$$E = pc \qquad [3]$$

Where $p = \hbar k$; is the momentum of massless particle. [4]

And $c = \lambda v$; is the velocity of particle

$\hbar$ = Reduce form of Planck's constant $= \frac{h}{2\pi}$ and $\lambda = \frac{2\pi}{\lambda}$; is the wave number [5]

$\therefore hv = pc = pv\lambda$; Where v is the frequency of propagating particle.

Finally, $h = p\lambda$ [6]

$\lambda = \frac{h}{n}$; Wave nature and $p = \frac{h}{\lambda}$; Particle nature [7]

Equation [7] is known as Einstein-de Broglie relation or *de-Broglie Hypothesis.*

Note from Equation [7]: [a] whenever we measure something that is of particle nature, wave nature disappears and vice-versa.

[b] A particle having more mass [massive] or momentum, it is less wave like and vice-versa.

1.15 de- Broglie Theory: de-Broglie wavelength of the matter waves

In Einstein's explanation of photo electric effect & later for the understanding of Compton scattering of X-rays, it was established that electromagnetic waves exhibit some properties were particle like & some that were wave like. Also when the radiation were emitted or absorbed, they did so in packets or quanta of hv.

According to Planck & Einstein, if E is the energy associated with the photon & the frequency v of the associated electromagnetic wave then they are related as

$$E = hv \quad [1]$$

According to the theory of relativity, the relation between the total relativistic energy E & the relativistic momentum P of a particle of rest mass m_0 is

$$E^2 = p^2c^2 + m_0^2c^4 \quad [2]$$

Since photons have zero rest mass. Therefore $E = PC$ or, $p = \frac{E}{C}$ [3]

From equation [1] and equation [3]

$$pc = hv = h\frac{c}{\lambda} \quad [\therefore c = \nu\lambda]$$

Or, $$\lambda = \frac{h}{p} \quad [4]$$

This relation represents the wave-particle relation for photon. In equation [4] , λ is the wavelength of the wave associated with photons of momentum p. Louis de- Broglie argued that the wave character is also associated with all particles[electrons, protons, neutrons, atoms, molecules etc] in motion and the wavelength and frequency associated with them are given by

$$\lambda = \frac{h}{p} = \frac{h}{mv}\ and\ E = h\nu \quad or, \nu = \frac{E}{h}\ and\ m = \frac{m_0}{\sqrt{1-\frac{v^2}{c^2}}} \qquad [5]$$

de-Broglie further argued that particle velocity is equal to the group velocity which we can see

$$V_g = \frac{d\omega}{dP}\ \ and\ V_p = \frac{dE}{dP} \qquad [6]$$

$$V_g = \frac{d(2\pi\nu)}{d\left(\frac{2\pi}{\lambda}\right)} = \frac{(d\nu)}{d\left(\frac{1}{\lambda}\right)} = \frac{d(h\nu)}{d\left(\frac{h}{\lambda}\right)} = \frac{dE}{dp} \quad \text{Or, } V_g = \frac{dE}{dP} \qquad [7]$$

From [7] we see that the wave packet moves as a classical particle.

Now $E = h\nu = \hbar\omega = (p^2c^2 + m_0^2c^4)^{\frac{1}{2}} = mc^2$ [8]

Now differentiating [8] with respect to k. [since ω is a function of k. So p must be the function of k]

$$\hbar\frac{d\omega}{dk} = \frac{1}{2}\,(p^2c^2 + m_0^2c^4)^{-\frac{1}{2}}\,(2pc^2\,\frac{dp}{dk})$$

Or, $$\hbar\frac{d\omega}{dk} = \frac{pc^2}{(p^2c^2+m_0^2c^4)^{\frac{1}{2}}}\cdot\frac{dp}{dk}$$

Or, $$\hbar\frac{d\omega}{dk} = \frac{pc^2}{mc^2}\cdot\frac{dp}{dk} = \frac{p}{m}\cdot\frac{dp}{dk} = \frac{mV_p}{m}\frac{dp}{dk}$$

Or, $\hbar v_g = v_p\frac{dp}{dk}$ Or, $\hbar = \frac{dp}{dk}$ [$\because$ and $v_g = v_p$]

Or, $\hbar\, dk = dp$

Now integrating on both sides

$p = \hbar k$ [taking is constant of integration to be zero]

Or, $p = \frac{h}{2\pi}\cdot\frac{2\pi}{\lambda}$ or, $\boxed{\lambda = \frac{h}{p}}$ [9]

This wavelength was called the de-Broglie wavelength of the matter waves associated with group velocity or particles velocity of waves. *Finally, de-Broglie argued that if radiation can act like a wave sometimes and like particle at the other times, then things like electrons should also exhibit wave properties under appropriate conditions.*

1.16 Physical meaning of Phase Velocity of de-Broglie waves

The Phase Velocity v_P of the de-Broglie wave associated with a particle in motion is not the same as the particle velocity v. Now from usual relation, the Phase velocity

$$v_P = \frac{\omega}{k} \quad [1]$$

Where k = propagation constant of the de-Broglie wave and ω = angular frequency

According to de-Broglie's Hypothesis, the momentum and the energy of the particle

$$p = \hbar k \quad [2]$$

$$E = h\nu = \hbar\omega \quad [3]$$

Now dividing [3] by [2]

$$\frac{E}{P} = \frac{\omega}{k} = V_P = \frac{h\nu}{\frac{h}{\lambda}} = \nu\lambda \qquad \boxed{\therefore\ v_P = \nu\lambda} \quad [4]$$

Now consider the following two cases:

When the particle velocity is small compared with the velocity of light, the energy E is

$$E = \frac{p^2}{2m}$$

Hence [4] becomes $v_P = \frac{1}{p}\frac{p^2}{2m} = \frac{p}{2m} = \frac{mv}{2m} = \frac{v}{2}$ [5]

When the particle velocity is comparable with the velocity of light, then E represents the total relativistic energy mc^2

Hence $E = hv = mc^2$

Equation [4] becomes $v_P = \frac{E}{p} = \frac{mc^2}{mv} = \frac{c^2}{v}$ [6]

Since the particle velocity v is always less than the velocity of light, the phase velocity V_P is greater than the velocity of light $i.e$ the de-Broglie waves always travel faster than light. But a velocity larger than the velocity of light cannot be measured. *Therefore the phase velocity has no physical meaning.*

1.17 de-Broglie waves move with the velocity of the particle

In order to understand this unexpected result we must look into the distinction between phase velocity & group velocity. Let us see the group velocity of de-Broglie waves, from de-Broglie relation

$$p = \hbar k \quad [1]$$

$$E = \hbar\omega \quad [2]$$

$$dp = \hbar dk \quad \text{Or,} \quad dk = \frac{dp}{\hbar} \qquad [3]$$

& $$dE = \hbar d\omega \quad \text{or,} \quad d\omega = \frac{dE}{\hbar} \qquad [4]$$

From [8] and [9] we write

$$\frac{d\omega}{dk} = \frac{dE}{dp} \quad \text{Or} \quad v_g = \frac{dE}{dp} \qquad [5]$$

Now consider the following two cases:

Free particle of classical mass: For such a particle the total classical energy is $E = \frac{p^2}{2m}$

Differentiating with respect to p on both sides

$$\frac{dE}{dP} = \frac{2p}{2m} = \frac{p}{m} = \frac{mv}{m} = v \qquad [6]$$

From [10] and [11], we have

$$v_g = v \qquad [7]$$

Free particle of relativistic mass and rest mass m_0, the total relativistic energy of the particle

$$E^2 = p^2c^2 + m_0^2c^4$$

Now differentiating with respect to p

$$2E\frac{dE}{dp} = c^2.2p \quad or, \quad \frac{dE}{dp} = \frac{Pc^2}{E} = \frac{mvc^2}{mc^2} = v \qquad [8]$$

From equation [10], we have

$$v_g = \frac{dE}{dp} = v \qquad [9]$$

Therefore, we concluded that the group velocity of the de-Broglie waves is the same as that of the particle. I.e., *de-Broglie waves move with the velocity of the particle.*

1.18 Wave-Particle duality of radiation and of matter

(a) Electromagnetic radiation exhibits well-know interference and diffraction phenomena which could only be explained assuming that electromagnetic radiation is a wave.

(b) When a radiation of high frequency, such as ultraviolet light or X-rays, is incident on a clean metal surface, electrons are emitted from the surface. Since the incident radiation causes the emission of the electrons, *the phenomenon is called photo-electric effect and emitted electrons are called the photo-electrons. The photo-electric effect was discovered by Heinrich Hertz in*

1887. The whole range of electro-magnetic radiation [a radiation of high frequency] from Gamma and X-rays to the ultra-violet, the visible and ifra-red rays produced this effect. This effect occurs also in solids, liquids and gases. Further experimental study was undertaken by Hallwachs in 1888. Then in 1899 Lenard showed that the carriers of electricity emitted from a metal surface, under the action of ultra-violet light, were electrons. Photoelectric Effect could not be explained assuming that electromagnetic radiation is wave.

Einstein could explain characteristics of photoelectric effect by suggesting that in inducing emission of photoelectrons, electromagnetic radiation does not act like a wave, it acts like a wave, it acts like a stream of discrete quanta or packets of energy or photons. The energy of each photon is hυ, where υ frequency of the radiation.

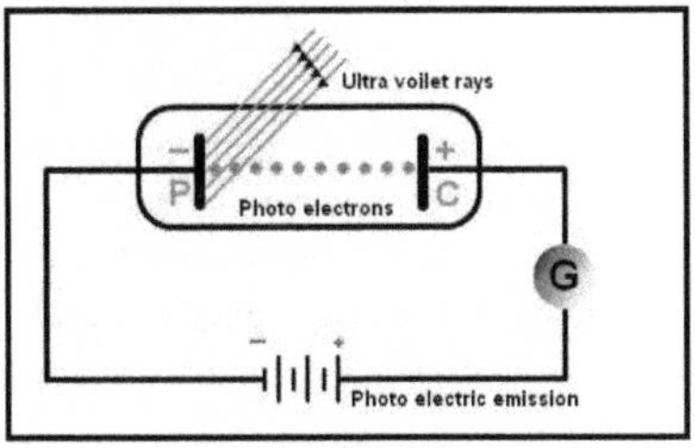

Fig.[1.18(a)]: Shows the Photoelectric Effect

Although this bears the essence that electromagnetic radiation consists of particles, it can be said that electromagnetic radiation is a wave that assumes particle characters in some unexplained manner at the instant of absorption by material.

(c) When a mono-chromatic beam of X-rays of wavelength λ is scattered by a light element like carbon, it is observed that the scattered X-rays have maximum intensities at two wave lengths. One maximum occurs at the same wavelength λ_i as that of the incident and other maximum occurs at a slightly longer wavelength λ_f. *This effect was discovered by A. H. Compton in 1922 and for the discovery and the explanation of the Effect, he was awarded the Nobel Prize in 1927. For this it is called the Compton Effect* and the change in wavelength, $\Delta\lambda = \lambda_f - \lambda_i$, called the Compton- shift of the wavelength of the scattered X-rays, from that of the incident beam. Compton Effect could not be explained assuming that x-ray is a wave; although X-ray diffraction can only be explained assuming X-ray is a wave. *Compton Effect* could only be explained assuming that X-ray of frequency υ consists of stream of particles (photons) of energy hυ each.

The success of the assumption in explaining the phenomenon makes it clear that electromagnetic radiation should be considered as a stream of particles (photons) even during propagation.

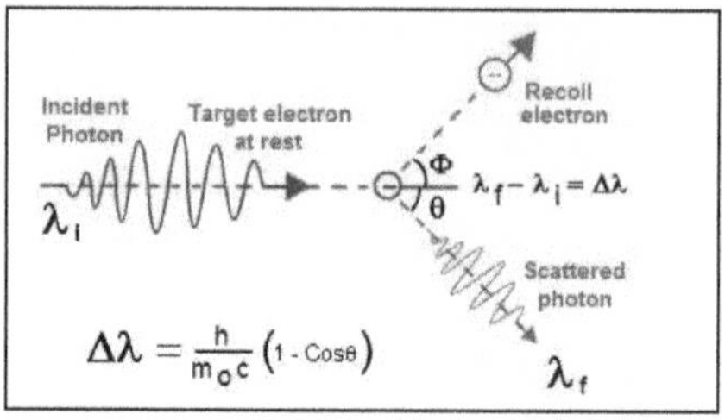

Fig.[1.18(b)]: Shows the Compton Scattering

Thus it is experimental fact that we need to consider electromagnetic radiation as a wave to explain some phenomena and we need to consider electromagnetic radiation as a stream of particles to explain some other phenomena. *This is called Wave-Particle duality of Electromagnetic radiation.*

Louis de-Broglie made a bold hypothesis that if electromagnetic radiation known as a wave can sometimes behave like particles, it should be possible for material particle (having non-zero rest mass) to show wave-like behavior. For free material particles, de-Broglie proposed that the associated matter wave has a wavelength λ and frequency υ related to momentum P and energy E by $\lambda = \frac{h}{P}$ and $\nu = \frac{E}{h}$ as is the case with photon. Here $\lambda = \frac{h}{P}$ is called de-Broglie wavelength of all particles. For material particles, one needs both the equations to obtain λ and υ. All material particles possess wave-like characteristic.

$$\lambda = \frac{h}{P} \quad \text{[Where } P = \hbar k \text{ and } \hbar = \frac{h}{2\pi} \text{ and wave number } k = \frac{2\pi}{\lambda}\text{]} \qquad [1]$$

$$\nu = \frac{E}{h} \quad \text{[Where } E = \hbar\omega \text{ and } \omega = 2\pi\nu\text{]} \qquad [2]$$

Experiments by Davission and Germer and by Kikuchi and Thomson demonstrated that

- A wave is associated with material particle
- The wavelength of the wave is $\lambda = \frac{h}{P}$

In these experiments, crystal diffraction pattern just like those produced by X-rays were obtained even when a beam of electrons was used instead of X-rays. Hence a wave is associated with material particle. In each of the experiments, the wavelength λ of the associated wave as determined from the deffraction pattern was found to be equal to $\frac{h}{p}$, in agreement with de-Broglie hypothesis. Electrons, Helium atoms, hydrogen molecules, neutrons etc. have been shown to exhibit wave property in agreement with de-Broglie hypothesis. It is an experimental fact that a material particle can manifest itself as having a wave associated with it. *This is the Wave-Particle duality of matter.*

1.19 Photo- electric Effect and Einstein Photo-electric equation

When a radiation of high frequency, such as ultraviolet light or X-rays, is incident on a clean metal surface, electrons are emitted from the surface. Since the incident radiation causes the emission of the electrons, *the phenomena are called Photo-electric Effect and emitted electrons are called the Photo-electrons.*

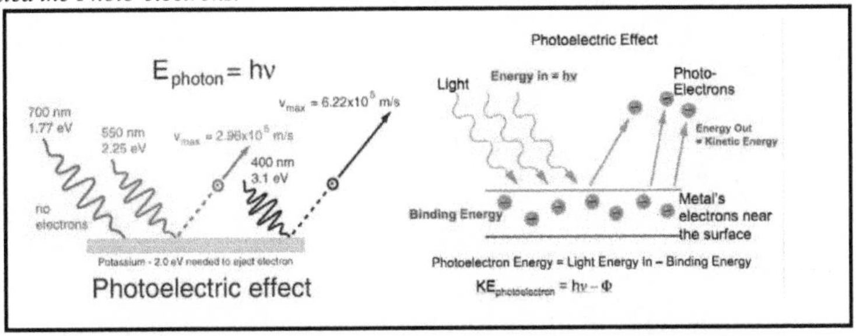

Fig.[1.19]: Shows Photo-Electric Effect and Einstein Photoelectron Energy

The Photo-electric Effect was discovered by Heinrich Hertz in 1887. In 1905 Albert Einstein applied Planck's quantum theory & made the following two assumptions:

- A radiation of frequency υ consists of a stream of discrete quanta each of energy$h\upsilon$; where h is Planck's constant. These quanta are called photons. The photons move through space with the speed of light
- When a photon of energy $h\upsilon$ is incident on a metal surface, the entire energy of the photon is absorbed by a single electron without any time lag

The energy $h\upsilon$ gained by the electron is used up in the following two ways:

- A part of the energy is used up by the electron to do a certain minimum amount of work to overcome the attractive forces of positive ions of the metal. This minimum amount of energy is called the photo-electric work function W_0 of the metal. The work function depends on the nature of the emitting surface
- The remaining energy $(h\upsilon - W_0)$ appears as the kinetic energy of the election. If the electron with this K.E comes from the surface, it does not suffer any collision with another electron

Therefore, it will be emitted from the surface into vacuum with this energy, as the maximum K.E= $\frac{1}{2}mV_{max}^2$.Where m is mass of the electron and V_{max} its maximum speed for the given frequency. Hence for such an electron

$\frac{1}{2}mV_{max}^2 = hv - W_0$ [1]; this is the Einstein Photo-electric equation.

Suppose the frequency of the incident radiation is reduced to a certain value v such that an electron is just emitted from the surface with zero K.E. Then we have

$\nu = \nu_0$ & $\frac{1}{2}mV_{max}^2 = 0$; From equation [1] $W_0 = h\nu_0$

Therefore [1] becomes $\frac{1}{2}mV_{max}^2 = h\nu - h\nu_0$ [2]

Equation [2] represents another form of Einstein equation. The frequency ν_0 is called the threshold frequency & it is defined as "It is the frequency of a photon which has just sufficient energy to liberate an election with zero K.E. from the emitting surface.'' From equation [2] it is evident that

- If $\nu < \nu_0$; no Photo -electrons are emitted
- If $\nu = \nu_0$; just emitted with Zero K.E.
- If $\nu > \nu_0$; Photo-electrons are emitted with K.E. [raising from zero to $\max^m$ value] or, intensity of the incident light is increased the number of photo- electrons emitted per-sec will increase

1.20 Comparative study of Photo Electric Effect and Compton Effect

When a radiation of high frequency, such as ultraviolet light or X-rays, is incident on a clean metal surface, electrons are emitted from the surface[26]. Since the incident radiation causes the emission of the electrons, *the phenomena are called Photo-electric Effect and emitted electrons are called the Photo-electrons.*

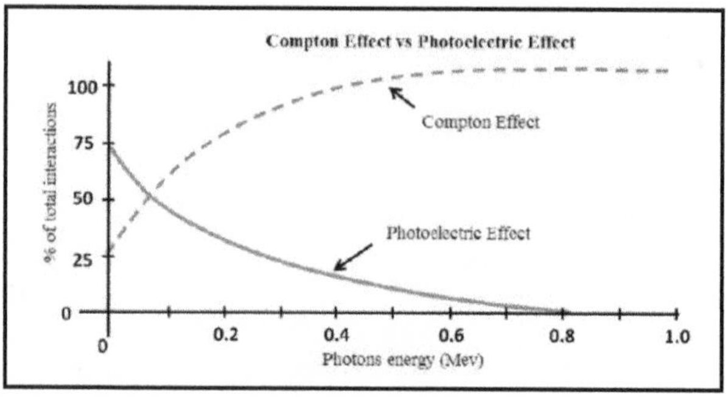

Fig.[1.20]: Shows the Photo Electric Effect and the Compton Effect

The Photo-electric Effect was discovered by Heinrich Hertz in 1887. When a mono-chromatic beam of X-rays of wavelength λ is scattered by a light element like carbon, it is observed that the scattered X-rays have maximum intensities at two wavelengths. One maximum occurs at the same wavelength λ_i as that of the incident and other maximum occurs at a slightly longer wavelength λ_f. *This effect was discovered by A. H. Compton in 1922 and for the discovery and the explanation of the Effect, he was awarded the Nobel Prize in 1927.* For this it is called the Compton Effect and the change in wavelength $\Delta\lambda = \lambda_f - \lambda_i$, called the Compton- shift of the wavelength of the scattered X-rays from that of the incident beam.

1.21 Physical significance Planck's constant[*h*]

- h (=$2\pi\hbar$) provides a measure of converting energy into frequency and vice-versa
- h is the smallest action in nature[$E = h\upsilon = h = ET$ [joule-sec means "action"]
- h related as a measure of the product of uncertainty in the measurement of two Canonically conjugate variables
- h^3 determines the smallest unit in phase space

1.22 Hilbert space

The Hilbert space provides, so to speak, the playgrounds for our analysis. To gain a deeper understanding of quantum mechanics, we will need a more solid mathematical basis for our discussion. We achieve this by studying more thoroughly the structure of the space that underlies our physical objects, which as so often, is a vector space, the Hilbert space. Hilbert space can be finite or infinite dimensional. *An infinite or finite dimensional vector space is called Hilbert Space*[5,7].

Important properties of the Hilbert space are:

- *A Hilbert space is a finite- or infinite-dimensional, linear vector space with scalar product in C*
- *A Hilbert space is a complete function space with scalar product in C*
- *A (function) space is called complete if every Cauchy-sequence has a limit in the space*

The basis vectors of Hilbert Space are $\Psi_1, \Psi_2, \Psi_3 ....\Psi_n$. An arbitrary vector can be written as column matrix as

$$|\Psi\rangle = \begin{bmatrix} \Psi_1 \\ \Psi_2 \\ . \\ . \\ . \\ \Psi_n \end{bmatrix} \quad \text{and} \quad \langle\Psi| = [\Psi_1^\times \ \Psi_2^\times \ \Psi_3^\times \ .... \ \Psi_n^\times] \qquad [1]$$

$$\langle\Psi|\Psi\rangle = \Psi_1^\times\Psi + \Psi_2^\times\Psi + \Psi_3^\times\Psi + ... + \Psi_n^\times\Psi \qquad [2]$$

Or, $\langle\Psi|\Psi\rangle=\sum_{i}^{n}\Psi_i^{\times}\Psi_i$ [3]

For Ψ's to be continuous. $\langle\Psi|\Psi\rangle=(\Psi,\Psi)=\int_{-\infty}^{\infty}\Psi^{\times}\Psi\partial\tau$ [4]

1.23 Phase Space

The state of a system of particles is completely specified classically at particular instant if the position and momentum of each of its constituent particles are known. Since position and momentum vectors with three components apiece, we must know six quantities[22]

x, y, z, p_x, p_y, p_z [1] ; for each particle

The position of a particle is a point having the coordinates x, y, z in ordinary three dimensional spaces. It is convenient to generalize this conception by imagining a six dimensional space in which a point has the six coordinates x, y, z, p_x, p_y, p_z *this combined position and momentum space is called Phase Space*. A point in Phase Space corresponds to particular position and momentum while a point in ordinary space corresponds to particular position only. The uncertainty principle compels us to elaborate what we mean by a point in Phase space. Let us divided Phase space into tiny six-dimensional cells whose sides are $dx, dy, dz, dp_x, dp_y, dp_z$ *respectively.* As we reduce the size of the cells we approach more and more closely to the limit of a point in Phase space. However, the volume of each of these cells is

$$\tau=dx, dy, dz, dp_x, dp_y, dp_z \qquad [2]$$

According to the uncertainty principle

$$dxdp_x\geq\hbar \quad , dydp_y\geq\hbar \quad and \quad dzdp_z\geq\hbar$$

Hence we see that $\tau=dx, dy, dz, dp_x, dp_y, dp_z=\hbar^3$ [3]

A point in phase space is actually a cell whose minimum volume is of the order of $\hbar^3$

1.24 Phase velocity and Group velocity

The wave velocity is known as phase velocity[16]. When a large number of plane waves with slightly different wavelengths and frequencies travel in the same direction, groups of waves are formed. These groups of waves are wave packets. The velocity of such group of waves is known as group velocity (V_g).

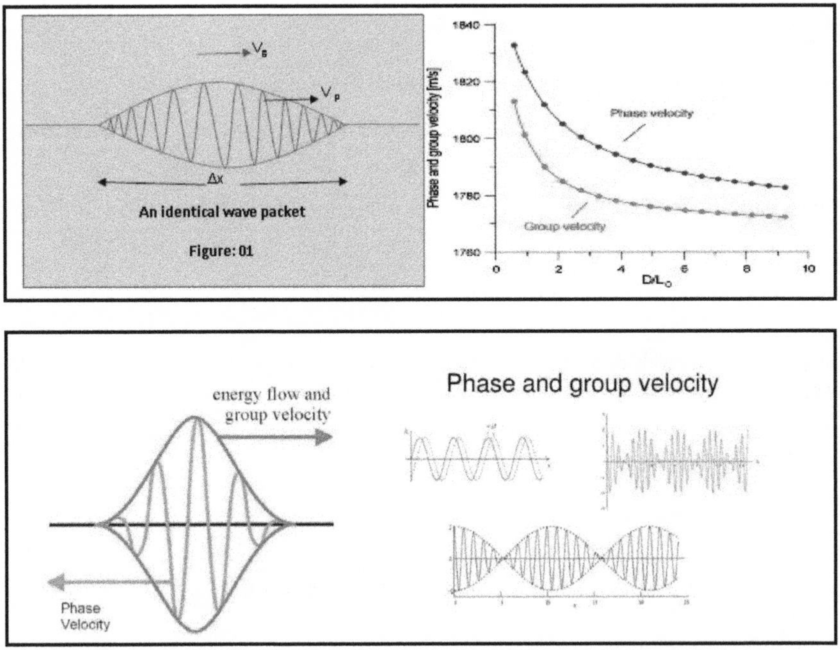

Fig.[1.24]: Shows Phase velocity and Group velocity

1.25 Relation between Phase Velocity and Group Velocity

A plain simple harmonic wave travelling in the positive X-direction

$$y = Asin\omega\left(t - \frac{x}{v_p}\right) = Asin(\omega t - \frac{\omega x}{v_p})$$

$$y = Asin(\omega t - kx) \qquad [1]$$

Where $$k = \frac{\omega}{v_p} = \frac{2\pi\upsilon}{\upsilon\lambda} = \frac{2\pi}{\lambda} \qquad [2]$$

Further $$\omega = 2\pi\upsilon = 2\pi\frac{v_p}{\lambda}$$

$$\frac{d\omega}{d\lambda} = 2\pi\left[-\frac{v_p}{\lambda^2} + \frac{1}{\lambda}\frac{dv_p}{d\lambda}\right] = -\frac{2\pi}{\lambda^2}\left[v_p - \lambda\frac{dv_p}{d\lambda}\right] \qquad [3]$$

Again $$\frac{dk}{d\lambda} = -\frac{2\pi}{\lambda^2} \qquad [4]$$

Dividing [3] by [4]

$$\frac{d\omega}{d\lambda} \cdot \frac{d\lambda}{dk} = v_p - \lambda \frac{dv_p}{d\lambda} \qquad [5]$$

From group velocity $$v_g = \frac{d\omega}{dk} = \frac{d\omega}{d\lambda} \cdot \frac{d\lambda}{dk} \qquad [6]$$

Or, $$v_g = v_p - \lambda \frac{dv_p}{d\lambda} \qquad [7]$$

In a medium in which there is no dispersion (dispersion means-if phase velocity varies with wave length, an effect is called dispersion) i.e. waves of all wave lengths travel with the same speed. Then

$$\frac{dv_p}{d\lambda} = 0 \qquad [8]$$

Therefore, in such a medium equation [7] reduces

$$v_g = v_p \qquad [9]$$

Equation [9] is true for electromagnetic waves in vacuum and elastic waves in a homogeneous medium.

1.26 Matter Wave

Planck- Einstein's light- quantum, Bohr model of the atom and de Broglie's matter waves have solid experimental support and each one of them is a convincing argument against the use of classical physics in the microscopic world. Yet this theory called the Quantum Theory of Plank, Einstein and Bohr is not really a comprehensive theory of Nature. Each of its successes makes it look sillier as Einstein remarked. What does one mean by quantum jumps? What is a matter wave precisely? In short, I have tried to explain the matter wave in the following:

- *Matter waves are not a physical phenomena but is a wave of probability*[6]
- *It is not electromagnetic waves but a new kind of waves that guide the material particle*
- *Matter waves can travel faster than light*
- *Larger the mass of particle, shorter the wave length associated with it*

- *Matter waves evidently have particle properties. By analogy with radiation matter should also have wave properties under suitable conditions [Interference, Diffraction]. This was first predicted by French theoretical Physicist Prince Louis de-Broglie in 1924. It was verified experimentally by C.J. Davisson and L.H. Germer in 1927*

A theory worth the name must not only answer such questions but it must also enable us to predict new phenomena. Such theory was indeed discovered during the unforgettable years of 1925-27 when Heisenberg, Schrödinger, Dirac, Pauli and many others laid the foundations of Quantum Mechanics.

Exercise

1. Why Quantum Mechanics is an important intellectual achievement of the 20th century?
2. Explain Quantum Mechanics is Insubstantial.
3. How Classical Mechanics can be explained by some of the followings Mathematical Preliminaries?
    a. Lagrangian Formulation
    b. Hamiltonian Formulation
    c. More on Hamiltonian
    d. Poisson Bracket
    e. Some useful knowledge of Matrix Algebra
    f. Identity, Hermitian, Symmetric, Inverse and Unitary Matrices
    g. Formulation of Planck's radiation law: Planck's hypothesis
    h. Formulation of another radiation laws from Planck's radiation law
4. Write preliminary remarks that help the introduction to Quantum Mechanics.
5. Show that Quantum Mechanics can be uniquely explained by
    a. Black Body Radiation
    b. Particle-Wave Duality of the Electron
6. Write down the difference between Quantum mechanics and Classical mechanics.
7. Mention the drawbacks of the Old Quantum Theory.
8. What are the basic Postulates of Quantum Mechanics?
9. Discuss that some Phenomena that could not be explained classically.
10. Give an idea of mass less Particles and the dual character of electromagnetic radiation.
11. Explain radiation possesses dual character.
12. What is the significance of Wave-Particle duality?
13. Derive the de-Broglie Hypothesis.
14. Write a brief description of de- Broglie wavelength of the matter waves.
15. What is the physical meaning of Phase Velocity of de-Broglie waves?
16. Show that the de-Broglie waves move with the velocity of the particle?
17. Discuss the Wave-Particle duality of radiation and of matter.
18. Explain Photo- electric Effect and find the Einstein Photo-electric equation.
19. Show a comparative study of Photo Electric Effect and Compton Effect.
20. Write down the Physical significance of Planck's constant[h].
21. Write a short note on:
    a. Hilbert space b. Phase Space c. Matter Wave d. Quanta/Photon e. Wave-Particle duality
22. Define the Phase velocity and the Group velocity. Establish the relation between Phase Velocity and Group Velocity.
23. Derive Wien's law $T\lambda_{max} = constant$ from Planck's radiation law.
24. Derive the Stefan constant from Planck's law and obtain its numerical value using

$k = 1.380662 \times 10^{-16} erg/°K \quad h = 6.626176 \times 10^{-27} erg.sec \quad c = 2.99742 \times 10^{10} cm/sec$

25. The maximum energy of photo-electrons from Al is 2.3eV for radiation of wavelength $2000A°$ and 0.9eV for radiation $3130A°$. Calculate Planck's constant and the work function for Al. Given $1\text{eV} = 1.60217 \times 10^{-12} erg$
26. If the stopping potential is 0.24 volts at wave length $2537A°$ for copper, what is the threshold frequency? Given 1 volt $= \frac{1}{300} erg/esu$
27. The photoelectric threshold of tungsten is $2300A°$. Determine the energy of the electrons ejected from the surface by ultraviolet light of wave length $1800A°$.
28. The work function for hydrogen is about 13.6 eV. For photons of wavelength $10A°$ incident on hydrogen. What is the maximum kinetic energy of the emitted photoelectrons? What is the ratio of the maximum speed of these photoelectrons to the speed of light?
29. What is the de Broglie wavelength of a 200 eV electron, a 20 MeV proton and a thermal neutron (neutron of energy kT)?
30. Find the de- Broglie wave length of an electron accelerated to a potential difference of Vvolts.
31. An electron microscope uses 1.25KeV electrons. Find its ultimate resolving power on the assumption that this is equal to the wave length of the electron.
32. Calculate the de- Broglie wave length associated with an electron of energy 5MeV.
33. The work- function of zinc is 3.6eV. What is the maximum energy of photon- electrons ejected by ultra violet- light of wave length $3000A°$?
34. Show that the wave length λ associated with an electron of mass m and kinetic energy T is $\lambda = \frac{h}{\sqrt{2mT}}$.
35. Show that for a non- relativistic free particle, the phase velocity is half of the group velocity.

Chapter 2 Basic Concepts of Quantum Mechanics

2.1 Waves and Wave function

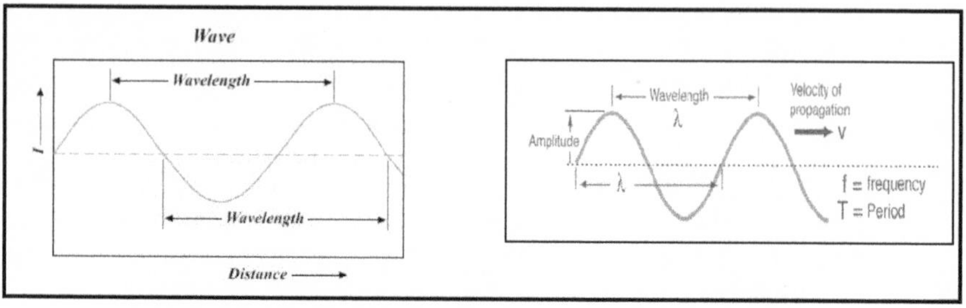

Fig.[2.1]: Shows Particle is propagating along the positive X-direction

$\Psi(x, t) = A \sin (kx - \omega t)$; is simple harmonic travelling along the positive X-direction.
$\Psi(x, t) = A \sin (kx + \omega t)$; is simple harmonic travelling along the negative X-direction.
$\Psi(x, t) = A \sin (kx) \cos(\omega t)$ is standing wave.

It can be represented for undamped (constant amplitude) and monochromatic (single frequency and single wave length) wave propagating in the positive X- direction

$$\Psi(x,t) = A exp\left(\frac{i}{\hbar}\right)(px - Et) \quad [1]$$

In three dimensions,

$$\Psi(\bar{r}, t) = A exp\left(\frac{i}{\hbar}\right)(\bar{p}.\bar{r} - Et) \quad [2]$$

The concept of a wave function is a fundamental postulate of quantum mechanics that defines the state of the system at each spatial position, and time. Using these postulates, Schrödinger's equation can be derived from the fact that the time-evolution operator must be unitary, and must therefore be generated by the exponential of a self-adjoint operator, which is the quantum Hamiltonian.

2.2 Physical significance [Interpretation] of Wave function

Let us consider a wave function [that is propagating from - ∞ to +∞]

$$\Psi = A + iB \quad [1]$$

Taking complex conjugate on both sides of equation [1]

$$\Psi^{\times} = A - iB \quad [2]$$

Therefore, $\Psi\Psi^{\times} = (A+iB)(A-iB) = = (A^2+B^2)$ [3]

From equation [3]; R.H.S is real whereas L.H.S is imaginary. So taking the absolute value of equation [3]

$$|\Psi\Psi^{\times}| = |\Psi|^2 = A^2 + B^2 \quad [4]$$

The wave function $|\Psi\Psi^{\times}|$ has no physical meaning but the probability of finding the electron in a certain region in space is proportional to the square of the wave function $|\Psi|^2$. This launches a new field called *Quantum Mechanics* and while it doesn't allow us to specify the exact location of an electron in an atom, it does define the region where the electron is most likely to be found at any given time. Equation [4] states that the probability of finding the particle at any instant in a certain region is maximum. Depending on the situation, according to Max Born, the L.H.S of equation [4] states that the square of the wave function is represented the probability of finding the electron at any instant in unit volume.

$$\therefore \quad |\Psi|^2 dv = \rho dv = 1 \quad [5]$$

Where $\rho = \Psi\Psi^{\times} = |\Psi|^2$; is the Probability density.

$$\therefore \int_{-\infty}^{\infty} \rho dv = \int_{-\infty}^{\infty} \Psi^{\times}\Psi dV = 1 \text{ or, } \int_{-\infty}^{\infty} \Psi^{\times}\Psi d\tau = 1 \text{ or, } \int_{-\infty}^{\infty} \Psi^{\times}\Psi dxdydz = 1 \text{ or, } \int_{-\infty}^{\infty} \Psi^{\times}\Psi d^3r = 1 \quad [6]$$

In any wave, a physical quantity varies with respect to both space [position] and time. There are some citations below that will be explained the physical meaning of wave function significantly.

- In case of electromagnetic waves, it is the electric and magnetic fields that vary with respect to position and time
- In case of sound wave, it is the longitudinal displacement of the particles

Of the transmitting medium from the equilibrium that varies with respect to both position (space) and time:

- In water wave, the physical quantity that varies with respect to position (space) and time is the vertical displacement of the molecules of the water surface from the equilibrium

- In case of matter wave, the quantity, whose variation constitutes the wave, is called the wave function and is denoted by Ψ

1. Ψ must in some sense be related to the presence of a particle.
2. Where $\Psi = 0$; we do not expect to find the particle in those regions of space.
3. If $\Psi \neq 0$; means there is a probability that the particle may be present there.

Although Ψ is related to the probability of finding a particle somewhere, Ψ itself is not taken as the probability. This is because Ψ can be positive or negative; moreover, we often write Ψ is a complex quantity. The probability of finding something somewhere at a given time can have values between 0 and 1. Zero [0] corresponds to certainty of its absence and 1 corresponds to certainty of its presence. A probability 0.25 somewhere means 25% chance of the presence of the particle there. But $\Psi^*\Psi = |\Psi|^2$ can be proportional to the probability of finding the particle somewhere at a given time. A larger value of $\Psi^*\Psi = |\Psi|^2$ means a greater probability of the particle's presence; a smaller value of $\Psi^*\Psi = |\Psi|^2$ means a smaller probability of its presence. So, probability is a physical quantity. If $\Psi^*\Psi = |\Psi|^2 \neq 0$ I.e., at any given instant, the particle is, in fact, at a particular point in space. This interpretation of wave function was made by Max Born.

2.3 Probability density

If we have a region of space containing a large number of non- interacting particles each having the same normalized wave function Ψ associated with it, so

$$\int_{-\infty}^{+\infty} \psi^{\times}(\bar{r},t)\psi(\bar{r},t) = |\psi(\bar{r},t)|^2 \qquad [1]$$

Equation [1] is proportional to the average number of particles per unit volume. Therefore

$\int_{-\infty}^{\infty} \Psi^{\times}\Psi d\tau = 1$; means the probability of presence of a particle in all space is maximum. Thus

$$\int_{-\infty}^{\infty} \left|\Psi^{\times}(\vec{r},t)\Psi(\vec{r},t)\right| d\tau = \left|\Psi(\vec{r},t)\right|^2 d\tau \qquad [2]$$

The above equation represents the probability that the particle is present in the volume $d\tau$ at time t. The state of a system is described by the wave function $\psi(x,t)$; and $|\psi(x,t)|^2 d^3x$

expresses the probability of finding the particle at any instant, somewhere, on the X-axis in a certain region i.e. in the volume element d^3x at time t.

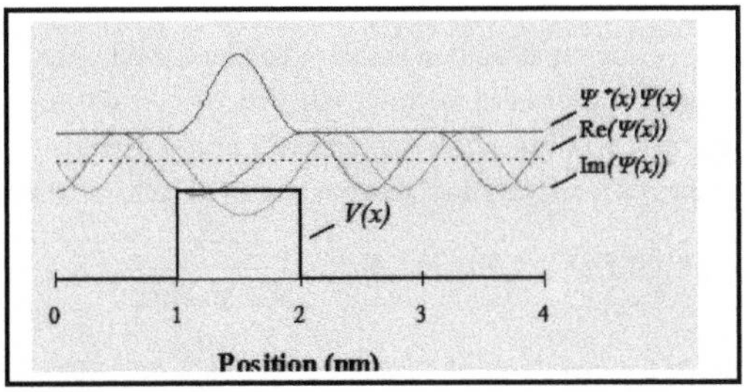

Fig.[2.3]: Shows $|\psi(x,t)|^2 d^3x$ expresses the probability density in the volume element d^3x at time t.

$$\therefore \quad \int_{-\infty}^{\infty} \left|\Psi^{\times}(x,t)\,\Psi(x,t)\right| d^3x = \left|\Psi(x,t)\right|^2 d^3x$$

Hence $\left|\Psi(\vec{r},t)\right|^2$ or $\left|\Psi(x,t)\right|^2$ is called the Probability density (ρ).

2.4 Limitations of Wave function

- Ψ must be finite for all values of x, y and z
- Ψ must be single-valued I.e., for each set of values of x ,y ,and z; Ψ must have one value only
- Ψ must be continuous in all regions except in those regions where the potential energy V(x, y, z) = ∞
- Ψ is analytic I.e., it possesses continuous first derivative
- Ψ vanishes at the boundaries

2.5 Normalization of Wave function and Normalizing constant

The probability of finding a particle in the volume element dτ is given by

$$\int_{-\infty}^{\infty} \left|\Psi^{\times}(\vec{r},t)\Psi(\vec{r},t)\right| d\tau = \left|\Psi(\vec{r},t)\right|^2 d\tau \qquad [1]$$

The total probability of finding the particle in the entire space is of course unity.

$$\int_{-\infty}^{\infty} \left|\Psi^{\times}(\vec{r},t)\Psi(\vec{r},t)\right| d\tau = \left|\Psi(\vec{r},t)\right|^2 d\tau = 1 \qquad [2]$$

Any wave function which satisfies the above condition is said to be normalized. *Usually ψ is not a normalized wave function.* In order to normalize the wave function, it is possible to multiply ψ by a constant A [say], to give a new wave function Aψ, which is also a solution of the wave function. The new wave function Aψ satisfying the above equation is said to be normalized.

$$\therefore \int_{-\infty}^{\infty} \left|(A\psi)^{\times}(\vec{r},t)\, A\Psi(\vec{r},t)\right| d\tau = |A|^2 \left|\Psi(\vec{r},t)\right|^2 d\tau = 1 \;\; or,\; |A|^2 = \frac{1}{\int \Psi\Psi^{\times} dxdydz}$$

Where $|A|$ is normalizing Constant. *From the normalization condition we can find the value of the normalizing constant and its sign.*

2.6 Normalization constant and Normalization wave function for a trial wave function [Trial wave function for Linear Harmonic Oscillator and Hydrogen atom]

[a] a trial wave function $\psi(x) = e^{\frac{-\alpha x^2}{2}}$ for Linear Harmonic Oscillator

Let the trial wave function $\Psi(x) = e^{-\alpha x^2/2}$ [1]

And the normalized wave function $\Psi(x) = A e^{-\alpha x^2/2}$ [2]

First of all, we have to find out the Normalization constant A. From the normalization condition

$$\int_{-\infty}^{\infty} \Psi^{\times}\Psi\, d\tau = 1 \quad or,\; \int_{-\infty}^{\infty} A^{\times} e^{-\alpha x^2/2} A e^{-\alpha x^2/2} dx = 1$$

$$Or, \quad |A|^2 \int_{-\infty}^{\infty} e^{-\alpha x^2} dx = 1 \qquad [3]$$

Let $Z = \alpha x^2 \quad or, x = \sqrt{\frac{z}{\alpha}} \quad or,\; dz = 2\alpha x dx \quad or,\; dx = \frac{dz}{2\alpha x} = \frac{dz}{2\alpha}\sqrt{\frac{\alpha}{z}} = \frac{z^{-\frac{1}{2}} dz}{2\sqrt{\alpha}}$

From equation [1]

$$or,\; |A|^2\, 2\int_0^{\infty} e^{-z} \frac{z^{-\frac{1}{2}} dz}{2\sqrt{\alpha}} = 1 \quad or,\; \frac{|A|^2}{\sqrt{\alpha}} \int_0^{\infty} e^{-z} z^{\frac{1}{2}-1} dz = 1$$

$$or,\; \frac{|A|^2}{\sqrt{\alpha}} \Gamma\left(\frac{1}{2}\right) = 1 \quad or,\; \frac{|A|^2}{\sqrt{\alpha}} \sqrt{\pi} = 1 \quad or,\; A = \left(\frac{\alpha}{\pi}\right)^{\frac{1}{4}}$$

Hence Normalization constant $A = \left(\frac{\alpha}{\pi}\right)^{\frac{1}{4}}$ [5]

Thus the normalized wave function is

$$\Psi(x)=\left(\frac{\alpha}{\pi}\right)^{\frac{1}{4}} e^{-\alpha x^2/2} \qquad [4]$$

[b] a trial wave function $\psi = e^{-\frac{\alpha r}{r_0}}$; for Hydrogen atom

The trial wave function $\psi = e^{-\frac{\alpha r}{r_0}}$

Let the normalized trial wave function $\psi = Ae^{-\frac{\alpha r}{r_0}}$ [1]

For Hydrogen atom the potential energy will be $V = -\frac{e^2}{r}$

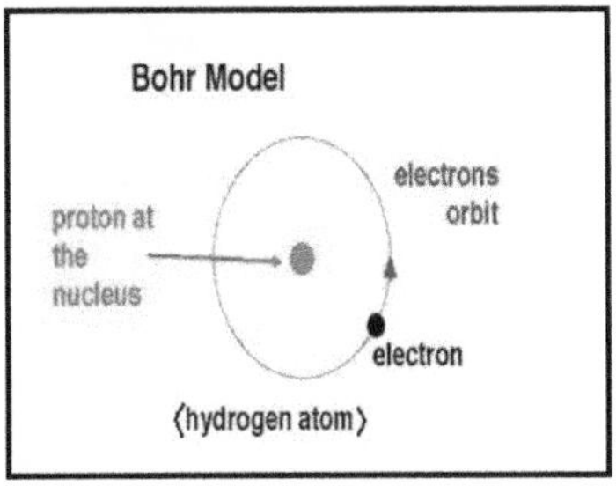

Fig.[2.6]: Two body system: Hydrogen atom

This is spherically Symmetric. First of all, we have to find out the Normalization constant A.

From the normalization condition

$$\int_{-\infty}^{\infty} \Psi^{\times}\Psi \, d\tau = 1$$

$$\int \Psi^{\times}\Psi d\tau = \int_0^{\infty}\int_0^{\pi}\int_0^{2\pi} A^{\times}e^{-\alpha r/r_0} \, Ae^{-\alpha r/r_0} \, r^2 \sin\theta \, d\theta \, d\phi \, dr$$

$$= |A|^2 (2\pi)[-\cos\theta]_0^{\pi} \int_0^{\infty} r^2 e^{-2\alpha r/r_0} dr = |A|^2 (4\pi)\int_0^{\infty} r^2 e^{-2\alpha r/r_0} dr$$

But we know from standard integral $\int_0^{\infty} x^n \, e^{-\alpha x} dx = \frac{n!}{\alpha^{n+1}}$

$$\therefore \int \psi^{\times}\psi dx = |A|^2 (4\pi)\frac{2!}{\left(\frac{2\alpha}{r_0}\right)^{2+1}} = |A|^2 (4\pi)\frac{2}{\frac{8\alpha^3}{r_0^3}} = |A|^2 \, 4\pi 2\frac{r_0^3}{8\alpha^3} = |A|^2 \frac{\pi r_0^3}{\alpha^3}$$

Or, $1 = |A|^2 \frac{\pi r_0^3}{\alpha^3}$ $\therefore |A| = \left(\frac{\alpha^3}{\pi r_0^3}\right)^{\frac{1}{2}}$

Hence Normalization constant $|A| = \left(\frac{\alpha^3}{\pi r_0^3}\right)^{\frac{1}{2}}$ [2]

Thus the normalized wave function

$$\Psi = Ae^{-\alpha r/r_0} = \left(\frac{\alpha^3}{\pi r_0^3}\right)^{\frac{1}{2}} e^{-\alpha r/r_0} \quad [3]$$

2.7 Normalization constant and Normalization wave function for one dimensional case

$$\psi(x,t) = sin\left(\frac{\pi x}{2a}\right)e^{\frac{-iEt}{\hbar}} \ ; -a \leq x \leq +a$$

The trial wave function $\psi(x,t) = sin\left(\frac{\pi x}{2a}\right)e^{\frac{-iEt}{\hbar}}$

Let the normalized wave function $\psi(x,t) = Asin\left(\frac{\pi x}{2a}\right)e^{\frac{-iEt}{\hbar}}$ [1]

First of all we have to find out the Normalization constant A. From the normalization condition

$$\int_{-\infty}^{+\infty} \psi^{\times} \psi dx = 1$$

$$\int_{-a}^{+a} A^{\times} sin\left(\frac{\pi x}{2a}\right)e^{\frac{+iEt}{\hbar}} Asin\left(\frac{\pi x}{2a}\right)e^{\frac{-iEt}{\hbar}} dx = 1$$

$$\int_{-a}^{+a} |A|^2 sin^2\left(\frac{\pi x}{2a}\right)dx = 1 \ or, \ \frac{|A|^2}{2}\int_{-a}^{+a} 2\,sin^2\left(\frac{\pi x}{2a}\right)dx = 1$$

$$\frac{|A|^2}{2}\int_{-a}^{+a}\left[1 - cos2\left(\frac{\pi x}{2a}\right)\right]dx = 1 \ or, \frac{|A|^2}{2}\left[x - \frac{a}{\pi}sin\frac{\pi x}{a}\right]_{-a}^{+a} = 1$$

$$\frac{|A|^2}{2}[a + a] = 1 \ or, \ \frac{|A|^2}{2}2a = 1 \ or, \ A = \frac{1}{\sqrt{a}}$$

Hence Normalization constant $A = \frac{1}{\sqrt{a}}$ [2]

Thus the normalized wave function $\psi(x,t) = \frac{1}{\sqrt{a}} sin\left(\frac{\pi x}{2a}\right)e^{\frac{-iEt}{\hbar}}$ [3]

2.8 Expectation value of dynamical variables

The expectation value is the mathematical expectation for the result of a single measurement or it is the average of the large number of measurements on independent systems. But in general, wave mechanics admits a fluctuation in these measurements, while in Classical Mechanics it is assumed that every variable is, in principle, absolutely determined.

The expectation value or the average value of any dynamical variable $\Omega(x)$ is defined as

$$\langle \Omega(x) \rangle = \int \Psi^{\times} \Omega(x) \Psi dx \qquad [1]$$

- The above equation is valid for one dimension
- The formula is applicable only if the wave function is normalized

If $\Psi(x)$ is not a normalized wave function then the expectation value or the average value of any dynamical variable $\Omega(x)$ is defined as

$$\langle \Omega(x) \rangle = \frac{\int \Psi^{\times} \Omega(x) \Psi dx}{\int \Psi^{\times} \Psi dx} \qquad [2]$$

$$\langle \Omega(\vec{r}) \rangle = \frac{\int \Psi^{\times} \Omega(\vec{r}) \Psi d\tau}{\int \Psi^{\times} \Psi d\tau} \qquad [3]$$

2.9 Expectation value of position, momentum and energy [Dynamical variables] for one dimensional wave function

$$\psi(x,t) = sin\left(\frac{\pi x}{2a}\right) e^{\frac{-iEt}{\hbar}} \ ; \ -a \le x \le +a$$

If $\Psi(x)$ is not a normalized wave function then the expectation value or the average value of any dynamical variable $\Omega(x)$ is defined as

$$\langle \Omega(x) \rangle = \frac{\int \Psi^{\times} \Omega(x) \Psi dx}{\int \Psi^{\times} \Psi dx} \qquad [1]$$

From the above equation, the expectation value of position can be written

$$\langle x \rangle = \frac{\int_{-a}^{+a} \psi^{\times} x \psi dx}{\int_{-a}^{+a} \psi^{\times} \psi dx} \qquad [2]$$

The above equation [2] is applicable only if the wave function is not normalized. The expectation value or the average value of any dynamical variable x is defined as

$$\langle x \rangle = \int \Psi^{\times} x \Psi dx \qquad [3]$$

The above equation [3] is applicable only if the wave function is normalized.

The trial wave function $\psi(x,t) = sin\left(\frac{\pi x}{2a}\right) e^{\frac{-iEt}{\hbar}}$

Let the normalized wave function $\psi(x,t) = Asin\left(\frac{\pi x}{2a}\right) e^{\frac{-iEt}{\hbar}}$ [4]

First of all, we have to find out the Normalization constant A. From the normalization condition

$$\int_{-\infty}^{+\infty} \psi^{\times} \psi dx = 1$$

$$\int_{-a}^{+a} A^{\times} sin\left(\frac{\pi x}{2a}\right) e^{\frac{+iEt}{\hbar}} Asin\left(\frac{\pi x}{2a}\right) e^{\frac{-iEt}{\hbar}} dx = 1$$

$$\int_{-a}^{+a} |A|^2 sin^2\left(\frac{\pi x}{2a}\right) dx = 1 \; or, \; \frac{|A|^2}{2}\int_{-a}^{+a} 2\,sin^2\left(\frac{\pi x}{2a}\right) dx = 1$$

$$\frac{|A|^2}{2}\int_{-a}^{+a}\left[1 - cos2\left(\frac{\pi x}{2a}\right)\right] dx = 1 \; or, \frac{|A|^2}{2}\left[x - \frac{a}{\pi} sin\frac{\pi x}{a}\right]_{-a}^{+a} = 1$$

$$\frac{|A|^2}{2}[a + a] = 1 \; or, \; \frac{|A|^2}{2} 2a = 1 \; or, \; A = \frac{1}{\sqrt{a}}$$

Hence Normalization constant $A = \frac{1}{\sqrt{a}}$ [5]

Thus the normalized wave function $\psi(x,t) = \frac{1}{\sqrt{a}} sin\left(\frac{\pi x}{2a}\right) e^{\frac{-iEt}{\hbar}}$ [6]

From equation [3], **the expectation value of position**

$$\langle x \rangle = \int_{-a}^{+a} \psi^{\times} x \psi dx = \int_{-a}^{+a}\left(\frac{1}{\sqrt{a}}\right)^{\times} sin\left(\frac{\pi x}{2a}\right) e^{\frac{+iEt}{\hbar}} \; x \frac{1}{\sqrt{a}} sin\left(\frac{\pi x}{2a}\right) e^{\frac{-iEt}{\hbar}} \quad [7]$$

$$= \int_{-a}^{+a}\left|\frac{1}{\sqrt{a}}\right|^2 x\, sin^2\left(\frac{\pi x}{2a}\right) dx = \frac{1}{2a}\int_{-a}^{+a} x\, 2sin^2\left(\frac{\pi x}{2a}\right) dx$$

$$= \frac{1}{2a}\int_{-a}^{+a} x\left[1 - cos2\left(\frac{\pi x}{2a}\right)\right] dx \quad or, \frac{1}{2a}\int_{-a}^{+a} x dx - \frac{1}{2a}\int_{-a}^{+a} x\, cos\frac{\pi x}{a} dx$$

$$= \frac{1}{2a}\left[\frac{x^2}{2}\right]_{-a}^{+a} - \frac{1}{2a}\left[x\,\frac{a}{\pi}\, sin\frac{\pi x}{a}\right]_{-a}^{+a} + \frac{1}{2a}\int_{-a}^{+a}\left[\frac{dx}{dx}\int_{-a}^{+a} cos\frac{\pi x}{a}\right] dx$$

$$= 0 - 0 + \frac{1}{2a}\int_{-a}^{+a}\frac{a}{\pi} sin\left(\frac{\pi x}{a}\right) d$$

$$\langle x \rangle \quad = \frac{1}{2a}.\frac{a}{\pi}.\frac{a}{\pi}. - \left[cos\frac{\pi x}{a}\right]_{-a}^{+a} = -\frac{a}{2\pi^2}[cos\pi - cos(-\pi)] = 0$$

From equation [3], **the expectation value of momentum**

$$\langle p \rangle = \int_{-a}^{+a} \psi^{\times} p \psi dx = \int_{-a}^{+a}\left(\frac{1}{\sqrt{a}}\right)^{\times} sin\left(\frac{\pi x}{2a}\right) e^{\frac{+iEt}{\hbar}} \; p\frac{1}{\sqrt{a}} sin\left(\frac{\pi x}{2a}\right) e^{\frac{-iEt}{\hbar}} dx$$

$$\langle p \rangle = \int_{-a}^{+a}\left(\frac{1}{\sqrt{a}}\right)^{\times} sin\left(\frac{\pi x}{2a}\right) e^{\frac{+iEt}{\hbar}}\left(-i\hbar\frac{d}{dx}\right)\frac{1}{\sqrt{a}} sin\left(\frac{\pi x}{2a}\right) e^{\frac{-iEt}{\hbar}} dx$$

$$\langle p \rangle = \frac{-i\hbar}{a}\int_{-a}^{+a} sin\left(\frac{\pi x}{2a}\right)\frac{d}{dx}\ sin\left(\frac{\pi x}{2a}\right)dx$$

$$\langle p \rangle = \frac{-i\hbar}{a}\int_{-a}^{+a} sin\left(\frac{\pi x}{2a}\right)\frac{\pi}{2a}\ cos\left(\frac{\pi x}{2a}\right)dx$$

$$\langle p \rangle = \frac{-i\hbar}{a}\frac{\pi}{2a}\frac{1}{2}\int_{-a}^{+a} 2sin\left(\frac{\pi x}{2a}\right)\ cos\left(\frac{\pi x}{2a}\right)dx$$

$$\langle p \rangle = \frac{-i\hbar\pi}{4a^2}\int_{-a}^{+a} sin2\left(\frac{\pi x}{2a}\right)\ dx$$

$$\langle p \rangle = \frac{-i\hbar\pi}{4a^2}\frac{a}{\pi}\left[-cos\frac{\pi x}{a}\right]_{-a}^{+a} \quad Or, \langle p \rangle = \frac{i\hbar}{4a}[cos\pi - cos\pi] = 0$$

From equation [3], **the expectation value of energy**

$$\langle E \rangle = \int_{-a}^{+a} \psi^{\times} E\psi dx = \int_{-a}^{+a}\left(\frac{1}{\sqrt{a}}\right)^{\times} sin\left(\frac{\pi x}{2a}\right)e^{\frac{+iEt}{\hbar}}\ E\frac{1}{\sqrt{a}}sin\left(\frac{\pi x}{2a}\right)e^{\frac{-iEt}{\hbar}}dx$$

$$\langle E \rangle = \int_{-a}^{+a}\left(\frac{1}{\sqrt{a}}\right)^{\times} sin\left(\frac{\pi x}{2a}\right)e^{\frac{+iEt}{\hbar}}\left(i\hbar\frac{d}{dt}\right)\frac{1}{\sqrt{a}}sin\left(\frac{\pi x}{2a}\right)e^{\frac{-iEt}{\hbar}}dx$$

$$\langle E \rangle = \frac{i\hbar}{a}\int_{-a}^{+a} sin^2\left(\frac{\pi x}{2a}\right)dx = \frac{-i^2\hbar E}{2a\hbar}\int_{-a}^{+a} 2sin^2\left(\frac{\pi x}{2a}\right)dx$$

$$\langle E \rangle = \frac{E}{2a}\int_{-a}^{+a}\left[1 - cos2\left(\frac{\pi x}{2a}\right)\right]dx = \frac{E}{2a}\int_{-a}^{+a}dx - \int_{-a}^{+a} cos\frac{\pi x}{a}dx$$

$$\langle E \rangle = \frac{E}{2a}[a + a] - \frac{E}{2a}\times 0 = \frac{E}{2a}2a \quad Or,\ \langle E \rangle = E$$

2.10 Orthogonal and Ortho-normal condition of Wave function

If the product of a function $\Psi_1(x)$ and the complex conjugate $\Psi^{\times}_2(x)$ of a function $\Psi_2(x)$ vanishes, when integrated with respect to x over the interval $a \le x \le b$.

$$\int_a^b \Psi_2^{\times}(x)\Psi_1(x)dx = 0 \quad [1]$$

Then $\Psi_1(x)$ and $\Psi_2(x)$ are said to be mutually orthogonal or simply orthogonal in the interval (a,b). If two wave functions Ψ_m and Ψ_n are said to be normalized if and only if

$$\left.\begin{aligned}\int_{-\infty}^{\infty}\Psi_m^{\times}(x)\Psi_n(x)dx = \delta_{mn} = 1\\ \int_{-\infty}^{\infty}\Psi_n^{\times}(x)\Psi_m(x)dx = \delta_{nm} = 1\end{aligned}\right\}\ \text{When } m = n;\quad \delta_{mn} = \delta_{nm} = 1 \quad [2]$$

These functions are said to be mutually orthogonal if

$$\left.\begin{aligned}\int_{-\infty}^{\infty}\Psi_m^{\times}(x)\Psi_n(x)dx = \delta_{mn} = 0\\ \int_{-\infty}^{\infty}\Psi_n^{\times}(x)\Psi_m(x)dx = \delta_{nm} = 0\end{aligned}\right\} \quad \text{When } m \neq n;\ \delta_{mn} = \delta_{nm} = 0 \qquad [3]$$

Thus *the functions which are orthogonal and also normalized are called ortho-normal wave functions.*

2.11 Bra and Ket notation and its property

The scalar product of two state vectors Ψ_m and Ψ_n can be written as

$$\left(\vec{\Psi}_m.\vec{\Psi}_n\right) = \int \Psi_m^{\times}\Psi_n\,\partial\tau \qquad [1]$$

And
$$\left(\vec{\Psi}_n.\vec{\Psi}_m\right) = \int \Psi_n^{\times}\Psi_m\,\partial\tau = (\Psi_m.\Psi_n)^{\times} \qquad [2]$$

This shows that
$$\left(\vec{\Psi}_n.\vec{\Psi}_m\right) \neq (\Psi_m.\Psi_n) \qquad [3]$$

Further,
$$\left.\begin{aligned}\left(\vec{\Psi}_m.C\vec{\Psi}_n\right) &= C\left(\vec{\Psi}_m.\vec{\Psi}_n\right)\\ \left(C\vec{\Psi}_m.\vec{\Psi}_n\right) &= C^{\times}\left(\vec{\Psi}_m.\vec{\Psi}_n\right)\end{aligned}\right\} \qquad [4]$$

The scalar product of two state vectors Ψ_m and Ψ_n can be written in the new notation

$$\left(\vec{\Psi}_m,\vec{\Psi}_n\right) = \langle\Psi_m|\Psi_n\rangle \qquad [5]$$

Where the symbol $\langle\Psi_m|$ is called the Bra vector corresponding to a Ket vector $|\Psi_n\rangle$. In this way Bra and Ket notations distinguish very clearly between pre-factor and post-factor wave vectors in a scalar product. Merely writing Ψ_m or Ψ_n one cannot know whether it is to be pre-factor or the post-factor. But writing $\langle\Psi_m|$ and $|\Psi_n\rangle$ clearly tells that Ψ_m will be a pre-factor and Ψ_n will be a post-factor. *We shall represents that the state vector of the wave functions Ψ by $|\Psi_n\rangle$ and its complex conjugate wave functions $\Psi^{\times}$ by $\langle\Psi_m|$*

Property of Bra and Ket notation

- Operation on a Ket vector from the left with an operator Ω_{op} produces another Ket vector I.e., $\hat{\Omega}|\Psi_n\rangle = |\Psi'\rangle$

- Operation on a Bra vector from the right with an operator Ω_{op} produces another Bra vector I.e., $\langle\Psi|\hat{\Omega}=\langle\Psi'|$
- The expectation value of an operator Ω_{op} in the state Ψ can be written as $\langle\hat{\Omega}\rangle=\int\Psi^{\times}\Omega\Psi\partial\tau=\int\Psi^{\times}\Psi'\partial\tau=(\vec{\Psi}.\vec{\Psi}')=\langle\Psi|\Psi'\rangle=\langle\Psi|\hat{\Omega}|\Psi\rangle$

2.12 de-Broglie's stationary Wave Quantized Orbits: Wave-mechanical concept of atom

The circumference must contain an integral number of waves.

I.e., $$2\pi r = n\lambda \quad [1]$$

Where r is the radius of the permitted orbit and n is a positive integer.

$$2\pi r = n\frac{h}{p} = n\frac{h}{mv} \quad [2]$$

$$\frac{2\pi rmv}{h} = n \quad or, \; mvr = \frac{nh}{2\pi} \quad [3]$$

For any permitted orbit; angular momentum

$$L = I\omega = mr^2\frac{v}{r}$$

The above equation can be written as

$$mvr = \frac{nh}{2\pi} = n\hbar \quad [4]$$

Or, $$mvr = L = n\hbar \quad [5]$$

Equation [5] represents the wave-mechanical concept of atom which also known as Bohr's quantization condition.

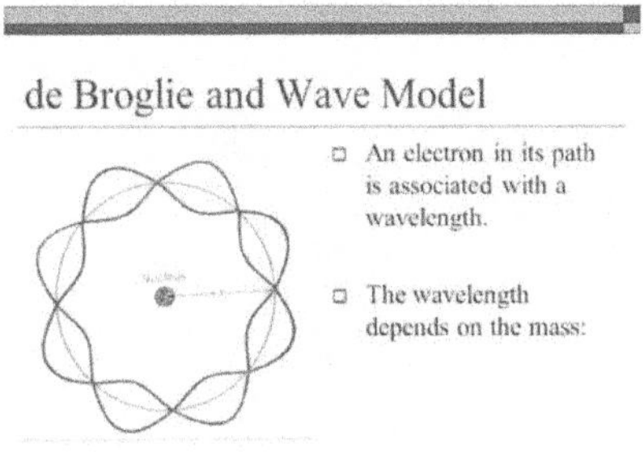
de Broglie and Wave Model
An electron in its path is associated with a wavelength.
The wavelength depends on the mass:

Fig.[2.12]: Shows Bohr's quantization condition

2.13 Brief description of Uncertainty Principle:

Section: 01 *A narrow de- Broglie wave group*, as shown in figure has a more precise particle's position that can be specified. However, the wavelength of the waves in a narrow packet is not well-defined; there are not enough waves to measure λ accurately.

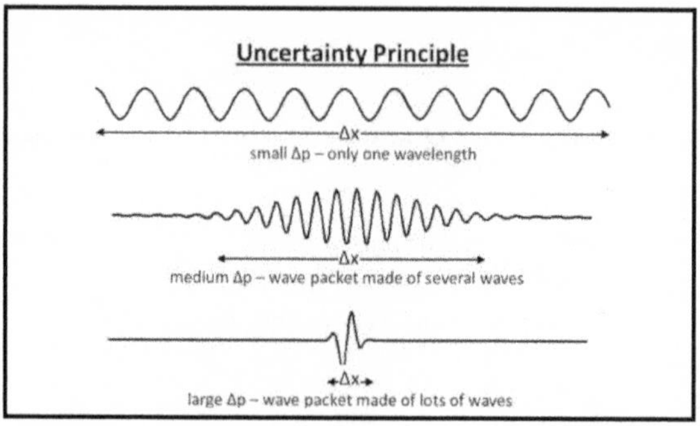

Fig.[2.13(a)]: Shows at the bottom $\Delta\, x\ small\ and\ \Delta\, p_x\ large$

This means that $[\lambda = \frac{h}{p} = \frac{h}{mv}]$ the particles momentum mv is not a precise quantity. If we make a series of momentum measurements we will find a broad range of values.

Section: 02 *A wide de-Broglie wave group* as shown in figure has a clearly defined wavelength. The momentum that corresponds to this wavelength is therefore, a precise quantity. A series of measurements will give a narrow range of values. But where is the particle located? The width $\Delta\, x$ of the group is too large for us to be able to say exactly where it is at a given time.

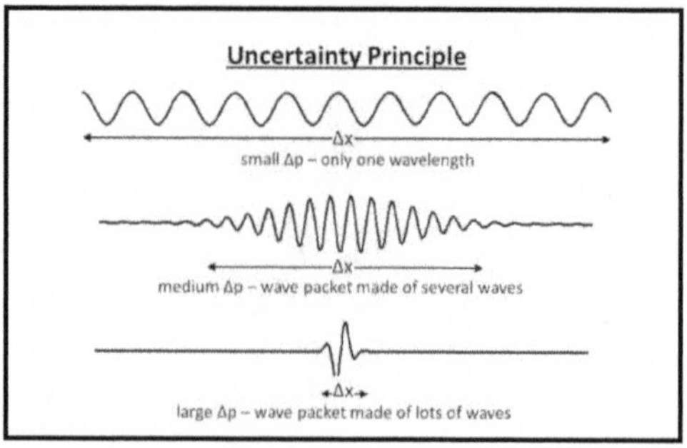

Fig.[2.13(b)]: Shows on the top $\Delta\, x\ large\ and\ \Delta\, p_x\ small$

Thus we have the Uncertainty Principle- *"It is impossible to know both the exact position & the exact momentum of an object at the same time"*.

On the basis of these considerations Werner Heisenberg in 1972, enunciated the principle of *unbestimmtheit. This term has been translated as uncertainty, Indeterminacy or Indefiniteness.* Werner Heisenberg has developed *unbestimmtheit as* [*uncertainty, indeterminacy or Indefiniteness*] "It is impossible to specify precisely & simultaneously the values of both members of particular pairs of physical variables that describe the behavior of an atomic system." The members of these pairs of variables are canonically conjugated with each other in the Hamiltonian sense.

More quantitatively, the Werner Heisenberg's Uncertainty Principle states that the order of magnitude of the product of the uncertainty in the knowledge of the two variables must be at least Planck's constant h divided by 2π. [$\hbar = \frac{h}{2\pi} = 1.0545 \times 10^{-27}\text{erg} - \text{sec or}, \hbar = \frac{6.6023}{2\pi} = 1.0545 \times 10^{-31}\text{joule} - \text{sec}$]. So, Heisenberg clarified Uncertainty Principle in different ways

$$\Delta x \, \Delta p_x \geq \hbar \qquad [1]$$

$$\Delta \varphi \, \Delta J_z \geq \hbar \qquad [2]$$

$$\Delta t \, \Delta E \geq \hbar \qquad [3]$$

Significance of equation [1]: The relation [1] means that a component of the momentum of a particle cannot be precisely specified without loss of all knowledge of the corresponding component of its position at that time. Similarly, a particle cannot be precisely localized in a particular direction without loss of all knowledge of its momentum component in that direction. So, the product of the uncertainties of the simultaneously measurable values of corresponding position & momentum components is at least of the order of magnitude of $\hbar$.

Significance of equation [2]: The relation [2] means that- the precise measurement of the angular position of a particle in an orbit carries with it the loss at that time of all knowledge of the component of angular momentum perpendicular to the place of the orbit.

Significance of equation [3]: The relation [3] means that an energy determination that has an accuracy ΔE must occupy at least a time interval $\Delta t \sim \frac{\hbar}{\Delta E}$. Thus if a system maintains a particular state of motion not longer than a time Δt , the energy of the system is that state is uncertain by at least the amount $\Delta t \sim \frac{\hbar}{\Delta E}$. Since Δt is the longest time interval available for the energy determination.

2.14 Physical significance of Heisenberg Uncertainty relation

If the position coordinate x of a particle in motion is accurately determined at some instant, so that $\Delta x = 0$ then at the same time [the uncertainty Δp_x in the determination of the momentum] Δp_x becomes infinite

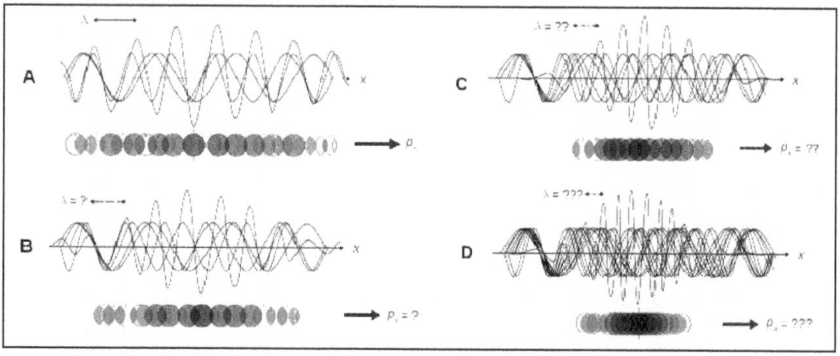

Fig.[2.14] Shows of Heisenberg Uncertainty relation

It the momentum p_x of a particle is accurately determined at some instant, so that $\Delta p_x = 0$, and then at the same instant the uncertainty Δx in the determination of the position coordinate becomes infinite.

Thus if an experiment is designed to measure x or p_x accurately, the other quantity will become completely uncertain. For a particle of mass m moving with velocity v, the product of the uncertainty Δx & the uncertainty Δv is

$$\Delta x \Delta v \geq \frac{\hbar}{m} \qquad [1]$$

Since for heavy particle $\frac{\hbar}{m}$ is very small. I.e., $\frac{\hbar}{m} \approx 0$. The uncertainties vanish & all quantities can be determined with perfect accuracy. This is the limiting case of Classical Mechanics. *Thus Classical Mechanics is true for heavy particles & the uncertainties are a characteristic of Quantum Mechanics which is applicable to light particles, such as elementary particles.* [Spin is a quantum mechanical property of an elementary particle by virtue of which it shows magnetic dipole moment.]

2.15 Heisenberg uncertainty principle on different view

If there are two canonically conjugated operators and if these operators do not commute with each other and if also in the measurement of the observables represented by one of these operators each of a certain amount, then no matter how accurately we measure the observable represented by the other operator, the accuracy in the later measurement cannot be better than Planck's constant divided by twice the uncertainty of the other experiment. [$\hbar = \frac{h}{2\pi}$]

x	p_x	$p_x = -i\hbar\frac{\partial}{\partial x}$	$\Delta x \, \Delta p_x \geq \hbar$
φ	L_x	$L_z = i\hbar\frac{\partial}{\partial \varphi}$	$\Delta \varphi \, \Delta L_z \geq \hbar$
t	E	$E = i\hbar\frac{\partial}{\partial t}$	$\Delta t \, \Delta E \geq \hbar$

2.16 Correspondence & Complementarity Principle

- ### Correspondence Principle:

Newtonian Mechanics & Classical Electrodynamics is based on thoroughly established experimental evidence. Therefore, it must be demanded that quantum theory yield, in every instance, results that become identical with those of classical physics if the mass & dimensions of the system under consideration are made to approach the mass and dimensions of classical systems.

This fundamental idea was stated by Niels Bohr in 1923 & is known as the correspondence principle. It is essential in the formulation of Quantum Mechanics. Quantum Mechanics must not only be consistent with Classical Physics but should yield the classical laws in a suitable approximation.

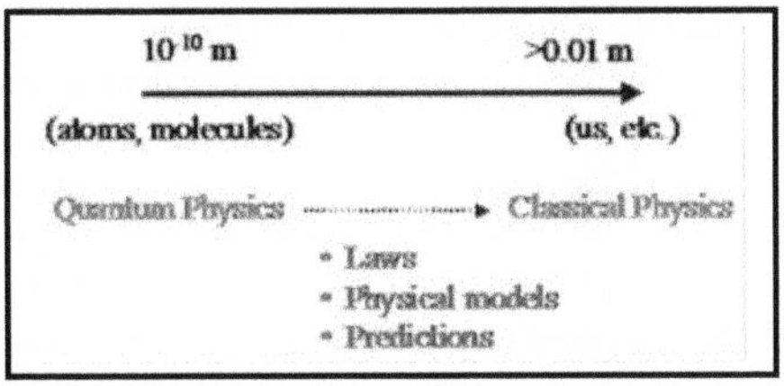

Fig.[2.15]: Shows the correspondence principle by Niels Bohr

When Quantum Physics is applied to macroscopic systems, it must reduce to the Classical Physics. Therefore, the non-classical Phenomena, such uncertainty and duality, must become undetectable. Niels Bohr codified this requirement into Correspondence principle.

- ### Complementarity Principle

Niels Bohr regarded the complementarity principle as the most fundamental principle of Quantum Mechanics. He introduced the concept of complementarity in his Lecture in 1927. The formal presentation was given in his book "*Atomic theory and the description of nature"* (Cambridge University Press, 1934). In quantum mechanics, however, evidence about atomic objects obtained by different experimental arrangements exhibits *"a novel kind of complementary relationship"*.

Specifically, what *"novel kind of complementary relationship"* is Bohr talking about? It is an often cited example is the complementarity of "particle" and "wave" views. Some experiments detect the wave nature of the elementary particles and other experiments detect the particle nature. Bohr is saying that both views are valid and complementary. *Bohr's "complementarity principle" has philosophical implications.*

According to some physicists, one of the philosophical implications of the "complementarity principle" is the view that elementary particles do not have intrinsic properties independent of the measuring device. I suppose the supporters of this view use the term "measuring device" in the most general sense. For example, any external (classical) field would be a measuring device. I suggest the term "measuring base" for this. If I go along with this line of thinking I should interpret the external field created by the attractive electric force between the protons in the nucleus and the orbiting electron as a measuring device. According to this view the measured properties of electrons in an atom such as charge, spin and mass are manifested by the local environment of the atom.

All electrons are identical because they all exhibit the exact same charge, spin and mass. It is true that one experimental setup may extract the wavelike behavior of electrons and another experimental setup may extract the particle behavior but other properties known as charge, spin and mass are still needed to explain the wavelike behavior as well as the particle behavior.

The particle & wave aspects of a physical entity are Complementarity & cannot be exhibited at the same time. *The inability to observe the wave & the particle aspects of matter at the same time is known as principle of* Complementarity. This is the Complementarity principle and was announced in 1928 by the Danish physicist Niels Bohr. Depending on the experimental arrangement, the behaviour of such phenomena as light and electrons is sometimes wavelike and sometimes particle-like; *I.e.,* such things have a wave-particle duality. The two aspects (wave & particle) of matter are necessary to understand all properties of matter but both aspects cannot be simultaneously observed.

In effect, the complementarity principle implies that phenomena on the atomic and subatomic scale are not strictly like large-scale particles or waves (*e.g.,* billiard balls and water waves). Such particle and wave characteristics in the same large-scale phenomenon are incompatible rather than complementary. Knowledge of a small-scale phenomenon, however, is essentially incomplete until both aspects are known.

2.17 Elementary proof of Heisenberg uncertainty principle (between energy and time)

Let us consider a wave-packet of width ≈ Δx. The packet advances along the positive X-axis with particle velocity v. The uncertainty in position of the particle is Δx, since the particle can be present anywhere within Δx. The uncertainty in time for the particle to cross a point say R on the X-axis is

$$\Delta t \approx \frac{\Delta x}{v} \qquad [1]$$

The uncertainty relation of position and momentum is

$$\Delta x\, \Delta p_x \geq h \qquad [2]$$

Therefore, $$\Delta t \geq \frac{h}{v}\frac{1}{\Delta p_x}$$

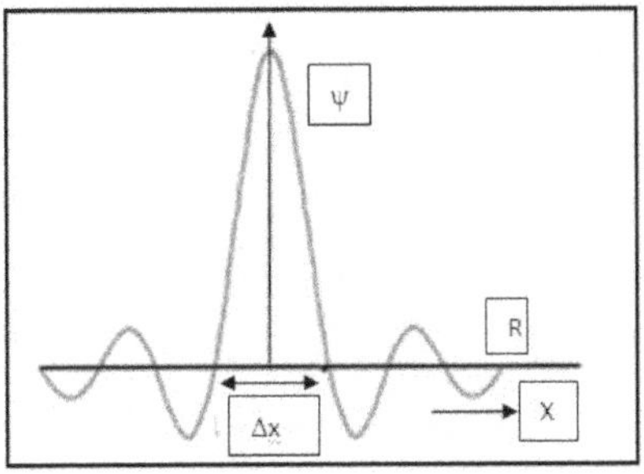

Fig.[2.17]: Shows Heisenberg uncertainty principle (between energy and time)

$$\Delta t\left(v\, \Delta p_x\right) \geq h \qquad [3]$$

For free particle, the kinetic energy

$$E_x = \frac{p_x^{\ 2}}{2m} \quad \text{or,} \quad \frac{dE_x}{dp} = \frac{2p_x}{2m} = \frac{p_x}{m} = v$$

$$dE_x = v\, dp_x \quad \text{or,} \quad \Delta E = v\, \Delta p_x \qquad [4]$$

Finally equation [3] becomes

$$\Delta t\, \Delta E \geq h \qquad [5]$$

$$\frac{1}{\Delta \upsilon}\,\Delta E \geq h \quad \text{or,} \quad \Delta E \geq h\Delta \upsilon \qquad [6]$$

Equation [6] can be written as $E = h\upsilon = n\,h\upsilon$; Where n is an integer.

2.18 Elementary proof of Heisenberg uncertainty principle (between position and momentum)

The de-Broglie wavelength of a particle of momentum p_x along the X-axis

$$\lambda = \frac{h}{p_x} \qquad [1]$$

And corresponding wave number is

$$k = \frac{2\pi}{\lambda} \qquad [2]$$

Therefore

$$p_x = \frac{h}{\lambda} = 2\pi\hbar.\frac{k}{2\pi} = \hbar k \qquad [3]$$

Hence an uncertainty $\Delta\, k$ in the wave number of the de-Broglie waves associated with the results in uncertainity $\Delta\, p$ in the particles momentum. Therefore, equation [3] becomes

$$\Delta\, p_x = \hbar\, \Delta\, k \qquad [4]$$

$$\Delta\, k = \frac{\Delta p_x}{\hbar} \qquad [5]$$

If $\Delta\, x$ is small, corresponding to narrow wave group then $\Delta\, p_x$ will be large. If $\Delta\, p_x$ is small, corresponding to wide wave group then $\Delta\, x$ will be large. So, if we reduce $\Delta\, p_x$ in some way, a broad wave group is inevitable and $\Delta\, x$ will be large.

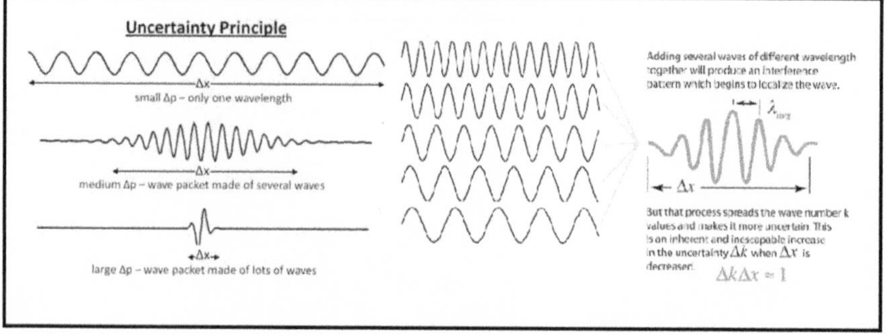

Fig.[2.18]: Shows Heisenberg uncertainty principle

But Fourier analysis of a single wave-packet in one dimension, it can be shown that the width $\Delta\, x$ of the wave-packet and the range $\Delta\, k$ of the propagating constants of the waves which gives rise to the wave-packet are given by

$$\Delta x \Delta k \geq \frac{1}{2} \ or, \ \Delta x \frac{\Delta p_x}{\hbar} \geq \frac{1}{2} \ or, \ \Delta x \Delta p_x \geq \frac{\hbar}{2} \quad [6]$$

Hence *"It is impossible to measure simultaneously the position of a particle along a particular direction and also its momentum in the same direction with unlimited accuracy*

2.19 Schwarz's inequalities:

Let us consider two sets of complex numbers

a_1, a_2, a_3 .... a_n

and b_1, b_2, b_3.. b_n.

The sum of the absolute squares is obviously non-negative. I.e.,

$$\sum_{i,j=1}^{n} \left|a_i b_j - a_j b_i\right|^2 \geq 0 \quad [1]$$

$$\Rightarrow \sum_{i,j=1}^{n} \left[\left(a_i^{\times} b^{\times}_j - a_j^{\times} b^{\times}_i\right)\left(a_i b_j - a_j b_i\right)\right] \geq 0$$

$$\Rightarrow \sum_{i,j=1}^{n} \left(a_i^{\times} a_i b^{\times}_j b_j - a_i^{\times} b_i a_j b^{\times}_j - a_i b^{\times}_i a_j^{\times} b_j + a_j^{\times} a_j b^{\times}_i b_i\right) \geq 0$$

$$But, \ \left(a_i^{\times} b_i a_j b^{\times}_j\right)^{\times} = \left(a_i b^{\times}_i a_j^{\times} b_j\right)$$

$$\therefore \sum_{i,j=1}^{n} \left(a_i^{\times} a_i b^{\times}_j b_j - a_i b^{\times}_i a_j^{\times} b_j - a_i b_i^{\times} a^{\times}_j b_j + a_j^{\times} a_j b^{\times}_i b_i\right) \geq 0$$

$$\Rightarrow \sum_{i,j=1}^{n} \left(a_i^{\times} a_i b^{\times}_j b_j + a_j^{\times} a_j b^{\times}_i b_i - 2 a_i b^{\times}_i a_j^{\times} b_j\right) \geq 0$$

$$\Rightarrow \sum_{i=j=1}^{n} \left(a_i^{\times} a_i b_i^{\times} b_i + a_i^{\times} a_i b_i^{\times} b_i - 2 a_i b_i^{\times} a_i^{\times} b_i\right) \geq 0$$

$$\Rightarrow \sum_{i=j=1}^{n} \left(2 a_i^{\times} a_i b_i^{\times} b_i - 2 a_i b_i^{\times} a_i^{\times} b_i\right) \geq 0 \quad Or, \quad 2\sum_i^{n} \left|a_i\right|^2 \left|b_i\right|^2 \geq 2\sum_i^{n} a_i^{\times} b_i \sum_i^{n} a_i b_i^{\times}$$

$$\Rightarrow \sum_i^{n} \left|a_i\right|^2 \left|b_i\right|^2 \geq \sum_i a_i b_i^{\times} \sum_i a^{\times}_i b_i \quad [2]$$

For two continuous functions a & b of x, equation [1] can also be written as

$$\int_{-\infty}^{+\infty} \left|a_i\right|^2 dx \int_{-\infty}^{+\infty} \left|b_i\right|^2 dx \geq \int_{-\infty}^{+\infty} \left|a^{\times} b dx\right|^2$$

$$\Rightarrow \int_{-\infty}^{+\infty} a^{\times} a \, dx \int_{-\infty}^{+\infty} b^{\times} b \, dx \ \geq \int_{-\infty}^{+\infty} \left|a^{\times} b dx\right|^2$$

2.20 Fundamental proof of Heisenberg uncertainty principle

Heisenberg uncertainty relation states that the product of the uncertainty Δx in the position x of a particle & the uncertainity ΔP_x in its momentum is of the order of reduce planks constant $\hbar$ or large

I.e., $$\Delta x \, \Delta P_x \geq \frac{\hbar}{2} \qquad [1]$$

Let's start from the uncertainty relation between x & P_x

$$[P_x, x] = \frac{\hbar}{i} \Rightarrow P_x x - xP_x = \frac{\hbar}{i} \qquad [2]$$

The mean square deviation of position from the average $<x>$ may be defined as

$$(\Delta x)^2 = \int_{-\infty}^{+\infty} \Psi^{\times} (\Delta x)^2 \, \Psi dx \qquad [3]$$

$$\Rightarrow (\Delta x)^2 = \int_{-\infty}^{+\infty} \Psi^{\times} (x - \langle x \rangle) \Psi dx$$

And the mean square deviation of momentum from the average $<P_x>$

$$(\Delta P_x)^2 = \int_{-\infty}^{+\infty} \Psi^{\times} (P_x - \langle P_x \rangle)^2 \, \Psi dx \qquad [4]$$

Let's define two new operators,

$$\alpha = x - \langle x \rangle \qquad [5]$$

$$\& \, \beta = P_x - \langle P_x \rangle \qquad [6]$$

Since, x and P_x are Hermitian. So x or $\langle x \rangle$ and P_x or $\langle P_x \rangle$ are also Hermitian.

$$\therefore \; [\alpha, \beta] = (\alpha\beta - \beta\alpha)$$

$$\Rightarrow [\alpha, \beta] = (x - \langle x \rangle)(P_x - \langle P_x \rangle) - (P_x - \langle P_x \rangle)(x - \langle x \rangle)$$

$$\Rightarrow [\alpha, \beta] = (xP_x + \langle x \rangle \langle P_x \rangle - \langle x \rangle P_x - \langle x \rangle \langle P_x \rangle) - (P_x x + \langle P_x \rangle \langle x \rangle - \langle P_x \rangle x - P_x \langle x \rangle)$$

$$\Rightarrow [\alpha, \beta] = (xP_x - P_x x)$$

$$\therefore \quad (\Delta x)^2 (\Delta P_x)^2 = \int_{-\infty}^{+\infty} \Psi^{\times} \alpha^2 \Psi \, dx \int_{-\infty}^{+\infty} \Psi^{\times} \beta^2 \Psi \, dx$$

$$= \int_{-\infty}^{+\infty} \Psi^{\times} \alpha\alpha \Psi \, dx \int_{-\infty}^{+\infty} \Psi^{\times} \beta\beta \Psi \, dx$$

Now apply the Hermitian property

$$\int_{-\infty}^{+\infty} g^{\times}(x) (\Omega f(x)) \, dx = \int_{-\infty}^{+\infty} (\Omega g(x))^{\times} f(x) dx$$

Or, $$(g, \Omega f) = (\Omega g, f)$$

Or, $$(\Delta\, x)^2(\Delta\, P_x)^2 = \int_{-\infty}^{+\infty}(\alpha\Psi)^{\times}(\alpha\Psi)\, \mathrm{dx} \int_{-\infty}^{+\infty}(\beta\Psi)^{\times}(\beta\Psi)\mathrm{dx}$$

$$= \int_{-\infty}^{+\infty}|\alpha\Psi|^2\, \mathrm{dx} \int_{-\infty}^{+\infty}|\beta\Psi|^2\mathrm{dx} \qquad [7]$$

We know from Schwarz's inequality

$$\int_{-\infty}^{+\infty} a^{\times}\, a\, dx \int_{-\infty}^{+\infty} b^{\times}\, b\, dx \geq \int_{-\infty}^{+\infty}|a^{\times} b\, dx|^2$$

Where $a = a_1, a_2, a_3, \ldots\ldots\ldots\ldots\ldots\ldots\ldots\ldots\ldots\ldots\ldots\ldots\ldots\ldots\ldots\ldots\ldots\ldots a_n$.

And $b = b_1, b_2, b_3, \ldots\ldots\ldots\ldots\ldots\ldots\ldots\ldots\ldots\ldots\ldots\ldots\ldots\ldots\ldots\ldots b_n$; are two sets of complex numbers.

Therefore, $$\int_{-\infty}^{+\infty} |a|^2\, dx \int_{-\infty}^{+\infty} |b|^2\, dx \geq \int_{-\infty}^{+\infty} |a^{\times} b\, dx|^2$$

So, equation [7] becomes,

$$(\Delta\, x)^2(\Delta\, P_x)^2 = \int_{-\infty}^{+\infty}|\alpha\Psi|^2\, \mathrm{dx} \int_{-\infty}^{+\infty}|\beta\Psi|^2\mathrm{dx} \geq \int_{-\infty}^{\infty}|(\alpha\Psi)^{\times}\, \beta\Psi\mathrm{dx}|^2$$

Or, $$(\Delta\, x)^2(\Delta\, P_x)^2 \geq \int_{-\infty}^{+\infty}|\Psi^{\times}\alpha\beta\, \Psi\mathrm{dx}|^2$$

Or, $$(\Delta\, x)^2(\Delta\, P_x)^2 \geq \left|\int_{-\infty}^{+\infty}\Psi^{\times}\frac{1}{2}(\alpha\beta + \beta\alpha)\, \Psi\mathrm{dx} + \int_{-\infty}^{+\infty}\Psi^{\times}\frac{1}{2}(\alpha\beta - \beta\alpha)\, \Psi\mathrm{dx}\right|^2$$

Or, $$(\Delta\, x)^2(\Delta\, P_x)^2 \geq \frac{1}{4}\left|\int_{-\infty}^{+\infty}\Psi^{\times}(\alpha\beta + \beta\alpha)\, \Psi\mathrm{dx}\right|^2 + \frac{1}{4}\left|\int_{-\infty}^{+\infty}\Psi^{\times}(\alpha\beta - \beta\alpha)\, \Psi\mathrm{dx}\right|^2 \qquad [8]$$

[the 1st term of R.H.S omitted because it vanishes]

Or, $$(\Delta\, x)^2(\Delta\, P_x)^2 \geq \frac{1}{4}\left|\int_{-\infty}^{\infty}\Psi^{\times}(\alpha\beta - \beta\alpha)\, \Psi\mathrm{dx}\right|^2$$

Or, $$(\Delta\, x)^2(\Delta\, P_x)^2 \geq \frac{1}{4}\left|\int_{-\infty}^{\infty}\Psi^{\times}[\alpha, \beta]\, \Psi\mathrm{dx}\right|^2$$

Or, $$(\Delta\, x)^2(\Delta\, P_x)^2 \geq \frac{1}{4}\left|\int_{-\infty}^{\infty}\Psi^{\times}(xP_x - P_x x)\, \Psi\mathrm{dx}\right|^2$$

Or, $$(\Delta\, x)^2(\Delta\, P_x)^2 \geq \frac{1}{4}\left|\left(-\frac{\hbar}{i}\right)\int_{-\infty}^{\infty}\Psi^{\times}\, \Psi\mathrm{dx}\right|^2$$

Or, $$(\Delta\, x)^2(\Delta\, P_x)^2 \geq \frac{1}{4}\left|\left(-\frac{\hbar}{i}\right)\right|^2 \left[\because \int_{-\infty}^{+\infty}\Psi^{\times}\, \Psi\mathrm{dx} = 1\right]$$

Or, $(\Delta x)^2(\Delta P_x)^2 \geq \frac{\hbar^2}{4}$ Finally, $(\Delta x)(\Delta P_x) \geq \frac{\hbar}{2}$

2.21 Ehrenfest's theorem

"*Quantum mechanics gives the same results as the classical mechanics for a particle for which the expectation values of dynamical quantities [i.e. position and momentum of a particle] are involved*". That is, if we mean by the "position" and "momentum" vectors of the wave packet the weighted average or expectation values of these quantities, we can show that the Classical and Quantum motions always agree[3].

For one dimensional motion of a particle by showing that a component of the "velocity" of the wave packet will be the time rate of change of the expectation value of that component of the

position; $< x >$ depends only on the time.

[1] $\frac{\partial}{\partial t} < x > = \frac{<P_x>}{m}$ And [2] $\frac{\partial}{\partial t} < P_x > = < F_x >$

The expectation value of position x of a particle of mass m at time t is

$$< x > = \int_{-\infty}^{\infty} \psi^{\times} x \, \psi \, dx \qquad [1]$$

The rate of change of the expectation value of position

$$\frac{\partial}{\partial t} < x > = \int_{-\infty}^{\infty} x \, \frac{\partial}{\partial t}(\psi^{\times} \psi) \, dx \qquad [2]$$

Because ψ and $\psi^{\times}$ is a function of time t. I.e., $\psi\,(t, x)$. The term $\frac{\partial}{\partial t}(\psi^{\times} \psi)$ in equation [2] which can be calculated from the equation of continuity.

$$\frac{\partial \rho}{\partial t} + \bar{\nabla}.\bar{J} = 0 \qquad [3]$$

$$\frac{\partial}{\partial t}(\psi^{\times} \psi) = -\bar{\nabla}.\bar{J} \qquad [4]$$

But the current density $\bar{J}$ is

$$\bar{J} = \frac{\hbar}{2im}[\psi^{\times}(\bar{\nabla}\psi) - (\bar{\nabla}\psi^{\times})\psi] \qquad [5]$$

Hence equation [4] becomes

$$\frac{\partial}{\partial t}(\psi^{\times} \psi) = -\frac{\hbar}{2im}[\bar{\nabla}.\{\psi^{\times}(\bar{\nabla}\psi) - (\bar{\nabla}\psi^{\times})\psi\}]$$

$$= -\frac{\hbar}{2\,i\,m}\left[\frac{\partial}{\partial x}\left\{\psi^\times\left(\frac{\partial\psi}{\partial x}\right) - \left(\frac{\partial\psi^\times}{\partial x}\right)\psi\right\}\right]$$

Therefore equation [2]

$$\frac{\partial}{\partial t} < x > = \frac{i\hbar}{2m}\int_{-\infty}^{\infty} x\frac{\partial}{\partial x}\left\{\psi^\times\left(\frac{\partial\psi}{\partial x}\right) - \left(\frac{\partial\psi^\times}{\partial x}\right)\psi\right\}\,dx$$

$$= \frac{i\hbar}{2m}\left[x\int_{-\infty}^{\infty}\frac{\partial}{\partial x}\left\{\psi^\times\left(\frac{\partial\psi}{\partial x}\right) - \left(\frac{\partial\psi^\times}{\partial x}\right)\psi\right\}\,dx - \int_{-\infty}^{\infty}\left\{\frac{\partial}{\partial x}(x)\int_{-\infty}^{\infty}\frac{\partial}{\partial x}\left(\psi^\times\frac{\partial\psi}{\partial x} - \frac{\partial\psi^\times}{\partial x}\psi\right)dx\right\}\right]$$

$$= \frac{i\hbar}{2m}\left[x\left(\psi^\times\left(\frac{\partial\psi}{\partial x}\right) - \left(\frac{\partial\psi^\times}{\partial x}\right)\psi\right)\right]_{-\infty}^{\infty} - \frac{i\hbar}{2m}\int_{-\infty}^{\infty}\left(\psi^\times\left(\frac{\partial\psi}{\partial x}\right) - \left(\frac{\partial\psi^\times}{\partial x}\right)\psi\right)\,dx \qquad [6]$$

For well-behaved wave functions ψ and $\frac{\partial\psi}{\partial x}$ approach zero as $x \rightarrow \pm\infty$. Hence the first term vanishes at both points.

$$\therefore \frac{\partial}{\partial t} < x > = -\frac{i\hbar}{2m}\int_{-\infty}^{\infty}\left(\psi^\times\frac{\partial\psi}{\partial x} - \frac{\partial\psi^\times}{\partial x}\psi\right)dx$$

$$= -\frac{i\hbar}{2m}\int_{-\infty}^{\infty}\psi^\times\frac{\partial\psi}{\partial x}dx + \frac{i\hbar}{2m}\int_{-\infty}^{\infty}\frac{\partial\psi^\times}{\partial x}\psi dx \qquad [7]$$

But the expectation value of $< P_x >$ is

$$< P_x > = \int_{-\infty}^{\infty}\psi^\times P_x\psi dx = \int_{-\infty}^{\infty}\psi^\times\frac{\hbar}{i}\frac{\partial\psi}{\partial x}\,dx$$

Or, $$\int_{-\infty}^{\infty}\psi^\times\frac{\partial\psi}{\partial x}\,dx = \frac{i}{\hbar} < P_x > \qquad [8]$$

Again $$< P_x > = \int_{-\infty}^{\infty}\psi^\times P_x\psi\,dx = \int_{-\infty}^{\infty}\psi^\times\frac{\hbar}{i}\frac{\partial\psi}{\partial x}\,dx \qquad [9]$$

But $\left[(\psi, p\psi = p\psi, \psi)\ or,\ \int_{-\infty}^{\infty}\psi^\times P_x\psi\,dx = \int_{-\infty}^{\infty}(P_x\psi)^\times\psi\,dx\right]$; therefore equation [9] becomes,

$$< P_x > = \int_{-\infty}^{\infty}\left(\frac{\hbar}{i}\frac{\partial\psi}{\partial x}\right)^\times\psi\,dx$$

$$= -\frac{\hbar}{i}\int_{-\infty}^{\infty}\frac{\partial\psi^\times}{\partial x}\psi\,dx$$

Or, $$\int_{-\infty}^{\infty}\frac{\partial\psi^\times}{\partial x}\psi\,dx = -\frac{i}{\hbar} < P_x > \qquad [10]$$

Finally [7] becomes

$$\frac{\partial}{\partial t} < x > = -\frac{i\hbar}{2m}\int_{-\infty}^{\infty}\psi^\times\frac{\partial\psi}{\partial x}dx + \frac{i\hbar}{2m}\int_{-\infty}^{\infty}\frac{\partial\psi^\times}{\partial x}\psi dx$$

$$= - \frac{i\hbar}{2m}\left(\frac{i}{\hbar}\right) < P_x > + \frac{i\hbar}{2m}(-\frac{i}{\hbar}) < P_x >$$

$$\boxed{\frac{\partial}{\partial t} < x > = \frac{< P_x >}{m}} \qquad [11]$$

Again the momentum expectation value in the integral form

$$< P_x > = \int_{-\infty}^{\infty} (\psi, P_x \psi)$$

The rate of change of momentum expectation value

$$\frac{\partial}{\partial t} < P_x > = \int_{-\infty}^{\infty} \frac{\partial \psi^\times}{\partial t} P\psi + \psi^\times \frac{\hbar}{i}\frac{\partial}{\partial t}\left(\frac{\partial \psi}{\partial x}\right); \quad where\ P_x = \frac{\hbar}{i}\frac{\partial}{\partial x}$$

$$= \frac{\hbar}{i}\int_{-\infty}^{\infty}\left[\frac{\partial \psi^\times}{\partial t}\frac{\partial \psi}{\partial x} + \psi^\times \frac{\partial^2 \psi}{\partial x\, \partial t}\right] dx$$

$$= \int_{-\infty}^{\infty} -i\hbar\left[\frac{\partial \psi^\times}{\partial t}\frac{\partial \psi}{\partial x} + \psi^\times(-i\hbar)\frac{\partial^2 \psi}{\partial x\, \partial t}\right] dx \qquad [12]$$

From time dependent Schrödinger equation in one dimension

$$i\hbar\frac{\partial \psi}{\partial t} = -\frac{\hbar^2}{2m}\frac{\partial^2 \psi}{\partial x^2} + V\psi \qquad [13]$$

Taking complex conjugate on both sides of above equation

$$-i\hbar\frac{\partial \psi^\times}{\partial t} = -\frac{\hbar^2}{2m}\frac{\partial^2 \psi^\times}{\partial x^2} + V\psi^\times \qquad [14]$$

Now differentiating equation [13] with respect to x

$$i\hbar\frac{\partial^2 \psi}{\partial x \partial t} = -\frac{\hbar^2}{2m}\frac{\partial^3 \psi}{\partial x^3} + \frac{\partial}{\partial x}(V\psi)$$

From equation [12]

$$\frac{\partial}{\partial t} < P_x > = \int_{-\infty}^{\infty} -i\hbar\left[\frac{\partial \psi^\times}{\partial t}\frac{\partial \psi}{\partial x} + \psi^\times(-i\hbar)\frac{\partial^2 \psi}{\partial x\, \partial t}\right] dx$$

$$\frac{\partial}{\partial t} < P_x > = \int_{-\infty}^{\infty}\left[\left(-\frac{\hbar^2}{2m}\frac{\partial^2 \psi^\times}{\partial x^2} + V\psi^\times\right)\frac{\partial \psi}{\partial x} - \psi^\times\left(-\frac{\hbar^2}{2m}\frac{\partial^3 \psi}{\partial x^3} + \frac{\partial}{\partial x}(V\psi)\right)\right] dx$$

$$= -\frac{\hbar^2}{2m}[\int_{-\infty}^{\infty}\left(\frac{\partial^2 \psi^\times}{\partial x^2}\frac{\partial \psi}{\partial x} - \psi^\times\frac{\partial^3 \psi}{\partial x^3}\right) dx + \int_{-\infty}^{\infty}\left(V\psi^\times\frac{\partial \psi}{\partial x} - \psi^\times\frac{\partial}{\partial x}(V\psi)\right) dx]$$

$$= -\frac{\hbar^2}{2m}\left[\int_{-\infty}^{\infty}\frac{\partial}{\partial x}\left(\frac{\partial \psi^\times}{\delta x}\frac{\partial \psi}{\partial x} - \psi^\times\frac{\partial^2 \psi}{\partial x^2}\right) dx + \int_{-\infty}^{\infty}\left(V\psi^\times\frac{\partial \psi}{\partial x} - \psi^\times\left(\psi\frac{\partial V}{\partial x} + V\frac{\partial \psi}{\partial x}\right)\right) dx\right]$$

$$= -\frac{\hbar^2}{2m}\left[\left(\frac{\partial \psi^\times}{\partial x}\frac{\partial \psi}{\partial x} - \psi^\times\frac{\partial^2 \psi}{\partial x^2}\right)\right]_{-\infty}^{+\infty} + \int_{-\infty}^{\infty}\left(V\psi^\times\frac{\partial \psi}{\partial x}dx\right) - \int_{-\infty}^{\infty}\left(\psi^\times\psi\frac{\partial V}{\partial x}dx\right) - \int_{-\infty}^{\infty}\psi^\times V\frac{\partial \psi}{\partial x}dx$$

$$= -\frac{\hbar^2}{2m}\left[\left(\frac{\partial\psi^\times}{\partial x}\frac{\partial\psi}{\partial x} - \psi^\times\frac{\partial^2\psi}{\partial x^2}\right)\right]_{-\infty}^{+\infty} - \int_{-\infty}^{\infty}\left(\psi^\times\psi\frac{\partial V}{\partial x}dx\right)$$

For well-behaved wave functions ψ, $\frac{\partial\psi}{\partial x}$ and $\frac{\partial^2\psi}{\partial x^2}$ approach zero as $x \to \pm\infty$

Hence the first term vanishes at both points.

$$\therefore \frac{\partial}{\partial t} < P_x > = -\int_{-\infty}^{\infty}\psi^\times\frac{\partial V}{\partial x}\psi\, dx \qquad [15]$$

But $< \frac{\partial V}{\partial x} > = \int_{-\infty}^{\infty}\psi^\times\frac{\partial V}{\partial x}\psi\, dx$

Hence from equation [15]

$$\frac{\partial}{\partial t} < P_x > = < \frac{\partial V}{\partial x} > \qquad [16]$$

In vector form, the above equation becomes

$$\frac{\partial}{\partial t}(< \bar{\mathrm{P}} >) = -< \bar{\nabla} V > \quad or, \frac{\partial}{\partial t} < \bar{\mathrm{P}} > = -< grad\, V > \quad or, \frac{\partial}{\partial t} < \bar{\mathrm{P}} > = -< \frac{\partial V}{\partial r} >$$

From the classical definition of potential I.e., $\bar{F} = -< \frac{\partial V}{\partial \bar{r}} >$

Finally, $\frac{\partial}{\partial t} < \bar{P} > = < \bar{F} >$ [17]

Equation [17] can be written in one dimension $\frac{\partial}{\partial t} < P_x > = < F_x >$

Finally ,

1. $\frac{\partial}{\partial t} < x > = \frac{<P_x>}{m}$ 2. $\frac{\partial}{\partial t} < P_x > = < F_x >$

2.22 Equation of Continuity

Let's consider a particle under the action of a conservative mechanical force. The wave equation of the associated matter wave is given by the Schrödinger equation. From time dependent Schrödinger equation

$$i\hbar\frac{\partial\psi}{\partial t} = -\frac{\hbar^2}{2m}\nabla^2\psi + V\psi \qquad [1]$$

And its complex conjugate

$$-i\hbar\frac{\partial\psi^\times}{\partial t} = -\frac{\hbar^2}{2m}\nabla^2\psi^\times + V\psi^\times \; ; \quad because\ V = V^\times \qquad [2]$$

Now multiplying equation [1] by $\psi^\times$ and equation [2] by ψ

$$i\hbar\, \psi^{\times} \frac{\partial \psi}{\partial t} = -\frac{\hbar^2}{2m} \psi^{\times}\, \nabla^2 \psi \, + \psi^{\times}\, V \psi \qquad [3]$$

and $$-\, i\hbar\, \psi\, \frac{\partial \psi^{\times}}{\partial t} = -\frac{\hbar^2}{2m} \psi\, \nabla^2 \psi^{\times} \, + \psi\, V \psi^{\times} \qquad [4]$$

Now subtracting [4] from [3]

$$i\hbar \left[\, \psi^{\times} \frac{\partial \psi}{\partial t} + \psi\, \frac{\partial\, \psi^{\times}}{\partial t} \right] = -\frac{\hbar^2}{2m} \left(\psi^{\times}\, \nabla^2 \psi - \psi\, \nabla^2 \psi^{\times}\right) + V(\psi^{\times}\psi - \psi\, \psi^{\times})$$

$$i\hbar\, \frac{\partial}{\partial t}(\psi^{\times}\psi) = -\frac{\hbar^2}{2m} [\psi^{\times}\overline{\nabla}.(\overline{\nabla}\psi) - \psi\, \overline{\nabla}.(\overline{\nabla}\psi^{\times})] + V(|\psi\,|^2 \, - \, |\psi\,|^2)$$

$$= -\frac{\hbar^2}{2m} [\psi^{\times}\overline{\nabla}.(\overline{\nabla}\psi) + \overline{\nabla}\psi^{\times}.\overline{\nabla}\psi - \overline{\nabla}\psi.\overline{\nabla}\psi^{\times} - \psi\, \overline{\nabla}.(\overline{\nabla}\psi^{\times})]$$

$$[Adding\ and\ subtracting\ \overline{\nabla}\psi^{\times}.\overline{\nabla}\psi]$$

$$= -\frac{\hbar^2}{2m} [\overline{\nabla}.(\psi^{\times}\overline{\nabla}\psi) - \overline{\nabla}.(\psi\, \overline{\nabla}\psi^{\times}\,)]$$

$$= -\, \overline{\nabla}.\ \left[\frac{\hbar^2}{2m}\{(\psi^{\times}\overline{\nabla}\psi) - (\psi\, \overline{\nabla}\psi^{\times})\}\right]$$

Or, $$\frac{\partial}{\partial t}(\psi^{\times}\psi) = -\, \overline{\nabla}.\ \left[\frac{\hbar}{2mi}\{(\psi^{\times}\overline{\nabla}\psi) - (\psi\, \overline{\nabla}\psi^{\times})\}\right]$$

Where $$\rho = \psi^{\times}\psi = |\psi\,|^2 \qquad [5]$$

And $$J = \frac{\hbar}{2mi}[(\psi^{\times}\overline{\nabla}\psi) - (\psi\, \overline{\nabla}\psi^{\times})] \qquad [6]$$

Or, $$\frac{\partial \rho}{\partial t} = -\overline{\nabla}\,.\bar{J} \quad ; \quad \text{Or,} \quad \boxed{\frac{\partial \rho}{\partial t} + \overline{\nabla}\,.\bar{J} = 0} \qquad [7]$$

Equation [7] represents the equation of continuity and equation [5] and [6] represent the probability density and probability current density.

2.23 Relation between Probability density [ρ] and Current density [J]

Consider a particle that moves freely in the positive x-direction. Its momentum

$$P_x = \frac{\hbar}{i}\frac{\partial}{\partial x}$$

Or, $$P_x\psi = \frac{\hbar}{i}\frac{\partial \psi}{\partial x} \quad and \quad \frac{\partial \psi}{\partial x} = \frac{i}{\hbar} P_x \psi$$

And $$P_x\psi^{\times} = -\frac{\hbar}{i}\frac{\partial \psi^{\times}}{\partial x} \quad and \quad \frac{\partial \psi^{\times}}{\partial x} = -\frac{i}{\hbar} P_x \psi^{\times}$$

So, the current density $J = \frac{\hbar}{2mi}\left[\left(\psi^{\times}\frac{\partial\psi}{\partial x}\right) - \left(\psi\ \frac{\partial\psi^{\times}}{\partial x}\right)\right] = \frac{\hbar}{2mi}\left[\left(\psi^{\times}\frac{i}{\hbar}P_x\psi\right) + \left(\psi\ \frac{i}{\hbar}P_x\psi^{\times}\ \right)\right]$

$$= \frac{\hbar}{2mi}\frac{i}{\hbar}[(\psi^{\times}P_x\psi) + (\psi\ P_x\psi^{\times})]\ = \frac{1}{2m}[P_x(\psi^{\times}\psi) + P_x(\psi\psi^{\times})]$$

$$= \frac{P_x}{2m}[(\psi^{\times}\psi) + (\psi\psi^{\times})]\ = \frac{P_x}{2\text{m}}\ [2|\psi\ |^2]\ = \frac{m\,V_x}{m}\rho$$

$$\boxed{J\ =\ \rho V_x} \qquad [1]$$

In vector form the above equation become

$$\bar{J} = \rho\bar{V} \qquad [2]$$

Equation [1] and [2] is the relation between Probability density [ρ] and Current density [J] in one and three dimensions respectively.

[NB: For well-behaved wave functions $\boldsymbol{\Psi}$, $\frac{d\psi}{dx}$ *and* $\frac{d^2\psi}{dx^2}$ vanishes at the boundaries:]

I.e., $x = \pm\infty[-\infty \le x \le +\infty]$

$$\Psi = A + iB = Ae^{i\alpha x} + Be^{-i\alpha x} = Ae^{-\alpha} + Be^{-0}\ \ [-\infty \le x \le 0]$$

$$= A\frac{1}{e^{\infty}} + \frac{B}{e^{0}} = A\frac{1}{\infty} + B = \mathbf{0 + B}$$

$$\Psi = A + iB = Ae^{i\alpha x} + Be^{-i\alpha x} = Ae^{0} + Be^{-\infty}\ \ [0 \le x \le \infty]$$

$$= A + B\frac{1}{e^{\infty}} = A + \frac{1}{\infty} = A + 0 = \mathbf{A + 0}$$

And $\left[\frac{d\psi}{dx}\right]_{-\infty}^{0} = i\alpha Ae^{i\alpha x} - i\alpha Be^{-i\alpha x} = i\alpha Ae^{-\infty} - i\alpha Be^{-0}$

$$= i\alpha A\frac{1}{e^{\infty}} - \frac{i\alpha B}{e^{0}} = i\alpha A\frac{1}{\infty} - i\alpha B = \mathbf{0 - i\alpha B}$$

$$\left[\frac{d\psi}{dx}\right]_{0}^{\infty} = i\alpha Ae^{i\alpha x} - i\alpha Be^{-i\alpha x} = i\alpha Ae^{0} - i\alpha Be^{-\infty} = i\alpha A - i\alpha B\frac{1}{e^{\infty}}$$

$$= i\alpha A - \frac{1}{\infty} = \boldsymbol{i\alpha A - 0}$$

$$\left[\frac{d^2\psi}{dx^2}\right]_{-\infty}^{0} = \frac{d}{dx}\left[i\alpha Ae^{i\alpha x} - i\alpha Be^{-i\alpha x}\right] = i^2\alpha^2Ae^{i\alpha x} + i^2\alpha^2Be^{-i\alpha x} = -\alpha^2Ae^{-\infty} - \alpha^2Be^{0}$$

$$= -\alpha^2Ae^{-\infty} - \alpha^2Be^{0} = -\alpha^2A\frac{1}{e^{\infty}} - \alpha^2B = -\alpha^2A\frac{1}{\infty} - \alpha^2B = \mathbf{0 - \boldsymbol{\alpha}^2B}$$

$$\left[\frac{d^2\psi}{dx^2}\right]_{0}^{\infty} = \frac{d}{dx}\left[i\alpha Ae^{i\alpha x} - i\alpha Be^{-i\alpha x}\right] = i^2\alpha^2Ae^{i\alpha x} + i^2\alpha^2Be^{-i\alpha x} = i^2\alpha^2Ae^{0} - \alpha^2Be^{-\infty}$$

$$= -\alpha^2A - \alpha^2Be^{-\infty} = -\alpha^2A - \alpha^2B\frac{1}{e^{\infty}} = -\alpha^2A - \alpha^2B\frac{1}{\infty} = \boldsymbol{-\alpha^2A - 0}$$

[Similarly, For well-behaved wave functions $\psi^{\times}$, $\frac{d\psi^{\times}}{dx}$ *and* $\frac{d^2\psi^{\times}}{dx^2}$ vanishes at the boundaries:
NB: Students should be done by themselves [Home Task]

Exercise

1. What is Wave and Wave function? Write physical significance of Wave function.
2. Define probability density and probability current density. What are the limitations of Wave function?
3. What are normalization of wave function and normalizing constant? Find normalization constant and normalization wave function for a trial wave function [Trial wave function for Linear Harmonic Oscillator and Hydrogen atom].
4. Find normalization constant and normalization wave function for one dimensional case
$$\psi(x,t) = sin\left(\frac{\pi x}{2a}\right) e^{\frac{-iEt}{\hbar}} \; ; \; -a \leq x \leq +a$$
5. What is expectation value of dynamical variables? Find Expectation value of position, momentum and energy [Dynamical variables] for one dimensional wave function
$$\psi(x,t) = sin\left(\frac{\pi x}{2a}\right) e^{\frac{-iEt}{\hbar}} \; ; \; -a \leq x \leq +a$$
6. Define orthogonal and ortho-normal condition of wave function. Write the property of Bra and Ket notation.
7. Explain the wave-mechanical concept of atom.
8. State and Brief description of Uncertainty Principle. What is the physical significance of Heisenberg Uncertainty relation?
9. State and explain the Bohr Correspondence & Complementarity Principle.
10. Show the elementary proof of Heisenberg uncertainty principle (between energy and time).
11. Show the elementary proof of Heisenberg uncertainty principle (between position and momentum).
12. Derive the Schwarz's inequalities.
13. Prove the Heisenberg uncertainty principle $\Delta x \, \Delta P_x \geq \frac{\hbar}{2}$
14. State and prove the Ehrenfest's theorem
$$[1]\ \frac{\partial}{\partial t} < x > = \frac{<P_x>}{m} \quad \text{And} \quad [2]\ \frac{\partial}{\partial t} < P_x > = < F_x >$$
15. Establish the following relation
$$\frac{\partial \rho}{\partial t} + \bar{\nabla}.\bar{J} = 0$$
Where, ρ is the probability density and $\bar{J}$ is the probability current density.
16. Write equation of Continuity. Establish the relation between Probability density [ρ] and Current density [J].
17. Calculate the uncertainty in the position of an electron of mass $m_e = 9 \times 10^{-28} gm$ moving with an uncertainty in speed of $3 \times 10^{10} cm/sec$.

18. An electron of energy 200eV is passed through a circular hole of radius $10^{-4}cm$. What is the uncertainty in the angle of emergences?
19. Calculate the uncertainty in the angle of emergences of 1MeV electron passing through a hole of diameter of 20 micron.
20. Write some application of Heisenberg uncertainty principle.
21. In a hydrogen atom, the electron is on the average at a distance of $0.5A°$ from the proton. Calculate (1) the minimum uncertainty in the momentum of the electron and (2) the minimum kinetic energy of the electron.
22. Find the uncertainty in the momentum of a particle when its position is determined within 0.01cm. Find also the uncertainty in the velocity of an electron and an alpha particle respectively, when they are located within $5 \times 10^{-8}cm$.
23. An electron of of mass $m_e = 9 \times 10^{-28} gm$ has a speed $2 \times 10^4 cm/sec$ accurate to 0.01%. What accuracy can be achieved in locating the position of the electron?
24. Obtain an expression for the current density from the equation.

$$\left[\frac{1}{c^2}\frac{\partial^2}{\partial t^2} - \frac{\partial^2}{\partial x^2} + \left(\frac{mc}{\hbar}\right)^2\right]\psi(x,t) = 0$$

25. The state of an oscillator of angular frequency ω is represented by

$$\psi(x) = e^{-m\omega x^2/\hbar}$$

What are the expectations of the momentum and the position?

26. Calculation the probability current density corresponding to the wave function $\psi(r) = \frac{e^{ikr}}{r}$; Where $r^2 = x^2 + y^2 + z^2$. Determine the current density for large values of r.
27. The wave function for a particle is given by $\psi(x) = Ae^{ikx} + Be^{-ikx}$. What current density or flux does this present?
28. What is the flux associated with a particle described by the wave function $\psi(x) = u(x)e^{ikx}$; where $u(x)$ is a real function.
29. Explain why it is plausible to choose the probability current density in Quantum Mechanics as $J = \frac{\hbar}{2mi}[\psi^{\times}\nabla\psi - \psi\nabla\psi^{\times}]$; the symbols have usual meaning.
30. Show that the electron cannot be present within the nucleus.
31. Can you measure the wave length of an electron ? If so but how ? If not, why ?

Chapter 3 Mathematical Structure of Quantum mechanics

3.1 Operator, Eigen function, Eigen value and Eigen value equation

- Operator: Any mathematical quantity which transforms one vector to another prescribed vector I.e., an operator is a rule which changes a function into another prescribed function. If a function f(x) is such that an operator Ω_{op} which operates on function *f(x)* gives

$$\Omega_{op} f(x) = \lambda f(x) \quad [1]$$

From the above equation we have noted that

- *f(x)* is called an Eigen function of the operator Ω_{op}
- Eigen function means an operator acting on a function reproduces the same function multiplied by a constant is called an Eigen function
- λ is called an Eigen value of the operator Ω_{op} belonging to the Eigen function *f(x)*
- Equation [1] represents the Eigen value equation

Eigen value equations play an important role in quantum mechanics. Remember that possible measurement outcomes are given by the Eigen values of Hermitian operators. This means, performing a measurement of an observable leaves the system in an Eigen state of the corresponding operator.

3.2 Operator for momentum and energy

For undamped [constant amplitude] monochromatic harmonics [constant angular momentum] wave propagating in the positive X- axis

$$y = A\exp\left(-i\omega\,[t-\frac{x}{v}]\right) \quad [1]$$

In Quantum Mechanics, the wave function corresponds to the displacement. Since y is not itself a measureable quantity and may be therefore be complex

$$\Psi(x,t) = A\exp\left(-i\omega\,[t-\frac{x}{v}]\right) \quad [2]$$

$$= A\exp\left(-i[2\pi\upsilon t - \frac{2\pi\upsilon}{\upsilon\lambda}x]\right) == A\exp\left(-i[\frac{E}{\hbar}t - \frac{2\pi}{\lambda}x]\right) == A\exp\left(-i[\frac{E}{\hbar}t - \frac{p}{\hbar}x]\right)$$

$$\Psi(x,t) = A\exp\left(-\frac{i}{\hbar}[Et - px]\right) \quad [3]$$

In three dimensions the above equation can be written as

$$\Psi(\vec{r},t)=A\exp\left(-\frac{i}{\hbar}[Et-\vec{p}.\vec{r}]\right) \qquad [4]$$

Now differentiating the equation [2] with respect to x

$$\frac{\partial\Psi(x,t)}{\partial x}=A(-p)(-\frac{i}{\hbar})\exp\left(-\frac{i}{\hbar}[Et-px]\right)=(\frac{i}{\hbar})p\Psi(x,t)$$

Or, $$\frac{\hbar}{i}\frac{\partial\Psi(x,t)}{\partial x}=p\Psi(x,t)$$

Or, $$\frac{\hbar}{i}\frac{\partial}{\partial x}=p_x \qquad \therefore \quad p_x=\frac{\hbar}{i}\frac{\partial}{\partial x}=-i\hbar\frac{\partial}{\partial x}$$

Similarly, $$p_y=\frac{\hbar}{i}\frac{\partial}{\partial y}=-i\hbar\frac{\partial}{\partial y} \qquad [5]$$

$$p_z=\frac{\hbar}{i}\frac{\partial}{\partial z}=-i\hbar\frac{\partial}{\partial z}$$

Equation [5] represents the x, y and z components of the momentum operators.

In three dimensions momentum operator can be written as

$$\hat{\vec{p}}=\hat{i}p_x+\hat{j}p_y+\hat{k}p_z=\hat{i}\left(-i\hbar\frac{\partial}{\partial x}\right)+\hat{j}\left(-i\hbar\frac{\partial}{\partial y}\right)+\hat{k}\left(-i\hbar\frac{\partial}{\partial z}\right)$$

$$\hat{\vec{p}}=-i\hbar[\left(\hat{i}\frac{\partial}{\partial x}\right)+\left(\hat{j}\frac{\partial}{\partial y}\right)+\left(\hat{k}\frac{\partial}{\partial z}\right)]$$

$$\hat{\vec{p}}=-i\hbar\vec{\nabla} \qquad [6]$$

Again differentiating the equation [2] with respect to t

$$\frac{\partial}{\partial t}\Psi(x,t)=-\left(\frac{i}{\hbar}E\right)A\exp\left(-\frac{i}{\hbar}[Et-px]\right)$$

$$\frac{\partial}{\partial t}\Psi(x,t)=-\left(\frac{i}{\hbar}E\right)\Psi(x,t)$$

$$\frac{\partial}{\partial t}\Psi(x,t)=\frac{1}{i\hbar}E\Psi(x,t)$$

$$\therefore \qquad E=\ i\hbar\frac{\partial}{\partial t} \qquad [7]$$

3.3 Hamiltonian operator and importance of Hamiltonian operator

When an operator acts on a function, it reproduces the same function; the operator is termed as Hamiltonian operator.

Importance of Hamiltonian Operator

- Hamiltonian operator acts on a function, it reproduces the same function. This function is said to be proper function or Eigen function
- Hamiltonian operator gives the total energy of the system
- It gives only the kinetic energy of the system for free particle
- An operator Ω is said to be constant in motion when it is commuted with Hamiltonian H I.e., $[\Omega, H] = 0$
- Hamiltonian operator can play a vital role for making the system is completely unchanged

3.4 Hermitian operator and its importance

In some sense, these operators play the role of the real numbers (being equal to their own "complex conjugate") and form a real vector space. They serve as the model of real-valued observables in quantum mechanics.

- Hermitian operator is widely used in Quantum Mechanics, because all eigen values of Hermitian operator are real. I.e., Hermatian operator gives real eigen values
- The sum of two Hermitian operators is Hermitian
- The product of the Hermitian Operator is Hermitian if and only if they commute
- If an operator say $\hat{\Omega}$ is Hermitian then it satisfies, the following condition

$$\left(\Psi, \hat{\Omega}\,\Psi\right) = \left(\hat{\Omega}\,\Psi, \Psi\right)$$

A linear operator which obeys the above rule is called Hermitian Operator.

3.5 Operator and Linear operator

Almost all operators encountered in quantum mechanics are *linear operators*. A linear operator is an operator which satisfies the following two conditions:

An entity $\hat{A}$ that relates every vector ψ in a linear vector space to another ϕ in this space by the equation

$$\phi = \hat{A}\psi \qquad [1]$$

is called an operator. The operator $\hat{A}$ is said to be linear if it possesses the properties

$$\hat{A}\,(\psi_a + \psi_b) = \hat{A}\;\psi_a + \hat{A}\,\psi_b \qquad [2]$$

And
$$\hat{A}\,(\lambda\psi) = \lambda\hat{A}\;\psi \qquad [3]$$

Where λ being a complex number.

3.6 Unitary operator and its importance

When the inverse and adjoint of an operator are identical then the operator is known as unitary operator. I.e.,

$$\Omega^{-1} = \Omega^{\times}$$
$$\Rightarrow \Omega\Omega^{-1} = \Omega^{\times}\Omega = 1$$

❖ Importance of unitary Operator

- Unitary operator makes unitary transformation and unitary transformation preserves scalar product
- Unitary operator leaves the system completely unchanged
- A matrix can be diagonilized by means of unitary transformation
- Hermitian operator can always be diagonalized by means of unitary transformation

3.7 Parity operator and Eigen values of Parity operator

The parity operator is defined by $\hat{\pi}$. When a parity operator $\hat{\pi}$ acts on a function $\Psi(x)$; [The operator gets its reflected in its co-ordinate] it reflects the function co-ordinate I.e.,

$$\hat{\pi}\,\Psi(x) = \Psi(-x)$$

From the Eigen value equation of parity operator

$$\hat{\pi}\Psi(x) = \lambda\Psi(x) \qquad [1]$$
$$\Rightarrow \hat{\pi}^2\Psi(x) = \hat{\pi}\left(\lambda\Psi(x)\right) = \lambda\hat{\pi}\,\Psi(x)$$
$$\Rightarrow \hat{\pi}^2\Psi(x) = \lambda\left(\lambda\Psi(x)\right) = \lambda^2\Psi(x) \qquad [2]$$

Again, operate equation [1] by $\hat{\pi}$

$$\Rightarrow \hat{\pi}\left(\hat{\pi}\,\Psi(x)\right) = \hat{\pi}\,\Psi(-x) = \Psi(x) \qquad [3]$$

Now compare [2] & [3]

$$\lambda^2 \Psi(x) = \Psi(x) \Rightarrow \lambda^2 = 1 \Rightarrow \lambda = \pm 1 \quad [4]$$

When $\lambda = +1$

The Eigen function of the Parity operator is even function,

$$\Psi_{even}(x) = \Psi_{even}(-x) \quad [5]$$

When $\lambda = -1$

The Eigen function of the Parity operators are odd function,

$$\Psi_{odd}(x) = -\Psi_{odd}(-x) \quad [6]$$

3.8 Conjugate Operator and Self Adjoint Operator

The adjoint of an operator $\widehat{\Omega}$ may also be called the Hermitian adjoint, Hermitian conjugate or Hermitian transpose *(after Charles Hermite)* of $\widehat{\Omega}$ and is denoted by $\Omega^{\times}$ or $\Omega^{dagger[\dagger]}$. It is the transpose of the complex conjugate of the operator $\widehat{\Omega}$ i.e., $\widetilde{\Omega}^{\times}$ is called adjoint or conjugate of the operator $\widehat{\Omega}$ and is denoted by $\Omega^{dagger[\dagger]}$.

$$\therefore \quad \widetilde{\Omega}^{\times} = \Omega^{dagger[\dagger]} \quad [1]$$

An operator $\widehat{\Omega}$ is said to be Hermitian or self adjoint if

$$\widehat{\Omega} = \Omega^{dagger[\dagger]} \quad [2]$$

3.9 Transpose operator [$\widetilde{\Omega}$] and its properties

The operator $\widetilde{\Omega}$ is said to be transpose of operator $\widehat{\Omega}$ if the rows and columns in their corresponding matrices are changed I.e.,

$$\widetilde{\Omega}_{nm} = \widehat{\Omega}_{mn}$$

Let us try to develop some example of Transpose operator

[1] $(a^{\times}, \widetilde{\Omega}\, b) = (b^{\times}, \Omega\, a)$

Take for two vectors **a** and **b** by two unit vectors e_i and e_k. $\therefore\ e_i^{\times} = e_i \ and\ e_k^{\times} = e_k$

Therefore, $(a^{\times}, \widetilde{\Omega}\, b) = (b^{\times}, \Omega\, a) = (e_i^{\times}, \widetilde{\Omega}\, e_k) = (e_k^{\times}, \Omega\, e_i)$

$$\text{Or, } (e_i^{\times}, \widetilde{\Omega}\, e_k) = (e_k^{\times}, \Omega\, e_i)$$

Or, $\left(e_i, \tilde{\Omega}\, e_k\right) = (e_k, \Omega\, e_i)$

Or, $\tilde{\Omega}(e_i\,,\, e_k) = \Omega(e_k, e_i)$ $\boxed{\tilde{\Omega}_{ik} = \Omega_{ki}}$

[2] $\Omega_{ik} = {\Omega^{\times}}_{ki}$

We know from Hermitian operator

$(a, \Omega\, b) = (\Omega a, b) = (b, \Omega a)^{\times} = (e_i, \Omega\, e_k) = (e_k, \Omega\, e_i)^{\times}$ [Take for two vectors **a** and **b** by two unit vectors e_i and e_k. $\therefore\ e_i^{\times} = e_i \ and\ e_k^{\times} = e_k$]

$\therefore$ $\Omega(e_i, e_k) = [\Omega(e_k, e_i)]^{\times}$ $\boxed{\Omega_{\mathrm{ik}} = \Omega_{\mathrm{ki}}^{\times}}$

3.10 Adjoint operator and its properties

If $\hat{\Omega}$ is such that$(\psi_1, \hat{\Omega}\psi_2) = (\hat{\Omega}\psi_1, \psi_2)$; $\hat{\Omega}$ is Hermitian operator. If $\hat{\Omega}$ is not Hermitian operator, then $(\psi_1, \hat{\Omega}\psi_2) \neq (\hat{\Omega}\psi_1, \psi_2)$. If two operators P and Q are such that $(\psi_1, \hat{\mathrm{P}}\psi_2) = (\hat{Q}\psi_1, \psi_2)$ I.e., if$\int \psi_1^{\times} \hat{\mathrm{P}}\psi_2 d\tau = \int (\hat{Q}\psi_1)^{\times} \psi_2 d\tau$, P and Q are said to be the adjoint of each other and we write $\hat{\mathrm{Q}} = \Omega^{dagger[\dagger]}$ or $\hat{\mathrm{P}} = \Omega^{dagger[\dagger]}$. Thus Hermitian operators are self-adjoint, I.e., $\hat{\Omega} = \Omega^{daggar[\dagger]}$

- $(\hat{\mathrm{P}} + \hat{\mathrm{Q}})^{\dagger} = \hat{P}^{\dagger} + \hat{Q}^{\dagger}$

$$(\psi_1, (P+Q)\psi_2) = \int \psi_1^{\times}(P+Q)\,\psi_2 d\tau = \int \psi_1^{\times}\,(P\psi_2 + Q\psi_2)d\tau$$

$$= \int \psi_1^{\times} P\psi_2 d\tau + \int \psi_1^{\times}\, Q\psi_2 d\tau = (\psi_1, \hat{\mathrm{P}}\psi_2) + (\psi_1, \hat{\mathrm{Q}}\psi_2)$$

$$= (\hat{P}^{\dagger}\psi_1, \psi_2) + (\hat{Q}^{\dagger}\psi_1, \psi_2) = \int (\hat{P}^{\dagger}\psi_1)^{\times}\psi_2 d\tau + \int (\hat{Q}^{\dagger}\psi_1)^{\times}\,\psi_2 d\tau$$

$$= \int \left[(\hat{P}^{\dagger}\psi_1)^{\times} + (\hat{Q}^{\dagger}\psi_1)^{\times}\right]\psi_2 d\tau$$

$$= (\hat{P}^{\dagger} + \hat{Q}^{\dagger}) \int \psi_1^{\times}\,\psi_2 d\tau = (\hat{P}^{\dagger} + \hat{Q}^{\dagger})(\psi_1, \psi_2)$$

Thus $\boxed{(\hat{\mathrm{P}} + \hat{\mathrm{Q}})^{\dagger} = \hat{P}^{\dagger} + \hat{Q}^{\dagger}}$

- If j is a complex number, $(j\,\hat{\mathrm{P}})^{\dagger} = j^{\times}\hat{P}^{\dagger}$

$$(\psi_1, jP\psi_2) = \int \psi_1^{\times} jP\,\psi_2 d\tau = j\int \psi_1^{\times} P\,\psi_2 d\tau = j(\psi_1, P\psi_2) = j(\hat{P}^{\dagger}\psi_1, \psi_2)$$

$$= j\int (\hat{P}^{\dagger}\psi_1)^{\times}\psi_2 d\tau = \int (j^{\times}\hat{P}^{\dagger}\psi_1)^{\times}\psi_2 d\tau = (j^{\times}\hat{P}^{\dagger}\psi_1, \psi_2)$$

Thus $(\psi_1, jP\psi_2) = (j^\times \hat{P}^\dagger \psi_1, \psi_2)$

$$\boxed{(j\hat{P})^\dagger = j^\times \hat{P}^\dagger}$$

- $(\hat{P}\hat{Q})^\dagger = Q^\dagger P^\dagger$

$$(\psi_1, PQ\psi_2) = (\psi_1, P(Q\psi_2)) = (\hat{P}^\dagger\psi_1, Q\psi_2) = \left((\hat{P}^\dagger\psi_1), Q\psi_2\right) = (\hat{Q}^\dagger(\hat{P}^\dagger\psi_1), \psi_2)$$

$$= (Q^\dagger P^\dagger \psi_1, \psi_2)$$

$\therefore (\hat{P}\hat{Q})^\dagger = Q^\dagger P^\dagger$ and $(\hat{P}\hat{Q})^\dagger = QP$ if P and Q are Hermitian or self- adjoint.

- $< P^\dagger P > = (\psi, P^\dagger P\psi) = \left(\psi, P^\dagger(P\psi)\right) = (P\psi, P\psi) = \int |P\psi|^2 d\tau = 0, Only\ if P\psi = 0$

3.11 Projection operator

A projection operator is that operator which on operating on any vector ψ , results the projection of it in some particular direction ψ_a [say] and the length of the projected vector ψ equal to the component ($\hat{\psi}_a\psi$). That is

$$P_a\psi = \hat{\psi}_a(\hat{\psi}_a, \psi) \quad and \quad P_a\hat{\psi}_a = \hat{\psi}_a \ \ or, P_a\,\psi_a = \psi_a$$

Hence for any vector ψ

$$P_a^2\psi = P_a(P_a\psi) = P_a\hat{\psi}_a(\hat{\psi}_a, \psi) = \hat{\psi}_a(\hat{\psi}_a, \psi) = P_a\psi\ ; \quad \boxed{P_a^2\psi = P_a\psi}$$

3.12 Degenerate and non-degenerate states

If the particle having the same energy in an excited state will have several different stationary states or different wave functions. I.e., *there are a number of independent quantum states of a system, each belonging to the same energy level. Such states and energy-levels are said to be degenerate.* If there are n linearly independent wave functions ψ_1, ψ_2, ψ_3 ……………………… ψ_n belonging to the same energy state then the energy level is said to be n-fold degenerate. It can be easily shown that any linearly combination of the degenerate wave functions $= C_1\psi_1 + C_2\psi_2 + C_3\psi_3 +$ ……………………… $C_n\psi_n$; is also an Eigen function belonging to the same energy Eigen value.

If there are a number of independent quantum states of a system, each belonging to the different energy level. Such states and energy-levels are said to be non-degenerate.

3.13 Unitary transformation preserves scalar product

Consider an operator $\hat{\Omega}$ belonging two vectors $\bar{a}\,\&\,\bar{b}$. The scalar product is

$$\hat{\Omega}\,\bar{a},\hat{\Omega}\,\bar{b}=\left(\bar{a},\hat{\Omega}^{\times}\hat{\Omega}\,b\right) \qquad \because\ \left[\left(\hat{\Omega}\,\bar{a},\bar{b}\right)=\left(\bar{a},\hat{\Omega}^{\times}\hat{\Omega}\,\bar{b}\right)\right]$$

$$\Rightarrow\left(\hat{\Omega}\,\bar{a},\hat{\Omega}\,\bar{b}\right)=\hat{\Omega}^{\times}\,\hat{\Omega}\left(\bar{a},b\right) \qquad \because\ \left[\left(\hat{\Omega}\,\bar{a},\bar{b}\right)=\left(\bar{a},\hat{\Omega}^{\times}\,\hat{\Omega}\,\bar{b}\right)\right]$$

$$\Rightarrow\left(\hat{\Omega}\,\bar{a},\hat{\Omega}\,\bar{b}\right)=\left(\bar{a},\bar{b}\right) \qquad \because\ \left[\hat{\Omega}^{\times}\,\hat{\Omega}=1\right]$$

3.14 Eigen values of Hermitian operator

Consider a wave function Ψ of an operator $\hat{\Omega}$, belonging to the Eigen value λ .Then,

$$\hat{\Omega}\Psi=\lambda\Psi \qquad [1]$$

If an operator $\hat{\Omega}$ is Hermitian operator then

$$(\Psi,\Omega\Psi)=(\Omega\Psi,\Psi)$$

$$or,(\Psi,\lambda\Psi)\ =(\lambda\Psi,\Psi)$$

$$or,\ \ \lambda(\Psi,\Psi)=\lambda^{\times}(\Psi,\Psi)$$

$$or,\lambda(\Psi,\Psi)-\lambda^{\times}(\Psi,\Psi)=0$$

$$or,\left(\lambda-\lambda^{\times}\right)(\Psi,\Psi)=0$$

$$or,\left(\lambda-\lambda^{\times}\right)=0$$

$$or,\ \ \lambda=\lambda^{\times}\left[Since,(\Psi,\Psi)\neq 0\right]$$

This is possible only when λ is a real number. Thus Hermitian Operator gives real Eigen value.

3.15 Sum and product of two Hermatian operators

If $\hat{A}$ is a Hermitian Operator, then

$$(\Psi_1,A\Psi_2)=(A\Psi_1,\Psi_2)$$

$$\Rightarrow\int\Psi_1^{\times}A\Psi_2dt=\int(A\Psi_1)^{\times}\Psi_2dt \qquad [1]$$

If $\hat{B}$ is Hermitian Operator

$$(\Psi_1,B\Psi_2)=(B\Psi_1,\Psi_2)$$

$$\Rightarrow\int\Psi_1^{\times}B\Psi_2dt=\int(B\Psi_1)^{\times}\Psi_2dt \qquad [2]$$

Adding equation [1] and [2]

$$\int\left(\Psi_1^{\times}\mathrm{A}\Psi_2+\Psi_1^{\times}\mathrm{B}\Psi_2\right)\mathrm{dt}=\int\left\{(\mathrm{A}\Psi_1)^{\times}\,\Psi_2+(\mathrm{B}\Psi_1)^{\times}\,\Psi_2\right\}\mathrm{dt}$$

$$\Rightarrow \int \Psi_1^{\times}(A+B)\Psi_2 dt = \int\{((A+B)\Psi_1)^{\times}\Psi_2 dt\}$$
$$\Rightarrow (\Psi_1^{\times}(A+B)\Psi_2) = ((A+B)\Psi_1, \Psi_2)$$

So, (A+B) is Hermitian, I.e., the sum of two Hermatian Operators is Hermatian.

Again, we have to prove that $(\Psi_1, AB\Psi_2) = (AB\Psi_1, \Psi_2)$ if A & B are Hermatian & if A & B commute with each other. Therefore, L.H.S

$$(\Psi_1, AB\Psi_2) = (\Psi_1, A(B\Psi_2) = (A\Psi_1, B\Psi_2) = ((A\Psi_1), B\Psi_2) = (B(A\Psi_1), \Psi_2) = (BA\Psi_1, \Psi_2)$$

$$= (AB\Psi_1, \Psi_2) = R.H.S$$

3.16 Hermitian operators form a linear independent set or orthogonal set

Let, $\hat{\Omega}$ be the Hermatian operator belonging two Eigen vectors u & v respectively. Then,

$$\hat{\Omega}u = \omega u \quad \& \quad \hat{\Omega}v = \omega' v$$

Now, take the scalar product of $\hat{\Omega}u$ *and* $\hat{\Omega}v$ which is $(\hat{\Omega}u, \hat{\Omega}v)$ can be written in the following

$$\therefore \quad (\hat{\Omega}u, \hat{\Omega}v) = u, \hat{\Omega}^2 v \qquad [\because \bar{a}, \hat{\Omega}\bar{b} = \hat{\Omega}\bar{a}, \bar{b}]$$
$$\Rightarrow (\hat{\Omega}\hat{\Omega}u, v) = (u, \hat{\Omega}\hat{\Omega}\, v)$$
$$\Rightarrow (\hat{\Omega}\omega u, v) = (u, \hat{\Omega}\omega' v)$$
$$\Rightarrow \omega(\hat{\Omega}u, v) = \omega'(u, \hat{\Omega}v) \qquad [\because (\bar{a}, \lambda\bar{b}) = \lambda(\bar{a}, \bar{b})] \text{ and } [(\lambda\bar{a}, \bar{b}) = \lambda^{\times}(\bar{a}, \bar{b})]$$
$$\Rightarrow \omega(u, \Omega v) = \omega'(u, \hat{\Omega}v)$$
$$\Rightarrow \omega(u, \omega' v) = \omega'(u, \omega' v)$$
$$\Rightarrow \omega\omega'(u, v) = \omega'^2(u, v)$$
$$\Rightarrow (\omega'^2 - \omega\omega')(u, v) = 0$$

If two states have the same Eigen values then they are degenerate. So (*u, v*) are non-degenerate. That is why $\omega'^2 - \omega\omega' \neq 0$ therefore, $(u, v) = 0$

So, the Eigen vectors u & v are orthogonal to each other.

3.17 Hermitian operator can be diagonalized by means of unitary transformation

Let us consider, $\Omega^{-1} H\, \Omega = D$. Taking Hermitian adjoint on the both sides

$$\left(\Omega^{-1}H\Omega\right)^{\times}=D^{\times}=D\quad[\because D^{\times}=D]$$

The Hermitian adjoint of the product of two matrices is equal to the product of their Hermitian adjoint taken in inverse order I.e.,

$$(AB)^{\times}=B^{\times}A^{\times}\qquad[1]$$

$$\Rightarrow\left(\Omega^{-1}H\Omega\right)^{\times}=(H\Omega)^{\times}\left(\Omega^{-1}\right)^{\times}$$

$$\therefore\ (AB)^{\times}=B^{\times}A^{\times}\qquad[2]$$

$$\Rightarrow\left(\Omega^{-1}H\Omega\right)^{\times}=(H\Omega)^{\times}\left(\Omega^{-1}\right)^{\times}$$

$$(H\Omega)^{\times}\left(S^{-1}\right)^{\times}=D\qquad[3]$$

$$\Rightarrow\Omega^{\times}H^{\times}\left(S^{-1}\right)^{\times}=D$$

$$\Rightarrow\Omega^{\times}H\left(S^{-1}\right)^{\times}=\Omega^{-1}H\Omega$$

$$\Rightarrow\Omega^{\times}H\left(\Omega^{\times}\right)^{\times}=\Omega^{-1}H\Omega$$

An operator is said to be unitary if its inverse is equal to its Hermitian adjoint.

$$\Omega^{\times}H\left(S^{-1}\right)^{\times}=\Omega^{-1}=\Omega^{\times}\quad\therefore\ \Omega^{\times}H\Omega=\Omega^{-1}H\Omega\quad\Rightarrow\Omega^{\times}=\Omega^{-1}$$

3.18 Momentum operator is Hermitian

The momentum operator in one dimension

$$P_x=\frac{\hbar}{i}\frac{d}{dx}$$

Momentum operator is Hermitian when it follows the condition

$$(\Psi,P_x\Psi)=(P_x\Psi,\Psi)$$

$$\Rightarrow\left(\Psi,\frac{\hbar}{i}\frac{d}{dx}\Psi\right)=\left(\frac{\hbar}{i}\frac{d}{dx}\Psi,\Psi\right)$$

$$\Rightarrow\frac{\hbar}{i}\left(\Psi,\frac{d}{dx}\Psi\right)=\left(\frac{\hbar}{i}\right)^{\times}\left(\frac{d}{dx}\Psi,\Psi\right)\qquad[1]$$

$$L.H.S=(\Psi,P_x\Psi)=\left(\Psi,\frac{\hbar}{i}\frac{d}{dx}\Psi\right)=\frac{\hbar}{i}\left(\Psi,\frac{d}{dx}\Psi\right)=\frac{\hbar}{i}\int_{-\infty}^{+\infty}\Psi^{\times}\frac{d}{dx}\Psi dx$$

$$=\left(\frac{\hbar}{i}\right)\left\{\Psi^{\times}\int_{-\infty}^{+\infty}\frac{d\Psi}{dx}dx-\int_{-\infty}^{+\infty}\frac{d\Psi^{\times}}{dx}\int_{-\infty}^{+\infty}\frac{d}{dx}\Psi\right\}dx$$

$$= \left(\frac{\hbar}{i}\right)\left\{\Psi^{\times}\left[\Psi\right]_{-\infty}^{+\infty} - \int_{-\infty}^{+\infty}\frac{d\Psi^{\times}}{dx}\Psi dx\right\}$$

For well-behaved wave functions Ψ approaches zero as $x \to \pm\infty$ I.e., Ψ vanishes at the boundaries.

$$= \frac{\hbar}{i}\left[0 - \int_{-\infty}^{+\infty}\frac{d\Psi^{\times}}{dx}\Psi dx\right] = -\frac{\hbar}{i}\left[\int_{-\infty}^{+\infty}\frac{d\Psi^{\times}}{dx}\Psi dx\right] = \int_{-\infty}^{+\infty}\left(\frac{\hbar}{i}\frac{d\Psi}{dx}\right)^{\times}\Psi dx = \left(\frac{\hbar}{i}\frac{d}{dx}\Psi, \Psi\right)$$

$$Therefore, \quad L.H.S = \frac{\hbar}{i}\left(\Psi, \frac{d}{dx}\Psi\right) = \frac{\hbar}{i}\int_{-\infty}^{+\infty}\Psi^{\times}\frac{d}{dx}\Psi dx = \int_{-\infty}^{+\infty}\left(\frac{\hbar}{i}\frac{d\Psi}{dx}\right)^{\times}\Psi dx = \left(\frac{\hbar}{i}\frac{d}{dx}\Psi, \Psi\right)$$

$$\therefore \quad \frac{\hbar}{i}\left(\Psi, \frac{d}{dx}\Psi\right) = \left(\frac{\hbar}{i}\frac{d}{dx}\Psi, \Psi\right)$$

$$\Rightarrow (\Psi, P_x \Psi) = (P_x \Psi, \Psi)$$

L.H.S = R.H.S; Hence *momentum operator is Hermitian.*

3.19 Necessary and sufficient condition for two operators to have common Eigen function

Let us consider two operators $\hat{\Omega}$ & $\hat{R}$ have the same Eigen vector $\bar{a}$ belonging to the Eigen values ω & ρ respectively.

$$\hat{\Omega}\,\bar{a} = \omega\,\bar{a} \qquad [1]$$

$$\&\ \hat{R}\,\bar{a} = \rho\,\bar{a} \qquad [2]$$

$$Now, \quad \hat{R}\hat{\Omega}\bar{a} = \hat{R}\omega\bar{a} = \omega\hat{R}\bar{a} = \omega\rho\,\bar{a} \quad \& \quad \hat{\Omega}\hat{R}\bar{a} = \hat{\Omega}\rho\bar{a} = \rho\,\hat{\Omega}\bar{a} = \rho\omega\bar{a}$$

$$\therefore \quad \hat{R}\,\hat{\Omega}\bar{a} = \hat{\Omega}\,\hat{R}\bar{a} \Rightarrow \left(\hat{R}\hat{\Omega} - \hat{\Omega}\hat{R}\right)\bar{a} = 0$$

$$\Rightarrow \left(\hat{R}\hat{\Omega} - \hat{\Omega}\hat{R}\right) = 0 \qquad [3]$$

$$\Rightarrow \left[\hat{R}, \hat{\Omega}\right] = 0 \qquad [4]$$

When two operators have common Eigen function then they commutes. This is the necessary and sufficient condition for two operators.

3.20 Parity operator and symmetric Hamiltonian

We know the Hamiltonian operator for one dimension

$$\therefore \quad \hat{H} = -\frac{\hbar^2}{2m}\frac{d^2}{dx^2} + V(x) \qquad [1]$$

$$\Rightarrow \hat{H}\Psi(x) = -\frac{\hbar^2}{2m}\frac{d^2\Psi}{dx^2} + V(x)\Psi(x) \qquad [2]$$

$$\Rightarrow \hat{\pi}\left(\hat{H}\Psi(x)\right) = -\frac{\hbar^2}{2m}\frac{d^2\Psi(x)}{dx^2} + V(-x)\Psi(-x) \qquad [3]$$

$$\Rightarrow \hat{\pi}\left(\hat{H}\Psi(x)\right) = -\frac{\hbar^2}{2m}\frac{d^2\Psi(-x)}{dx^2} + V(x)\Psi(-x) \qquad [4]$$

Because $\hat{H}$ is symmetric, so $V(-x) = V(x)$. Similarly,

$$\hat{H}\left(\hat{\pi}\,\Psi(x)\right) = \left(-\frac{\hbar^2}{2m}\frac{d^2}{dx^2} + V(x)\right)\Psi(-x)$$

$$\Rightarrow \hat{H}\left(\hat{\pi}\,\Psi(x)\right) = -\frac{\hbar^2}{2m}\frac{d^2\Psi(x)}{dx^2} + V(-x)\Psi(-x) \qquad [5]$$

From equation [5] & [4], we have

$$\hat{H}\left(\hat{\pi}\,\Psi(x)\right) - \hat{\pi}\left(\hat{H}\Psi(x)\right) = 0$$

$$\Rightarrow \left(\hat{H}\hat{\pi} - \hat{\pi}\hat{H}\right)\Psi(x) = 0$$

$$\Rightarrow \left(\hat{H}\hat{\pi} - \hat{\pi}\hat{H}\right) = 0$$

$$\Rightarrow \left[\hat{H}, \hat{\pi}\right] = 0$$

So, parity operator commutes with symmetric Hamiltonian

3.21 Parity operator is Harmitian

Let us consider two functions Ψ_1 & Ψ_2 and also consider the scalar product of $\hat{\pi}\,\Psi_1$ & Ψ_2 is

$$(\hat{\pi}\,\Psi_1, \Psi_2) = \int (\hat{\pi}\,\Psi_1(x))^{\times}\Psi_2(x)dx = \int \Psi^{\times}_1(-x)\Psi_2(x)dx$$

$$\Rightarrow (\hat{\pi}\,\Psi_1, \Psi_2) = \int \Psi^{\times}_1(x')\Psi_2(-x')dx$$

Where, $-x = x'$ *and* $x = -x'$; *on renaming the variable, the value of the* integral is unaffected.

$$\Rightarrow (\hat{\pi}\,\Psi_1, \Psi_2) = \int \Psi^{\times}_1(x)\hat{\pi}\,\Psi_2(x)dx$$

$$\Rightarrow (\hat{\pi}\Psi_1, \Psi_2) = (\Psi_1, \hat{\pi}\Psi_2) \qquad [1]$$

Hence, parity operator is Hermatian.

3.22 Matrix representation of an operator

We know the operator equation

$$f = \hat{Q}\varphi \qquad [1]$$

Let $$\varphi = \sum_{n=1}^{n} a_n \psi_n^{(r)} \text{ and } f = \sum_{k=1}^{n} b_k \psi_k^{(r)} \qquad [2]$$

Here, ψ_n forms an orthogonal set of wave functions. Now put the value of φ *and* f in [1]

$$\sum_{k=1}^{n} b_k \psi_k^{(r)} = \hat{Q} \sum_{n=1}^{n} a_n \psi_n^{(r)} \qquad [3]$$

Scalar multiplication on both sides of equation [3] by $\psi_m^{(r)}$

$$\psi_m^{(r)} \sum_{k=1}^{n} b_k \psi_k^{(r)} = \hat{Q}\psi_m^{(r)} \sum_{n=1}^{n} a_n \psi_n^{(r)} \qquad [4]$$

$$\sum_{k=1}^{n} b_k \left(\psi_m^{(r)}, \psi_k^{(r)}\right) = \sum_{n=1}^{n} a_n \left(\psi_m^{(r)}, \hat{Q}\psi_n^{(r)}\right) \qquad [5]$$

$$b_m = \sum_{n=1}^{n} a_n \hat{Q}_{mn} \qquad [6]$$

Here, $\left(\psi_m^{(r)}, \psi_k^{(r)}\right) = \delta_{mk} = 1$; when $m = k$. And $\hat{Q}_{mn} = \left(\psi_m^{(r)}, \hat{Q}\psi_n^{(r)}\right)$ [7]

The set of quantities $\hat{Q}_{mn}$ they represent the operator. From equation [6] we write

$$b = \begin{pmatrix} b_1 \\ b_2 \\ \vdots \\ b_m \end{pmatrix} = \begin{pmatrix} Q_{11} & Q_{12} & Q_{13} & \cdots & Q_{1n} \\ Q_{21} & Q_{22} & Q_{23} & \cdots & Q_{2n} \\ Q_{31} & Q_{32} & Q_{33} & \cdots & Q_{3n} \\ \vdots & \vdots & \vdots & \cdots & \vdots \\ Q_{n1} & Q_{n2} & Q_{n3} & \cdots & Q_{nn} \end{pmatrix} \begin{pmatrix} a_1 \\ a_2 \\ \vdots \\ a_n \end{pmatrix} \qquad [8]$$

Therefore, the operator equation has been converted into a matrix equation. *So, an operator can be represented by a matrix.*

3.23 Operator formalism of Quantum Mechanics

We now introduce commutator and describe how to deal with them. If *A and B* are two operators, *(AB - BA)* is called the commutator of *A and B* and it is denoted by *[A, B] = (AB- BA).* If two operators commute then *[A, B] = 0* and if the two operators do not commute, *[A, B] ≠ 0.* Now we evaluate commutator of *x and* p_x ; That is

(1) $[x, P_x] = -\frac{\hbar}{i} = i\hbar$

$L.H.S = [x, P_x] = xP_x - P_x x$

$$Or, \quad [x, P_x] = x\left(\frac{\hbar}{i}\frac{d}{dx}\right) - \left(\frac{\hbar}{i}\frac{d}{dx}\right)x = \frac{\hbar}{i}\left(x\frac{d}{dx} - \frac{d}{dx}x\right)$$

$$\Rightarrow [x, P_x]\Psi(x) = \frac{\hbar}{i}\left(x\frac{d}{dx} - \frac{d}{dx}x\right)\Psi(x)$$

$$\Rightarrow [x, P_x]\Psi(x) = \frac{\hbar}{i}\left(x\frac{d\Psi(x)}{dx} - \Psi(x)\frac{dx}{dx} - x\frac{d\Psi(x)}{dx}\right)$$

$$\Rightarrow [x, P_x]\Psi(x) = -\frac{\hbar}{i}\Psi(x) \Rightarrow [x, P_x] = -\frac{\hbar}{i} = i\hbar \qquad (Proved)$$

(ii) $\left[x^n, P_x\right] = ni\hbar x^{n-1}$

$$We\ know\ [x, P_x] = i\hbar \quad \therefore \quad \left[x^2, P_x\right] = [xx, P_x]$$

$$[ab, c] = a[b, c] + [a, c]b = x[x, P_x] + [x, P_x]x$$

$$\Rightarrow a[b, c] + [a, c]b = xi\hbar + i\hbar x$$

$$\Rightarrow a[b, c] + [a, c]b = 2i\hbar x$$

$$\Rightarrow a[b, c] + [a, c]b = 2i\hbar x^{2-1}$$

$$Similarly, \left[x^3, P_x\right] = \left[x^2 x, P_x\right]$$

$$\Rightarrow \left[x^3, P_x\right] = x^2[x, P_x] + \left[x^2, P_x\right]x$$

$$\Rightarrow \left[x^3, P_x\right] = x^2 i\hbar + 2i\hbar x^2$$

$$\Rightarrow \left[x^3, P_x\right] = 3i\hbar x^2$$

$$\Rightarrow \left[x^3, P_x\right] = 3i\hbar x^{3-1}$$

$$Finally, \left[x^n, P_x\right] = i\hbar n x^{n-1}$$

(iii) $\left[x, P_x^n\right] = ni\hbar P_x^{n-1}$

We know, $[x, P_x] = i\hbar$

$$\therefore \left[x, P^2{}_x\right] = [x, P_x P_x]$$

$$[a, bc] = b[a, b] + [a, c]b = P_x[x, P_x] + [x, P_x]P_x = P_x i\hbar + i\hbar P_x = 2i\hbar P_x = 2i\hbar P_x{}^{2-1}$$

Similarly,

$$\left[x, P_x^n\right] = ni\hbar P_x^{n-1}$$

(iv) $\left[\hat{x}_i, \hat{P}_j\right] = i\hbar\delta_{ij}$

From the operator equation

$$\hat{x}_i \Psi(x) = x_i \Psi(x)$$

& the momentum operator acts on a function $\Psi(x)$

$$\hat{P}_j \Psi(x) = -i\hbar \frac{d\Psi(x)}{dx_j}$$

$$Now,\ \hat{x}_i \hat{P}_j \Psi(x) = \hat{x}_i\ \hat{P}_j \Psi(x) = -i\hbar \hat{x}_i \frac{d\Psi(x)}{dx_j}$$

$$\&\ \hat{P}_j \hat{x}_i \Psi(x) = -i\hbar \frac{d}{dx_j}\left(\hat{x}_i \Psi(x)\right)$$

$$\Rightarrow \quad \hat{P}_j \hat{x}_i \Psi(x) = -i\hbar \hat{x}_i \frac{d\Psi(x)}{dx_j} - i\hbar \Psi(x) \frac{dx_i}{dx_j}$$

$$\therefore \quad \hat{x}_i \hat{P}_j \Psi(x) - \hat{P}_j \hat{x}_i \Psi(x) = i\hbar \Psi(x) \frac{dx_i}{dx_j}$$

$$\Rightarrow \quad \left(\hat{x}_i \hat{P}_j - \hat{P}_j \hat{x}_i\right) \Psi(x) = i\hbar \Psi(x) \frac{dx_i}{dx_j}$$

$$\Rightarrow \quad \left[\hat{x}_i, \hat{P}_j\right] \Psi(x) = i\hbar \delta_{ij} \Psi(x)$$

$$\Rightarrow \quad \left[\hat{x}_i, \hat{P}_j\right] = i\hbar \delta_{ij}$$

3.24 The Schrödinger wave equation

Schrödinger equation is a mathematical equation that describes the changes over time of a physical system in which quantum effects, such as wave–particle duality, are significant.

These systems are referred to as quantum (mechanical) systems. The equation is considered a central result in the study of quantum systems, and its derivation was a significant landmark in the development of the theory of quantum mechanics. It was named after Erwin Schrödinger, who derived the equation in 1925, and published it in 1926, forming the basis for his work that resulted in his being awarded the Nobel Prize in Physics in 1933.

The concept of a wave function is a fundamental postulate of quantum mechanics that defines the state of the system at each spatial position, and time. Using these postulates, Schrödinger's equation can be derived from the fact that the time-evolution operator must be unitary, and must therefore be generated by the exponential of a self-adjoint operator, which is the quantum Hamiltonian. This derivation is explained below.

In the Copenhagen interpretation of quantum mechanics, the wave function is the most complete description that can be given of a physical system. Solutions to Schrödinger's equation describe not only molecular, atomic, and subatomic systems, but also macroscopic systems, possibly even the whole universe. Schrödinger's equation is central to all applications of quantum mechanics including quantum field theory which combines special relativity with quantum mechanics. Theories of quantum gravity, such as string theory, also do not modify Schrödinger's equation.

The Schrödinger equation is not the only way to study quantum mechanical systems and make predictions, as there are other quantum mechanical formulations such as matrix mechanics, introduced by Werner Heisenberg, and path integral formulation, developed chiefly by Richard Feynman. Paul Dirac incorporated matrix mechanics and the Schrödinger equation into a single formulation.

3.25 Wave equation for free and non- free particle: Schrödinger Time dependent equation

For undamped [constant amplitude] monochromatic harmonics [constant angular momentum] wave propagating in the positive X- axis

$$y = A\exp\left(-i\omega[t-\frac{x}{v}]\right) \qquad [1]$$

In Quantum Mechanics the wave function corresponds to the displacement. Since y is not itself a measureable quantity and it may therefore, be complex.

$$\Psi(x,t) = A\exp\left(-i\omega\,[t-\frac{x}{v}]\right) = A\exp\left(-i[2\pi\upsilon t - \frac{2\pi\upsilon}{\upsilon\lambda}x]\right)$$

$$= A\exp\left(-i[\frac{E}{\hbar}t - \frac{2\pi}{\lambda}x]\right) = A\exp\left(-i[\frac{E}{\hbar}t - \frac{p}{\hbar}x]\right)$$

$$\Psi(x,t) = A\exp\left(-\frac{i}{\hbar}[Et - px]\right) \qquad [2]$$

In three dimensions the above equation can be written as

$$\psi(\bar{r},t) = Aexp\left(\frac{i}{\hbar}\right)(\bar{p}.\bar{r} - Et) \;\; or, \overline{\nabla}\,\psi(\bar{r},t) = A\overline{\nabla}[exp\left(\frac{i}{\hbar}\right)(\bar{p}.\bar{r} - Et)]$$

$$\overline{\nabla}\,\psi(\bar{r},t) = A\left[\hat{\imath}\frac{\partial}{\partial x} + \hat{\jmath}\frac{\partial}{\partial y} + \hat{k}\frac{\partial}{\partial z}\right][exp\left(\frac{i}{\hbar}\right)(p_x x + p_y y + p_z z - Et)]$$

$$= \left(\frac{i}{\hbar}\right)A\left(\hat{\imath}p_x + \hat{\jmath}p_y + \hat{k}p_z\right)\left[exp\left(\frac{i}{\hbar}\right)\left(p_x x + p_y y + p_z z - Et\right)\right] = \left(\frac{i}{\hbar}\right)\bar{p}\psi(\bar{r},t)$$

$$\overline{\nabla}\,\psi(\bar{r},t) = -\frac{1}{i\hbar}\bar{p}\psi(\bar{r},t) \;or, -i\hbar\overline{\nabla}\,\psi(\bar{r},t) = \bar{p}\psi(\bar{r},t) \qquad \boxed{\bar{p} = -i\hbar\overline{\nabla}} \qquad [3]$$

Again, differentiating equation [2] with respect to t

$$\frac{\partial\psi(\bar{r},t)}{\partial t} = \left(-\frac{i}{\hbar}\right) EAexp\left(\frac{i}{\hbar}\right)(\bar{p}.\bar{r} - Et) = \frac{1}{i\hbar}E\psi(\bar{r},t)$$

Or, $$E = i\hbar\frac{\partial}{\partial t} \qquad [4]$$

For free particle, the total energy is equal to the kinetic energy

$$E = T + V = T \quad [Here, V = 0]$$

Or, $$E = \frac{P^2}{2m}$$

Or, $$E\Psi(\overline{r},t) = \frac{1}{2m}P^2\Psi(\overline{r},t)$$

Or, $$i\hbar\frac{d\Psi(\overline{r},t)}{dt} = \frac{1}{2m}(-i\hbar\overline{\nabla})^2\Psi(\overline{r},t)$$

Or, $$i\hbar\frac{d\Psi(\overline{r},t)}{dt} = -\frac{\hbar^2}{2m}\nabla^2\Psi(\overline{r},t)$$

Or, $$-\frac{\hbar^2}{2m}\nabla^2\Psi(\overline{r},t) = i\hbar\frac{d\Psi(\overline{r},t)}{dt} \qquad [5]$$

The equation [5] represents the time dependent Schrödinger equation for free particle in three dimensions.

If the particle is propagated by an external force & its potential energy is $V(\overline{r},t)$; then total energy of the particle

$$E = \frac{P^2}{2m} + V(\overline{r},t)$$

Or, $$E\Psi(\overline{r},t) = \frac{1}{2m}P^2\Psi(\overline{r},t) + V(\overline{r},t)\Psi(\overline{r},t)$$

Or, $$i\hbar\frac{d\Psi(\overline{r},t)}{dt} = \frac{-\hbar^2}{2m}\nabla^2\Psi(\overline{r},t) + V(\overline{r},t)\Psi(\overline{r},t) \qquad [6]$$

If the potential energy does not depend on time I.e., $V = V(\overline{r})$ then equation [6] becomes,

$$i\hbar\frac{d\Psi(\overline{r},t)}{dt} = \left[\frac{-\hbar^2}{2m}\nabla^2 + V(\bar{r})\right]\Psi(\overline{r},t) \qquad [7]$$

Equation [7] represents the three dimensional time dependent Schrödinger equation. Equation [7] can be written as

$$i\hbar\frac{d\Psi(\overline{r},t)}{dt} = H\Psi(\overline{r},t) \qquad [8]$$

Where $H = -\frac{\hbar^2}{2m}\nabla^2 + V(\bar{r})$; is called the Hamiltonian operator. [9]

Again, let $\Psi(\bar{r},t) = C\Psi(\bar{r})e^{-i/_{\hbar}Et}$ [10]

Now from equation [7]

$$i\hbar\frac{d}{dt}\left[C\Psi(r)e^{-i/_{\hbar}Et}\right] = \left(-\frac{\hbar^2}{2m}\right)\nabla^2 C\Psi(r)e^{-i/_{\hbar}Et} + V(r)C\Psi(r)e^{-i/_{\hbar}Et}$$

Or, $$i\hbar\left(-\frac{i}{\hbar}\right)EC\Psi(r)e^{-i/_{\hbar}Et} = \left(-\frac{\hbar^2}{2m}\right)\nabla^2 C\Psi(r)e^{-i/_{\hbar}Et} + V(r)C\Psi(r)e^{-i/_{\hbar}Et}$$

Or, $$E\Psi(\bar{r},t) = -\frac{\hbar^2}{2m}\nabla^2\Psi(\bar{r},t) + V(r)\Psi(\bar{r},t)$$

Or, $$\nabla^2\Psi(\bar{r},t) + \frac{2m}{\hbar^2}\big(E - V(r)\big)\Psi(\bar{r},t) = 0 \qquad [11]$$

Equation [11] represents the time dependent Schrödinger equation, if the particle is propagated by an external force.

3.26 Time dependent Schrödinger equation and its solution: Under Conservative force

Schrödinger time dependent equation for a particle under conservative force is

$$-\frac{\hbar^2}{2m}\nabla^2\,\psi(\bar{r},t) + V(\bar{r})\psi(\bar{r},t) = i\hbar\frac{\partial\psi(\bar{r},t)}{\partial t} \qquad [1]$$

Let us try as a solution of equation [1]

$$\psi(\bar{r},t) = u(\bar{r})\,f(t) \qquad [2]$$

Where u a function of space coordinates only and f is a function of time only. From equation [1]

$$-\frac{\hbar^2}{2m}\nabla^2[\,u(\bar{r})f(t)] + V(\bar{r})u(\bar{r})f(t) = i\hbar\frac{\partial}{\partial t}\,[u(\bar{r})f(t)]$$

$$-\frac{\hbar^2}{2m}\,f(t)\nabla^2 u(\bar{r}) + V(\bar{r})u(\bar{r})f(t) = i\hbar\,u(\bar{r})\frac{\partial f(t)}{\partial t}$$

Dividing both sides by $u(\bar{r})f(t)$.

$$-\frac{\hbar^2}{2m}\frac{\nabla^2 u(\bar{r})}{u(\bar{r})} + V(\bar{r}) = \frac{i\hbar}{f(t)}\frac{\partial f(t)}{\partial t} \qquad [3]$$

L.H.S of equation [3] is a function of position only whereas R.H.S is a function of time only. To solve the equation [3]; if both sides of equation [3] are equal to a constant say C.

$$-\frac{\hbar^2}{2m}\frac{\nabla^2 u(\bar{r})}{u(\bar{r})} + V(\bar{r}) = C \quad [4]$$

And $$\frac{i\hbar}{f(t)}\frac{\partial f(t)}{\partial t} = C \quad [5]$$

From equation [4]

$$-\frac{\hbar^2}{2m}\nabla^2 u(\bar{r}) + V(\bar{r})u(\bar{r}) = C\,u(\bar{r})$$

$$\left[-\frac{\hbar^2}{2m}\nabla^2 + V(\bar{r})\right]u(\bar{r}) = C\,u(\bar{r}) \quad [6]$$

$$H_{op}u(\bar{r}) = C\,u(\bar{r}) \quad [7]$$

Where, $H_{op} = \left[-\frac{\hbar^2}{2m}\nabla^2 + V(\bar{r})\right]$; is Hamiltonian operator. Equation [7] represents the Eigen value equation, C is an Eigen value and it is real and $u(\bar{r})$ is the Eigen function. Also, equation [7] represents the time independent Schrödinger equation. Let us denote C by E, because C indicates the Eigen value of total energy.

$$H_{op}u(\bar{r}) = E\,u(\bar{r}) \quad [8]$$

Equation [6] becomes

$$\left[-\frac{\hbar^2}{2m}\nabla^2 + V(\bar{r})\right]u(\bar{r}) = E\,u(\bar{r}) \quad [9]$$

Equation [9] is called time independent Schrödinger equation. Again, from equation [5]

$$\frac{i\hbar}{f(t)}\frac{\partial f(t)}{\partial t} = C \;\; or,\;\; i\hbar\frac{\partial f(t)}{\partial t} = C\,f(t) = E\,f(t) \;\; or, \frac{\partial f(t)}{\partial t} = \frac{E\,f(t)}{i\hbar}$$

Let $f(t) = e^{nt}$ [10]

$\therefore r\,e^{nt} = \frac{E}{i\hbar}e^{nt} \quad or, (n - \frac{E}{i\hbar})\,e^{nt} = 0$

$\therefore (n - \frac{E}{i\hbar}) = 0$ And $e^{nt} \neq 0$ Therefore $n = \frac{E}{i\hbar} = \frac{-i\,E}{\hbar} = -i\,\omega$

$$\therefore f(t) = e^{-i\,\omega\,t} = e^{-\frac{i}{\hbar}E\,t} \;;\;\; E = h\nu = 2\pi\hbar\nu = \frac{E}{\hbar} = 2\pi\,\nu\left(\omega = \frac{2\pi}{T} = 2\pi\,\nu\right)$$

Finally equation [2] becomes

$$\boxed{\psi(\bar{r},t) = u(\bar{r})\,f(t) = u(\bar{r})\,e^{-\frac{i}{\hbar}E\,t} = u(\bar{r})\,e^{-i\,\omega\,t}} \quad [11]$$

Equation [11] represents the solution of Schrödinger equation under conservative force.

3.27 Time independent Schrödinger equation and Stationary State

Schrödinger time dependent equation for a particle under conservative force is

$$-\frac{\hbar^2}{2m}\nabla^2 \psi(\bar{r},t) + V(\bar{r})\psi(\bar{r},t) = i\hbar\frac{\partial\psi(\bar{r},t)}{\partial t} \qquad [1]$$

Let us try as a solution of equation [1]

$$\psi(\bar{r},t) = u(\bar{r})\, f(t) \qquad [2]$$

Where u a function of space coordinates only and f is a function of time only. From equation [1]

$$-\frac{\hbar^2}{2m}\nabla^2[\, u(\bar{r})f(t)] + V(\bar{r})u(\bar{r})f(t) = i\hbar\frac{\partial}{\partial t}\,[u(\bar{r})f(t)]$$

$$-\frac{\hbar^2}{2m}\, f(t)\nabla^2 u(\bar{r}) + V(\bar{r})u(\bar{r})f(t) = i\hbar\, u(\bar{r})\frac{\partial f(t)}{\partial t}$$

Dividing both sides by $u(\bar{r})f(t)$.

$$-\frac{\hbar^2}{2m}\frac{\nabla^2 u(\bar{r})}{u(\bar{r})} + V(\bar{r}) = \frac{i\hbar}{f(t)}\frac{\partial f(t)}{\partial t} \qquad [3]$$

L.H.S of equation [3] is a function of position only whereas R.H.S is a function of time only. To solve the equation [3]; if both sides of equation [3] are equal to a constant say C.

$$-\frac{\hbar^2}{2m}\frac{\nabla^2 u(\bar{r})}{u(\bar{r})} + V(\bar{r}) = C \qquad [4]$$

And

$$\frac{i\hbar}{f(t)}\frac{\partial f(t)}{\partial t} = C \qquad [5]$$

From equation [4]

$$-\frac{\hbar^2}{2m}\nabla^2 u(\bar{r}) + V(\bar{r})u(\bar{r}) = C\, u(\bar{r})$$

$$\left[-\frac{\hbar^2}{2m}\nabla^2 + V(\bar{r})\right]u(\bar{r}) = C\, u(\bar{r}) \qquad [6]$$

$$H_{op}u(\bar{r}) = C\, u(\bar{r}) \qquad [7]$$

Where, $H_{op} = \left[-\frac{\hbar^2}{2m}\nabla^2 + V(\bar{r})\right]$; is Hamiltonian operator. Equation [7] represents the Eigen value equation, C is an Eigen value and it is real and $u(\bar{r})$ is the Eigen function. Also, equation [7] represents the time independent Schrödinger equation. Let us denote C by E, because C indicates the Eigen value of total energy.

$$H_{op}u(\bar{r}) \ = E\,u(\bar{r}) \qquad [8]$$

Equation [6] becomes

$$\left[-\frac{\hbar^2}{2m}\nabla^2 + V(\bar{r})\right]u(\bar{r}) \ = E\,u(\bar{r}) \qquad [9]$$

Equation [9] is called time independent Schrödinger equation.

- **Stationary State**

$$|\psi(\bar{r},t)\,|^2 = \psi(\bar{r},t)\,\psi^{\times}(\bar{r},t)$$

$$= u(\bar{r})\,e^{-\frac{i}{\hbar}E\,t}u^{\times}(r)\,e^{-\frac{i}{\hbar}E^{\times}\,t}$$

$$= u(\bar{r})\,e^{-\frac{i}{\hbar}E\,t}u^{\times}(r)\,e^{+\frac{i}{\hbar}E\,t} \quad [\because E\ is\ real\ \therefore\ E^{\times} = E\,]$$

$$|\psi(\bar{r},t)\,|^2 = u(\bar{r})u^{\times}(\bar{r}) = |u(\bar{r})|^2 \qquad [10]$$

Thus we find $|\psi(\bar{r},t)\,|^2$is independent of time. So, the states given by equation [10] are stationary states.

3.28 Matrix form of Schrödinger equation

From time dependent Schrödinger equation

$$i\hbar\frac{\partial}{\partial t}\Psi(\bar{r},t) = -\frac{\hbar^2}{2m}\nabla^2\Psi(\bar{r},t) + V(\bar{r})\Psi(\bar{r},t) \qquad [1]$$

If we use $\Psi(\bar{r},t) = <\bar{r}|\Psi(t)>$ in equation [1] then

$$i\hbar\frac{\partial}{\partial t} < \bar{r}|\Psi(t) = -\frac{\hbar^2}{2m}\nabla^2 < \bar{r}|\Psi(t) > +V(\bar{r}) < \bar{r}|\Psi(t) >$$

$$< \bar{r}\left|i\hbar\frac{\partial}{\partial t}\right|\Psi(t) > \ = \ < \bar{r}\left|\frac{p^2}{2m}\right|\Psi(t) > + < \bar{r}|V(\bar{r})|\Psi(t) >$$

$$[\because\ \bar{p} = -i\hbar\bar{\nabla}\ and < \bar{r}|\hat{p}^n|\gamma > \ = (-i\hbar)^{\mathrm{n}}\bar{\nabla}^n < \bar{r}|\gamma >]$$

$$i\hbar\frac{\partial}{\partial t}|\Psi(t) > \ = \left[\left|\frac{p^2}{2m}\right| + V(\bar{r})\right]|\Psi(t) >$$

$$i\hbar\frac{\partial}{\partial t}|\Psi(t) > \ = H|\Psi(t) >$$

$$i\hbar\frac{\partial}{\partial t}|\Psi(t) > \ = \left[\frac{\hat{p}^2}{2m} + V(\bar{r})\right]|\Psi(t) > \quad \boxed{i\hbar\frac{\partial}{\partial t}|\Psi(t) >= \hat{H}|\Psi(t) >} \qquad [2]$$

Generally, equation [2] is known as Schrödinger equation for $|\Psi(t)>$. Again $|\Psi(t)>$ can be expanded by the of the Hamiltonian's Eigen ket $|\text{n}>$ which is not explicitly time dependent.

$$|\Psi(t) = \sum_n C_n(t)|\text{n}> \quad [3]$$

From equation [2] and [3]

$$i\hbar \sum_n \frac{\partial}{\partial t} C_n(t)|\text{n}> \; = \sum_n C_n \hat{H}\,|\text{n}> \quad [4]$$

Multiplying both sides on the left by the $< k|$

$$i\hbar \sum_n \frac{\partial}{\partial t} C_n(t) < k|\text{n}> \; = \sum_n C_n < k|\hat{H}\,|\text{n}>$$

$$i\hbar \frac{\partial C_k}{\partial t} = \sum_n H_{kn} C_n \quad [5]$$

Equation [5] is the matrix form of the Schrödinger equation. It is applicable in all respects. The term can be written in matrix form

$$H_{kn} = < k|\hat{H}|\text{n}> \quad [6]$$

Sets of equation can be written from equation [5]

$$\begin{aligned} i\hbar\frac{\partial C_1}{\partial t} &= H_{11}C_1 + H_{12}C_2 + H_{13}C_3 + \cdots \cdots \cdots\cdots \cdots\cdots\cdots\cdots\cdots\cdots\cdots\cdots\cdots \\ i\hbar\frac{\partial C_2}{\partial t} &= H_{21}C_1 + H_{22}C_2 + H_{23}C_3 + \cdots \cdots \cdots\cdots \cdots\cdots\cdots\cdots\cdots\cdots\cdots\cdots\cdots \\ i\hbar\frac{\partial C_3}{\partial t} &= H_{31}C_1 + H_{32}C_2 + H_{33}C_3 + \cdots \cdots \cdots\cdots \cdots\cdots\cdots\cdots\cdots\cdots\cdots\cdots\cdots \\ &\vdots \qquad\quad \vdots \qquad\quad \vdots \end{aligned}$$

Finally,

$$i\hbar\frac{\partial}{\partial t}\begin{pmatrix} C_1 \\ C_2 \\ C_3 \end{pmatrix} = \begin{pmatrix} H_{11} & H_{12} & H_{13}\cdots \\ H_{21} & H_{22} & H_{23}\cdots \\ H_{31} & H_{32} & H_{33}\cdots \end{pmatrix}$$

$$i\hbar\frac{\partial}{\partial t}\begin{pmatrix} C_1 \\ C_2 \\ C_3 \\ \vdots \end{pmatrix} = \begin{pmatrix} H_{11} & H_{12} & H_{13} & \cdots \\ H_{21} & H_{22} & H_{23} & \cdots \\ H_{31} & H_{32} & H_{33} & \cdots \\ \vdots & \vdots & \vdots & \ddots \end{pmatrix} \quad [7]$$

Equation [7] represents the matrix form of Schrödinger equation and it is consistence with the equation [2].

3.29 Momentum representation of Schrödinger equation

For undammed monochromatic wave function

$$\psi(\bar{r},t) = A\, e^{i\,(\bar{k}.\bar{r}-\omega_k t)} = Ae^{i\,(\bar{P}.\bar{r}-Et)/\hbar} \quad [1]$$

$$[\because P = \hbar\, k\,,\ k = \frac{P}{\hbar}\,; \quad \omega = \frac{E}{\hbar}\ and\ \hbar = \frac{h}{2\pi}]$$

According to Fourier transformation, equation [1] can be written as

$$\psi(\bar{r},t) = (2\pi\hbar)^{-\frac{3}{2}} \int \psi(p)\, e^{\frac{i}{\hbar}(\bar{k}.\bar{r}-\omega_k t)} d^3k \quad [2]$$

Where momentum wave function

$$\psi(p) = (2\pi\hbar)^{-\frac{3}{2}} \int \psi(\bar{r},t)\ e^{-\frac{i}{\hbar}(\bar{k}.\bar{r}-\omega_k t)} d^3r \quad [3]$$

From the time Dependent Schrödinger equation

$$i\hbar \frac{\partial\psi(\bar{r},t)}{\partial t} = -\frac{\hbar^2}{2m}\nabla^2\psi(\bar{r},t)\ + V(r)\psi(\bar{r},t)$$

For the time dependent wave function

$$\psi(\bar{r},t) = \psi(\bar{r})e^{-\frac{i}{\hbar}(Et)} \quad [4]$$

Equation [4] can be written in momentum form

$$\psi(\bar{p},t) = \psi(\bar{p})e^{-\frac{i}{\hbar}(Et)} \quad [5]$$

$$= (2\pi\hbar)^{-\frac{3}{2}} \int \psi(\bar{r},t)\ e^{-\frac{i}{\hbar}(\bar{k}.\bar{r})} e^{\frac{i}{\hbar}(\omega t)} d^3r\ e^{-\frac{i}{\hbar}(Et)}$$

$$= (2\pi\hbar)^{-\frac{3}{2}} \int \psi(\bar{r},t)\ e^{-\frac{i}{\hbar}(\bar{P}.\bar{r})} d^3r$$

Now, multiplying both side by $i\hbar$ and differentiating with respect to time.

$$i\hbar \frac{\partial\psi(\bar{P},t)}{\partial t} = (2\pi\hbar)^{-\frac{3}{2}} \int i\hbar \frac{\partial\psi(\bar{r},t)}{\partial t} e^{-\frac{i}{\hbar}(\bar{P}.\bar{r})} d^3r$$

$$i\hbar \frac{\partial\psi(\bar{P},t)}{\partial t} = (2\pi\hbar)^{-\frac{3}{2}} \int \left[-\frac{\hbar^2}{2m}\nabla^2 + V(\bar{r})\right]\psi(\bar{r},t)\ e^{-\frac{i}{\hbar}(\bar{P}.\bar{r})} d^3r$$

$$= (2\pi\hbar)^{-\frac{3}{2}} \int -\frac{\hbar^2}{2m}\nabla^2\psi(\bar{r},t)\ e^{-\frac{i}{\hbar}(\bar{P}.\bar{r})} d^3r + (2\pi\hbar)^{-\frac{3}{2}} \int V(\bar{r})\, \psi(\bar{r},t)\ e^{-\frac{i}{\hbar}(\bar{P}.\bar{r})} d^3r$$

$$= (2\pi\hbar)^{-\frac{3}{2}} \int -\frac{\hbar^2}{2m}\nabla^2 e^{-ikr}\psi(\bar{r},t)\ d^3r + (2\pi\hbar)^{-\frac{3}{2}} \int V(\bar{r})\, e^{-\frac{i}{\hbar}(\bar{P}.\bar{r})}\psi(\bar{r},t)\ d^3r$$

$$= (2\pi\hbar)^{-\frac{3}{2}} \int -\frac{\hbar^2}{2m}(-\mathrm{ik})^2 e^{-ikr}\psi(\bar{r},t)\ d^3r + (2\pi\hbar)^{-\frac{3}{2}} \int V(\bar{r})\, e^{-\frac{i}{\hbar}(\bar{P}.\bar{r})}\psi(\bar{r},t)\ d^3r$$

$$= \frac{\hbar^2 k^2}{2m}(2\pi\hbar)^{-\frac{3}{2}} \int \psi(\bar{r},t)\ e^{-\frac{i}{\hbar}(\bar{P}.\bar{r})} d^3r + V(\bar{r})(2\pi\hbar)^{-\frac{3}{2}} \int \psi(\bar{r},t)\ e^{-\frac{i}{\hbar}(\bar{P}.\bar{r})}\ d^3r$$

$$i\hbar\frac{\partial\psi(\bar{P},t)}{\partial t} = \frac{P^2}{2m}\psi(\bar{p},t) + V(\bar{r})\psi(\bar{p},t)\ [\psi(\bar{p},t) = (2\pi\hbar)^{-\frac{3}{2}} \int \psi(\bar{r},t)\ e^{-\frac{i}{\hbar}(\bar{P}.\bar{r})} d^3r\ ] \qquad [6]$$

Equation [6] represents the momentum representation of Schrödinger equation.

[Note: Fourier transformation of wave function $\psi(\bar{r},t) = (2\pi)^{-\frac{3}{2}} \int A(k)\, e^{i(\bar{k}.\bar{r}-\omega_k t)}\, dk$

Where $A(k)$ is the momentum wave function $A(k) = (2\pi)^{-\frac{3}{2}} \int \psi(\bar{r},t)\ e^{-i(\bar{k}.\bar{r}-\omega_k t)}$

$E = \hbar\,\omega \leftrightarrow \omega = \frac{E}{\hbar}$; $P = \hbar\, k \leftrightarrow k = \frac{P}{\hbar} = -i\bar{\nabla}$; $P = -i\,\hbar\bar{\nabla}$]

3.30 Momentum Eigen function & Eigen values

According to de-Broglie hypothesis the linear momentum and wave number is related by

$$p = \hbar k \qquad [1]$$

The momentum Eigen value equation for the Eigen function $\Phi_k(r)$

$$\hat{\mathrm{p}}\Phi_k(r) = \hbar k\Phi_k(r) \qquad [2]$$

The Eigen function $\Phi_k(r)$ can be written as the separation of variables

$$\Phi_k(r) = \Phi_{k_x}(x)\Phi_{k_y}(y)\Phi_{k_z}(z) \qquad [3]$$

Therefore, equation [2] becomes

$$(\hat{\imath}\mathrm{p_x} + \hat{\jmath}\mathrm{p_y} + \hat{k}\mathrm{p_z})\,\Phi_{k_x}(x)\Phi_{k_y}(y)\Phi_{k_z}(z) = \hbar(\hat{\imath}\mathrm{k_x} + \hat{\jmath}\mathrm{k_y} + \hat{k}\mathrm{k_z})\Phi_{k_x}(x)\Phi_{k_y}(y)\Phi_{k_z}(z)$$

So,

$$\mathrm{p_x}\Phi_{k_x}(x) = \hbar\mathrm{k_x}\Phi_{k_x}(x) \qquad [4]$$

$$\mathrm{p_y}\Phi_{k_y}(x) = \hbar\mathrm{k_y}\Phi_{k_y}(y) \qquad [5]$$

$$\mathrm{p_z}\Phi_{k_z}(x) = \hbar\mathrm{k_z}\Phi_{k_z}(z) \qquad [6]$$

Now introducing the operator $\mathrm{p_x} = -i\hbar\frac{\partial}{\partial \mathrm{x}}$ in equation [4]

$$-\mathrm{i}\hbar\frac{\partial}{\partial \mathrm{x}}\Phi_{k_x}(x) = \hbar\mathrm{k_x}\Phi_{k_x}(x)$$

$$\int \frac{\partial\Phi_{k_x}(x)}{\Phi_{k_x}(x)} = \mathrm{ik_x}\int \mathrm{dx}\ \text{ or, } \ln\Phi_{k_x}(x) = \mathrm{ik_x x}\ \text{ or, } \Phi_{k_x}(x) = \mathrm{N_1 exp(ik_x x)} \qquad [7]$$

Similarly, using the operators $p_y = -i\hbar\frac{\partial}{\partial y}$ and $p_z = -i\hbar\frac{\partial}{\partial z}$ in equation [5] and [6] and finally we can write

$$\Phi_{k_y}(y) = N_2 \exp(ik_y y) \quad [8]$$

$$\Phi_{k_z}(z) = N_3 \exp(ik_z z) \quad [9]$$

Equations [4], [5] and [6] represent the momentum Eigen value equations and their corresponding Eigen values are $\hbar k_x$, $\hbar k_y$ and $\hbar k_z$ respectively.

Equation [7], [8] and [9] represents the momentum Eigen functions. From equation [3]

$$\Phi_k(r) = \Phi_{k_x}(x)\Phi_{k_y}(y)\Phi_{k_z}(z)$$

$$\Phi_k(\boldsymbol{r}) = N_1N_2N_3\exp(ik_x x + ik_y y + ik_z z) = N\exp(i\bar{k}.\bar{r}) \quad [10]$$

Equation [10] indicates the momentum Eigen function in three dimensions. $N = N_1N_2N_3$ is the normalization constant. Equation [10] can also be written as

$$\Phi_p(\boldsymbol{r}) = Nexp(i\bar{p}.\bar{r}/\hbar)$$

From the above equation the momentum Eigen value equation will be

$$\hat{p}\Phi_p(\boldsymbol{r}) = \hbar k\Phi_p(\boldsymbol{r})\ ;\ \text{where } \bar{p} = \hbar\bar{k} \quad [11]$$

3.31 Normalization of momentum Eigen function [Box normalization]

Let the origin at the centre of a square of arm L of an inertial frame. Also consider the length of the cube is parallel to three axes of the coordinate system. Consider the x-axis and according to the boundary condition we can write

$$[\Phi_k(\boldsymbol{r})]_{x=-L/2} = [\Phi_k(\boldsymbol{r})]_{x=+L/2}$$

$$exp(-ik_xL/2) = exp(+ik_xL/2)\ ;\ [\Phi_k(\boldsymbol{r}) = N\exp(i\bar{k}.\bar{r}) = N_1\exp(ik_x x)]$$

$$exp(-ik_xL/2)exp(+ik_xL/2) = exp(+ik_xL/2)exp(+ik_xL/2)$$

[Multiplying both sides by $exp(+ik_xL/2)$]

Therefore, $1 = exp(2ik_xL/2) = exp(i2\pi n_x)$; where $n_x = 0, \pm1, \pm2, \cdots\cdots\cdots$

Here, $k_x = \frac{2\pi n_x}{L}$ $or, \bar{k}\frac{2\pi n}{L}$

The value of the wave number is not absolutely independent. For that reason, from the condition of box normalization

$$\int \Phi_k^\times(\vec{r})\,\Phi_k(\vec{r})\,d^3\vec{r} = 1 \quad Or, |N|^2 \int d^3 r = 1$$

$$Or, |N|^2 \int_{-L/2}^{L/2}\int_{-L/2}^{L/2}\int_{-L/2}^{L/2} dxdydz = 1$$

$$Or, \quad |N|^2 L^3 = 1, \text{ where } N = L^{-3/2}$$

[Where $\Phi_k(\boldsymbol{r}) = \mathrm{N}\exp(\mathrm{i}\bar{\mathrm{k}}.\bar{\mathrm{r}})$]

The *momentum Eigen function in three dimensions*

$$\Phi_k(\mathrm{r}) = \mathrm{N_1 N_2 N_3} \exp\left[i\left(k_x x + k_y y + k_z z\right)\right] = \mathrm{N}\exp\left(i\vec{k}.\vec{r}\right) = L^{-3/2} e^{i\vec{k}.\vec{r}}$$

$$\Phi_k(\mathrm{r}) = L^{-3/2} e^{i\vec{k}.\vec{r}}$$

$$\boxed{\Phi_k(\mathrm{r}) = L^{-3/2} e^{i\vec{k}.\vec{r}}} \qquad [1]$$

3.32 Momentum Eigen function using Box Normalization

For Box normalization

$$\phi_k(\bar{r}) = (L)^{-\frac{3}{2}} e^{i\,(\bar{k}.\bar{r})}$$

From wave vector $k = \frac{P}{\hbar} = -i\bar{\nabla}$ $or,\ k\,\phi_k(\bar{r}) = -i\,\bar{\nabla}\phi_k(\bar{r})$ $or,\ \bar{\nabla}\,\phi_k(\bar{r}) = i\,k\,\phi_k(\bar{r})$

$$\frac{\partial}{\partial x}\phi_k(\bar{r}) = i\,k_x\phi_k(\bar{r})$$

Similarly, $$\frac{\partial}{\partial y}\phi_k(\bar{r}) = i\,k_y\phi_k(\bar{r})$$

And $$\frac{\partial}{\partial z}\phi_k(\bar{r}) = i\,k_z\phi_k(\bar{r})$$

$\therefore$ $$\phi_k(\bar{r}) = \phi_k(x,y,z) = X(x)\,Y(y)\,Z(z)$$

$\therefore$ $$\frac{\partial}{\partial x}\phi_k(\bar{r}) = \frac{\partial}{\partial x}X(x)\,Y(y)\,Z(z) = Y(y)\,Z(z)\frac{\partial X(x)}{\partial x}$$

$$= Y(y)\,Z(z)\,i\,k_x\,X(x) \quad or, \frac{\partial X(x)}{X(x)} = i\,k_x\partial x$$

Or, $$\int\frac{d\,X(x)}{X(x)} = \int ik_x\partial\,x \rightarrow \ln X(x) = i\,k_x\,x + C_1$$

$\therefore$ $$X(x) = C_1 e^{\,i\,K_x\,x} \ similarly,\ Y(y) = C_1 e^{\,i\,K_y\,y}\ and\ Z(z) = C_1 e^{\,i\,K_z\,z}$$

$$\phi_K(\bar{r}) = X(x)\,Y(y)\,Z(z) = C_1C_2C_3\ e^{\,i(\,K_x\,x + K_y\,y + K_z\,z)}$$

$$\boxed{\therefore\ \phi_k(\bar{r}) = N\,e^{i\,(\bar{k}.\bar{r})}} \quad ; \quad \text{Where } N = C_1C_2C_3$$

3.33 Physical Significance of wave function in momentum space

From the expectation value of momentum [dynamical variable]

$$\langle p \rangle = \int_{-\infty}^{\infty} \psi^{\times} \frac{\hbar}{i} \frac{\partial \psi}{\partial x} dx \qquad [1]$$

Where $$\psi(x,t) = Ae^{\frac{i}{\hbar}(px-Et)} = \frac{1}{\sqrt{2\pi\hbar}} \int_{-\infty}^{\infty} C(p)\, e^{\frac{i}{\hbar}(px-Et)} dp \qquad [2]$$

Therefore, $$\langle p \rangle = \int_{-\infty}^{\infty} \psi^{\times} \frac{\hbar}{i} \frac{\partial \psi}{\partial x} dx = \int_{-\infty}^{\infty} \psi^{\times} \frac{\hbar}{i} \frac{1}{\sqrt{2\pi\hbar}} \int_{-\infty}^{\infty} \frac{i}{\hbar} C(p) p e^{\frac{i}{\hbar}(px-Et)} dp dx$$

$$= \frac{1}{\sqrt{2\pi\hbar}} \int_{-\infty}^{\infty} \int_{-\infty}^{\infty} \psi^{\times} e^{\frac{i}{\hbar}(px-Et)} dx\, p\, C(p)\, dp$$

$$= \int_{-\infty}^{\infty} c^{\times}(p)\, p\, C(p)\, dp \qquad [3]$$

Where $$C(p) = \frac{1}{\sqrt{2\pi\hbar}} \int_{-\infty}^{\infty} \psi(x,t) e^{-\frac{i}{\hbar}(px-Et)}\, dx \qquad [4]$$

$$\langle p \rangle = \int_{-\infty}^{\infty} p |C(p)|^2 dp \qquad [5]$$

$\therefore$ $|C(p)|^2$ is probability density function of p whereas $|\psi(x,t)|^2$ is probability density function of x. Comparison of

$$\langle x \rangle = \int_{-\infty}^{\infty} x |\psi(x,t)|^2 dx \qquad [6]$$

$$\langle p \rangle = \int_{-\infty}^{\infty} p |c(p)|^2 dp \qquad [7]$$

It is clear that $C(p)$ plays the same formula role for p that $\psi(x)$ does for x. $C(p)$ is called wave function in momentum space whereas $\psi(x,t)$ is called wave function in coordinate space or real space.

Parseval's formula gives $$\int_{-\infty}^{\infty} |\psi(x,t)|^2 dx = \int_{-\infty}^{\infty} |c(p)|^2 dp \qquad [8]$$

From the above equation, both sides are independent of time. Thus if we have a normalized $\psi(x,t)$, it will remain normalized although $\psi(x,t)$ changes with time.

3.34 Closure property of the momentum Eigen functions

We know, the normalized momentum Eigen function in three dimension

$$\langle r \mid p \rangle = (2\pi\hbar)^{-3/2} \exp(i\vec{p}.\vec{r}/\hbar)$$

The above equation can be written

$$\langle p \mid \vec{r}\,' \rangle = (2\pi\hbar)^{-3/2} \exp\left(\frac{-i\vec{p}.\vec{r}\,'}{\hbar}\right)$$

Integrating the product of the above two equations

$$\int \langle \vec{r}\,|\,p\rangle\langle p\,|\,\vec{r}\,'\rangle\, d^3p = \frac{1}{(2\pi\hbar)^3}\int \exp\left(\frac{i\vec{p}.(\vec{r}-\vec{r}')}{\hbar}\right) d^3p$$

$$= \frac{1}{(2\pi)^3}\int \exp\left(i\vec{k}.[\vec{r}-\vec{r}']\right) d^3k \;\; ; \; because,\; \vec{p} = \hbar\vec{k}$$

$$= \delta^3(\vec{r}-\vec{r}\,') = \langle \vec{r}\,|\,\vec{r}\,'\rangle$$

Where, $\delta^3(\vec{r}-\vec{r}\,') = \delta(x-x')\delta(y-y')\delta(z-z')$

Finally we can write

$$\int \langle \vec{r}\,|\,\vec{p}\rangle\langle \vec{p}\,|\,\vec{r}\,'\rangle\, d^3p = \delta^3(\vec{r}-r') = \langle \vec{r}\,|\,\vec{r}\,'\rangle$$

$$\therefore \int |\,\vec{p}\rangle\langle \vec{p}\,|\, d^3p = 1$$

3.35 Wave packet & probability can be calculated from coordinate momentum

We know, from closure property of momentum $\int d^3p\,|\,p\rangle\langle p\,| = 1$

$$Therefore, \langle r\,|\,\alpha\rangle = \int d^3p \langle r\,|\,p\rangle\langle p\,|\,\alpha\rangle$$

$$= \int d^3p\,(2\pi\hbar)\, e^{ip.r/\hbar}\langle p\,|\,\alpha\rangle \quad [\because \langle r\,|\,p\rangle = (2\pi\hbar)^{-3/2}\exp(^{ip.r/\hbar}]$$

$$\therefore \langle r\,|\,\alpha\rangle = (2\pi\hbar)^{-3/2}\int \langle p\,|\,\alpha\rangle e^{ip.r/\hbar}\, d^3p$$

Therefore, $\psi_a(r) = (2\pi\hbar)^{-3/2}\int \Phi_a(p)\, e^{ip.r/\hbar}\, d^3p$ [1]

$$\because \int d^3r\;|\,r\rangle\langle r\,| = 1 \quad ; \; we\; can\; write$$

$$\langle p\,|\,\alpha\rangle = \int d^3r \langle p\,|\,r\rangle\langle r\,|\,\alpha\rangle$$

$$= (2\pi\hbar)^{-3/2}\int d^3r \langle r\,|\,\alpha\rangle\, e^{-ip.r/\hbar} \quad [\because \langle r\,|\,p\rangle = (2\pi\hbar)^{-3/2}\exp(^{ip.r/\hbar}]$$

$$\Phi_a(p) = (2\pi\hbar)^{-3/2}\int \psi_a(r)\, e^{-ip.r/\hbar}\, d^3r \qquad [2]$$

Generally, according to equation [1] the above equation can be written

$$\psi(r) = (2\pi\hbar)^{-3/2}\int \Phi(p)\, e^{i\,p.r/\hbar}\, d^3p \qquad [3]$$

Generally, according to equation [1] the momentum wave function can be written

$$\Phi_a(p) = (2\pi\hbar)^{-3/2}\int \psi(\vec{r})\, e^{-i\vec{p}.\vec{r}/\hbar}\, d^3r \qquad [4]$$

So, time dependent wave function $\psi(r,t) = \psi(r)e^{-iEt/\hbar}$

Therefore,

$$\psi(\vec{r},t) = (2\pi\hbar)^{-3/2}\int \Phi(p)\, e^{i(\vec{p}.\vec{r}-Et)/\hbar}\, d^3r = (2\pi\hbar)^{-3/2}\int \Phi(p,t)\, e^{i\vec{p}.\vec{r}/\hbar}\, d^3r \qquad [5]$$

Equation [5] represents wave packet for plane wave function. So momentum wave function

$$\Phi(\text{p},\text{t}) = \Phi(\text{p})\,e^{-iEt/\hbar}$$

$$= (2\pi\hbar)^{-3/2}\int \psi(r)\,e^{-iEt/\hbar}\,e^{-i\vec{p}.\vec{r}/\hbar}\,d^3r$$

$$\textit{Therefore,}\ \ \Phi(\text{p},\text{t}) = (2\pi\hbar)^{-3/2}\int \psi(\vec{r},t)\,e^{-i\vec{p}.r/\hbar}\,d^3r \qquad [6]$$

Equation [6] represents wave packet for plane wave function. Finally,

$$\psi(\vec{\text{r}},\text{t}) = (2\pi\hbar)^{-3/2}\int \Phi(\text{p})\,e^{i(\vec{p}.\vec{r}-Et)/\hbar}\,d^3r = (2\pi\hbar)^{-3/2}\int \Phi(\text{p},\text{t})\,e^{i\vec{p}.\vec{r}/\hbar}\,d^3r$$

$$\textit{Therefore,}\ \ \Phi(\text{p},\text{t}) = (2\pi\hbar)^{-3/2}\int \psi(\vec{r},t)\,e^{-i\vec{p}.r/\hbar}\,d^3r$$

Therefore, $\psi(\vec{\text{r}},\text{t})$ and $\Phi(\vec{\text{p}},\text{t})$ are mutually Fourier- transform with respect to each other.

The equation [5] and [6] can be written as

$$\psi(\vec{\text{r}},\text{t}) = (2\pi)^{-3/2}\int \Phi(\text{k},\text{t})\,e^{i\vec{k}.\vec{r}/\hbar}\,d^3k \quad [\because\ p = \hbar k\,]$$

And *Therefore,*

$$\Phi(\vec{\text{k}},\text{t}) = (2\pi)^{-3/2}\int \psi(\vec{r},t)\,e^{-i\vec{k}.r/\hbar}\,d^3r$$

3.36 Probability for momentum and space coordinates from closure property

We know, from closure property of momentum $\int d^3p\,|\,\text{p}\rangle\langle p\,| = 1$

and closure property of space $\int d^3r\,|\,\text{r}\rangle\langle r\,| = 1;$

Therefore, $\langle\alpha\,|\,\alpha\rangle = \langle\alpha\,|\,\alpha\rangle = 1$; *from this we can write*

$$\int d^3p\langle\alpha\,|\,\vec{\text{p}}\rangle\langle\vec{\text{p}}|\alpha\rangle = 1 \ \ \textit{and} \int d^3r\langle\alpha\,|\,\vec{\text{r}}\rangle\langle\vec{\text{r}}\,|\,\alpha\rangle = 1 \ \ \textit{Or,}\ \int |\langle\vec{p}\,|\,\alpha\rangle|^2\,d^3p = \int |\langle\vec{r}\,|\,\alpha\rangle|^2\,d^3r = 1$$

$$\therefore \int |\Phi_a(\vec{\text{p}})|^2\,d^3\vec{p} = \int |\psi_a(\vec{\text{r}})|^2\,d^3\vec{r} = 1$$

According to our need the above equations can be written in different forms

$$\int |\Phi(\text{p})|^2\,d^3p = \int |\psi(\text{r})|^2\,d^3r = 1 \qquad \textit{Or,}\ \int |\Phi(\text{p})|^2\,dp = \int |\psi(\text{x})|^2\,dx = 1$$

3.37 Fourier transformation and inverse Fourier transformation

This is an important integral transformation used in all branches of Physics. It is used, for example, *to determine the frequency spectrum of time dependent signal*[29]. The Fourier transform $\tilde{f} \equiv \mathcal{F}[f]$ of a function $f(x)$ is a function of parameter k defined as follows:

$$\tilde{f}(k) = \mathcal{F}[f]f(x) = \frac{1}{\sqrt{2\pi}}\int_{-\infty}^{+\infty} e^{-ikx} f(x)\, dx \qquad [1]$$

The *inverse Fourier transform* $\mathcal{F}^{-1}[g]$ of a function $f(k)$ is a function of parameter x defined as follows:

$$\tilde{g}(x) = \mathcal{F}^{-1}[g](x) = \frac{1}{\sqrt{2\pi}}\int_{-\infty}^{+\infty} e^{+ikx} g(k)\, dk \qquad [2]$$

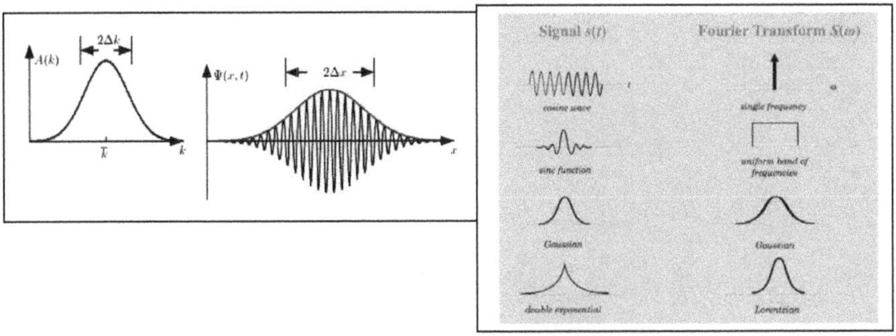

Fig. [3.37]: Plots of a typical wave packet $\psi(x)$ and its Fourier transform

3.38 Fourier transformations: Superposition of monochromatic waves

Fourier transform can be written as

$$\psi(x,t) = \frac{1}{\sqrt{2\pi}} \int_{-\infty}^{\infty} a(k)\, e^{i(kx-\omega t)}\, dk \qquad [1]$$

$$a(k) = \frac{1}{\sqrt{2\pi}} \int_{-\infty}^{\infty} \psi(x,t) e^{-i(kx-\omega t)}\, dx \qquad [2]$$

Here,
$$\begin{aligned} a(k)e^{i(kx-\omega t)} &= a(k)\cos(kx-\omega t) + ia(k)\sin(kx-\omega t) \\ &= C_1\cos(kx-\omega t) + C_2\sin(kx-\omega t) \\ &= \mathrm{C}\cos(\mathrm{kx}-\omega \mathrm{t}+\varphi) = \mathrm{C}\cos(\frac{2\pi}{\lambda}\mathrm{x} - 2\pi\upsilon\,\mathrm{t}+\varphi) \end{aligned}$$

Thus $a(k)e^{i(kx-\omega t)}$ is monochromatic wave of wave length λ, frequency υ and amplitude C~ a(k) which is independent of time although t explicitly appears in equation [2]. *The equation [1] gives superposition of monochromatic waves* of a continuous range of wave number k for which a(k) ≠ 0

3.39 Normalization of Delta function

We know, $\Phi_{p_x}(x) = A_1 \exp(ip_x x/\hbar)$

Therefore, $\langle x | p_x \rangle = A_1 \exp(ip_x x/\hbar) \quad \therefore \quad \langle p_x | x' \rangle = A_1^{\times} \exp(ip_x x'/\hbar)$

Therefore, $\langle x | p_x \rangle \langle p_x | x' \rangle$ *Or,* $\langle x | p_x \rangle \langle p_x | x' \rangle = |A_1|^2 \exp\left[ip_x (x - x')/\hbar\right]$ [1]

Now integrating the equation [1]

$$|A_1|^2 \int \exp\left[ip_x (x-x')/\hbar\right] dp_x = \int \langle x | p_x \rangle \langle p_x | x' \rangle dp_x = \langle x | \int | p_x \rangle \langle p_x | dp_x | x' \rangle = \langle x | x' \rangle$$

$$[\because \int | p_x \rangle \langle p_x | dp_x = 1]$$

$$|A_1|^2 \int \exp\left[ip_x (x-x')/\hbar\right] dp_x = \delta(x-x') = \frac{1}{2\pi} \int \exp\left[ik_x (x-x')\right] dk_x$$

$$|A_1|^2 \int \exp\left[ip_x (x-x')/\hbar\right] dp_x = \frac{1}{2\pi\hbar} \int \exp\left[ip_x (x-x')/\hbar\right] dp_x \quad ;where,\ p_x = \hbar k_x$$

$$|A_1|^2 = \frac{1}{2\pi\hbar} \quad \text{Or,} \quad A_1 = \frac{1}{\sqrt{2\pi\hbar}}$$

$$\therefore \qquad \langle x | p_x \rangle = \frac{1}{\sqrt{2\pi\hbar}} \exp(ip_x x/\hbar)$$

Similarly, $\langle y | p_y \rangle = \frac{1}{\sqrt{2\pi\hbar}} \exp(ip_y y/\hbar)$ *and* $\langle z | p_z \rangle = \frac{1}{\sqrt{2\pi\hbar}} \exp(ip_z z/\hbar)$

In three dimensions,

$$\langle r | p \rangle = \langle (x,y,z) | (p_x, p_y, p_z) \rangle = \langle x | p_x \rangle \langle y | p_y \rangle \langle z | p_z \rangle = (2\pi\hbar)^{-3/2} \exp\left[i\left(p_x x + p_y y + p_z z\right)/\hbar\right]$$

$$\boxed{\langle r | p \rangle = (2\pi\hbar)^{-3/2} \exp(i\vec{p}.\vec{r}/\hbar)}$$

3.40 Dirac-Delta function

My students always asked me what a Dirac delta function really is. I tried to explain it but eventually it is really tuff to make them understand. I have seen approximation of the Dirac delta function as an infinitely peaked Gaussian. I have also seen interpretation of Dirac delta function as a Fourier Transform which is widely adopted in study of quantum theory. So what really is a Dirac delta function? Is it a function? Or is it some kind of limit for a large class of functions? It is really confusing. But for the sake of my students it can be defined as *the delta function is not a function on the real line because we need to define its values in a way that has nothing to do with the real line and everything to do with what occurs if we integrate it against another function.*

$$\int_{-\infty}^{\infty} f(x)\delta(x-x')dx = f(x') \quad \textit{and}\ \text{hence} \quad \int_{-\infty}^{\infty} \delta(x-x')dx = 1$$

The usual 'intuitive' picture of a Dirac delta function is a function that is zero everywhere except for one point, where it is infinite. Thus Dirac delta is defined as a function that satisfies these Constraints.

$$\delta(x-x')=0 \quad if \;\; x \neq x' \qquad and \;\; \delta(x-x')=\infty \;\; if \;\; x=x'$$

Kronecker delta (named after Leopold Kronecker) is a piecewise function δ_{ij} of two variables i and j usually just non-negative integers. The function is 1 if the variables are equal and 0 otherwise. Therefore,

$$\delta_{ij}=\begin{pmatrix} 1 \;\; if \;\; i=j \\ 0 \;\; if \;\; i\neq j \end{pmatrix}$$ For example, $\delta_{12}=0$ whereas $\delta_{33}=1$

The Kronecker delta appears naturally in many areas of mathematics, physics and engineering, as a means of compactly expressing its definition above. The delta function belongs to the class of so-called *generalized functions.* [That can be defined as the limit of a class of delta sequences] This means that it is meaningful only as a part of an integral expression. I.e., for a smooth function $f(x)$, this takes zero values at $\pm\infty$,

$$\int_{-\infty}^{+\infty}\left[\frac{d}{dx}\delta(x)\right]f(x)dx=-\left[\frac{df}{dx}\right]_{x=0} \qquad [1]$$

$$\left[\frac{d}{dx}\delta(x)\right]f(x)=-\delta(x)\frac{df}{dx} \qquad [2]$$

In other words, the expression $\frac{d}{dx}\delta(x)$ can be seen as an operator acting on functions

$$\frac{d}{dx}\delta(x)=-\delta(x)\frac{d}{dx} \qquad [3]$$

In Quantum Mechanics we use $\delta(x)$ to write, for example, the wave function of a state with a well-defined position. We now introduce closure.

If $$\psi(x,t)=\sum_i C_i\psi_i(x,t) \qquad [4]$$

$$C_i=(\psi_i,\psi)\int_{-\infty}^{+\infty}\psi_i^{\times}(x,t)\psi(x,t)dx=\int_{-\infty}^{+\infty}\psi_i^{\times}(x',t)\psi(x',t)dx' \qquad [5]$$

$$\therefore \psi(x,t)=\sum_i\left(\int_{-\infty}^{+\infty}\psi_i^{\times}(x',t)\psi(x',t)dx'\right)\psi_i(x,t) \qquad [6]$$

$$=\int_{-\infty}^{+\infty}\left[\sum_i\psi_i^{\times}(x',t)\psi_i(x,t)\right]\psi(x',t)dx' \qquad [7]$$

If we compare equation [7] with

$$f(x)=\int_{-\infty}^{\infty}f(x')\delta(x-x')\,dx' \qquad [8]$$

Comparing equation [7] and [8], finally we have

$$\delta(x-x')=\sum_i\psi_i^{\times}(x',t)\psi_i(x,t) \qquad [9]$$

The property that Eigen functions obey equation [9] is called closure. The delta function is sometimes called "Dirac's delta function" or the "impulse symbol". *The delta function also obeys the so-called sifting property.*

3.41 Shifting property of Dirac-Delta function

An ordinary function $x(t)$ has the property that for $t = t_0$ the value of the function is given by $x(t_0)$. In contrast, the Delta function is a generalized function or distribution defined as

$$\delta(t) = 0 \quad for\ t \neq 0 \qquad [1]$$

$$\int_{-\infty}^{+\infty} \delta(t - t_0)\, x(t)dt = x(t_0) \qquad [2]$$

For any function $x(t)$ that is continuous at $t = t_0$ [See Fig. 3.42]. In particular,

$$\int_{-\infty}^{+\infty} \delta(t)\, x(t)dt = x(0) \quad \text{And} \quad \int_{-\infty}^{+\infty} \delta(t)\, dt = 1 \qquad [3]$$

The Delta function represents an idealized pulse that in practice can only be approximated. Its width approaches zero as its amplitude approaches infinity while the area under the curve of the function describing the pulse remains constant. The operation on $x(t)$ indicated on the left side of equation [2] yields the value of $x(t)$ at the singularity $t = t_0$ of the Delta function. This is called the "shifting property" of the Delta function [28].

The Delta function is not a function in the usual sense. Its definition contains a contradiction with respect to Riemann integration, because the integral of a function whose value is non-zero only at a finite number of points should be equal to zero. *Thus the Delta function is an abstract,* but useful tool for carrying out certain calculations based on mathematical rules derived from the definition in equation [2]. The sifting property of the Dirac delta function is clearly shown in the following figure 3.42.

Some property of Dirac-Delta Function:

[1] $\delta(ax) = \frac{1}{a}\delta(x)$ [2] $\delta(-x) = \delta(x)$

[3] $\int_{-\infty}^{\infty} \delta(x)dx = 1$ [4] $x\delta(x) = 0$ [5] $\delta'(x) = -\frac{1}{x}\delta(x)$ [6] $\delta(x - x') = \langle x|x'\rangle$

[7] $\langle r|r'\rangle = \delta^3(r - r') = (2\pi)^{-3}\int d^3k e^{ik.(r-r')}$ [8] $\delta'(x) = -\delta'(-x)$

[9] $f(x)\delta(x - a) = f(a)\delta(x - a)$ [10] $\delta(a - b) = \int \delta(a - x)\,\delta(x - b)dx$

[11] $\int_{-\infty}^{\infty} f(x)\delta(x - a)dx = f(a)$ [12] $\int (r')\,\delta^3(r - r')d^3r' = f(r)$

[13] $\delta(x) = \lim_{g\to\infty}\frac{1}{\pi}\frac{sin^2 gx}{gx^2}$ [14] $\delta(x^2 - a^2) = \frac{1}{2a}[\delta(x - a) + \delta(x + a)]$

[15] $\delta(x) = \lim_{g\to\infty} = \frac{singx}{\pi x}$ [16] $f(x) = \int_{-\infty}^{\infty} \delta(x - x')\, f(x')dx'$

[17] $\delta(x - x') = \frac{1}{2\pi}\int_{-\infty}^{\infty} e^{ik(x-x')}dk \to \delta(x) = \frac{1}{2\pi}\int_{-\infty}^{\infty} e^{ikx}\, dk$ [when $x' = 0$]

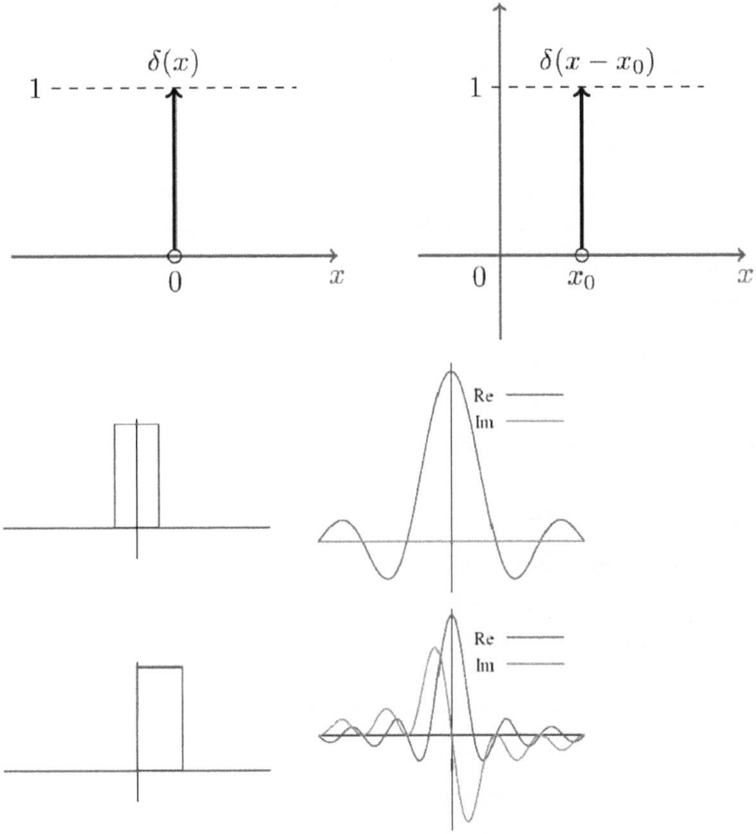

Fig. [3.41]: Illustration of the sifting property of the Dirac delta function.

3.42 Kronecker Delta [named after Leopold Kronecker] Function δ_{ij} and Levi-Civita [Epsilon] Symbol ε_{ijk}

1. Definitions

$$\delta_{ij} = \begin{pmatrix} 1 & if \ i = j \\ 0 & if \ i \neq j \end{pmatrix} \qquad \varepsilon_{ijk} = \{+1 \ \ if \ \ ijk = 123, 312, or\ 231\}$$

$$= \{-1 \quad if \quad ijk = 213, 321, or\ 132\}$$

$$= \{0 \quad all\ other\ cases\ [i.e., any\ two\ equal]\}$$

- So, for example, $\varepsilon_{112} = \varepsilon_{313} = \varepsilon_{222} = 0$
- The $+1$ $[or\ even]$ permutations are related by rotating the numbers around; think of starting with 123 and moving [in your mind] the 3 to the front of the line, to get 312. Do it again with the 2 and you get 231.The $-1[or\ odd]$ permutations starting with 231 are related to each other the same way; they are related to 123 by interchanging just two of the numbers [e.g., switch the 1 and 3 to get 321]

2. Applying δ_{ij} and ε_{ijk} to Vectors in Cartesian coordinates

- Instead of using $x, y, and\ z$ to label the components of a vector, we use 1, 2, and 3.
- Then the letters i, j, k ... can be used as summation variables, running from 1 to 3. [We could use any other letters, a, b, ...; it is merely a convention]
- Don't confuse the use of the dummy summation variables i, j, k, each of which can be 1, 2, or 3, with the unit vectors $\boldsymbol{i}, \boldsymbol{j}, \boldsymbol{k}$. These are independent notations.
- The dot product of two vectors **A.B** in this notation is

$$A.B = A_1B_1 + A_2B_2 + A_3B_3 = \sum_{i=1}^{3} A_iB_i = \sum_{i=1}^{3}\sum_{j=1}^{3} A_iB_j\,\delta_{ij}.$$

Note that there are nine terms in the final sums, but only three of them are non-zero.

- The ith component of the cross product of two vectors $\bar{A} \times \bar{B}$

$$(\bar{A} \times \bar{B})_i = \sum_{j=1}^{3}\sum_{k=1}^{3} \varepsilon_{ijk}\, A_jB_k$$

Again, there are nine terms in the sum, but this time only two of them are non-zero. Note also that this expression summarizes three equations, namely for $i = 1, 2, 3$.

3. Einstein Summation Convention

- We might notice that the summations in the expressions for $\boldsymbol{A}.\boldsymbol{B}\ and\ \bar{A} \times \bar{B}$ are redundant, because they only appear when an index like $i\ or\ j$ appears twice on one side of an equation. So we can omit them. thus

$$\sum_{i=1}^{3}\sum_{j=1}^{3} A_iB_j\,\delta_{ij} \rightarrow A_iB_j\delta_{ij} = A_iB_i \quad \text{And} \quad \sum_{j=1}^{3}\sum_{k=1}^{3} \varepsilon_{ijk}\, A_jB_k \rightarrow \varepsilon_{ijk}\, A_jB_k$$

- Rules: If an index appears [exactly] twice, then it is summed over and appears only on one side of an equation. A single index [called a free index] appears once on each side of the equation. So

  Valid: $A_i = A_j\delta_{ij}$, $B_k = \varepsilon_{ikl}\, A_iC_l$; Invalid: $A_i = B_iC_i$, $A_i = \varepsilon_{ijk}B_iC_j$

- When you have a Kronecker Delta δ_{ij} and one of the indices is repeated [say i], then you simplify it by replacing the other i index on that side of the equation by j and removing the δ_{ij} . For example

  $A_j\delta_{ij} = A_i,$ $B_{ij}C_{jk}\delta_{ik} = B_{kj}C_{jk} = B_{ij}C_{ji}$

Note that in the second case we had two choices of how to simplify the equation; use either one.

- The triple or box product $\bar{A}.(\bar{B}\times\bar{C})$ can be written

  $\bar{A}.(\bar{B}\times\bar{C}) = \varepsilon_{ijk}A_iB_jC_k = \varepsilon_{kij}A_iB_jC_k = \varepsilon_{kij}C_kA_iB_j = \bar{C}.(\bar{A}\times\bar{B})$

Where we have used the properties of ε_{ijk} to prove a relation among triple products with the vectors in a different order.

A very useful identity [if the repeated index is not first in both $\varepsilon's$, permute until it is]: $\varepsilon_{ijk}\varepsilon_{ilm} = \delta_{jl}\delta_{km} - \delta_{jm}\delta_{kl}$

4. Example: Proving a Vector Identity

We will write the ith Cartesian component of the gradient operator $\boldsymbol{\nabla}$ as $\partial_i \left(cf.\ \frac{\partial}{\partial x_i}\right)$. Let's simplify $\boldsymbol{\nabla}\times\left(\boldsymbol{\nabla}\times\mathbf{A}(\mathrm{x})\right)$. We start by considering the ith component and then we use our expression for the cross product

$$\left(\boldsymbol{\nabla}\times(\boldsymbol{\nabla}\times\mathbf{A})\right)_{\mathrm{i}} = \varepsilon_{ijk}\partial_j(\boldsymbol{\nabla}\times\mathbf{A})_k.$$

Next we replace the remaining cross product, making sure to introduce new dummy summation variables l *and* m.

$$\left(\boldsymbol{\nabla}\times(\boldsymbol{\nabla}\times\mathbf{A})\right)_{\mathrm{i}} = \varepsilon_{ijk}\partial_j\varepsilon_{klm}\partial_l\mathrm{A}_m = \varepsilon_{ijk}\varepsilon_{klm}\partial_j\partial_l\mathrm{A}_m$$

[The partial derivatives act only on the components of **A,** so we can pull out the $\varepsilon's$]. We rotated the indices in one of the $\varepsilon's$ in the last step so that we can now directly apply our very useful identity [and simplify]

$$\left(\boldsymbol{\nabla}\times(\boldsymbol{\nabla}\times\mathbf{A})\right)_{\mathrm{i}} = \left(\delta_{il}\delta_{jm} - \delta_{im}\delta_{lj}\right)\partial_j\partial_l\mathrm{A}_m = \partial_m\partial_i\,\mathrm{A}_m - \partial_l\partial_l\,\mathrm{A}_i = \partial_i(\partial_m\,\mathrm{A}_m) - (\partial_l\partial_l\mathrm{A})_i$$

or, finally $\boldsymbol{\nabla}\times(\boldsymbol{\nabla}\times\mathbf{A}) = \boldsymbol{\nabla}(\boldsymbol{\nabla}.\mathbf{A}) - \nabla^2$

Exercise

1. Define Eigen function, Eigen value and Eigen value equation.
2. What is an Operator? When an Operator is said to be a constant of motion?
3. Write an Operator for momentum and energy.
4. What is Hamiltonian Operator? Mention the importance of Hamiltonian Operator.
5. Define Hermitian Operator. Explain its importance.
6. Write a short note on:
    a. Linear Operator b.Unitary Operator c. Parity Operator d. Conjugate Operator and Self Adjoint Operator e. Projection Operator f. Transpose Operator $[\tilde{\Omega}]$
7. Define Unitary operator? What is the significance of Unitary operator?
8. Find out the Eigen values of Parity operator.
9. Give a brief description of degenerate and non-degenerate states.
10. Show that Unitary transformation preserves scalar product.
11. Show that the Eigen values of Hermitian operators are real.
12. Prove that the sum and product of two Hermatian operators are Hermatian.
13. Show that Hermitian operators form a linear independent set or orthogonal set.
14. Show that Hermitian operator can be diagonalized by means of unitary transformation.
15. Show that Momentum operator and Parity operator are Hermitian.
16. Find the Necessary and sufficient condition for two operators to have common Eigen function.
17. Define Parity operator? Show that parity operator commutes with symmetric Hamiltonian.
18. Show that an Operator can be represented by a matrix.
19. Find the following Operator formalism of Quantum Mechanics:

    (1) $[x, P_x] = -\frac{\hbar}{i} = i\hbar$ (2) $\left[x^n, P_x\right] = ni\hbar x^{n-1}$ (3) $\left[x, P_x^{\ n}\right] = ni\hbar P_x^{n-1}$ (4) $\left[\hat{x}_i, \hat{P}_j\right] = i\hbar\delta_{ij}$
20. Write about the Schrödinger wave equation. Find Schrödinger time dependent Wave equation for free and non- free particle.
21. Find the time dependent Schrödinger equation and its solution under Conservative force.
22. Find the time independent Schrödinger equation and find the Stationary State.
23. Established the Matrix form of Schrödinger equation.
24. Write down the Momentum representation of Schrödinger equation.
25. Define Momentum Eigen function & Eigen values. Find Momentum Eigen function using Box Normalization.
26. Find normalization of momentum Eigen function using Box normalization.
27. Show the physical Significance of wave function in momentum space.
28. Explain the closure property of the momentum Eigen functions.
29. What is Wave packet? Show that the probability can be calculated from coordinate momentum.

30. Explain the Probability for momentum and space coordinates from closure property.
31. Define the Fourier transformation and the inverse Fourier transformation. Explain Fourier transformations for superposition of monochromatic waves.
32. Explain Normalization of Delta function.
33. Define Dirac-Delta function δ_{ij} and Levi-Civita [Epsilon] Symbol ε_{ijk}. Explain the Shifting property of Dirac-Delta function. Find the normalization of Delta function.
34. Let the function $\psi(\theta)$ of the angular variable $\theta(-\pi \leq \theta \leq \pi)$ satisfy the condition $\psi(\pi) = \psi(-\pi)$. Show that the operator $L = -i\hbar\frac{\partial}{\partial\theta}$ is Hermitian.
35. What is the necessary and sufficient condition that the operator say A is a constant of motion?
36. What is the significance of commutation relation between position and momentum operator?
37. What do you mean by the commutation of operators ? Discuss why the Hamiltonian operator in Quantum Mechanics is represented by Hermitian opertor.
38. Show that the sum of two Hermitian operators is Hermitian.
39. Show that the product of the Hermitian opertors is Hermitian if and only if they commute.
40. Write a short note on:
    a. Bra and Ket notation b. Linear operator c. Hermitian operator d. Identity operator e. Inverse operator f. Unitary operator g. Parity operator h. Schwartz's inequality
41. Show that the commutator of two Hermitian operators is anti Hermitian.
42. Consider the operators $A_1\psi(x) = x^3\psi(x)$ *and* $A_2\psi(x) = x\frac{\partial}{\partial x}\psi(x)$. Calculate the commutator $[A_1, A_2]$
43. Show that the expectation value of position and momentum [the dynamical variables] is not real.
44. Show that the energy of a particle in a three dimensional cubical box of side L is $$E = \frac{\pi^2\hbar^2}{3mL^3}(n_1^2 + n_2^2 + n_3^2);$$ Where n_1, n_2, n_3 is a set of integers.
45. What is Delta function ? Write down its integral representation. Prove that if a is a real constant but $a \neq 0$ then (1) $\delta(xa) = \frac{1}{|a|}\delta(x)$
    (2) $f(x)\delta(x-a) = f(a)\delta(x-a)$
46. Normalize the wave function $\psi(x) = e^{-|x|}\sin ax$; in the range $-\infty \leq x \leq \infty$
47. Write down the Schrödinger equation for a particle moving in a periodic potential $V(x) = V(x+a)$, in momentum representation.
48. Derive the Schrödinger wave equation in momentum representation. The Gaussian function is represented by

$$\psi(x) = \frac{1}{(\sigma\sqrt{\pi})^{1/2}} e^{-\frac{x^2}{2\sigma^2}};$$

where σ is the width of wave packet. Calculate the amplitude function $A(k)$

Chapter 4 Matrix formulation of Quantum Mechanics

4.1 Importance of matrix formulation in Quantum Mechanics

Dynamical quantities whose expectation values are real are called observables. If A is an observable

$\langle A\rangle=\langle A\rangle^{\times}$; Expectation value of A is real.

Observables are usually functions of position [x] and momentum [P_x]. The differential form of position and momentum operators are x and $\frac{\hbar}{i}\frac{\partial}{\partial x}$ respectively. Thus operators of observables can be written as differential operators.

$$\hat{\text{E}}=\frac{\text{P}^2}{2m}+V(x) \quad [1]$$

$$\hat{\text{E}}=-\frac{\hbar^2}{2m}\frac{\partial^2}{\partial x^2}+V(x) \quad [2]$$

There are some observables which are not possible to be written as a function of x and P_x. Also, they cannot be written as the differential operators for them. In such a case, we can use matrix formulation of Quantum Mechanics. Hence it is more general formulation in Quantum Mechanics

4.2 Unitary operator and unitary transformation

An operator is said to be unitary operator if $\text{U}^{\text{dagger}}=\text{U}^{-1}$; The adjoint [U^{dagger}] and the inverse of [U^{-1}] the operator are the same.

$$\text{U}^{\text{dagger}}\text{U}^{-1}=\text{U}^{-1}\text{U}=\text{I}, \quad \text{U}\text{U}^{\text{dagger}}=\text{U}\text{U}^{-1}=\text{I} \quad \therefore\ \text{U}\text{U}^{\text{dagger}}=\text{U}^{\text{dagger}}\text{U}=\text{I}$$

When a unitary operator acts on a wave function to get a new wave function is called a unitary transformation. Since, *a unitary transformation leaves the system completely unchanged.*

Let Ψ be a wave function and Ω be an operator, such that $\Omega\psi=X$ where X is a new wave function. Let U be a unitary operator, such that $U\psi=\Psi'$, $UX=X'$; where Ψ' *and* X' are the unitary transforms of Ψ and X respectively. Again $\Omega'\psi'=X'$, where Ω' defined the transformed operator.

$$\Omega'\Psi'=X'\ or, \Omega'U\Psi=UX=U\Omega\Psi \quad ; \text{ Because } \Omega\psi=X$$

$$\therefore\quad \Omega'\text{U}=\text{U}\Omega \text{ or, } \Omega'\text{U}\text{U}^{\text{dagger}}=\text{U}\Omega\text{U}^{\text{dagger}}. \text{ Hence } \Omega'=\text{U}\Omega\text{U}^{\text{dagger}}$$

We have $U\psi = \Psi'$

or, $U^{dagger}\Psi' = U^{dagger}U\Psi = I\Psi = \Psi.$

Thus $U^{dagger}\Psi' = \Psi$

Again we have $U\Omega U^{dagger} = \Omega'$

or, $U\Omega U^{dagger}U = \Omega' U$

or, $U\Omega I = \Omega' U$

or, $U\Omega = \Omega' U$

or, $U^{dagger}U\Omega = U^{dagger}\Omega' U$

or, $\Omega = U^{dagger}\Omega' U$

Thus a converse of $U\Psi = \Psi'$ and $\Omega' = U\Omega U^{dagger}$ respectively, We have $U^{dagger}\Psi' = \Psi$ and $\Omega = U^{dagger}\Omega' U$ respectively; because of unitary transformation, operator Ω becomes $\Omega' = U\Omega U^{dagger}$ This means change of basis from one orthonomal set to another orthonormal set and operator changes from Ω to $\Omega' = U\Omega U^{dagger}$

4.3 Advantage of unitary transformation

- Unitary transformation leaves the system completely unchanged
- Matrix should be diagonalized by means of unitary transformation
- Expectation values of an observables remain unchanged under unitary transformation
- Unitary transformation preferable to use that set of orthonormal functions as basis which are simple and simplifies the calculation of elements of matrix operator is given by

$$\Omega_{mn} = \langle \Psi_m | \Omega_{mn} | \Psi_n \rangle = (\Psi_m , \Omega_{mn} \Psi_n) = \int_{-\infty}^{\infty} \Psi_m \Omega_{mn} \Psi_n \partial\tau \qquad [1]$$

- The Eigen values of observables(Ω) and transformed observables(Ω') are the same under unitary transformation

4.4 Some properties of unitary transformation

(A) If an observable Ω is Hermitian, then Ω'[its transformed operator] is also Hermitian.

We know that $\Omega' = U\Omega U^{dagger}$

$\therefore (\Omega')^{\dagger} = (U\Omega U^{\dagger})^{\dagger} = (U^{\dagger})^{\dagger}\,\Omega^{\dagger}\,U^{\dagger}$ [$\therefore (XYZ)^{\dagger} = X^{\dagger}\,Y^{\dagger}\,Z^{\dagger}$] $= U\,\Omega^{\dagger}\,U^{\dagger}$ [$(U^{\dagger})^{\dagger} = U$] $= U\Omega U^{\dagger}$ [$\Omega^{\dagger} = \Omega$]

Since Ω is Hermitian, so $\Omega^{\dagger} = \Omega$

Or, $(\Omega')^{\dagger} = U\Omega\, U^{\dagger}$

So, $(\Omega')^{\dagger} = \Omega'$ i.e. Ω' is Hermitian. The Eigen values of transformed operator Ω' are real. And its Eigen functions are orthogonal to each other.

(B) Expectation values of transformed observables remain the same

$$\langle\Omega'\rangle = (\Psi', \Omega'\Psi)$$

Or, $\langle\Omega'\rangle = (U\Psi, U\Omega U^{dagger} U\,\Psi)$ $[\because U\Psi = \Psi']$

Or, $\langle\Omega'\rangle = (\Psi, U^{dagger} U\Omega U^{dagger} U\,\Psi)$

Or, $\langle\Omega'\rangle = (\Psi, I\Omega I\,\Psi)$ $[\because U^{dagger} U = I]$

Or, $\langle\Omega'\rangle = (\Psi, \Omega\,\Psi)$

Or, $\langle\Omega'\rangle = \langle\Omega\rangle$

Thus $\langle\Omega'\rangle = \langle\Omega\rangle$. That is *the Expectation values of transformed observables remain the same (unchanged) in unitary transformation.*

(C) The Eigen values of observables [Ω] and transformed observables [Ω′] are the same

Let us consider an operator equation

$$\Omega\Psi_n = a_n\Psi_n \qquad [1]$$

$$\Omega U^{dagger} U\,\Psi_n = a_n\ UU^{dagger}\ \Psi_n \qquad [\because U^{dagger} U = I]$$

$$\Omega U^{dagger}\Psi'_n = a_n\ U^{dagger}\ \Psi'_n \qquad [\because U\Psi = \Psi']$$

$$U\,\Omega U^{dagger}\Psi'_n = U(a_n\ U^{dagger}\ \Psi'_n)$$

$$\Omega'\Psi'_n = a_n\ UU^{dagger}\ \Psi'_n \qquad [\because \Omega' = U\,\Omega U^{dagger}]$$

$$\Omega'\Psi'_n = a_n\ \Psi'_n \quad [\ \therefore\ UU^{dagger} = I\,] \qquad [2]$$

Comparing equation [1] and [2] we find that the Eigen values of observables Ω and transformed observables [$\Omega' = U\Omega U^{dagger}$] are the same.

4.5 Some properties of matrices

- If two matrices (*Ω and Φ*) have the same number of rows and columns then the matrices are said to be equal. i.e. $\Omega_{mn} = \Phi_{mn}$
- Two matrices (*Ω and Φ*) can be added if and only if they have the same number of rows and columns. Therefore, the sum of two matrices (*Ω andΦ*) is given by $P_{mn} = \Omega_{mn} + \Phi_{mn}$

  $P_{mn} = (\Omega + \Phi)_{mn}$
- The product of two matrices (*Ω andΦ*) is given by $P_{mn} = (\Omega\Phi)_{mn}$

  Where, $P_{mn} = \sum_k \Omega_{mk}\ \phi_{kn} = \sum_i \Omega_{mi}\ \phi_{in}$

- Matrix multiplication obeys [To carry out the multiplication, number of columns of matrix Ω must be equal to number of rows of matrix Φ]

  (1) Law of distribution $\Omega(\Phi + P) = \Omega\Phi + \Omega P$

  (2) Law of association $\Omega(\Phi P) = (\Omega\Phi)P$
- Inverse of matrix Ω denoted by Ω^{-1}. Such that $\Omega\Omega^{-1}=\Omega^{-1}\Omega = I$; where I is the unit matrix having elements $I_{mn} = \delta_{mn}$
- A matrix which possesses an inverse is said to be non-singular
- Transpose of matrix Ω^T such that $\Omega^T{}_{mn} = \Omega_{nm}$
- Adjoint of matrix Ω^{dagger} such that $\Omega^{dagger}{}_{mn} = \Omega^{\times}{}_{nm}$
- Self-adjoint or Hermitian of matrix Ω^{dagger} such that $\Omega^{dagger} = \Omega$ and hence

  $\Omega^{dagger}{}_{mn} = \Omega_{mn}$ or, $\Omega^{\times}{}_{nm} = \Omega_{mn}$
- If $\Omega\Phi = \Phi\Omega$, then two matrices are said to commute. But in general

  $\Omega\Phi \neq \Phi\Omega$,
- For unitary matrix, the matrix is said to be unitary if the inverse of matrix $[\Omega^{-1}]$ is equal to its adjoint $[\Omega^{dagger}]$, such that
- $\Omega^{-1} = \Omega^{dagger}$ or, $\Omega\Omega^{dagger} = \Omega^{dagger}\Omega = I$; where I is the unit matrix having elements

  $(\Omega\Omega^{dagger})_{mn} = (\Omega^{dagger}\Omega)_{mn} = \delta_{mn} = I_{mn}$

4.6 Some properties of matrix Operators

- **Hermitian operator is a Hermitian matrix**

Let us consider a Hermitian operator Ω. Elements of matrix operator Ω are given by

$$\Omega_{mn} = (\Psi_m, \Omega\,\Psi_n)$$

Now we can write $\Omega_{mn} = (\Psi_m, \Omega\,\Psi_n) = (\Omega\Psi_n, \Psi_m)^{\times} = (\Psi_n, \Omega\Psi_m)^{\times} = \Omega^{\times}{}_{nm}$

Since Ω is Hermitian operator such that $\Omega_{mn} = \Omega^{\times}{}_{nm}$

I.e., operator of an observable, $\Omega_{mn} = \Omega^{\times}{}_{nm}$ which means matrix, Ω is Hermitian matrix. Thus *if an operator is Hermitian ,its matrix operator is Hermitian matrix.*

- **Unitary operator is a Unitary matrix**

If U is a unitary operator and it can be written as

$$U^{-1} = U^{dagger} \text{ or, } U^{dagger}U = UU^{dagger} = I\text{; Where I is unit operator.}$$

Matrix elements of UU^{daggar} are given by

$$(UU^{dagger})_{mn} = (\Psi_m , UU^{dagger}\,\Psi_n) = (\Psi_m , I\,\Psi_n) = (\Psi_m , \Psi_n) = \delta_{mn}$$

And $$(U^{dagger}U)_{mn} = (\Psi_m , U^{dagger}U\,\Psi_n) = (\Psi_m , I\,\Psi_n) = (\Psi_m , \Psi_n) = \delta_{mn}$$

Hence $(UU^{dagger})_{mn} = (U^{dagger}U)_{mn} = \delta_{mn}$; which means unitary U is unitary matrix. Thus *unitary operator is a unitary matrix.*

- **An adjoint operator is a adjoint matrix**

Elements of matrix operator Ω are given by

$$\Omega_{mn} = (\Psi_m, \Omega\,\Psi_n)$$

Elements of adjoint of the matrix operator Ω are given by

$$(\Omega^{dagger})_{mn} = (\Psi_m, \Omega^{dagger}\,\Psi_n) = (\Omega\Psi_m , \Psi_n) = (\Psi_n , \Omega\Psi_m)^{\times} = (\Omega^{\times})_{nm}$$

Thus *An adjoint operator(Ω^{dagger}) is a adjoint matrix*

- **The sum of two operators is the sum of two matrix operators**

Let us consider two operators *Ω and Φ* respectively. Matrix elements of (*Ω*+*Φ*) are given by

$$\begin{aligned}(\Omega+\Phi)_{mn} &= (\Psi_m ,(\Omega+\Phi)\Psi_n) \\ &= \int \Psi_m^{\times}(\Omega+\Phi)\Psi_n\, d\tau \\ &= \int \Psi_m^{\times}(\Omega\Psi_n + \Phi\Psi_n) d\tau \\ &= \int \Psi_m^{\times}\Omega\Psi_n\, d\tau + \int \Psi_m^{\times}\Phi\Psi_n d\tau \\ &= (\Psi_m , \Omega\Psi_n) + (\Psi_m , \Phi\Psi_n) \\ (\Omega+\Phi)_{mn} &= (\Omega_{mn} + \Phi_{mn})\end{aligned}$$

Thus the sum of two operators is the sum of two matrix operators.

- **The product of two operators is the product of two matrix operators**

Let us consider two operators Ω and Φ respectively. Matrix elements of the product of two operators (ΩΦ) are given by

$$(\Omega\,\Phi)_{mn} = (\Psi_m ,(\Omega\Phi)\Psi_n) = (\Psi_m , \Omega\,(\Phi\,\Psi_n)) = (\Psi_m\, \Omega \sum b_{kn}\,\Psi_k)$$

Where $$\Phi\Psi_n = \sum b_{kn}\,\Psi_k \qquad [1]$$

Where b_{kn}'s are expansion coefficients.

$$(\Omega\,\Phi)_{mn} = \sum_k (\Psi_m\,, \Omega\Psi_k)\, b_{kn} = \sum_k \Omega_{mk}\, b_{kn} \qquad [2]$$

If b_{kn}'s are the elements of operator Φ , then equation [2] means that *The product of two operators is the product of two matrix operators*

4.7 Diagonalization of a matrix

Let Ψ be a wave function and Ω be an operator, matrix elements of an operator Ω_{mn} is given by

$$\Omega_{mn} = (\Psi_m\,, \Omega_{mn}\Psi_n) = (\Psi_m\,, a_n\Psi_n) = a_n\,(\Psi_m, \Psi_m) = a_n\delta_{mn} \qquad [1]$$

Here we noted that *the basic argument and drawbacks* of diagonalization of a matrix are:

- The matrix elements of an operator Ω_{mn} is a diagonal matrix with Eigen values as diagonal elements
- If the basis functions used are not Eigen functions, matrix elements of an operator Ω_{mn} will not be diagonalised Matrix

If the matrix elements of an operator Ω_{mn} are not be diagonalised Matrix, we will understand that basis functions used in calculating the matrix elements are not Eigen functions. We can change the basis and use another set of functions as basis. If the new basis functions are Eigen functions, the new matrix operator will become diagonal matrix and Eigen values will be diagonal elements of the new matrix operator. This is called the *diagonalisation of a matrix*[20,21].

4.8 Matrix representation of wave function and operator

Let $\Psi_1, \Psi_2, \Psi_3, \ldots\ldots\ldots\ldots\ldots\ldots\ldots\ldots\ldots\ldots \Psi_n$ be the known wave functions of a dynamical observables. An arbitrary state Ψ_c can be written as

$$\Psi_C = \sum_n C_n \psi_n \qquad [1]$$

If the set of numbers C_n's are known, Ψ_c is known. Because Ψ_n's are known.
Suppose an operator Ω acts on Ψ_c and get

$$\Psi_d = \Omega\Psi_c \qquad [2]$$

Let

$$\Omega_d = \sum_m d_m \psi_m \qquad [3]$$

If the set of numbers d_m's are known, Ψ_d is known, because Ψ_m's are known.

$$d_m = (\Psi_m\,, \Psi_d) = (\Psi_m\,, \Omega\Psi_c) = \left(\Psi_m\,, \Omega\sum_n C_n\,\Psi_n\right) = \left(\Psi_m\,, \sum_n C_n\Omega\Psi_n\right) = \sum_n (\Psi_m\,, \Omega\Psi_n)\,C_n$$

Thus $$d_m = \sum_n \Omega_{mn} C_n \quad [4]$$

Where $$\Omega_{\text{mn}} = (\Psi_{\text{m}}, \Omega\Psi_{\text{n}}) \quad [5]$$

Equation [4] can be written as

$$\begin{pmatrix} d_1 \\ d_2 \\ d_3 \\ . \\ . \\ . \\ d_n \end{pmatrix} = \begin{pmatrix} \Omega_{11} & \Omega_{12} & \Omega_{13} \ldots\ldots\ldots\ldots & \Omega_{1n} \\ \Omega_{21} & \Omega_{22} & \Omega_{23} \ldots\ldots\ldots\ldots & \Omega_{2n} \\ \Omega_{31} & \Omega_{32} & \Omega_{33} \ldots\ldots\ldots\ldots & \Omega_{3n} \\ . & . & \ldots\ldots\ldots\ldots & . \\ . & . & \ldots\ldots\ldots\ldots & . \\ . & . & \ldots\ldots\ldots\ldots & . \\ \Omega_{n1} & \Omega_{n2} & \Omega_{n3} \ldots\ldots\ldots & \Omega_{nn} \end{pmatrix} \begin{pmatrix} C_1 \\ C_2 \\ C_3 \\ . \\ . \\ . \\ C_n \end{pmatrix} \quad [6]$$

This is equivalent to $\Psi_d = \Omega\Psi_c$. Elements of square matrix are given by $\Omega_{\text{mn}} = (\Psi_{\text{m}}, \Omega\Psi_{\text{n}})$. Thus the square matrix Ω_{mn} represents an operator Ω.The column matrices represents Ψ_C and Ψ_d respectively.

$$\Psi_C = \begin{pmatrix} C_1 \\ C_2 \\ C_3 \\ . \\ . \\ . \\ C_n \end{pmatrix} \quad and \quad \Psi_d \begin{pmatrix} d_1 \\ d_2 \\ d_3 \\ . \\ . \\ . \\ d_n \end{pmatrix} \quad [7]$$

$$(\Psi_C, \Psi_d) = \left(\sum_i C_i \Psi_i, \sum_j d_j \Psi_j \right) \quad [8]$$

$$(\Psi_C, \Psi_d) = \sum_i \sum_j C_i^{\times} d_j (\Psi_i, \Psi_j)$$

$$= \sum_i \sum_j C_i^{\times} d_j \delta_{ij} = \sum_i C_i^{\times} d_i$$

$$(\Psi_C, \Psi_d) = \begin{pmatrix} C_1^{\times} & C_2^{\times} & C_3^{\times} \ldots\ldots\ldots\ldots C_n^{\times} \end{pmatrix} \begin{pmatrix} d_1 \\ d_2 \\ d_3 \\ . \\ . \\ d_n \end{pmatrix} \quad [9]$$

Thus $\Psi_C^{\times}$ is represented by the row matrix $\left(C_1^{\times} \quad C_2^{\times} \quad C_3^{\times} \ldots\ldots\ldots\ldots C_n^{\times}\right)$. Thus complex conjugate of wave function is represented by row matrix

4.9 Technique of diagonalization of a matrix

Let us consider a matrix elements of an operator Ω_{mn} (2×2) which is not a diagonalised matrix and can be written as

$$\Omega_{mn} = \begin{pmatrix} 0 & -i\lambda \\ i\lambda & 0 \end{pmatrix} \qquad [1]$$

If the matrix elements of an operator Ω_{mn} are not be diagonalized matrix, we will understand that basis functions used in calculating the matrix elements are not Eigen functions. We can change the basis and use another set of functions as basis. If the new basis functions are Eigen functions, the new matrix operator will become diagonal matrix and Eigen values will be diagonal elements of the new matrix operator. *This is the Technique of Diagonalization of a Matrix*

4.10 Matrix form of Eigen value equation

Consider a matrix elements of an operator Ω_{mn} (3×3) can be written as

$$\Omega_{mn} = \begin{pmatrix} \Omega_{11} & \Omega_{12} & \Omega_{13} \\ \Omega_{21} & \Omega_{22} & \Omega_{23} \\ \Omega_{31} & \Omega_{32} & \Omega_{33} \end{pmatrix} \qquad [1]$$

If λ be the Eigen values of the observable Ω_{mn} (3×3); the secular equation can be written as

$$\Omega_{mn} = \begin{vmatrix} \Omega_{11}-\lambda & \Omega_{12} & \Omega_{13} \\ \Omega_{21} & \Omega_{22}-\lambda & \Omega_{23} \\ \Omega_{31} & \Omega_{32} & \Omega_{33}-\lambda \end{vmatrix} \qquad [2]$$

From the above equation we can write

$$\left.\begin{aligned} (\Omega_{11}-\lambda)C_1 + \Omega_{12}C_2 + \Omega_{13}C_3 &= 0 \\ \Omega_{21}C_1 + (\Omega_{22}-\lambda)C_2 + \Omega_{23}C_3 &= 0 \\ \Omega_{31}C_1 + \Omega_{32}C_2 + (\Omega_{33}-\lambda)C_3 &= 0 \end{aligned}\right\} \qquad [3]$$

Or

$$\left.\begin{aligned} \Omega_{11}C_1 + \Omega_{12}C_2 + \Omega_{13}C_3 &= \lambda C_1 \\ \Omega_{21}C_1 + \Omega_{22}C_2 + \Omega_{23}C_3 &= \lambda C_2 \\ \Omega_{31}C_1 + \Omega_{32}C_2 + \Omega_{33}C_3 &= \lambda C_3 \end{aligned}\right\} \qquad [4]$$

Or $$\left.\begin{pmatrix} \Omega_{11} & \Omega_{12} & \Omega_{13} \\ \Omega_{21} & \Omega_{22} & \Omega_{23} \\ \Omega_{31} & \Omega_{32} & \Omega_{33} \end{pmatrix}\begin{pmatrix} C_1 \\ C_2 \\ C_3 \end{pmatrix} = \lambda\begin{pmatrix} C_1 \\ C_2 \\ C_3 \end{pmatrix}\right\} \quad [5]$$

Equation [5] represents the matrix form of an Eigen value equation.

4.11 Eigen value, Eigen function and Diagonalization of some matrices [20.21]

(a) $\Omega = \begin{pmatrix} 0 & i & 0 \\ -i & 0 & 0 \\ 0 & 0 & 0 \end{pmatrix}$ (b) $\Omega = \begin{pmatrix} 0 & 1 & 0 \\ 1 & 0 & 0 \\ 0 & 0 & 1 \end{pmatrix}$ (c) $\Omega = \begin{pmatrix} 5 & 0 & \sqrt{3} \\ 0 & 3 & 0 \\ \sqrt{3} & 0 & 3 \end{pmatrix}$

(d) $\Omega = \begin{pmatrix} 0 & i \\ -i & 0 \end{pmatrix}$ (e) $\Omega = \begin{pmatrix} 1 & i \\ -i & 1 \end{pmatrix}$ (f) $\Omega = \begin{pmatrix} 3 & 2 \\ 2 & 1 \end{pmatrix}$

(g) $\Omega = \begin{pmatrix} 0 & 1 \\ 1 & 0 \end{pmatrix}$ (h) $\Omega = \begin{pmatrix} 1 & \frac{3}{2} \\ \frac{3}{2} & 4 \end{pmatrix}$ (i) $\Omega = \begin{pmatrix} 1 & 0 & 0 \\ 0 & 0 & 1 \\ 0 & 1 & 0 \end{pmatrix}$

[b]Given $\Omega = \begin{pmatrix} 0 & 1 & 0 \\ 1 & 0 & 0 \\ 0 & 0 & 1 \end{pmatrix}$

Let us write the Secular equation by the given matrix

$$|\Omega - I\lambda| = 0 \quad [1]$$

Or, $$\begin{vmatrix} 0 & 1 & 0 \\ 1 & 0 & 0 \\ 0 & 0 & 1 \end{vmatrix} - \lambda\begin{vmatrix} 1 & 0 & 0 \\ 0 & 1 & 0 \\ 0 & 0 & 1 \end{vmatrix} = 0 \quad or, \quad \begin{vmatrix} 0 & 1 & 0 \\ 1 & 0 & 0 \\ 0 & 0 & 1 \end{vmatrix} - \begin{vmatrix} \lambda & 0 & 0 \\ 0 & \lambda & 0 \\ 0 & 0 & \lambda \end{vmatrix} = 0 \quad [2]$$

Or, $$\begin{vmatrix} -\lambda & 1 & 0 \\ 1 & -\lambda & 0 \\ 0 & 0 & 1-\lambda \end{vmatrix} = 0 \quad Or, \quad (1-\lambda)(\lambda^2 - 1) = 0 \quad \therefore \quad \lambda = 1, \pm 1$$

Therefore, The Eigen values are $\lambda = (1,\ +1,\ -1)$

The Characteristic Eigen vector equation of the given matrix

$$(\Omega - I\lambda)(X_1) = 0 \quad [4]$$

Or $$\begin{pmatrix} -\lambda & 1 & 0 \\ 1 & -\lambda & 0 \\ 0 & 0 & 1-\lambda \end{pmatrix}\begin{pmatrix} x_1 \\ x_2 \\ x_3 \end{pmatrix} = 0 \qquad [5]$$

Where x_1 , x_2 and x_3 are the components of the Eigen vector X.

$\therefore$ $$\left.\begin{aligned} -\lambda x_1 + x_2 &= 0 \\ x_1 - \lambda x_2 &= 0 \\ (1-\lambda)\, x_3 &= 0 \end{aligned}\right\} \qquad [6]$$

- ***Case -01: When $\lambda = -1$***

And $$\left.\begin{aligned} x_1 + x_2 &= 0 \\ x_1 + x_2 &= 0 \\ 2x_3 &= 0 \end{aligned}\right\} \qquad [7]$$

Therefore, $x_3 = 0$, $x_1 + x_2 = 0$ or, $x_1 = -x_2 = C$ [let $x_2 = -C$]

$\therefore$Eigen vector $$(X_1) = \begin{pmatrix} x_1 \\ x_2 \\ x_3 \end{pmatrix} = \begin{pmatrix} +C \\ -C \\ 0 \end{pmatrix}$$

From normalization condition $X_1(X_1)^\times = 1$

or, $\begin{pmatrix} +C \\ -C \\ 0 \end{pmatrix}(+C \quad -C \quad 0) = 1 \therefore C = \pm\frac{1}{\sqrt{2}}$ $Or, C = \frac{1}{\sqrt{2}}$; Hence the normalized Eigen vector $(X_1) = \begin{pmatrix} \frac{1}{\sqrt{2}} \\ -\frac{1}{\sqrt{2}} \\ 0 \end{pmatrix}$

- ***Case – 02: When $\lambda = 1$***

$\therefore$ $(1-\lambda)\, x_3 = 0$ or, x_3 is arbitrary ; And $x_1 - \lambda x_2 = 0$ or, $x_1 = x_2$. Again $-\lambda x_1 + x_2 = 0$ or, $x_1 = x_2$.

So that $x_1 = x_2 = C$ (say) ; any real value.

Eigen vector $$(X_2) = \begin{pmatrix} x_1 \\ x_2 \\ x_3 \end{pmatrix} = \begin{pmatrix} C \\ C \\ 0 \end{pmatrix}$$

From normalization condition $X_2\,(X_2)^\times = 1$ or, $\begin{pmatrix} C \\ C \\ 0 \end{pmatrix}(C \quad C \quad 0) = 1$; $C = \pm\frac{1}{\sqrt{2}} = \frac{1}{\sqrt{2}}$. Hence the normalized Eigen vector $(X_2)=\begin{pmatrix} \frac{1}{\sqrt{2}} \\ \frac{1}{\sqrt{2}} \\ 0 \end{pmatrix}$

The Eigen vector X_3 to be perpendicular on both X_1 and X_2. So the Eigen vector X_3 will be

$$X_3 = X_1 \times X_2 = \begin{vmatrix} \hat{i} & \hat{j} & \hat{k} \\ \frac{1}{\sqrt{2}} & -\frac{1}{\sqrt{2}} & 0 \\ \frac{1}{\sqrt{2}} & \frac{1}{\sqrt{2}} & 0 \end{vmatrix} = 0\hat{i} + 0\hat{j} + 1\hat{k} = \begin{pmatrix} \hat{i} & \hat{j} & \hat{k} \end{pmatrix}\begin{pmatrix} 0 \\ 0 \\ 1 \end{pmatrix} \therefore X_3 = \begin{pmatrix} 0 \\ 0 \\ 1 \end{pmatrix}$$

Hence the vectors of the matrix $(X_1)=\begin{pmatrix} \frac{1}{\sqrt{2}} \\ -\frac{1}{\sqrt{2}} \\ 0 \end{pmatrix}$, $(X_2)=\begin{pmatrix} \frac{1}{\sqrt{2}} \\ \frac{1}{\sqrt{2}} \\ 0 \end{pmatrix}$, $(X_3)=\begin{pmatrix} 0 \\ 0 \\ 1 \end{pmatrix}$ [8]

A matrix can be diagonalized by means of unitary transformation.

I.e., $$\Omega' = U^{-1}\Omega U \qquad [9]$$

Now we shall check whether Ω is Hermitian or not.

$$\Omega = \begin{pmatrix} 0 & 1 & 0 \\ 1 & 0 & 0 \\ 0 & 0 & 1 \end{pmatrix} \qquad \tilde{\Omega} = \begin{pmatrix} 0 & 1 & 0 \\ 1 & 0 & 0 \\ 0 & 0 & 1 \end{pmatrix}^{\times}$$; By inspection Ω is Hermitian.

$$U = \begin{pmatrix} \frac{1}{\sqrt{2}} & \frac{1}{\sqrt{2}} & 0 \\ -\frac{1}{\sqrt{2}} & \frac{1}{\sqrt{2}} & 0 \\ 0 & 0 & 1 \end{pmatrix} \qquad or,\ U^{\times} = \begin{pmatrix} \frac{1}{\sqrt{2}} & -\frac{1}{\sqrt{2}} & 0 \\ \frac{1}{\sqrt{2}} & \frac{1}{\sqrt{2}} & 0 \\ 0 & 0 & 1 \end{pmatrix}$$

Now we shall check whether U is unitary or not. From the unitary condition

$$UU^{\times} = 1 \qquad [10]$$

$$\text{L.H.S} = \begin{pmatrix} \frac{1}{\sqrt{2}} & \frac{1}{\sqrt{2}} & 0 \\ -\frac{1}{\sqrt{2}} & \frac{1}{\sqrt{2}} & 0 \\ 0 & 0 & 1 \end{pmatrix}\begin{pmatrix} \frac{1}{\sqrt{2}} & -\frac{1}{\sqrt{2}} & 0 \\ \frac{1}{\sqrt{2}} & \frac{1}{\sqrt{2}} & 0 \\ 0 & 0 & 1 \end{pmatrix} = \begin{pmatrix} \frac{1}{2}+\frac{1}{2}+0 & -\frac{1}{2}+\frac{1}{2}+0 & 0+0+0 \\ -\frac{1}{2}+\frac{1}{2}+0 & \frac{1}{2}+\frac{1}{2}+0 & 0+0+0 \\ 0+0+0 & 0+0+0 & 0+0+1 \end{pmatrix} = \begin{pmatrix} 1 & 0 & 0 \\ 0 & 1 & 0 \\ 0 & 0 & 1 \end{pmatrix}$$

Thus $$UU^{\times} = 1 = \begin{pmatrix} 1 & 0 & 0 \\ 0 & 1 & 0 \\ 0 & 0 & 1 \end{pmatrix}$$ [11]

Thus U is unitary. So the matrix can be diagonalized.

$$\Omega' = U^{-1}\Omega U = U^{\times}\Omega U \quad [\because U^{-1} = U^{\times}]$$

$$\begin{pmatrix} \frac{1}{\sqrt{2}} & -\frac{1}{\sqrt{2}} & 0 \\ \frac{1}{\sqrt{2}} & \frac{1}{\sqrt{2}} & 0 \\ 0 & 0 & 1 \end{pmatrix} \begin{pmatrix} 0 & 1 & 0 \\ 1 & 0 & 0 \\ 0 & 0 & 1 \end{pmatrix} \begin{pmatrix} \frac{1}{\sqrt{2}} & \frac{1}{\sqrt{2}} & 0 \\ -\frac{1}{\sqrt{2}} & \frac{1}{\sqrt{2}} & 0 \\ 0 & 0 & 1 \end{pmatrix} = \begin{pmatrix} \frac{1}{\sqrt{2}} & -\frac{1}{\sqrt{2}} & 0 \\ \frac{1}{\sqrt{2}} & \frac{1}{\sqrt{2}} & 0 \\ 0 & 0 & 1 \end{pmatrix} \begin{pmatrix} -\frac{1}{\sqrt{2}} & \frac{1}{\sqrt{2}} & 0 \\ \frac{1}{\sqrt{2}} & \frac{1}{\sqrt{2}} & 0 \\ 0 & 0 & 1 \end{pmatrix}$$

$$= \begin{pmatrix} -\frac{1}{2}-\frac{1}{2}+0 & \frac{1}{2}-\frac{1}{2}+0 & 0+0+0 \\ -\frac{1}{2}+\frac{1}{2}+0 & \frac{1}{2}+\frac{1}{2}+0 & 0+0+0 \\ 0+0+0 & 0+0+0 & 0+0+1 \end{pmatrix} = \begin{pmatrix} -1 & 0 & 0 \\ 0 & 1 & 0 \\ 0 & 0 & 1 \end{pmatrix}$$

We see that the diagonal elements are the Eigen values of the matrix. *So the Matrix has been Diagonilized.*

(G) Diagonalize the Matrix $\begin{pmatrix} 0 & 1 \\ 1 & 0 \end{pmatrix}$

$$\text{Let } \Omega = \begin{pmatrix} 0 & 1 \\ 1 & 0 \end{pmatrix}$$

We know, the secular equation $|\Omega - \lambda I| = 0 \quad or, \begin{vmatrix} 0 & 1 \\ 1 & 0 \end{vmatrix} - \lambda \begin{vmatrix} 1 & 0 \\ 0 & 1 \end{vmatrix} = 0$

$$or, \begin{vmatrix} 0 & 1 \\ 1 & 0 \end{vmatrix} - \begin{vmatrix} \lambda & 0 \\ 0 & \lambda \end{vmatrix} = 0 \quad or, \begin{vmatrix} -\lambda & 1 \\ 1 & -\lambda \end{vmatrix} = 0$$

$$or,\ \lambda^2 - 1 = 0 \quad or, \lambda = \pm 1$$

So the Eigen values are $\lambda = (+1, -1)$

We know, the characteristic equation of Eigen Vector

$$(\Omega - \lambda I)(a) = 0 \ or, \begin{pmatrix} -\lambda & 1 \\ 1 & -\lambda \end{pmatrix} \begin{pmatrix} x_1 \\ x_2 \end{pmatrix} = 0$$

Where x_1& $x_2$are the components of Eigen Vectors.

$$-\lambda x_1 + x_2 = 0$$ [1]

And $$x_1 + \lambda x_2 = 0$$ [2]

Case-01: When $\lambda = 1$

$$\therefore -x_1 + x_2 = 0 \Rightarrow x_1 = x_2$$

And

$$+x_1 - x_2 = 0 \Rightarrow x_1 = x_2$$

Let $x_2 = C$; Then, $x_1 = C$ $\therefore$ The Eigen vector $a = \begin{pmatrix} x_1 \\ x_2 \end{pmatrix} = \begin{pmatrix} C \\ C \end{pmatrix}$

From the normalization condition

$$aa^{\times} = 1 \Rightarrow \begin{pmatrix} C \\ C \end{pmatrix}\begin{pmatrix} C & C \end{pmatrix} = 1 \Rightarrow C^2 + C^2 = 1 \Rightarrow 2C^2 = 1 \therefore C = \frac{1}{\sqrt{2}}$$

Therefore, normalized Eigen vector

$$a = \begin{pmatrix} \frac{1}{\sqrt{2}} \\ \frac{1}{\sqrt{2}} \end{pmatrix} = \frac{1}{\sqrt{2}}\begin{pmatrix} 1 \\ 1 \end{pmatrix}$$

Case-02: When $\lambda = -1$

$$x_1 + x_2 = 0 \Rightarrow x_1 = -x_2 \Rightarrow x_1 = -C[say]$$

Then, the Eigen vector

$$b = \begin{pmatrix} x_1 \\ x_2 \end{pmatrix} = \begin{pmatrix} -C \\ C \end{pmatrix} \quad \& \quad b^{\times} = \begin{pmatrix} -C & C \end{pmatrix}$$

From the normalization condition

$$bb^{\times} = 1 \Rightarrow \begin{pmatrix} -C \\ C \end{pmatrix}\begin{pmatrix} -C & C \end{pmatrix} = 1 \Rightarrow C^2 + C^2 = 1 \Rightarrow 2C^2 = 1 \therefore C = \frac{1}{\sqrt{2}}$$

Therefore, normalized Eigen vector

$$b = \begin{pmatrix} x_1 \\ x_2 \end{pmatrix} = \begin{pmatrix} -\frac{1}{\sqrt{2}} \\ \frac{1}{\sqrt{2}} \end{pmatrix} = \frac{1}{\sqrt{2}}\begin{pmatrix} -1 \\ 1 \end{pmatrix}$$

A matrix can be diagonalized by means of unitary transformation. I.e.,

$$\Omega' = S^{-1}\Omega S$$

First we observe that Ω is Hermatian or not.

Here, $\Omega = \begin{pmatrix} 0 & 1 \\ 1 & 0 \end{pmatrix}$ $\therefore \tilde{\Omega} = \begin{pmatrix} 0 & 1 \\ 1 & 0 \end{pmatrix}^{\times}$ $[\Omega = \tilde{\Omega}]$

By Inspection Ω is Hermatian.

Now,

$$S = \begin{pmatrix} \frac{1}{\sqrt{2}} & -\frac{1}{\sqrt{2}} \\ \frac{1}{\sqrt{2}} & \frac{1}{\sqrt{2}} \end{pmatrix} \qquad \therefore S^{\times} = \begin{pmatrix} \frac{1}{\sqrt{2}} & \frac{1}{\sqrt{2}} \\ -\frac{1}{\sqrt{2}} & \frac{1}{\sqrt{2}} \end{pmatrix}$$

Now we shall check whether S is unitary or not. From the unitary condition

$$SS^{\times} = 1$$

$$\text{L.H.S} = \begin{pmatrix} \frac{1}{\sqrt{2}} & -\frac{1}{\sqrt{2}} \\ \frac{1}{\sqrt{2}} & \frac{1}{\sqrt{2}} \end{pmatrix}\begin{pmatrix} \frac{1}{\sqrt{2}} & \frac{1}{\sqrt{2}} \\ -\frac{1}{\sqrt{2}} & \frac{1}{\sqrt{2}} \end{pmatrix} = \begin{pmatrix} \frac{1}{2}+\frac{1}{2} & \frac{1}{2}-\frac{1}{2} \\ \frac{1}{2}-\frac{1}{2} & \frac{1}{2}+\frac{1}{2} \end{pmatrix} = \begin{pmatrix} 1 & 0 \\ 0 & 1 \end{pmatrix} = 1$$

$\therefore SS^{\times} = 1$ Hence, S is unitary; so the matrix can be diagonalized.

$$\therefore \Omega' = S^{-1}\Omega S = S^{\times}\Omega S \; [\because S^{\times} = S^{-1} = \tfrac{1}{S}]$$

$$= \begin{pmatrix} \frac{1}{\sqrt{2}} & \frac{1}{\sqrt{2}} \\ -\frac{1}{\sqrt{2}} & \frac{1}{\sqrt{2}} \end{pmatrix}\begin{pmatrix} 0 & 1 \\ 1 & 0 \end{pmatrix}\begin{pmatrix} \frac{1}{\sqrt{2}} & -\frac{1}{\sqrt{2}} \\ \frac{1}{\sqrt{2}} & \frac{1}{\sqrt{2}} \end{pmatrix} = \begin{pmatrix} \frac{1}{\sqrt{2}} & \frac{1}{\sqrt{2}} \\ -\frac{1}{\sqrt{2}} & \frac{1}{\sqrt{2}} \end{pmatrix}\begin{pmatrix} \frac{1}{\sqrt{2}} & \frac{1}{\sqrt{2}} \\ \frac{1}{\sqrt{2}} & -\frac{1}{\sqrt{2}} \end{pmatrix} = \begin{pmatrix} \frac{1}{2}+\frac{1}{2} & \frac{1}{2}-\frac{1}{2} \\ -\frac{1}{2}+\frac{1}{2} & -\frac{1}{2}-\frac{1}{2} \end{pmatrix}$$

$$= \begin{pmatrix} 1 & 0 \\ 0 & -1 \end{pmatrix}$$

We see that the diagonal elements are the Eigen values of the matrix. So the Matrix has been diagonalized.

(a) ***Let*** $\Omega = \begin{pmatrix} 0 & i & 0 \\ -i & 0 & 0 \\ 0 & 0 & 0 \end{pmatrix}$

We know the secular equation $|\Omega - I\lambda| = 0$

$$\Rightarrow \begin{vmatrix} 0 & i & 0 \\ -i & 0 & 0 \\ 0 & 0 & 0 \end{vmatrix} - \lambda \begin{vmatrix} 1 & 0 & 0 \\ 0 & 1 & 0 \\ 0 & 0 & 1 \end{vmatrix} = 0 \quad \Rightarrow \begin{vmatrix} 0 & i & 0 \\ -i & 0 & 0 \\ 0 & 0 & 0 \end{vmatrix} - \begin{vmatrix} \lambda & 0 & 0 \\ 0 & \lambda & 0 \\ 0 & 0 & \lambda \end{vmatrix} = 0$$

$$\Rightarrow \begin{vmatrix} -\lambda & i & 0 \\ -i & -\lambda & 0 \\ 0 & 0 & -\lambda \end{vmatrix} = 0 \quad \Rightarrow \quad -\lambda(\lambda^2 - 0) + i(0 - i\lambda) = 0$$

$$\Rightarrow -\lambda^3 - i^2\lambda = 0 \Rightarrow -\lambda^3 + \lambda = 0 \Rightarrow \lambda(1-\lambda^2) = 0 \therefore \lambda = 0, \pm 1$$

Therefore, the Eigen values are $\lambda = (0, +1, -1)$

We know, the characteristic equation of Eigen vector

$$(\Omega - \lambda I)(X_1) = 0$$

$$\Rightarrow \begin{pmatrix} -\lambda & i & 0 \\ -i & -\lambda & 0 \\ 0 & 0 & -\lambda \end{pmatrix}\begin{pmatrix} x_1 \\ x_2 \\ x_3 \end{pmatrix} = \begin{pmatrix} 0 \\ 0 \\ 0 \end{pmatrix}$$

Where x_1, x_2, x_3 are the components of the eigen vectors (X_1).

$$\Rightarrow -\lambda x_1 + i x_2 = 0 \qquad [1]$$

$$-ix_1 - \lambda x_2 = 0 \quad [2]$$
$$-\lambda x_3 = 0 \quad [3]$$

Case -01: When, $\lambda = 0$

From equation [3] x_3 is orbitary ($x_3 = t_3$[say]). From equation [1] $x_2 = 0$ and from equation [2] $x_1 = 0$. Therefore, the Eigen vector

$$X_1 = \begin{pmatrix} x_1 \\ x_2 \\ x_3 \end{pmatrix} \Rightarrow X_1 = \begin{pmatrix} 0 \\ 0 \\ t_3 \end{pmatrix}$$

From the normalization condition

$$X_1 X_1^{\times} = \begin{pmatrix} 0 \\ 0 \\ t_3 \end{pmatrix} (0 \quad 0 \quad t_3) \Rightarrow t_3{}^2 = 1 \therefore t_3 = \pm 1$$

Hence, the normalized Eigen vector $X_1 = \begin{pmatrix} 0 \\ 0 \\ 1 \end{pmatrix}$

Case-02: When, $\lambda = 1$

From equation [3] $x_3 = 0$

From equation [1]; $x_1 = \mathrm{i}x_2$ And From equation [2]; $x_2 = -\mathrm{i}x_1$; let $x_1 = \mathrm{C}$ $\therefore$ $x_2 = -\mathrm{iC}$

Therefore, the Eigen vector

$$X_2 = \begin{pmatrix} x_1 \\ x_2 \\ x_3 \end{pmatrix} \begin{pmatrix} \mathrm{c} \\ -\mathrm{ic} \\ 0 \end{pmatrix}$$

From the normalization condition

$$X_2 X_2^{\times} = \begin{pmatrix} \mathrm{c} \\ -\mathrm{ic} \\ 0 \end{pmatrix} (\mathrm{c} \quad +\mathrm{ic} \quad 0) = 1 \Rightarrow \mathrm{c}^2 + \mathrm{c}^2 = 1 \Rightarrow 2\mathrm{c}^2 = 1 \therefore \mathrm{c} = \pm\frac{1}{\sqrt{2}}$$

Hence, the normalized Eigen vector $X_2 = \begin{pmatrix} \frac{1}{\sqrt{2}} \\ -\frac{\mathrm{i}}{\sqrt{2}} \\ 0 \end{pmatrix}$

Case-03: When, $\lambda = -1$

From equation [1]; $x_1 = \mathrm{i}x_2 = \beta$ [Say $x_1 = \beta$] & from equation [3] $x_3 = 0$ & from [2] $x_2 = \mathrm{i}x_1 = \mathrm{i}\beta$

Therefore, the Eigen vector $X_3 = \begin{pmatrix} x_1 \\ x_2 \\ x_3 \end{pmatrix} \begin{pmatrix} \beta \\ \mathrm{i}\beta \\ 0 \end{pmatrix}$

From the normalization condition

$$X_3 X_3^{\times} = \begin{pmatrix} \beta \\ \mathrm{i}\beta \\ 0 \end{pmatrix} (\beta \quad -\mathrm{i}\beta \quad 0) = 1 \Rightarrow \beta^2 + \beta^2 = 1 \Rightarrow 2\beta^2 = 1 \therefore \beta = \pm\frac{1}{\sqrt{2}} = \frac{1}{\sqrt{2}}$$

Hence, the normalized Eigen vector $X_3 = \begin{pmatrix} \frac{1}{\sqrt{2}} \\ \frac{\mathrm{i}}{\sqrt{2}} \\ 0 \end{pmatrix}$

The Eigen values of the given matrix are $(0 \quad -1 \quad +1)$ & the corresponding Eigen vectors are

$$X_1 = \begin{pmatrix} 0 \\ 0 \\ 1 \end{pmatrix}, X_2 = \begin{pmatrix} \frac{1}{\sqrt{2}} \\ -\frac{i}{\sqrt{2}} \\ 0 \end{pmatrix} \; and \; X_3 = \begin{pmatrix} \frac{1}{\sqrt{2}} \\ \frac{i}{\sqrt{2}} \\ 0 \end{pmatrix}$$

A matrix can be diagonalized by means of unitary transformation, I.e., $\Omega' = U^{-1}\Omega U$

First we shall observe that Ω is Hermitian or not.

Here, $$\Omega = \begin{pmatrix} 0 & i & 0 \\ -i & 0 & 0 \\ 0 & 0 & 0 \end{pmatrix} \quad \& \quad \tilde{\Omega} = \begin{pmatrix} 0 & -i & 0 \\ i & 0 & 0 \\ 0 & 0 & 0 \end{pmatrix}^{\times}$$

By inspection, the Matrix is Hermitian.

$$U = \begin{pmatrix} 0 & \frac{1}{\sqrt{2}} & \frac{1}{\sqrt{2}} \\ 0 & -\frac{i}{\sqrt{2}} & \frac{i}{\sqrt{2}} \\ 1 & 0 & 0 \end{pmatrix} \; and \; U^{\times} = \begin{pmatrix} 0 & 0 & 1 \\ \frac{1}{\sqrt{2}} & -\frac{i}{\sqrt{2}} & 0 \\ \frac{1}{\sqrt{2}} & \frac{i}{\sqrt{2}} & 0 \end{pmatrix}^{\times} = \begin{pmatrix} 0 & 0 & 1 \\ \frac{1}{\sqrt{2}} & \frac{i}{\sqrt{2}} & 0 \\ \frac{1}{\sqrt{2}} & -\frac{i}{\sqrt{2}} & 0 \end{pmatrix}$$

Now we shall check whether U is unitary or not. From the unitary condition $UU^{\times} = 1$

$$\begin{pmatrix} 0 & \frac{1}{\sqrt{2}} & \frac{1}{\sqrt{2}} \\ 0 & -\frac{i}{\sqrt{2}} & \frac{i}{\sqrt{2}} \\ 1 & 0 & 0 \end{pmatrix} \begin{pmatrix} 0 & 0 & 1 \\ \frac{1}{\sqrt{2}} & \frac{i}{\sqrt{2}} & 0 \\ \frac{1}{\sqrt{2}} & -\frac{i}{\sqrt{2}} & 0 \end{pmatrix} = \begin{pmatrix} 0+\frac{1}{2}+\frac{1}{2} & 0+\frac{i}{2}-\frac{i}{2} & 0+0+0 \\ 0-\frac{i}{2}+\frac{i}{2} & 0-\frac{i^2}{2}-\frac{i^2}{2} & 0+0+0 \\ 0+0+0 & 0+0+0 & 1+0+0 \end{pmatrix} = \begin{pmatrix} 1 & 0 & 0 \\ 0 & 1 & 0 \\ 0 & 0 & 1 \end{pmatrix}$$

So that, U is unitary. Therefore, the matrix can be diagonalized. $\Omega' = U^{-1}\Omega U = U^{\times}\Omega U$

$$= \begin{pmatrix} 0 & 0 & 1 \\ \frac{1}{\sqrt{2}} & \frac{i}{\sqrt{2}} & 0 \\ \frac{1}{\sqrt{2}} & -\frac{i}{\sqrt{2}} & 0 \end{pmatrix} \begin{pmatrix} 0 & i & 0 \\ -i & 0 & 0 \\ 0 & 0 & 0 \end{pmatrix} \begin{pmatrix} 0 & \frac{1}{\sqrt{2}} & \frac{1}{\sqrt{2}} \\ 0 & -\frac{i}{\sqrt{2}} & \frac{i}{\sqrt{2}} \\ 1 & 0 & 0 \end{pmatrix} = \begin{pmatrix} 0 & 0 & 1 \\ \frac{1}{\sqrt{2}} & \frac{i}{\sqrt{2}} & 0 \\ \frac{1}{\sqrt{2}} & -\frac{i}{\sqrt{2}} & 0 \end{pmatrix} \begin{pmatrix} 0 & \frac{1}{\sqrt{2}} & -\frac{1}{\sqrt{2}} \\ 0 & -\frac{i}{\sqrt{2}} & \frac{i}{\sqrt{2}} \\ 0 & 0 & 0 \end{pmatrix}$$

$$= \begin{pmatrix} 0 & 0 & 0 \\ 0 & 1 & 0 \\ 0 & 0 & -1 \end{pmatrix}$$

We see that the diagonal elements are the Eigen values of the Matrix[20,21]. So the matrix has been diagonalized.

4.12 Eigen vectors of an operator in the basis of Eigen functions

The Eigen value equation $A\Psi_n = a_n\Psi_n$ becomes matrix Eigen value equation on the basis of Eigen functions Ψ_1, Ψ_2, Ψ_3, Ψ_4,

$$\begin{pmatrix} a_1 & 0 & 0 & 0 \\ 0 & a_2 & 0 & 0 \\ 0 & 0 & a_3 & 0 \\ 0 & 0 & 0 & a_4 \end{pmatrix} \begin{pmatrix} c_1 \\ c_2 \\ c_3 \\ c_4 \end{pmatrix} = a_n \begin{pmatrix} c_1 \\ c_2 \\ c_3 \\ c_4 \end{pmatrix} \qquad [1]$$

The Eigen vectors $\begin{pmatrix} c_1 \\ c_2 \\ c_3 \\ c_4 \end{pmatrix}$ can be obtained by solving the above Eigen value equation [1].

For $a_n = a_1, a_2, a_3, a_4$; The Eigen value equation [1] becomes

$$\begin{pmatrix} a_1 & 0 & 0 & 0 \\ 0 & a_2 & 0 & 0 \\ 0 & 0 & a_3 & 0 \\ 0 & 0 & 0 & a_4 \end{pmatrix} \begin{pmatrix} c_1 \\ c_2 \\ c_3 \\ c_4 \end{pmatrix} = a_1 \begin{pmatrix} c_1 \\ c_2 \\ c_3 \\ c_4 \end{pmatrix} \qquad \text{Or,} \quad \begin{pmatrix} a_1c_1 \\ a_2c_2 \\ a_3c_3 \\ a_4c_4 \end{pmatrix} = \begin{pmatrix} a_1c_1 \\ a_1c_2 \\ a_1c_3 \\ a_1c_4 \end{pmatrix}$$

$\therefore$ $a_1c_1 = a_1c_1$ [Insignificant or negligible]

$a_2c_2 = a_1c_2 \ Or, (a_1 - a_2)c_2 = 0 \ \therefore c_2 = 0 \ [\because a_1 - a_2 \neq 0 \ and \ a_1 \neq a_2]$

$a_3c_3 = a_1c_3 \ Or, (a_1 - a_3)c_3 = 0 \ \therefore c_3 = 0 \ [\because a_1 - a_3 \neq 0 \ and \ a_1 \neq a_3]$

$a_4c_4 = a_1c_4 \ Or, (a_1 - a_4)c_4 = 0 \ \therefore c_4 = 0 \ [\because a_1 - a_4 \neq 0 \ and \ a_1 \neq a_4]$

$\therefore$ The Eigen vector $a_1 = \begin{pmatrix} c_1 \\ 0 \\ 0 \\ 0 \end{pmatrix}$

From the Normalization condition

$$c_1^{\times} c_1 = 1;$$

$(c_1^{\times} \ \ 0 \ \ 0 \ \ 0) \begin{pmatrix} c_1 \\ 0 \\ 0 \\ 0 \end{pmatrix} = 1 \ \therefore c_1 = 1e^{i\delta}$; Where δ is anything real.

If $\delta = 0$ then $c_1 = 1$. The normalized Eigen vector $a_1 = \begin{pmatrix} 1 \\ 0 \\ 0 \\ 0 \end{pmatrix}$. And Eigen value equation is

$$\begin{pmatrix} a_1 & 0 & 0 & 0 \\ 0 & a_2 & 0 & 0 \\ 0 & 0 & a_3 & 0 \\ 0 & 0 & 0 & a_4 \end{pmatrix} \begin{pmatrix} 1 \\ 0 \\ 0 \\ 0 \end{pmatrix} = a_1 \begin{pmatrix} 1 \\ 0 \\ 0 \\ 0 \end{pmatrix} \qquad [2]$$

It is clear that L.H.S. of equation [2] is $\begin{pmatrix} a_1 \\ 0 \\ 0 \\ 0 \end{pmatrix}$ and R.H.S. of equation [2] is also $\begin{pmatrix} a_1 \\ 0 \\ 0 \\ 0 \end{pmatrix}$.

Similarly, for $a_n = a_2, a_3, a_4,$ the normalized Eigenvectors are

$$a_2 = \begin{pmatrix} 0 \\ 1 \\ 0 \\ 0 \end{pmatrix},\ a_3 = \begin{pmatrix} 0 \\ 0 \\ 1 \\ 0 \end{pmatrix} \text{ and } a_4 = \begin{pmatrix} 0 \\ 0 \\ 0 \\ 1 \end{pmatrix} \text{ respectively.}$$

These Eigen vectors are Ortho-normal. Because the scalar product of the Eigen vectors a_2 and a_3 corresponding to Eigen values

$$(0 \quad 1^{\times} \quad 0 \quad 0)\begin{pmatrix} 0 \\ 0 \\ 1 \\ 0 \end{pmatrix} \quad Or, (0 \quad 1 \quad 0 \quad 0)\begin{pmatrix} 0 \\ 0 \\ 1 \\ 0 \end{pmatrix} = 0$$

That is, normalized Eigen vectors are column matrices having one element 1 and the others 0 on the basis of Eigen functions.

4.13 Matrix operators of angular momentum components for $j = \frac{1}{2}$

No differential operators for J^2, J_x, J_y, J_z are available for non-integer values of j. In case of orbital angular momentum, there is no Eigen function like spherical harmonics $Y_{\ell,m}$ and there is no differential operators that we use in case of integral values of $j\ (= \ell)$. To use square matrices as an operators and column matrices for Eigen functions. For integer values of j, we do not have to use matrix formulation[24].

For $j = 1/2$ and hence $m_l = \pm 1/2$, we want to calculate matrix operators of J^2, J_x, J_y, J_z
We have

$$J^2\Psi_{j,m} = j(j+1)\hbar^2\Psi_{j,m} = \frac{1}{2}\left(\frac{1}{2}+1\right)\hbar^2\Psi_{j,m} = \frac{3}{4}\hbar^2\Psi_{j,m}$$

$$J_z\Psi_{j,m} = m_l\hbar\Psi_{j,m} = \pm\frac{1}{2}\hbar\Psi_{j,m}$$

$$J_+\Psi_{j,m} = \sqrt{j(j+1) - m(m+1)}\hbar\Psi_{j,m+1}$$

$$J_-\Psi_{j,m} = \sqrt{j(j+1) - m(m-1)}\,\hbar\Psi_{j,m-1}$$

J^2 and J_z have common normalized Eigen functions $\Psi_{j,m}$. So $\Psi_1 = \Psi_{\frac{1}{2},\frac{1}{2}}$ and $\Psi_2 = \Psi_{\frac{1}{2},-\frac{1}{2}}$ are only two allowed $\Psi_{j,m}$'s which are ortho-normal. Because they have common Eigen functions belonging to different Eigen values. Since Ψ_1 and the Ψ_2 as ortho-normal basis functions and use the Ψ_1 and the Ψ_2 to construct matrix operators of J^2, J_x, J_y, J_z. The matrix elements are given by

$(J^2)_{11} = (\Psi_1, J^2\Psi_1)$	$(J_z)_{11} = (\Psi_1, J_z\Psi_1)$
$(J^2)_{12} = (\Psi_1, J^2\Psi_2)$	$(J_z)_{12} = (\Psi_1, J_z\Psi_2)$
$(J^2)_{21} = (\Psi_2, J^2\Psi_1)$	$(J_z)_{21} = (\Psi_2, J_z\Psi_1)$
$(J^2)_{22} = (\Psi_2, J^2\Psi_2)$	$(J_z)_{22} = (\Psi_2, J_z\Psi_2)$

$$(J_+)_{11} = (\Psi_1, J_+\Psi_1) \qquad (J_-)_{11} = (\Psi_1, J_-\Psi_1)$$
$$(J_+)_{12} = (\Psi_1, J_+\Psi_2) \qquad (J_-)_{12} = (\Psi_1, J_-\Psi_2)$$
$$(J_+)_{21} = (\Psi_2, J_+\Psi_1) \qquad (J_-)_{21} = (\Psi_2, J_-\Psi_1)$$
$$(J_+)_{22} = (\Psi_2, J_+\Psi_2) \qquad (J_-)_{22} = (\Psi_2, J_-\Psi_2)$$

And the matrices are given by

$$J^2 = \begin{pmatrix} (J^2)_{11} & (J^2)_{12} \\ (J^2)_{21} & (J^2)_{22} \end{pmatrix} \text{ and } J_z = \begin{pmatrix} (J_z)_{11} & (J_z)_{12} \\ (J_z)_{21} & (J_z)_{22} \end{pmatrix}$$

$$J_+ = \begin{pmatrix} (J_+)_{11} & (J_+)_{12} \\ (J_+)_{21} & (J_+)_{22} \end{pmatrix} \text{ and } J_- = \begin{pmatrix} (J_-)_{11} & (J_-)_{12} \\ (J_-)_{21} & (J_-)_{22} \end{pmatrix}$$

$$J_x = \frac{1}{2}(J_+ + J_-) \text{ and } \qquad J_y = \frac{1}{2i}(J_+ - J_-)$$

Matrix elements of J^2 can be obtained as follows:

$$(J^2)_{11} = (\Psi_1, J^2\Psi_1) = \left(\Psi_{\frac{1}{2},\frac{1}{2}},\ J^2\Psi_{\frac{1}{2},\frac{1}{2}}\right) = \left(\Psi_{\frac{1}{2},\frac{1}{2}},\ \frac{3}{4}\hbar^2\Psi_{\frac{1}{2},\frac{1}{2}}\right)$$
$$= \frac{3}{4}\hbar^2\left(\Psi_{\frac{1}{2},\frac{1}{2}},\ \Psi_{\frac{1}{2},\frac{1}{2}}\right) = \frac{3}{4}\hbar^2(1) = \frac{3}{4}\hbar^2$$

$$(J^2)_{12} = (\Psi_1,\ J^2\Psi_2) = \left(\Psi_{\frac{1}{2},\frac{1}{2}},\ J^2\Psi_{\frac{1}{2},-\frac{1}{2}}\right) = \left(\Psi_{\frac{1}{2},\frac{1}{2}},\ \frac{3}{4}\hbar^2\Psi_{\frac{1}{2},-\frac{1}{2}}\right)$$
$$= \frac{3}{4}\hbar^2\left(\Psi_{\frac{1}{2},\frac{1}{2}},\ \Psi_{\frac{1}{2},-\frac{1}{2}}\right) = \frac{3}{4}\hbar^2(0) = 0$$

$$(J^2)_{21} = (\Psi_2,\ J^2\Psi_1) = \left(\Psi_{\frac{1}{2},-\frac{1}{2}},\ J^2\Psi_{\frac{1}{2},\frac{1}{2}}\right) = \left(\Psi_{\frac{1}{2},-\frac{1}{2}},\ \frac{3}{4}\hbar^2\Psi_{\frac{1}{2},\frac{1}{2}}\right)$$
$$= \frac{3}{4}\hbar^2\left(\Psi_{\frac{1}{2},-\frac{1}{2}}, \Psi_{\frac{1}{2},\frac{1}{2}}\right) = \frac{3}{4}\hbar^2(0) = 0$$

$$(J^2)_{22} = (\Psi_2,\ J^2\Psi_2) = \left(\Psi_{\frac{1}{2},-\frac{1}{2}},\ J^2\Psi_{\frac{1}{2},-\frac{1}{2}}\right) = \left(\Psi_{\frac{1}{2},-\frac{1}{2}},\ \frac{3}{4}\hbar^2\Psi_{\frac{1}{2},-\frac{1}{2}}\right)$$
$$= \frac{3}{4}\hbar^2\left(\Psi_{\frac{1}{2},-\frac{1}{2}},\ \Psi_{\frac{1}{2},-\frac{1}{2}}\right) = \frac{3}{4}\hbar^2(1) = \frac{3}{4}\hbar^2$$

Thus $$J^2 = \begin{pmatrix} (J^2)_{11} & (J^2)_{12} \\ (J^2)_{21} & (J^2)_{22} \end{pmatrix} = \begin{pmatrix} \frac{3}{4}\hbar^2 & 0 \\ 0 & \frac{3}{4}\hbar^2 \end{pmatrix} = \frac{3}{4}\hbar^2\begin{pmatrix} 1 & 0 \\ 0 & 1 \end{pmatrix}$$

Matrix elements of J_z can be obtained as follows:

$$(J_z)_{11} = (\Psi_1, J_z\Psi_1) = \left(\Psi_{\frac{1}{2},\frac{1}{2}}, J_z\Psi_{\frac{1}{2},\frac{1}{2}}\right) = \left(\Psi_{\frac{1}{2},\frac{1}{2}}, \frac{1}{2}\hbar\Psi_{\frac{1}{2},\frac{1}{2}}\right)$$

$$= \frac{1}{2}\hbar\left(\Psi_{\frac{1}{2},\frac{1}{2}}, \Psi_{\frac{1}{2},\frac{1}{2}}\right) = \frac{1}{2}\hbar(1) = \frac{1}{2}\hbar$$

$$(J_z)_{12} = (\Psi_1, J_z\Psi_2) = \left(\Psi_{\frac{1}{2},\frac{1}{2}}, J_z\Psi_{\frac{1}{2},-\frac{1}{2}}\right) = \left(\Psi_{\frac{1}{2},\frac{1}{2}}, -\frac{1}{2}\hbar\Psi_{\frac{1}{2},-\frac{1}{2}}\right)$$

$$= -\frac{1}{2}\hbar\left(\Psi_{\frac{1}{2},\frac{1}{2}}, \Psi_{\frac{1}{2},-\frac{1}{2}}\right) = -\frac{1}{2}\hbar(0) = 0$$

$$(J_z)_{21} = (\Psi_2, J_z\Psi_1) = \left(\Psi_{\frac{1}{2},-\frac{1}{2}}, J_z\Psi_{\frac{1}{2},\frac{1}{2}}\right) = \left(\Psi_{\frac{1}{2},-\frac{1}{2}}, \frac{1}{2}\hbar\Psi_{\frac{1}{2},\frac{1}{2}}\right)$$

$$= \frac{1}{2}\hbar\left(\Psi_{\frac{1}{2},-\frac{1}{2}}, \Psi_{\frac{1}{2},\frac{1}{2}}\right) = \frac{1}{2}\hbar(0) = 0$$

$$(J_z)_{22} = (\Psi_2, J_z\Psi_2) = \left(\Psi_{\frac{1}{2},-\frac{1}{2}}, J_z\Psi_{\frac{1}{2},-\frac{1}{2}}\right) = \left(\Psi_{\frac{1}{2},-\frac{1}{2}}, -\frac{1}{2}\hbar\Psi_{\frac{1}{2},-\frac{1}{2}}\right)$$

$$= -\frac{1}{2}\hbar\left(\Psi_{\frac{1}{2},-\frac{1}{2}}, \Psi_{\frac{1}{2},-\frac{1}{2}}\right) = -\frac{1}{2}\hbar(1) = -\frac{1}{2}\hbar$$

Thus $$J_z = \begin{pmatrix} (J_z)_{11} & (J_z)_{12} \\ (J_z)_{21} & (J_z)_{22} \end{pmatrix} = \begin{pmatrix} \frac{1}{2}\hbar & 0 \\ 0 & -\frac{1}{2}\hbar \end{pmatrix}; \; So,\; J_z = \frac{1}{2}\hbar\begin{pmatrix} 1 & 0 \\ 0 & -1 \end{pmatrix}$$

Matrix elements of J_+ can be obtained as follows:

$$(J_+)_{11} = (\Psi_1, J_+\Psi_1) = \left(\Psi_{\frac{1}{2},\frac{1}{2}}, J_+\Psi_{\frac{1}{2},\frac{1}{2}}\right) = \left(\Psi_{\frac{1}{2},\frac{1}{2}}, \sqrt{\frac{1}{2}\left(\frac{1}{2}+1\right) - \frac{1}{2}\left(\frac{1}{2}+1\right)}\;\hbar\Psi_{\frac{1}{2},\frac{1}{2}+1}\right)$$

$$= \sqrt{\frac{1}{2}\left(\frac{1}{2}+1\right) - \frac{1}{2}\left(\frac{1}{2}+1\right)}\;\hbar\left(\Psi_{\frac{1}{2},\frac{1}{2}}, \Psi_{\frac{1}{2},\frac{3}{2}}\right) = (0)(\hbar)\left(\Psi_{\frac{1}{2},\frac{1}{2}}, \Psi_{\frac{1}{2},\frac{3}{2}}\right) = 0$$

$$(J_+)_{12} = (\Psi_1, J_+\Psi_2) = \left(\Psi_{\frac{1}{2},\frac{1}{2}}, J_+\Psi_{\frac{1}{2},-\frac{1}{2}}\right) = \left(\Psi_{\frac{1}{2},\frac{1}{2}}, \sqrt{\frac{1}{2}\left(\frac{1}{2}+1\right) - \left(-\frac{1}{2}\right)\left(-\frac{1}{2}+1\right)}\;\hbar\Psi_{\frac{1}{2},-\frac{1}{2}+1}\right)$$

$$= \sqrt{\frac{1}{2}\left(\frac{1}{2}+1\right) - \left(-\frac{1}{2}\right)\left(-\frac{1}{2}+1\right)}\;\hbar\left(\Psi_{\frac{1}{2},\frac{1}{2}}, \Psi_{\frac{1}{2},\frac{1}{2}}\right) = \sqrt{\frac{1}{2}\left(\frac{3}{2}\right) + \frac{1}{2}\left(\frac{1}{2}\right)}\;\hbar(1) = \hbar$$

$$(J_+)_{21} = (\Psi_2, J_+\Psi_1) = \left(\Psi_{\frac{1}{2},-\frac{1}{2}}, J_+\Psi_{\frac{1}{2},\frac{1}{2}}\right) = \left(\Psi_{\frac{1}{2},-\frac{1}{2}}, \sqrt{\frac{1}{2}\left(\frac{1}{2}+1\right) - \frac{1}{2}\left(\frac{1}{2}+1\right)}\;\hbar\Psi_{\frac{1}{2},\frac{1}{2}+1}\right)$$

$$= \sqrt{\frac{1}{2}\left(\frac{1}{2}+1\right) - \frac{1}{2}\left(\frac{1}{2}+1\right)}\;\hbar\left(\Psi_{\frac{1}{2},-\frac{1}{2}}, \Psi_{\frac{1}{2},\frac{3}{2}}\right) = (0)\hbar\left(\Psi_{\frac{1}{2},-\frac{1}{2}}, \Psi_{\frac{1}{2},\frac{3}{2}}\right) = 0$$

$$(J_+)_{22} = (\Psi_2, J_+\Psi_2) = \left(\Psi_{\frac{1}{2},-\frac{1}{2}}, J_+\Psi_{\frac{1}{2},-\frac{1}{2}}\right) = \left(\Psi_{\frac{1}{2},-\frac{1}{2}}, \sqrt{\frac{1}{2}\left(\frac{1}{2}+1\right)-\frac{1}{2}\left(-\frac{1}{2}+1\right)}\,\hbar\Psi_{\frac{1}{2},-\frac{1}{2}+1}\right.$$

$$= \sqrt{\frac{1}{2}\left(\frac{1}{2}+1\right)-\left(-\frac{1}{2}\right)\left(\frac{1}{2}+1\right)}\,\hbar\left(\Psi_{\frac{1}{2},-\frac{1}{2}}, \Psi_{\frac{1}{2},\frac{1}{2}}\right) = \sqrt{\frac{1}{2}\left(\frac{3}{2}\right)+\frac{1}{2}\left(\frac{1}{2}\right)}\,\hbar(0) = 0$$

Thus $J_+ = \begin{pmatrix} (J_+)_{11} & (J_+)_{12} \\ (J_+)_{21} & (J_+)_{22} \end{pmatrix} = \begin{pmatrix} 0 & \hbar \\ 0 & 0 \end{pmatrix}$

Matrix elements of J_- can be obtained as follows:

$$(J_-)_{11} = (\Psi_1, J_-\Psi_1)\left(\Psi_{\frac{1}{2},\frac{1}{2}}, J_-\Psi_{\frac{1}{2},\frac{1}{2}}\right) = \left(\Psi_{\frac{1}{2},\frac{1}{2}}, \sqrt{\frac{1}{2}\left(\frac{1}{2}+1\right)-\frac{1}{2}\left(\frac{1}{2}-1\right)}\,\hbar\Psi_{\frac{1}{2},\frac{1}{2}-1}\right)$$

$$= \sqrt{\frac{1}{2}\left(\frac{1}{2}+1\right)-\frac{1}{2}\left(\frac{1}{2}-1\right)}\,\hbar\left(\Psi_{\frac{1}{2},\frac{1}{2}}, \Psi_{\frac{1}{2},-\frac{1}{2}}\right) = \sqrt{\frac{3}{4}+\frac{1}{4}}\,\hbar(0) = 0$$

$$(J_-)_{12} = (\Psi_1, J_-\Psi_2) = \left(\Psi_{\frac{1}{2},\frac{1}{2}}, J_-\Psi_{\frac{1}{2},-\frac{1}{2}}\right) = \left(\Psi_{\frac{1}{2},\frac{1}{2}}, \sqrt{\frac{1}{2}\left(\frac{1}{2}+1\right)-\left(-\frac{1}{2}\right)\left(-\frac{1}{2}-1\right)}\,\hbar\Psi_{\frac{1}{2},-\frac{1}{2}-1}\right)$$

$$= \sqrt{\frac{1}{2}\left(\frac{1}{2}+1\right)-\left(-\frac{1}{2}\right)\left(-\frac{1}{2}-1\right)}\,\hbar\left(\Psi_{\frac{1}{2},\frac{1}{2}}, \Psi_{\frac{1}{2},-\frac{3}{2}}\right) = (0)\hbar\left(\Psi_{\frac{1}{2},\frac{1}{2}}, \Psi_{\frac{1}{2},-\frac{3}{2}}\right) = 0$$

$$(J_-)_{21} = (\Psi_2, J_-\Psi_1) = \left(\Psi_{\frac{1}{2},-\frac{1}{2}}, J_-\Psi_{\frac{1}{2},\frac{1}{2}}\right) = \left(\Psi_{\frac{1}{2},-\frac{1}{2}}, \sqrt{\frac{1}{2}\left(\frac{1}{2}+1\right)-\frac{1}{2}\left(\frac{1}{2}-1\right)}\,\hbar\Psi_{\frac{1}{2},\frac{1}{2}-1}\right)$$

$$= \sqrt{\frac{1}{2}\left(\frac{1}{2}+1\right)-\frac{1}{2}\left(\frac{1}{2}-1\right)}\,\hbar\left(\Psi_{\frac{1}{2},-\frac{1}{2}}, \Psi_{\frac{1}{2},-\frac{1}{2}}\right) = \sqrt{\frac{1}{2}\left(\frac{3}{2}\right)-\frac{1}{2}\left(-\frac{1}{2}\right)}\,\hbar(1) = \hbar$$

$$(J_-)_{22} = (\Psi_2, J_-\Psi_2) = \left(\Psi_{\frac{1}{2},-\frac{1}{2}}, J_-\Psi_{\frac{1}{2},-\frac{1}{2}}\right) = \left(\Psi_{\frac{1}{2},-\frac{1}{2}}, \sqrt{\frac{1}{2}\left(\frac{1}{2}+1\right)-\left(-\frac{1}{2}\right)\left(-\frac{1}{2}-1\right)}\,\hbar\Psi_{\frac{1}{2},-\frac{1}{2}-1}\right)$$

$$= \sqrt{\frac{1}{2}\left(\frac{1}{2}+1\right)-\left(-\frac{1}{2}\right)\left(-\frac{1}{2}-1\right)}\,\hbar\left(\Psi_{\frac{1}{2},-\frac{1}{2}}, \Psi_{\frac{1}{2},-\frac{3}{2}}\right) = (0)\hbar\left(\Psi_{\frac{1}{2},-\frac{1}{2}}, \Psi_{\frac{1}{2},-\frac{3}{2}}\right) = 0$$

Thus $J_- = \begin{pmatrix} (J_-)_{11} & (J_-)_{12} \\ (J_-)_{21} & (J_-)_{22} \end{pmatrix} = \begin{pmatrix} 0 & 0 \\ \hbar & 0 \end{pmatrix}$

$$J_x = \frac{1}{2}(J_+ + J_-) = \frac{1}{2}\left[\begin{pmatrix} 0 & \hbar \\ 0 & 0 \end{pmatrix} + \begin{pmatrix} 0 & 0 \\ \hbar & 0 \end{pmatrix}\right] = \frac{1}{2}\begin{pmatrix} 0 & \hbar \\ \hbar & 0 \end{pmatrix} = \frac{1}{2}\hbar\begin{pmatrix} 0 & 1 \\ 1 & 0 \end{pmatrix}$$

$$J_y = \frac{1}{2i}(J_+ - J_-) = \frac{1}{2i}\left[\begin{pmatrix} 0 & \hbar \\ 0 & 0 \end{pmatrix} - \begin{pmatrix} 0 & 0 \\ \hbar & 0 \end{pmatrix}\right]$$

$$= \frac{1}{2i}\begin{pmatrix} 0 & \hbar \\ -\hbar & 0 \end{pmatrix} = -\frac{i\hbar}{2}\begin{pmatrix} 0 & 1 \\ -1 & 0 \end{pmatrix} = \frac{\hbar}{2}\begin{pmatrix} 0 & -i \\ i & 0 \end{pmatrix}$$

4.14 Commutation relations of the matrix operators of angular momentum components for $j = 1/2$

We have the angular momentum matrices for $j = 1/2$,

$$J_x = \frac{1}{2}\hbar\begin{pmatrix}0 & 1\\1 & 0\end{pmatrix}, \qquad J_y = \frac{\hbar}{2}\begin{pmatrix}0 & -i\\i & 0\end{pmatrix}$$

$$J_z = \frac{1}{2}\hbar\begin{pmatrix}1 & 0\\0 & -1\end{pmatrix}, \qquad J^2 = \frac{3}{4}\hbar^2\begin{pmatrix}1 & 0\\0 & 1\end{pmatrix}$$

For $= {}^1\!/_2$; the matrix operators of J_x, J_y, J_z, J^2 obey the following commutation relations

$$[J_x,J_y] = i\hbar J_z, \quad [J_y,J_z] = i\hbar J_x \quad ,[J_z,J_x] = i\hbar J_y$$

$$[J^2,J_x] = [J^2,J_y] = [J^2,J_z] = 0$$

So, the commutation rule of the matrix operators of J_x and J_y as

$$[J_x,J_y] = (J_xJ_y - J_yJ_x) = \frac{1}{2}\hbar\begin{pmatrix}0 & 1\\1 & 0\end{pmatrix}\frac{1}{2}\hbar\begin{pmatrix}0 & -i\\i & 0\end{pmatrix} - \frac{1}{2}\hbar\begin{pmatrix}0 & -i\\i & 0\end{pmatrix}\frac{1}{2}\hbar\begin{pmatrix}0 & 1\\1 & 0\end{pmatrix}$$

$$= \left(\frac{1}{2}\hbar\right)^2\begin{pmatrix}0 & 1\\1 & 0\end{pmatrix}\begin{pmatrix}0 & -i\\i & 0\end{pmatrix} - \left(\frac{1}{2}\hbar\right)^2\begin{pmatrix}0 & -i\\i & 0\end{pmatrix}\begin{pmatrix}0 & 1\\1 & 0\end{pmatrix}$$

$$= \left(\frac{1}{2}\hbar\right)^2\begin{pmatrix}i & 0\\0 & -i\end{pmatrix} - \left(\frac{1}{2}\hbar\right)^2\begin{pmatrix}-i & 0\\0 & i\end{pmatrix}$$

$$= \left(\frac{1}{2}\hbar\right)^2\begin{pmatrix}2i & 0\\0 & -2i\end{pmatrix} = \left(\frac{1}{2}\hbar\right)^2 2i\begin{pmatrix}1 & 0\\0 & -1\end{pmatrix}$$

$$\therefore \quad [J_x,J_y] = (J_xJ_y - J_yJ_x) = i\hbar\frac{1}{2}\hbar\begin{pmatrix}1 & 0\\0 & -1\end{pmatrix} = i\hbar J_z$$

The matrix operators of J_x and J_y satisfy the commutation relation $[J_x,J_y] = i\hbar J_z$.

Again, the commutation rule of the matrix operators of of J^2 and J_z

$$[J^2,J_z] = (J^2J_z - J_zJ^2)$$

$$= \frac{3}{4}\hbar^2\begin{pmatrix}1 & 0\\0 & 1\end{pmatrix}\frac{1}{2}\hbar\begin{pmatrix}1 & 0\\0 & -1\end{pmatrix} - \frac{1}{2}\hbar\begin{pmatrix}1 & 0\\0 & -1\end{pmatrix}\frac{3}{4}\hbar^2\begin{pmatrix}1 & 0\\0 & 1\end{pmatrix}$$

$$= \frac{3}{4}\hbar^2\frac{1}{2}\hbar\left\{\begin{pmatrix}1 & 0\\0 & 1\end{pmatrix}\begin{pmatrix}1 & 0\\0 & -1\end{pmatrix} - \begin{pmatrix}1 & 0\\0 & -1\end{pmatrix}\begin{pmatrix}1 & 0\\0 & 1\end{pmatrix}\right\}$$

$$= \frac{3}{8}\hbar^3\left\{\begin{pmatrix}1 & 0\\0 & -1\end{pmatrix} - \begin{pmatrix}1 & 0\\0 & -1\end{pmatrix}\right\} = \begin{pmatrix}0 & 0\\0 & 0\end{pmatrix}$$

The matrix operators of J^2 and J_z satisfy the commutation relation $[J^2,J_z] = 0$. Similarly, the matrix operators obey all the other commutation relations

$$[J_y,J_z] = i\hbar J_x \quad , [J_z,J_x] = i\hbar J_y, \quad [J^2,J_x] = [J^2,J_y] = 0.$$

4.15 Diagonalization of the angular momentum matrices for j $= \frac{1}{2}$

The matrix operators of angular momentum [components] calculated in the basis of common Eigen functions of J^2 and J_z for $= 1/2$,

$$J_x = \frac{1}{2}\hbar\begin{pmatrix}0 & 1\\ 1 & 0\end{pmatrix}, \qquad J_y = \frac{\hbar}{2}\begin{pmatrix}0 & -i\\ i & 0\end{pmatrix}$$

$$J_z = \frac{1}{2}\hbar\begin{pmatrix}1 & 0\\ 0 & -1\end{pmatrix}, \qquad J^2 = \frac{3}{4}\hbar^2\begin{pmatrix}1 & 0\\ 0 & 1\end{pmatrix}$$

The basis functions used are Eigen functions of J^2 and J_z so the matrices of J^2 and J_z are already diagonal. Hence their diagonal elements are Eigen values. The basis functions used are not Eigen functions of J_x and J_y so the matrices for J_x and J_y are not diagonal. The matrices of J_x and J_y, can be diagonalise by using Secular equation. The Secular equation for J_x will be

$$\begin{vmatrix}0-\lambda & \hbar/2\\ \hbar/2 & 0-\lambda\end{vmatrix} = 0$$

Or, $\lambda^2 - (\hbar/2)^2 = 0 \; or, \lambda^2 = (\hbar/2)^2 \;\therefore \lambda = \pm\hbar/2$

Thus Eigen values of J_x are $\pm\hbar/2$, same as those of J_z. The diagonalized matrix of J_x will be

$$J_x = \begin{pmatrix}\hbar/2 & 0\\ 0 & -\hbar/2\end{pmatrix};$$

And the Secular equation for J_y is

$$\begin{vmatrix}0-\lambda & -i\hbar/2\\ i\hbar/2 & 0-\lambda\end{vmatrix} = 0$$

Or, $\lambda^2 - (-i\hbar/2)(i\hbar/2) = 0$ or, $\lambda^2 - (\hbar/2)^2 = 0 \; or, \lambda^2 = (\hbar/2)^2 \;\therefore \lambda = \pm\hbar/2.$

Thus Eigen values of J_y are $\pm\hbar/2$, same as those of J_z. The diagonalized matrix of J_y will be

$$J_y = \begin{pmatrix}\hbar/2 & 0\\ 0 & -\hbar/2\end{pmatrix}.$$

Thus the Eigen values of J_x, J_y, J_z are the same.

4.16 Pauli spin matrices

The components of generalized angular momentum $\vec{J}[J_x, J_y J_z]$ have the following matrix operators for j $= \frac{1}{2}$

$$J_x = \frac{\hbar}{2}\begin{pmatrix}0 & 1\\ 1 & 0\end{pmatrix} = \frac{\hbar}{2}\sigma_x$$

$$J_y = \frac{\hbar}{2}\begin{pmatrix}0 & -i\\ i & 0\end{pmatrix} = \frac{\hbar}{2}\sigma_y$$

$$J_z = \frac{\hbar}{2}\begin{pmatrix}1 & 0\\ 0 & -1\end{pmatrix} = \frac{\hbar}{2}\sigma_z$$

Since J^2 and J_z have the common Eigen functions. So in the basis of the common Eigen functions of J^2 and J_z, the matrices are

$$\sigma_x = \begin{pmatrix} 0 & 1 \\ 1 & 0 \end{pmatrix}$$

$$\sigma_y = \begin{pmatrix} 0 & -i \\ i & 0 \end{pmatrix}$$

$$and\ \sigma_z = \begin{pmatrix} 1 & 0 \\ 0 & -1 \end{pmatrix}$$

The above are called *Pauli spin matrices*

4.17 For = 1/2 ; the Eigen vectors of generalized angular momentum and its components J^2, J_x, J_y, J_z

We want to calculate the Eigen vectors of generalized angular momentum and its components J^2, J_x, J_y, J_z for $= \frac{1}{2}$. Matrix form of Eigen value equation

$$J^2\Psi = \frac{3}{4}\hbar^2\Psi$$

Or, $$\frac{3}{4}\hbar^2 \begin{pmatrix} 1 & 0 \\ 0 & 1 \end{pmatrix}\begin{pmatrix} C_1 \\ C_2 \end{pmatrix} = \frac{3}{4}\hbar^2 \begin{pmatrix} C_1 \\ C_2 \end{pmatrix}$$

Or, $\begin{pmatrix} 1 & 0 \\ 0 & 1 \end{pmatrix}\begin{pmatrix} C_1 \\ C_2 \end{pmatrix} = \begin{pmatrix} C_1 \\ C_2 \end{pmatrix}$ Or, $\begin{pmatrix} C_1 \\ C_2 \end{pmatrix} = \begin{pmatrix} C_1 \\ C_2 \end{pmatrix}$, $\therefore c_1 = c_1$ and $c_2 = c_2$[Insignificant or negligible]

Matrix form of Eigen value equation

$$J_z\Psi = +\frac{\hbar}{2}\Psi$$

$$\frac{1}{2}\hbar \begin{pmatrix} 1 & 0 \\ 0 & -1 \end{pmatrix}\begin{pmatrix} C_1 \\ C_2 \end{pmatrix} = +\frac{1}{2}\hbar \begin{pmatrix} C_1 \\ C_2 \end{pmatrix},$$

Or, $\begin{pmatrix} C_1 \\ -C_2 \end{pmatrix} = \begin{pmatrix} C_1 \\ C_2 \end{pmatrix}$ $\therefore c_1 = c_1$(trivial) and $c_2 = -c_2$, or, $2c_2 = 0$, $\therefore c_2 = 0$

J^2 and J_z commute and hence have the same common Eigen vector. So, we have used operator obtained in the basis of common Eigen functions of J^2 and J_z

$\therefore$ The Eigen vector is $\begin{pmatrix} C_1 \\ 0 \end{pmatrix}$.

From Normalization condition

$$(c_1^\times \quad 0)\begin{pmatrix} C_1 \\ 0 \end{pmatrix} = 1$$

$c_1^\times c_1 = 1\ or, c_1 = 1e^{i\delta}$, Where δ is anything real. Thus $c_1 = 1$ taking $\delta = 0$.

Hence Normalized Eigen vector is $\begin{pmatrix}1\\0\end{pmatrix}$ for eigenvalue of J_z equal to $+\frac{\hbar}{2}$.

Again, matrix form of Eigen value equation

$$J_z\Psi = -\frac{\hbar}{2}\Psi \text{ for } J_z$$

$$\frac{1}{2}\hbar\begin{pmatrix}1 & 0\\0 & -1\end{pmatrix}\begin{pmatrix}c_1\\c_2\end{pmatrix} = -\frac{1}{2}\hbar\begin{pmatrix}c_1\\c_2\end{pmatrix}$$

Or, $\begin{pmatrix}1 & 0\\0 & -1\end{pmatrix}\begin{pmatrix}c_1\\c_2\end{pmatrix} = -\begin{pmatrix}c_1\\c_2\end{pmatrix}$, or, $\begin{pmatrix}c_1\\-c_2\end{pmatrix} = \begin{pmatrix}-c_1\\-c_2\end{pmatrix}$, $\therefore c_1 = -c_1$, or, $2c_1 = 0$

$\therefore c_1 = 0$, and $c_2 = c_2$ (trivial)

$\therefore$ The Eigen vector is $\begin{pmatrix}0\\c_2\end{pmatrix}$.

From Normalization condition

$$(0 \quad c_2^\times)\begin{pmatrix}0\\c_2\end{pmatrix} = 1$$

$c_2^\times c_2 = 1 \therefore c_2 = 1e^{i\delta}$; Where δ is anything real. Thus $c_2 = 1$ taking $\delta = 0$.

Hence Normalized Eigen vector is $\begin{pmatrix}0\\1\end{pmatrix}$ for eigenvalue of J_z equal to $-\frac{\hbar}{2}$.

For $=\frac{1}{2}$: [1] The Normalized Eigen vector is $\begin{pmatrix}1\\0\end{pmatrix}$ for eigenvalue of J_z equal to $+\frac{\hbar}{2}$.

[2] The Normalized Eigen vector is $\begin{pmatrix}0\\1\end{pmatrix}$ for eigenvalue of J_z equal to $-\frac{\hbar}{2}$.

For The Normalized Eigen vector is $\begin{pmatrix}1\\0\end{pmatrix}$:

We have $\frac{3}{4}\hbar^2\begin{pmatrix}1 & 0\\0 & 1\end{pmatrix}\begin{pmatrix}1\\0\end{pmatrix} = \frac{3}{4}\hbar^2\begin{pmatrix}1\\0\end{pmatrix}$. [$\therefore \begin{pmatrix}1\\0\end{pmatrix}$ is an Eigen vector of J^2 also.]

Again, For The Normalized Eigen vector is $\begin{pmatrix}0\\1\end{pmatrix}$:

We have $\frac{3}{4}\hbar^2\begin{pmatrix}1 & 0\\0 & 1\end{pmatrix}\begin{pmatrix}0\\1\end{pmatrix} = \frac{3}{4}\hbar^2\begin{pmatrix}0\\1\end{pmatrix}$. [$\therefore \begin{pmatrix}0\\1\end{pmatrix}$ is an Eigen vector of J^2 also.]

The normalized common Eigen vectors of J^2 and J_z which are $\begin{pmatrix}1\\0\end{pmatrix}$ and $\begin{pmatrix}0\\1\end{pmatrix}$ are called *spinors.*

Again Matrix form of Eigen value equation of J_x for Eigen value $= +\hbar/2$,

$$J_x\Psi = +\frac{\hbar}{2}\Psi$$

$$\frac{1}{2}\hbar\begin{pmatrix}0 & 1\\1 & 0\end{pmatrix}\begin{pmatrix}c_3\\c_4\end{pmatrix} = +\frac{1}{2}\hbar\begin{pmatrix}c_3\\c_4\end{pmatrix}.$$

Thus we have $\frac{\hbar}{2}c_4 = \frac{\hbar}{2}c_3$ *and* $\frac{\hbar}{2}c_3 = \frac{\hbar}{2}c_4$. $\therefore c_4 = c_3$ and $c_3 = c_4 = c$. [Using operator obtained in the basis of common Eigen functions of J^2 and J_z.]

$\therefore$ The Eigen vector of J_x is $\begin{pmatrix}c\\c\end{pmatrix}$.

From Normalization condition

$$(c^{\times} \quad c^{\times})\begin{pmatrix} c \\ c \end{pmatrix} = 1$$

$or,\ \ c^{\times}c + c^{\times}c = 1\ or,\ \ 2|c|^2 = 1\ \ or,\ \ c = \frac{1}{\sqrt{2}}e^{i\delta}$

Where δ is anything real. Thus $c = \frac{1}{\sqrt{2}}$ taking $\delta = 0$.

The normalized Eigen vector of J_x is $\begin{pmatrix} c \\ c \end{pmatrix} = \begin{pmatrix} \frac{1}{\sqrt{2}} \\ \frac{1}{\sqrt{2}} \end{pmatrix} = \frac{1}{\sqrt{2}}\begin{pmatrix} 1 \\ 1 \end{pmatrix}$ for eigenvalue of J_x equal to $+\frac{\hbar}{2}$

Matrix form of Eigen value equation of J_x for Eigen value $= -\hbar/2$,

$$J_x\Psi = -\frac{\hbar}{2}\Psi$$

$$\frac{1}{2}\hbar\begin{pmatrix} 0 & 1 \\ 1 & 0 \end{pmatrix}\begin{pmatrix} c_3 \\ c_4 \end{pmatrix} = -\frac{1}{2}\hbar\begin{pmatrix} c_3 \\ c_4 \end{pmatrix} \quad or,\ \frac{1}{2}\hbar\begin{pmatrix} c_3 \\ c_4 \end{pmatrix} = -\frac{1}{2}\hbar\begin{pmatrix} c_3 \\ c_4 \end{pmatrix}$$

Or, $\ c_4 = -c_3$ and $c_3 = -c_4$. Let $c_3 = -c_4 = c$. [say]

The Eigen vector of J_x is $\begin{pmatrix} -c \\ c \end{pmatrix}$. We can also take $\begin{pmatrix} c \\ -c \end{pmatrix}$.

From Normalization condition

$$(-c^{\times} \quad c^{\times})\begin{pmatrix} -c \\ c \end{pmatrix} = 1 \ \ or, \qquad c^{\times}c + c^{\times}c = 1 \ \ \ or, 2|c|^2 = 1$$

$\therefore c = \frac{1}{\sqrt{2}}e^{i\delta}$. Where δ is anything real. Thus $c = \frac{1}{\sqrt{2}}$ taking $\delta = 0$.

The normalized Eigen vector of J_x is

$$\begin{pmatrix} -c \\ c \end{pmatrix} = \begin{pmatrix} -\frac{1}{\sqrt{2}} \\ \frac{1}{\sqrt{2}} \end{pmatrix} = \frac{1}{\sqrt{2}}\begin{pmatrix} -1 \\ 1 \end{pmatrix} \ ; \ \text{for Eigen value } -\hbar/2.$$

We can also take The normalized Eigen vector of J_x is

$$\begin{pmatrix} c \\ -c \end{pmatrix} = \frac{1}{\sqrt{2}}\begin{pmatrix} 1 \\ -1 \end{pmatrix}; \ \text{for Eigen value } -\hbar/2.$$

Matrix form of Eigen value equation of J_y for Eigen value $= +\hbar/2$,

$$J_y\Psi = +\frac{\hbar}{2}\Psi$$

$$\frac{1}{2}\hbar\begin{pmatrix} 0 & -i \\ i & 0 \end{pmatrix}\begin{pmatrix} c_5 \\ c_6 \end{pmatrix} = \pm\frac{1}{2}\hbar\begin{pmatrix} c_5 \\ c_6 \end{pmatrix}$$

Or, $\qquad -ic_6 = c_5$ and $ic_5 = c_6$.

Thus Eigen vector of $\begin{pmatrix} c_5 \\ c_6 \end{pmatrix}$ of J_y is $\begin{pmatrix} c_5 \\ ic_5 \end{pmatrix}$.

From Normalization condition

$$(c_5^{\times} \quad -ic_5^{\times})\begin{pmatrix} c_5 \\ ic_5 \end{pmatrix} = 1 \ \text{ or, } \ c_5^{\times}c_5 + c_5^{\times}c_5 = 1$$

$\therefore c_5 = \frac{1}{\sqrt{2}} e^{i\delta}$ where δ is real. Thus $c_5 = \frac{1}{\sqrt{2}}$ taking $\delta = 0$.

Thus a normalized Eigen vector of J_y is $\begin{pmatrix} \frac{1}{\sqrt{2}} \\ i\frac{1}{\sqrt{2}} \end{pmatrix} = \frac{1}{\sqrt{2}}\begin{pmatrix} 1 \\ i \end{pmatrix}$ belonging to the Eigen value $\hbar/2$.

Matrix form of Eigen value equation of J_y for Eigen value $= -\hbar/2$,

$$J_y \Psi = -\frac{\hbar}{2}\Psi$$

$$\frac{1}{2}\hbar\begin{pmatrix} 0 & -\mathrm{i} \\ \mathrm{i} & 0 \end{pmatrix}\begin{pmatrix} c_5 \\ c_6 \end{pmatrix} = -\frac{1}{2}\hbar\begin{pmatrix} c_5 \\ c_6 \end{pmatrix}, \quad \text{or, } -\mathrm{i}c_6 = -c_5 \text{ and } \mathrm{i}c_5 = -c_6.$$

Thus Eigen vector $\begin{pmatrix} c_5 \\ c_6 \end{pmatrix}$ of J_y is $\begin{pmatrix} \mathrm{i}c_6 \\ c_6 \end{pmatrix}$.

From Normalization condition

$$(-\mathrm{i}c_6^{\times} \quad c_6^{\times})\begin{pmatrix} \mathrm{i}c_6 \\ c_6 \end{pmatrix} = 1$$

Or, $c_6^{\times}c_6 + c_6^{\times}c_6 = 1$ Or, $c_6 = \frac{1}{\sqrt{2}}e^{\mathrm{i}\delta}$, where δ is anything real. Taking $\delta = 0$; $c_6 = \frac{1}{\sqrt{2}}$

Thus a normalized Eigen vector of J_y is $\begin{pmatrix} \mathrm{i}c_6 \\ c_6 \end{pmatrix} = \begin{pmatrix} \mathrm{i}\frac{1}{\sqrt{2}} \\ \frac{1}{\sqrt{2}} \end{pmatrix} = \frac{1}{\sqrt{2}}\begin{pmatrix} \mathrm{i} \\ 1 \end{pmatrix}$; belonging to the Eigen value $-\hbar/2$.

4.18 Spin up and Spin down States

For $j = \frac{1}{2}$, we have $m = +\frac{1}{2}$ and $-\frac{1}{2}$. Thus $J_z = m\hbar = \pm\frac{1}{2}\hbar = \pm 0.5\hbar$. Therefore

$J = \sqrt{j(j+1)}\hbar = \sqrt{\frac{1}{2}\left(\frac{1}{2}+1\right)}\,\hbar = \frac{\sqrt{3}}{2}\hbar \approx 0.86\hbar$. Thus there can be only two allowed orientations of j at an angle

$\cos^{-1}\left(\frac{\frac{1}{2}\hbar}{\frac{\sqrt{3}}{2}\hbar}\right) = \cos^{-1}\left(\frac{1}{\sqrt{3}}\right) = 54^{\circ}$ with +ve z-axis and with -ve z-axis.

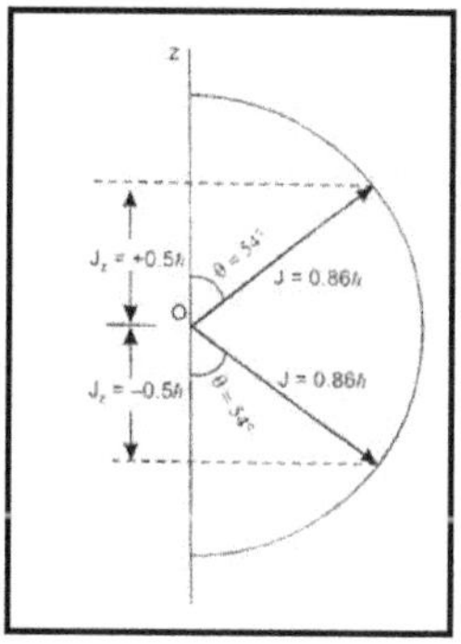

Although $\vec{J}$ cannot actually point along +ve or -ve z-axis, the two allowed orientations are still referred to as "spin-up" and "spin-down" states[22,23].

4.19 Arbitrary spin state

For a particle having $j = \frac{1}{2}$ and we do not know if it is in spin up state Ψ_1 or in spin down state Ψ_2. Thus we can write its state Ψ as

$$\Psi = c_1\Psi_1 + c_2\Psi_2 = c_1\begin{pmatrix}1\\0\end{pmatrix} + c_2\begin{pmatrix}0\\1\end{pmatrix}, \qquad \text{[Where } \Psi_1 = \begin{pmatrix}1\\0\end{pmatrix}\ and\ \Psi_2 = \begin{pmatrix}0\\1\end{pmatrix}]$$

Where Eigen vectors of J_z are $\begin{pmatrix}1\\0\end{pmatrix}$ and $\begin{pmatrix}0\\1\end{pmatrix}$; belonging to the Eigen value $+\frac{1}{2}\hbar$ and $-\frac{1}{2}\hbar$ respectively. Thus

$$\Psi = \begin{pmatrix}c_1\\0\end{pmatrix} + \begin{pmatrix}0\\c_2\end{pmatrix} = \begin{pmatrix}c_1\\c_2\end{pmatrix}.$$

From Normalization condition:

$$(c_1^\times \quad c_2^\times)\begin{pmatrix}c_1\\c_2\end{pmatrix} = 1\ or, |c_1|^2 + |c_2|^2 = 1$$

Where $|c_1|^2$ is probability that J_z is $+\frac{1}{2}\hbar$, I.e., the particle is in spin up state.

And $|c_2|^2$ is probability that J_z is $-\frac{1}{2}\hbar$, I.e., the particle is in spin down state. The expectation value of $\langle J_z\rangle$

$$\langle J_z\rangle = (\Psi, J_z\Psi) = (c_1^\times \quad c_2^\times)\frac{1}{2}\hbar\begin{pmatrix}1 & 0\\0 & -1\end{pmatrix}\begin{pmatrix}c_1\\c_2\end{pmatrix}$$

$$= \frac{1}{2}\hbar(c_1^\times \quad c_2^\times)\begin{pmatrix}1 & 0\\0 & -1\end{pmatrix}\begin{pmatrix}c_1\\c_2\end{pmatrix}$$

$$= \frac{1}{2}\hbar(c_1^\times \quad c_2^\times)\begin{pmatrix}c_1\\-c_2\end{pmatrix}$$

$$= \frac{1}{2}\hbar(c_1^\times c_1 - c_2^\times c_2)$$

$$= \frac{1}{2}\hbar|c_1|^2 + \left(-\frac{1}{2}\hbar\right)|c_2|^2$$

$$\langle J_z\rangle = \left(\frac{1}{2}\hbar\right)(probability\ of\ spin\ up) + \left(-\frac{1}{2}\hbar\right)(probability\ of\ spin\ down).$$

This is a weighted average.

Again the expectation of the generalized angular momentum

$$\langle J^2\rangle = (\Psi, J^2\Psi)$$

$$= (c_1^\times \quad c_2^\times)\frac{3}{4}\hbar^2\begin{pmatrix}1 & 0\\0 & 1\end{pmatrix}\begin{pmatrix}c_1\\c_2\end{pmatrix}$$

$$= \frac{3}{4}\hbar^2(c_1^\times \quad c_2^\times)\begin{pmatrix}1 & 0\\0 & 1\end{pmatrix}\begin{pmatrix}c_1\\c_2\end{pmatrix}$$

$$= \frac{3}{4}\hbar^2(c_1^\times \quad c_2^\times)\begin{pmatrix}c_1\\c_2\end{pmatrix}$$

$$= \frac{3}{4}\hbar^2(c_1^\times c_1 - c_2^\times c_2)$$

$$= \frac{3}{4}\hbar^2(|c_1|^2 + |c_2|^2)$$

$$= \frac{3}{4}\hbar^2(1) = \frac{3}{4}\hbar^2$$

Finally, $\frac{3}{4}\hbar^2$ *is the only possible value of* J^2 *for* $j = \frac{1}{2}$

Exercise

1. Explain the importance of matrix formulation in Quantum Mechanics.
2. Define unitary operator and unitary transformation. What is the advantage of unitary transformation?
3. Prove Some properties of unitary transformation:
   (a) If an observable Ω is Hermitian, then Ω′[its transformed operator] is also Hermitian.
   (b) Expectation values of transformed observables remain the same.
   (c) The Eigen values of observables [Ω] and transformed observables [Ω′] are the same.
4. Show the following arguments:

(a) If two matrices (Ω *and* Φ) have the same number of rows and columns then the matrices are said to be equal. I.e., $\Omega_{mn} = \Phi_{mn}$

(b) Two matrices (Ω *and* Φ) can be added if and only if they have the same number of rows and columns. $P_{mn} = \Omega_{mn} + \Phi_{mn}$ *Or*, $P_{mn} = (\Omega + \Phi)_{mn}$

(c) The product of two matrices (Ω *and* Φ) is given by $P_{mn} = (\Omega\Phi)_{mn}$
Where, $P_{mn} = \sum_k \Omega_{mk}\, \phi_{kn} = \sum_i \Omega_{mi}\, \phi_{in}$

5. Write some properties of matrices and properties of matrix Operators.
6. How can be diagonalized of a matrix?
7. Establish the Matrix representation of wave function and operator.
8. Write the technique of diagonalization of a matrix.
9. Write the Matrix form of Eigen value equation.

10. Find the Eigen value, Eigen function and Diagonalization of some given matrices:

(a) $\Omega = \begin{pmatrix} 0 & i & 0 \\ -i & 0 & 0 \\ 0 & 0 & 0 \end{pmatrix}$ (b) $\Omega = \begin{pmatrix} 0 & 1 & 0 \\ 1 & 0 & 0 \\ 0 & 0 & 1 \end{pmatrix}$ (c) $\Omega = \begin{pmatrix} 5 & 0 & \sqrt{3} \\ 0 & 3 & 0 \\ \sqrt{3} & 0 & 3 \end{pmatrix}$

(d) $\Omega = \begin{pmatrix} 0 & i \\ -i & 0 \end{pmatrix}$ (e) $\Omega = \begin{pmatrix} 1 & i \\ -i & 1 \end{pmatrix}$ (f) $\Omega = \begin{pmatrix} 3 & 2 \\ 2 & 1 \end{pmatrix}$

11. Show that the Eigen vectors of an operator in the basis of Eigen functions.

12. Find the Matrix operators of angular momentum components for $j = \frac{1}{2}$.

13. Establish the Commutation relations of the matrix operators of angular momentum components for $j = 1/2$.

14. Write down the Diagonalization of the angular momentum matrices for $j = \frac{1}{2}$.

15. Define Pauli spin matrices.

16. For $= 1/2$; Find the Eigen vectors of generalized angular momentum and its components J^2, J_x, J_y, J_z.

17. Explain the following states:

(a) Arbitrary spin state (b) Spin up and Spin down States

18. Show that the Hermitian nature of a matrix is invariant under unitary transformation.

19. Show that the Eigen values of a Hermitian matrix are real and Eigen vectors corresponding to different Eigen values are orthogonal.

Chapter 5 Some application of Quantum mechanics: Solution of simple one dimensional problems

5.1 Potential barrier

If the force field acting on a particle is zero or nearly zero everywhere except in a limited region,

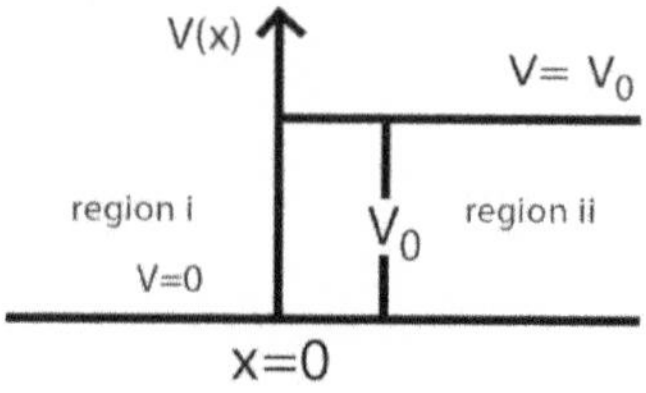

Fig.[5.1]: Shows a one dimensional single step Potential barrier

it is said to be a potential. A single step potential barrier is as shown in figure. At *x=0*, the face field action or a particle is v_0 is $F = -\frac{dV_0}{dx}$. Where v_0 is the height of the potential barrier.

$$V(x) = 0 \quad ; \quad \text{When} \quad x < 0.$$
$$V(x) = V_0 \quad ; \quad \text{When} \quad x > 0.$$

5.2 Quantum theory for single step potential barrier

Suppose a beam of particles travelling parallel to the x - axis from the left to the right is incident or the potential barrier as shown in figure.

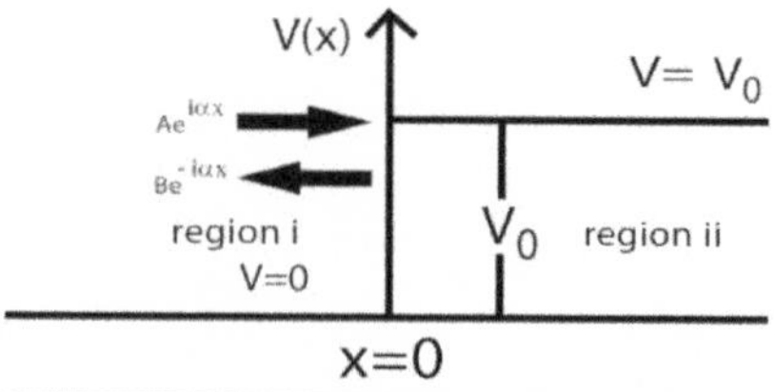

Fig.[5.2]: Shows Quantum theory for one dimensional single step Potential barrier

When this wave packet or beam of particles reaches the barrier, a part is reflected & a part is transmitted. The reflected part of the wave packet would yield the probability that the particles were reflected while transmitted part would yield the probability that the particles are transmitted[17].

Case-A [E>V$_0$]: The Schrödinger equation for the first region; Where E > V$_0$ is

$$\frac{d^2\psi_1}{dx^2} + \frac{2mE}{\hbar^2}\psi_1 = 0 \quad [1]$$

Or, $$\frac{d^2\psi_1}{dx^2} + \alpha^2 \psi_1 = 0 \quad [2]$$

Where $$\alpha^2 = \frac{2mE}{\hbar^2} \quad [3]$$

The general solution of Schrödinger equation is

$$\psi_1 = Ae^{i\alpha x} + Be^{-i\alpha x} \quad [4]$$

Where A and B are constants of integration.

Case-B [Where $x > 0$ & V(x) =V$_0$]: The Schrödinger equation for the second region;

$$\frac{d^2\psi_2}{dx^2} + \frac{2m}{\hbar^2}(E - V_0)\psi_2 = 0 \quad [5]$$

$$or,\ \frac{d^2\psi_2}{dx^2} + \beta^2 \psi_2 = 0 \quad [6]$$

Where $$\beta^2 = \frac{2m}{\hbar^2}(E - V_0) \quad [7]$$

The general solution of equation [6] is

$$\psi_2 = Ce^{i\beta x} + De^{-i\beta x} \quad [8]$$

Where A and B are constants of integration.

We note from equation [4]

Where $e^{i\alpha x}$ represents the incident wave & $e^{-i\alpha x}$ represents the reflected wave.

And we note from equation [8]

Where $e^{i\beta x}$ represents the transmitted wave & $e^{-i\beta x}$ represents a wave moving in the negative x direction i.e. reflected wave.

But at $x = 0$ & after; there is no discontinuity in second region. Hence there should not at all arise a question of reflection in this region. Due to this fact, the term $e^{-i\beta x}$ is discarded from equation [8], hence equation [8] becomes

$$\psi_2 = Ce^{i\beta x} \quad [9]$$

Since ψ is continuous; we have the following boundary conditions

$(\psi_1)_{x=0} = (\psi_2)_{x=0}$ [10]

$(\frac{d\psi_1}{dx})_{x=0} = (\frac{d\psi_2}{dx})_{x=0}$ [11]

Apply these boundary conditions in equation [4] & [9], we have

$\mathrm{A + B = C}$ [12]

& $\alpha A - \alpha B = C\beta = (A+B)\beta$ [13]

Or, $\alpha A - A\beta = B\beta + \alpha B \quad or, \quad A(\alpha-\beta) = B(\alpha+\beta)$

Or, $B = \frac{(\alpha-\beta)}{(\alpha+\beta)} A$ [14]

Putting the value of B in equation [12]

$$\mathrm{A + B = C}$$

Or, $\mathrm{A} + \frac{(\alpha-\beta)}{(\alpha+\beta)} A = \mathrm{C}$

Or $C = A[\,1 + \frac{(\alpha-\beta)}{(\alpha+\beta)}] = [\frac{2\alpha}{(\alpha+\beta)}]A$

From equation [3] and [7] We have

$$\alpha = \left(\frac{2mE}{\hbar^2}\right)^{\frac{1}{2}} = \left(\frac{2m.\frac{1}{2}mv^2}{\hbar^2}\right)^{\frac{1}{2}} = \frac{mv}{\hbar}$$

Or, $v = \frac{\alpha\hbar}{m}$ [15]

Where v is the velocity of incident wave packets.

& $\beta = \left[\frac{2m\,(E-V_0)}{\hbar^2}\right]^{\frac{1}{2}} = \left[\frac{2m.\frac{1}{2}mv_1{}^2}{\hbar^2}\right]^{\frac{1}{2}}$

Or, $\beta = \frac{mv_1}{\hbar^2}$ $or\ v_1 = \frac{\beta\hbar}{m}$ [16]

From $= \left(\frac{2mE}{\hbar^2}\right)^{\frac{1}{2}}$ $and\ \beta = \left[\frac{2m\,(E-V_0)}{\hbar^2}\right]^{\frac{1}{2}}$; we easily write $\beta < \alpha$. So that $v_1 < v$. Thus the velocity of the transmitted particles in such a region is less than the incident velocity. Whereas the velocity of the reflected particles are the same as that of the incident beam. If N_i be the incident flux or current density is defined as

$$N_i = (A^{\times}A)v = |A|^2 v$$

Or $$N_i = \frac{\alpha\hbar}{m}|A|^2 \quad [17]$$

[From the relation of current density & the probability density I.e., $J = \rho v = (\psi^{\times}\psi)v = |\psi|^2 v$]

& the reflected flux N_r *is*

$$N_r = |B|^2 v = \left|\frac{\alpha-\beta}{\alpha+\beta} A\right|^2 v = \left|\frac{\alpha-\beta}{\alpha+\beta}\right|^2 |A|^2 v$$

Or, $$N_r = \left|\frac{\alpha-\beta}{\alpha+\beta}\right|^2 - \left(\frac{\alpha\hbar}{m}\right)|A|^2$$

Or, $$N_r = \left|\frac{\alpha-\beta}{\alpha+\beta}\right|^2 N_i \quad \text{[From equation17]} \quad [18]$$

& the transmitted flux is

$$N_T = |C|^2 v_1 = \left|\frac{2\alpha}{\alpha+\beta}\right|^2 |A|^2 v_1$$

Or, $$N_T = \left|\frac{2\alpha}{\alpha+\beta}\right|^2 |A|^2 \frac{\beta\hbar}{m} = \frac{4\alpha}{(\alpha+\beta)^2}\frac{\alpha\beta\hbar}{m}|A|^2$$

Or, $$N_T = \frac{4\alpha\beta}{(\alpha+\beta)^2}\frac{\alpha\hbar}{m}|A|^2$$

Or, $$N_T = \frac{4\alpha\beta}{(\alpha+\beta)^2} N_i \text{ (Form equation 17)} \quad [19]$$

Now from equation [18] & [19]

$$N_r + N_T = N_i\left[\frac{(\alpha-\beta)^2}{(\alpha+\beta)^2} + \frac{4\alpha\beta}{(\alpha+\beta)^2}\right]$$

$$= N_i\left[\frac{(\alpha-\beta)^2 + 4\alpha\beta}{(\alpha+\beta)^2}\right]$$

$$= N_i\left[\frac{(\alpha+\beta)^2}{(\alpha+\beta)^2}\right]$$

Or, $$N_i = N_r + N_T \quad [20]$$

Equation [20] represents that the total number of particles is conserved. Therefore the reflection coefficient

$$R = \frac{\textit{Reflected Flux}}{\textit{Incident Flux}} = \frac{N_r}{N_i} = \frac{(\alpha-\beta)^2}{(\alpha+\beta)^2} \quad [21]$$

& Transmission coefficient

$$T = \frac{\text{Transmission } Flux}{Incident\ Flux} = \frac{N_T}{N_i} = \frac{4\alpha\beta}{(\alpha+\beta)^2} \qquad [22]$$

Now adding equations [21] & [22], we have

$$T + R = \frac{4\alpha\beta}{(\alpha+\beta)^2} + \frac{(\alpha-\beta)^2}{(\alpha+\beta)^2}$$

$$= \frac{4\alpha\beta + (\alpha-\beta)^2}{(\alpha+\beta)^2} = \frac{(\alpha+\beta)^2}{(\alpha+\beta)^2}$$

Finally, $$\boxed{\mathbf{T + R = 1}} \qquad [23]$$

5.3 Penetration of a potential barrier: TUNNEL EFECT [17]

A rectangular potential barrier of height V_0 & width 'a' for a particle has shown in figure. It extends over the region II from $x = 0\ to\ x = a$ in which the potential energy V of the particle will be constant & equal to V_0. On the both side of the barrier $I.e.$, in region I & III, the potential energy $V = 0$; It means that when the particle is in these regions no forces act on it.

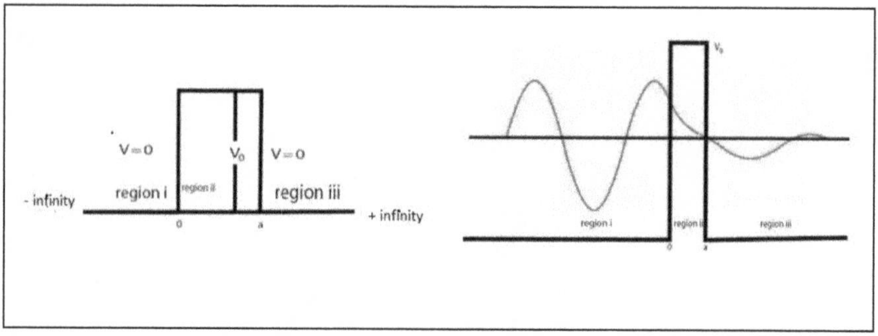

Fig.[5.3]: Shows A rectangular Potential barrier

Suppose a beam of particles travelling parallel to the X- axis from left to right is incident or the potential barrier. In the region I & III, the energy E of a particle is fully kinetic & in the region II it is partly kinetic & partly potential. If $< V_0$; *then according to classical mechanics , the probability of any particle reaching the region III after crossing the region II is Zero. However, according to Q.M the transmission probability has a small but definite value.* Thus if a particle impinging on the barrier with energy certainly less than the height of potential barrier $[E < V_0]$, it will not necessarily be reflected by the barrier but there is always a probability that it may cross the barrier & continue its forward motion. *This probability of crossing the barrier is called the Tunnel effect* [17].

Case – A [$-\infty < x < 0$]: The Schrödinger equation for the first region, where $V = 0$.

$\therefore$ $$\frac{d^2\psi_1}{dx^2} + \frac{2mE}{\hbar^2}\psi_1 = 0 \qquad [1]$$

Or $$\frac{d^2\psi_1}{dx^2} + \alpha^2\psi_1 = 0 \qquad [2]$$

Where $$\alpha^2 = \frac{2mE}{\hbar^2} \qquad [3]$$

The solution of equation [2] is

$$\psi_1 = Ae^{i\alpha x} + Be^{-i\alpha x} \qquad [4]$$

Where A & B are constant of integrations.

Case-B [$0 \le x \le a$]: The Schrödinger equation for the second region, where $V = V_0$.

$\therefore$ $$\frac{d^2\psi_2}{dx^2} - \frac{2m}{\hbar^2}(V_0 - E)\psi_2 = 0 \qquad [5]$$

Or, $$\frac{d^2\psi_2}{dx^2} - \beta^2\psi_2 = 0 \qquad [6]$$

Where $$\beta^2 = \frac{2m}{\hbar^2}(V_0 - E) \qquad [7]$$

Equation [6] has the solution

$$\psi_2 = Ce^{\beta x} + De^{-\beta x} \qquad [8]$$

Case –C [$a \le x \le \infty$]: The Schrödinger equation for the third region, where $V = 0$.

$\therefore$ $$\frac{d^2\psi_3}{dx^2} + \frac{2mE}{\hbar^2}\psi_3 = 0 \qquad [9]$$

Or, $$\frac{d^2\psi_3}{dx^2} - \alpha^2\psi_3 = 0 \qquad [10]$$

Equation [10] has the general solution

$$\psi_3 = Fe^{i\alpha x} + Ge^{-i\alpha x} \qquad [11]$$

Since there is no reflection in region III, hence the terms $Ge^{-i\alpha x}$ is discarded.

$$\psi_3 = Fe^{i\alpha x} \qquad [12]$$

Where, F is a constant of integration. We note from equation [4];

$e^{i\alpha x}$ is the incident wave & $e^{-i\alpha x}$ is the reflected wave.

And from equation [8];

$e^{-\beta x}$ is an exponentially decreasing wave function representing a non-oscillatory disturbance which more through the barrier in the positive X-direction.

And $e^{\beta x}$ is reflected disturbance within the barrier.

In order that ψ & $\frac{d\psi}{dx}$ be continuous at $x = 0$, we have

$$(\psi_1)_{x=0} = (\psi_2)_{x=0}$$

And $$(\frac{d\psi_1}{dx})_{x=0} = (\frac{d\psi_2}{dx})_{x=0}$$

Apply these boundary conditions in equations [4] & [(8]

$$A + B = C + D \quad [13]$$

And $$i\alpha A - i\alpha B = \beta C - \beta D$$

Or, $$i\alpha\,(A - B) = \beta C - \beta D$$

Or $$(A - B) = \frac{\beta}{i\alpha}(C - D) \quad [14]$$

Now $$A + B = C + D$$

And $$A - B = \frac{\beta}{i\alpha}(C - D)$$

∴ $$2A = C + D + \frac{\beta}{i\alpha}C - \frac{\beta}{i\alpha}D = C + D - \frac{i\beta}{\alpha}C + \frac{i\beta}{\alpha}D$$

$$= \left(1 - \frac{i\beta}{\alpha}\right)C + (1 + \frac{i\beta}{\alpha})D$$

∴ $$A = \left(1 - \frac{i\beta}{\alpha}\right)\frac{C}{2} + (1 + \frac{i\beta}{\alpha})\frac{D}{2} \quad [15]$$

Again in order that ψ & $\frac{d\psi}{dx}$ be continuous at $x = a$, we must have

$$(\psi_2)_{x=a} = (\psi_3)_{x=a}$$

And $$(\frac{d\psi_2}{dx})_{x=a} = (\frac{d\psi_3}{dx})_{x=a}$$

∴ $$Ce^{\beta a} + De^{-\beta a} = Fe^{i\alpha a} \quad [16]$$

And $$\beta C\,e^{\beta a} - \beta De^{-\beta a} = i\alpha Fe^{i\alpha a} \quad [17]$$

Or, $$\beta(C\,e^{\beta a} - De^{-\beta a}) = i\alpha Fe^{i\alpha a}$$

Or, $$C\,e^{\beta a} - De^{-\beta a} = \frac{i\alpha}{\beta}Fe^{i\alpha a} \qquad [18]$$

From equations [16] & [18]

$$2C\,e^{\beta a} = (1+\frac{i\alpha}{\beta})Fe^{i\alpha a}$$

Or, $$C = (1+\frac{i\alpha}{\beta})\frac{F}{2}e^{(i\alpha-\beta)a} \qquad [19]$$

And subtracting

$$2D\,e^{-\beta a} = (1-\frac{i\alpha}{\beta})Fe^{i\alpha a}$$

Or, $$D = (1-\frac{i\alpha}{\beta})\frac{F}{2}e^{(i\alpha+\beta)a} \qquad [20]$$

From equation [15] we rewrite

$$2A = \left(1+\frac{\beta}{i\alpha}\right)c + (1-\frac{\beta}{i\alpha})D$$

$$= [\left(1+\frac{\beta}{i\alpha}\right)\left(1+\frac{i\alpha}{\beta}\right)\frac{1}{2}e^{-\beta a} + \left(1-\frac{\beta}{i\alpha}\right)\left(1-\frac{i\alpha}{\beta}\right)\frac{1}{2}e^{\beta a}]Fe^{i\alpha a}$$

$$= [\frac{1}{2}\left(1+\frac{i\alpha}{\beta}+\frac{\beta}{i\alpha}+1\right)e^{-\beta a} + \frac{1}{2}\left(1-\frac{i\alpha}{\beta}-\frac{\beta}{i\alpha}+1\right)e^{\beta a}]Fe^{i\alpha a}$$

$$= [\left\{1+\frac{1}{2}\left(\frac{i\alpha}{\beta}+\frac{\beta}{i\alpha}\right)\right\}e^{-\beta a} + \left\{1-\frac{1}{2}\left(\frac{i\alpha}{\beta}+\frac{\beta}{i\alpha}\right)\right\}e^{\beta a}]Fe^{i\alpha a}$$

$$= [e^{-\beta a} + \frac{1}{2}\left(\frac{i\alpha}{\beta}+\frac{\beta}{i\alpha}\right)e^{-\beta a} + e^{\beta a} - \frac{1}{2}\left(\frac{i\alpha}{\beta}+\frac{\beta}{i\alpha}\right)e^{\beta a}]Fe^{i\alpha a}$$

$$= [e^{\beta a} + e^{-\beta a} - \frac{1}{2}\left(\frac{i\alpha}{\beta}+\frac{\beta}{i\alpha}\right)(e^{\beta a} - e^{-\beta a})]Fe^{i\alpha a}$$

$$\mathrm{A} = \left[\frac{e^{\beta a}+e^{-\beta a}}{2} - \frac{1}{2}\left(\frac{i\alpha}{\beta}+\frac{\beta}{i\alpha}\right)\left(\frac{e^{\beta a}-e^{-\beta a}}{2}\right)\right]Fe^{i\alpha a}$$

$$\mathrm{A} = \left[\cosh\beta a - \frac{1}{2}\left(\frac{i\alpha}{\beta}+\frac{\beta}{i\alpha}\right)\sinh\beta a\right]Fe^{i\alpha a}$$

$$\mathrm{A} = \left[\cosh\beta a - \frac{1}{2}\left(\frac{i\alpha}{\beta}+\frac{i\beta}{\alpha}\right)\sinh\beta a\right]Fe^{i\alpha a}$$

$$\mathrm{A} = \left[\cosh\beta a + \frac{i}{2}\left(\frac{\beta}{\alpha}-\frac{\alpha}{\beta}\right)\sin h\beta a\right]Fe^{i\alpha a}$$

Or, $$\frac{\mathrm{A}}{F} = \left[\cos h\beta a + \frac{i}{2}\left(\frac{\beta}{\alpha}-\frac{\alpha}{\beta}\right)\sinh\beta a\right]e^{i\alpha a} \qquad [21]$$

And $\left(\frac{A}{F}\right)^{\times} = \left[\cos h\beta a - \frac{i}{2}\left(\frac{\beta}{\alpha} - \frac{\alpha}{\beta}\right)\sin h\beta a\right] e^{-i\alpha a}$ [22]

Similarly,

$$2B = C + D + \frac{i\beta}{\alpha}c - \frac{i\beta}{\alpha}D$$

$$= \left(1 + \frac{i\beta}{\alpha}\right)C + \left(1 - \frac{i\beta}{\alpha}\right)D$$

$$= \left[\left(1 + \frac{i\beta}{\alpha}\right)\frac{1}{2}\left(1 + \frac{i\alpha}{\beta}\right)e^{-\beta a} + \left(1 - \frac{i\beta}{\alpha}\right)\frac{1}{2}\left(1 - \frac{i\alpha}{\beta}\right)e^{-\beta a}\right]Fe^{i\alpha a}$$

$$= \left[\left(1 - \frac{\beta}{i\alpha}\right)\frac{1}{2}\left(1 + \frac{i\alpha}{\beta}\right)e^{-\beta a} + \left(1 + \frac{\beta}{i\alpha}\right)\frac{1}{2}\left(1 - \frac{i\alpha}{\beta}\right)e^{\beta a}\right]Fe^{i\alpha a}$$

$$= \left[\frac{1}{2}\left(1 + \frac{i\alpha}{\beta} - \frac{\beta}{i\alpha} - 1\right)e^{-\beta a} + \frac{1}{2}\left(1 - \frac{i\alpha}{\beta} + \frac{\beta}{i\alpha} - 1\right)e^{\beta a}\right]Fe^{i\alpha a}$$

$$= \left[-\frac{1}{2}\left(\frac{\beta}{i\alpha} - \frac{i\alpha}{\beta}\right)e^{-\beta a} + \frac{1}{2}\left(\frac{\beta}{i\alpha} - \frac{i\alpha}{\beta}\right)e^{\beta a}\right]Fe^{i\alpha a}$$

$$= \left[\left(\frac{\beta}{i\alpha} - \frac{i\alpha}{\beta}\right)\left(\frac{e^{\beta a} - e^{-\beta a}}{2}\right)\right]Fe^{i\alpha a}$$

$$= \left(\frac{\beta}{i\alpha} - \frac{i\alpha}{\beta}\right)\sin h\beta a\ Fe^{i\alpha a}$$

$$= \left(-\frac{i\beta}{\alpha} - \frac{i\alpha}{\beta}\right)\sin h\beta a\ Fe^{i\alpha a}$$

$$= -\left(\frac{i\beta}{\alpha} + \frac{i\alpha}{\beta}\right)\sin h\beta a\ Fe^{i\alpha a}$$

$\therefore$ $B = \frac{i}{2}\left(\frac{\beta}{\alpha} + \frac{\alpha}{\beta}\right)\sinh \beta a\ Fe^{i\alpha a}$ [23]

Therefore, $\frac{1}{T} = \left(\frac{A}{F}\right)\left(\frac{A}{F}\right)^{\times}$

$$= \left[\cosh \beta a + \frac{i}{2}\left(\frac{\beta}{\alpha} - \frac{\alpha}{\beta}\right)\sin h\beta a\right]e^{i\alpha a}\left[\cosh \beta a - \frac{i}{2}\left(\frac{\beta}{\alpha} - \frac{\alpha}{\beta}\right)\sinh \beta a\right]e^{-i\alpha a}$$

$$= Cosh^2\beta a + \frac{1}{4}\left(\frac{\beta}{\alpha} - \frac{\alpha}{\beta}\right)^2 Sinh^2\beta a$$

$$= 1 + Sinh^2\beta a + \frac{1}{4}\left(\frac{\beta}{\alpha} - \frac{\alpha}{\beta}\right)^2 Sinh^2\beta a = 1 + \left[1 + \frac{1}{4}\left(\frac{\beta}{\alpha} - \frac{\alpha}{\beta}\right)^2\right]Sinh^2\beta a$$

$$= 1 + \frac{1}{4}\left[\left(\frac{\beta}{\alpha} - \frac{\alpha}{\beta}\right)^2 + 4\right]Sinh^2\beta a$$

$$\frac{1}{T} = 1 + \frac{1}{4}\left[\left(\frac{\beta}{\alpha} + \frac{\alpha}{\beta}\right)^2\right] Sinh^2\beta a \qquad [24]$$

$$T = \frac{1}{1+\frac{1}{4}\left[\left(\frac{\beta}{\alpha}+\frac{\alpha}{\beta}\right)^2\right]Sinh^2\beta a} \qquad [25]$$

Again dividing equation [23] by [21]

$$\frac{B}{A} = \frac{\left[-\frac{i}{2}\left(\frac{\beta}{\alpha}+\frac{\alpha}{\beta}\right)\sinh\beta a\right]Fe^{i\alpha a}}{\left[\cos\beta a+\frac{i}{2}\left(\frac{\beta}{\alpha}-\frac{\alpha}{\beta}\right)\sin\beta a\right]Fe^{i\alpha a}}$$

Or, $$\frac{B}{A} = \frac{-\frac{i}{2}\left(\frac{\beta}{\alpha}+\frac{\alpha}{\beta}\right)\sinh\beta a}{\cos\beta a+\frac{i}{2}\left(\frac{\beta}{\alpha}-\frac{\alpha}{\beta}\right)\sin\beta a}$$

And $$\left(\frac{B}{A}\right)^{\times} = \frac{\frac{i}{2}\left(\frac{\beta}{\alpha}+\frac{\alpha}{\beta}\right)\sinh\beta a}{\cos\beta a-\frac{i}{2}\left(\frac{\beta}{\alpha}-\frac{\alpha}{\beta}\right)\sin\beta a}$$

Hence the reflection coefficient

$$R = \left(\frac{B}{A}\right)\left(\frac{B}{A}\right)^{\times}$$

$$= \frac{-\frac{i}{2}\left(\frac{\beta}{\alpha}+\frac{\alpha}{\beta}\right)\sinh\beta a}{\cos\beta a+\frac{i}{2}\left(\frac{\beta}{\alpha}-\frac{\alpha}{\beta}\right)\sin\beta a} \times \frac{\frac{i}{2}\left(\frac{\beta}{\alpha}+\frac{\alpha}{\beta}\right)\sinh\beta a}{\cos\beta a-\frac{i}{2}\left(\frac{\beta}{\alpha}-\frac{\alpha}{\beta}\right)\sin\beta a}$$

$$= \frac{\frac{1}{4}\left(\frac{\beta}{\alpha}+\frac{\alpha}{\beta}\right)^2 \text{Sinh}^2\beta a}{\text{Cos}^2\,\beta a+\frac{1}{4}\left(\frac{\beta}{\alpha}-\frac{\alpha}{\beta}\right)\text{Sinh}^2\,\beta a}$$

$$= \frac{\frac{1}{4}\left(\frac{\beta}{\alpha}+\frac{\alpha}{\beta}\right)^2 \text{Sinh}^2\beta a}{1+\text{Sinh}^2\,\beta a+\frac{1}{4}\left(\frac{\beta}{\alpha}-\frac{\alpha}{\beta}\right)^2\text{Sinh}^2\,\beta a} = \frac{\frac{1}{4}\left(\frac{\beta}{\alpha}+\frac{\alpha}{\beta}\right)^2 \text{Sinh}^2\beta a}{1+\left[1+\frac{1}{4}\left(\frac{\beta}{\alpha}-\frac{\alpha}{\beta}\right)^2\right]\text{Sinh}^2\,\beta a}$$

$$R = \frac{\frac{1}{4}\left(\frac{\beta}{\alpha}+\frac{\alpha}{\beta}\right)^2 \text{Sinh}^2\beta a}{1+\frac{1}{4}\left[\left(\frac{\beta}{\alpha}-\frac{\alpha}{\beta}\right)^2+4\right]\text{Sinh}^2\,\beta a} = \frac{\frac{1}{4}\left(\frac{\beta}{\alpha}+\frac{\alpha}{\beta}\right)^2 \text{Sinh}^2\beta a}{1+\frac{1}{4}\left[\left(\frac{\beta}{\alpha}+\frac{\alpha}{\beta}\right)^2\right]\text{Sinh}^2\,\beta a}$$

Now adding equation [25] and [26], we have

$$\mathbf{T+R} = \frac{1}{1+\frac{1}{4}\left[\left(\frac{\beta}{\alpha}+\frac{\alpha}{\beta}\right)^2\right]Sinh^2\beta a} + \frac{\frac{1}{4}\left(\frac{\beta}{\alpha}+\frac{\alpha}{\beta}\right)^2 \text{Sinh}^2\beta a}{1+\frac{1}{4}\left[\left(\frac{\beta}{\alpha}+\frac{\alpha}{\beta}\right)^2\right]\text{Sinh}^2\,\beta a} = \frac{1+\frac{1}{4}\left[\left(\frac{\beta}{\alpha}+\frac{\alpha}{\beta}\right)^2\right]\text{Sinh}^2\,\beta a}{1+\frac{1}{4}\left[\left(\frac{\beta}{\alpha}+\frac{\alpha}{\beta}\right)^2\right]\text{Sinh}^2\,\beta a}$$

$$\mathbf{T+R=1} \qquad [27]$$

5.4 Energy levels for one dimensional square-well potential of finite depth or finite potential well: **Bound State problems** [17,26].

Let the potential be equal to zero within a distance 'a' on either side of the origin & equal to $+V$ elsewhere.

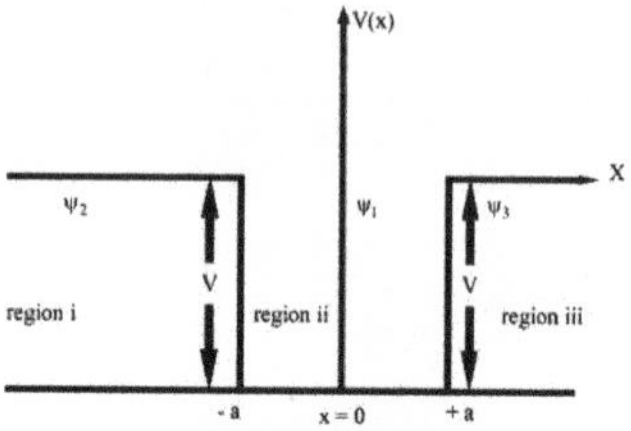

Fig.[5.4]: Shows Energy levels for one dimensional square-well potential of finite depth

Energy levels for one dimensional square-well potential of finite depth or finite potential well $i.e., V = 0$; for $-a < x < a$ and $V = +V$; for $-a > x > a$ *; Such a potential energy function is called a square-will potential.*

Case - A [$-\boldsymbol{a} < \boldsymbol{x} < \boldsymbol{a}(\boldsymbol{I.e.}, |\boldsymbol{x}|$ $< a)$]: For the motion of a particle along the x-axis inside the box; where $V(x) = 0$. The Schrödinger equation for the second region, where $V = 0$.

$$\frac{d^2\psi_1}{dx^2} + \frac{2mE}{\hbar^2}\psi_1 = 0 \qquad [1]$$

Or,

$$\frac{d^2\psi_1}{dx^2} + \alpha^2\psi_1 = 0 \qquad [2]$$

Where

$$\alpha^2 = \frac{2mE}{\hbar^2} \qquad [3]$$

The solution of equation [2] is

$$\psi_1 = Acos\alpha x + Bsin\alpha x \qquad [4]$$

Case - B [$-\boldsymbol{\alpha} > \boldsymbol{x} > \boldsymbol{a}$ $(\boldsymbol{I.e.}, |\boldsymbol{x}|$ $> a)$]: The Schrödinger equation for the first and third regions I.e., outside of the square well regions, where $V = +V$.

$$\frac{d^2\psi_2}{dx^2} - \frac{2m}{\hbar^2}(V - E)\psi_2 = 0 \qquad [5]$$

Or, $$\frac{d^2\psi_2}{dx^2} - \beta^2\psi_2 = 0 \quad [6]$$

Where $$\beta^2 = \frac{2m}{\hbar^2}(V - E\,) \quad [6a]$$

And the wave equation for third region

$$\frac{d^2\psi_3}{dx^2} - \beta^2\psi_3 = 0 \quad [7]$$

For Bound States $i.e\ \ E < V$; equation [6] and equation [7] have the general solutions

$$\psi_2 = Ce^{\beta x} + De^{-\beta x} \quad [8a]$$

And $$\psi_3 = Ce^{\beta x} + De^{-\beta x} \quad [8b]$$

Now Applying the boundary Conditions at $x = \pm a$ require that; in the region $x > a$; $C = 0$; and in the region $x < -a$; $D = 0$

Thus the solutions of the wave equations in different regions are

$$\psi_1 = Acos\alpha x + Bsin\alpha x \quad ; \quad -a < x < a \quad [9]$$

$$\psi_2 = Ce^{\beta x} \quad ; \quad x < -a \quad [10]$$

$$\psi_3 = De^{-\beta x} \quad ; \quad x > a \quad [11]$$

In order that ψ & $\frac{d\psi}{dx}$ be continuous at $x = +a$; we must have

$$(\psi_1)_{x=a} = (\psi_3)_{x=a}$$

Or, $$Acos\alpha a + Bsin\alpha a = De^{-\beta a} \quad [12]$$

And $$\left(\frac{d\psi_1}{dx}\right)_{x=a} = \left(\frac{d\psi_3}{dx}\right)_{x=a}$$

Or, $$-A\alpha\, sin\alpha a + B\alpha\, cos\alpha a = -D\beta e^{-\beta a} \quad [13]$$

Again in order that ψ & $\frac{d\psi}{dx}$ be continuous at $x = -a$; we must have

$$(\psi_1)_{x=-a} = (\psi_2)_{x=-a}$$

Or, $$Acos\alpha a - Bsin\alpha a = Ce^{-\beta a} \quad [14]$$

And $$\left(\frac{d\psi_1}{dx}\right)_{x=-a} = \left(\frac{d\psi_2}{dx}\right)_{x=-a}$$

Or, $$A\alpha\, sin\alpha a - B\alpha\, cos\alpha a = C\beta e^{-\beta a} \quad [15]$$

From equations [12] & [14]; we have

$$2A cos\alpha a = (C + D)\, e^{-\beta a} \qquad [16]$$

$$2B\ sin\alpha a = (D - C)\, e^{-\beta a} \qquad [17]$$

Similarly, from equations [13] and [15]; we have

$$2B\alpha\ cos\alpha a = (C - D)\beta e^{-\beta a} = -(D - C)\, \beta e^{-\beta a} \qquad [18]$$

$$2A\alpha\ sin\alpha a = (C + D)\, \beta e^{-\beta a} \qquad [19]$$

From equations [16] & [19]; we have

$$\alpha \tan \alpha a = \beta \qquad [20]$$

And from equations [16] & [18]; we have

$$\alpha \cot \alpha a = -\beta \qquad [21]$$

Now consider equation [20] & put

$$\xi = \alpha a \quad \& \quad \beta a = \eta$$

$$\alpha \tan \xi = \eta \qquad [22]$$

And $\xi^2 + \eta^2 = \alpha^2 a^2 + \beta^2 a^2 = a^2(\alpha^2 + \beta^2) = a^2\left(\frac{2mE}{\hbar^2} + \frac{2m}{\hbar^2}(V - E)\right) = \frac{2m}{\hbar^2}\, V a^2$ [23]

Now consider a special case in which V approaches infinity *I.e.*, the potential barrier is infinitely high.

$$\therefore\ \tan \alpha a = \frac{\beta}{\alpha} = \infty \text{ or, } \alpha a = (2n + 1)\frac{\pi}{2}$$

$$\frac{\beta}{\alpha} = \frac{\frac{2m}{\hbar^2}(V-E)}{\frac{2m}{\hbar^2}E} = \frac{(V-E)}{E} = \infty$$

$$\sqrt{\left(\frac{2mE}{\hbar^2}\right)}\ a = (2n + 1)\frac{\pi}{2} \quad or, \quad \frac{2mE}{\hbar^2}\ a^2 = \frac{(2n+1)^2}{4}\,\pi^2$$

The Characteristic energy values

$$E_n = \frac{(2n+1)^2}{8ma^2}\pi^2\hbar^2 \qquad [24]$$

And for the second group solution is

$$\cot \alpha a = -\frac{\beta}{\alpha} \quad and \ \tan \alpha a = -\frac{\alpha}{\beta} \approx 0 \ \ or, \alpha a = n\pi$$

$\frac{\alpha}{\beta} = \frac{E}{V-E} \approx \frac{1}{V} = \frac{1}{\infty} = 0$; When V > E

$$\left(\frac{2mE}{\hbar^2}\right)^{\frac{1}{2}} a = n\pi \quad or, \ \frac{2mE}{\hbar^2}\ a^2 = n^2\,\pi^2 \ \ or, \quad E_n = \frac{n^2\,\pi^2\hbar^2}{2ma^2}\,\pi^2$$

$$or, \quad E_n = \frac{(2n)^2 \pi^2 \hbar^2}{8ma^2} \pi^2 \qquad [25]$$

5.5 Energy levels of a particle enclosed with one dimensional rigid wall with infinite potential well

Consider a square-well with infinitely high sides as shown in figure. A particle bounded by impenetrable walls of width $2a$. From Figure

$$V(x) = 0\,; for - a < x < a \;\; or, \; |x| < a$$

And $$V(x) = \infty; \;\; for \;\; -a > x > a \;\; or, \; |x| > a$$

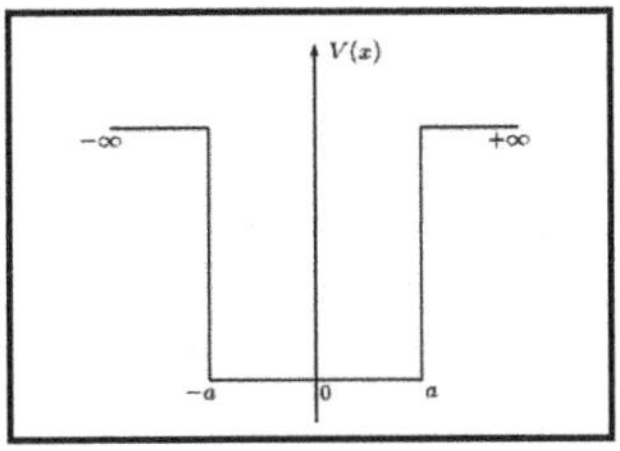

Fig.[5.5]: Shows a square-well with infinitely high sides

The proper boundary condition to be imposed is that the wave functions vanish at the walls.

$$(\psi)_{x=a} = 0 \;\; \& \;\; (\psi)_{x=-a} = 0\,; \; I.e., \qquad \psi_{(x)=0} \;\; when \; x = \pm a$$

Case- A $[-a < x < a\,; or\, |x| < a\,]$: For the Motion of the particle along the x-axis in the region $-a < x < a \;\; or, |x| < a$; I.e., inside the square-well or Box where $V(x) = 0$. Therefore the Schrödinger equation

$$\frac{d^2\psi(x)}{dx^2} + \frac{2m}{\hbar^2} E\psi(x) = 0 \qquad [1]$$

Or, $$\frac{d^2\psi(x)}{dx^2} + \alpha^2\psi(x) = 0 \qquad [2]$$

Where $$\alpha^2 = \frac{2m}{\hbar^2} E \qquad [3]$$

The general solution of equation [2] is

$$\psi(x) = A\; Cos\alpha x + B\; Sin\alpha x \qquad [4]$$

Where A & B are constants of integration.

Since ψ is Continuous; we have the following boundary conditions at $x = \pm a$. The proper boundary conditions to be imposed is that the wave functions vanish at the wall I.e.,

$$\psi(x)_{x=a} = 0 \quad \& \quad \psi(x)_{x=-a} = 0$$

Apply these boundary conditions in equation [4]

$$A\ Cos\alpha a + B\ Sin\alpha a = 0 \qquad [5]$$

$$A\ Cos(-\alpha a) + B\ Sin(-\alpha a) = 0$$

Or, $$A\ Cos\alpha a - B\ Sin\alpha a = 0 \qquad [6]$$

Now adding equations [5] & [6]; we have

$$2A\ Cos\alpha a = 0$$

$\therefore$ $$A = 0\ or,\ \ Cos\alpha a = 0 \qquad [7]$$

Again subtracting equations [5] & [6]; we have

$$2B\ Sin\alpha a = 0$$

$$B = 0\ or\ \ Sin\alpha a = 0 \qquad [8]$$

From equation [4]; it is clear that both constants A & B cannot be zero. Since this would give the physically uninteresting solutions $\psi = 0$. Also it is not possible to make both $Cos\alpha a$ & $Sin\alpha a$ zero for a given value of α & E. So there are two possible classes of solutions.

(1) For the first class solution:

$$A = 0\ \ \&\ Sin\alpha a = 0 = sin\ \frac{n\pi}{2}$$

$$\alpha a = \frac{n\pi}{2} \qquad [9]$$

Where n is even integer $I.e.,\ n = 2, 4, 6, \dots \dots \dots ..$ ....

(2) For the second class solution:

$$B = 0\ \ \&\ \ Cos\alpha a = 0 = Cos\ \frac{n\pi}{2}$$

Or, $$\alpha a = \frac{n\pi}{2} \qquad [10]$$

Where n is odd integer $i.e\ \ n = 1, 3, 5 \dots \dots \dots ....$ ....

Therefore from equation [3]; the energy again value

$$\alpha^2 = \frac{2mE}{\hbar^2} \qquad Or, \ \left(\frac{n\pi}{2a}\right)^2 = \frac{2m}{\hbar^2} E_n$$

$$E_n = \frac{\pi^2\hbar^2}{8ma^2} n^2 \tag{11}$$

Where $n = 1, 2, 3 \ldots \ldots \ldots \ldots \ldots \ldots \ldots \ldots \ldots \ldots \ldots \ldots \ldots \ldots \ldots \ldots \ldots \ldots \ldots$

Therefore, the general form of wave functions

$$\psi_n = A\, Cos\alpha x$$

$$= A\, Cos\frac{n\pi}{2a}x \qquad [when\, n\, is\, odd]$$

$$\psi_n = B\, Sin\alpha x$$

$$= B\, Sin\frac{n\pi}{2a}x \qquad [when\, n\, is\, even]$$

Form the normalization condition

$$\int_{-\infty}^{\infty} \psi^{\times}\psi dx = 1 \tag{12}$$

But in our problem equation [12] can be written as

$$\int_{-a}^{a} \psi^{\times}\psi dx = 1 \tag{13}$$

$$\int_{-a}^{+a} A^{\times} Cos\frac{n\pi x}{2a} A\, Cos\frac{n\pi x}{2a} dx = 1$$

$$\frac{|A|^2}{2}\int_{-a}^{+a} 2\, Cos^2\frac{n\pi x}{2a} dx = 1$$

$$\frac{|A|^2}{2}\int_{-a}^{+a} (1 + Cos\, 2\, \frac{n\pi x}{2a}) dx = 1$$

$$\frac{|A|^2}{2}\left[[x]_{-a}^{+a} + \frac{sin(2\frac{n\pi x}{2a}}{\frac{2n\pi}{2a}}\right]_{-a}^{+a} = 1$$

$$\frac{|A|^2}{2}[2a + sinn\pi] = 1$$

$$\frac{|A|^2}{2} 2a + 0 = 1 \qquad [\because Sin\pi,\ sin2\pi \ldots \ldots \ldots = 0]$$

$$\therefore \qquad |A|^2 = \frac{1}{a}\ and\quad A = \frac{1}{\sqrt{a}} \tag{14}$$

Similarly

$$B = \frac{1}{\sqrt{a}} \tag{15}$$

Hence the normalized wave function

$$\psi_n = A\, Cos \frac{n\pi}{2a} x$$

$$\psi_n = \frac{1}{\sqrt{a}}\, Cos \frac{n\pi}{2a} x \qquad [\, where\, n = 1, 3, 5 \ldots \ldots \ldots]$$

And $$\psi_n = \frac{1}{\sqrt{a}}\, Sin \frac{n\pi}{2a} x \qquad [\, where\, n = 2, 4, 6 \ldots \ldots \ldots]$$

5.6 The energy levels and corresponding normalized Eigen function of a particle in one dimensional potential-well of the form

$$V(x) = \infty ;\ for\ x < 0\ \ \&\ \ for\ \ x > 0$$

And $$V(x) = 0\ ;\ for\ 0 < x < a$$

For the motion of the particle along the x-axis in the region $0 < x < a$ $I.e.$, inside the square-well or box where $V(x) = 0$. The particle bounded by impenetrable walls of width 'a'. From figure or given

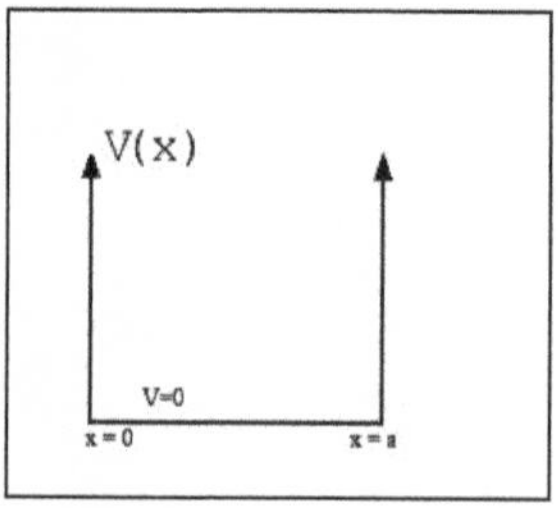

Fig.[5.6]: Shows one dimensional potential well

$$V(x) = \infty ;\ for\ \ x < 0\ \ \&\ \ for\ \ x > 0.$$

$$\&\ \ V(x) = 0\ ;\ for\ \ 0 < x < a.$$

The Schrödinger equation is

$$\frac{d^2\psi}{dx^2} + \frac{2m}{\hbar^2} E\psi = 0 \qquad [1]$$

$$\frac{d^2\psi}{dx^2} + \alpha^2\psi = 0 \qquad [2]$$

Where $$\alpha^2 = \frac{2mE}{\hbar^2} \qquad [3]$$

The general solution of equation [2] is

$$\psi = A\, Cos\alpha x + B\, Sin\alpha x \qquad [4]$$

Where A & B are Constants of integration.

Since ψ is Continuous; we have the following boundary conditions are

$$(i)\ \psi = 0 \quad at \quad x = 0$$

$$(ii)\ \psi = 0 \quad at \quad x = a$$

Apply these boundary conditions in equation [4]

$$0 = A\ Cos0 + B\ Sin0 \qquad \therefore\ A = 0$$

And $$0 = A\ Cos\alpha a + B\ Sin\alpha a$$

$$0 = B\ Sin\alpha a \qquad [\therefore A = 0]$$

$$B\ Sin\alpha a = 0 \quad But \quad B \neq 0$$

$\therefore$ $$Sin\alpha a = 0$$

Or, $$sin\alpha a = sin\, n\pi \quad or \quad \alpha a = n\pi \quad where\ \ n\ is\ even\ integer$$

Or, $$\alpha = \frac{n\pi}{a}$$

Therefore, the energy Eigen value is

$$\alpha^2 = \frac{2mE}{\hbar^2}$$

$$\left(\frac{n\pi}{a}\right)^2 = \frac{2mE_n}{\hbar^2}$$

$$E_n = \frac{\pi^2\hbar^2}{2ma^2}\, n^2 \qquad [5]$$

The corresponding Eigen function

$$\psi_n = B\ Sin\alpha x \qquad [6]$$

From the normalization condition

$$\int_{-\infty}^{\infty} \psi^{\times}\psi dx = 1 \qquad [7]$$

But in our problem equation [7] as

$$\int_0^a B^{\times} sin\alpha x\ Bsin\alpha x\ \ dx = 1$$

Or, $$|B|^2 \int_0^a sin^2\alpha x\ \ dx = 1$$

Or, $\frac{|B|^2}{2}\int_0^a 2sin^2\alpha x\ dx = 1$

Or, $\frac{|B|^2}{2}\int_0^a (1-cos2\alpha x)\ dx = 1$

Or, $\frac{|B|^2}{2}\left[x-\frac{sin2\alpha x}{2\alpha}\right]_0^a = 1$

Or, $\frac{|B|^2}{2}[x-0]_0^a = 1 \quad or,\ \frac{|B|^2}{2}a = 1 \quad or, B = \sqrt{\frac{2}{a}}$

$$\frac{\sin 2\frac{n\pi}{a}x}{\frac{2n\pi}{a}} = \frac{\sin(2n\pi)}{\frac{2n\pi}{a}} = \frac{a}{2n\pi}.sin2n\pi = 0$$

Hence the normalized wave function

$$\psi_n = Bsin\alpha x \quad Or, \quad \psi_n = \sqrt{\frac{2}{a}}\, sin\frac{n\pi}{a}x \qquad [8]$$

5.7 A particle of mass m is confined within one dimensional potential well defined by

$$V = -V_0\ ;\ \text{for } |x| < a$$

$$V = 0\ ;\ \text{for } |x| > a$$

The total energy of the particle is $E = -W$. Where W is positive. Assuming that the wave function of the particle in the lowest state(W_0) is an even function of x. Show that

$$\tan\beta a = \left(\frac{W_0}{V_0-W_0}\right)^{1/2};\ \text{Where } \beta^2 = \frac{8\pi^2 m}{\hbar^2}(V_0-W_0)$$

Consider the particle in the lowest state. For the lowest state $V = W_0$ & $E = 0$.The Schrödinger equation in the 1st region

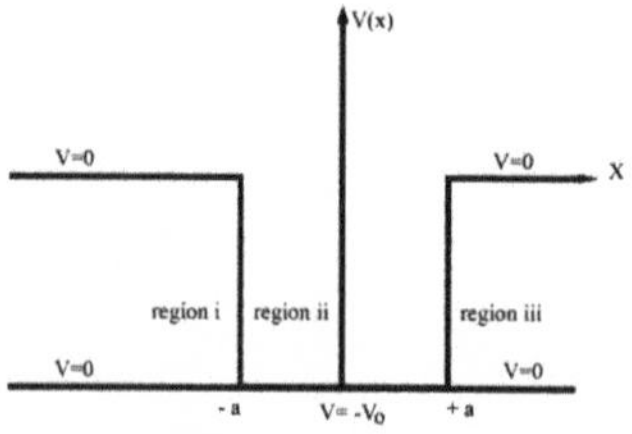

Fig.[5.7]: Shows a particle of mass m is confined within one dimensional potential well

$$\frac{d^2\psi_1}{dx^2}+\frac{2m}{\hbar^2}(E-V)\psi_1=0$$

Or,
$$\frac{d^2\psi_1}{dx^2}-\frac{2m}{\hbar^2}V\psi_1=0$$

Or,
$$\frac{d^2\psi_1}{dx^2}-\frac{2m}{\hbar^2}W_0\psi_1=0 \qquad [1]$$

$$\frac{d^2\psi_1}{dx^2}-\alpha^2\psi_1=0 \qquad [2]$$

Where
$$\alpha^2=\frac{2m}{\hbar^2}W_0 \qquad [3]$$

The wave equation for the second region is

$$\frac{d^2\psi_2}{dx^2}+\frac{2m}{\hbar^2}(E-V)\psi_2=0$$

$$\frac{d^2\psi_2}{dx^2}+\frac{2m}{\hbar^2}\big(-W_0-(-V_0)\big)\psi_2=0 \quad [\because E=-W]$$

$$\frac{d^2\psi_2}{dx^2}+\frac{2m}{\hbar^2}(V_0-W_0)\psi_2=0$$

$$\frac{d^2\psi_2}{dx^2}+\beta^2\psi_2=0$$

Where
$$\beta^2=\frac{2m}{\hbar^2}(V_0-W_0) \qquad [4]$$

The wave equation for the third region

$$\frac{d^2\psi_3}{dx^2}-\alpha^2\psi_3=0 \qquad [5]$$

The solution of wave equation for different regions is

$$\psi_1(x)=Ae^{\alpha x}+Be^{-\alpha x} \qquad [6]$$

$$\psi_1(x)=Ae^{\alpha x} \qquad [7]$$

[For the lowest state the term $Be^{-\alpha x}$ *can be dropped*]

$$\psi_2=Be^{i\beta x}+ce^{-i\beta x} \qquad [8]$$

$$\psi_3=Fe^{\alpha x}+Ge^{-\alpha x}$$

$$\psi_3=Ge^{-\alpha x} \qquad [9]$$

[According to the given condition, the 1st term is dropped]

In the lowest state, the wave function is an even function of x I.e., $\psi(x) = \psi(-x)$

[Note: $if\ f(-x) = f(x)\ then\ f(x)$ is called an even function.]

Hence, we have from equation [7] & [9] $A = G$

And from equation [8] $B = C$

$$\psi_2 = Be^{i\beta x} + ce^{-i\beta x} = Be^{i\beta x} + Be^{-i\beta x}$$

$$= B\left(e^{i\beta x} + e^{-i\beta x}\right) = 2B\left(\frac{e^{i\beta x} + e^{-i\beta x}}{2}\right)$$

$$\psi_2 = 2B\cos\beta x$$

$$\psi_2(x) = Be^{i\beta x} + ce^{-i\beta x}$$

$$\psi_2(-x) = Be^{-i\beta x} + ce^{i\beta x}$$

$$\psi_2(x) = \psi_2(-x)$$

$$Be^{i\beta x} + ce^{-i\beta x} = Be^{-i\beta x} + ce^{i\beta x}$$

Now we apply the following boundary conditions

$$(\psi_1)_{x=-a} = (\psi_2)_{x=-a} \quad ; \quad (\psi_2)_{x=a} = (\psi_3)_{x=a}$$

And $$\left(\frac{d\psi_1}{dx}\right)_{x=-a} = \left(\frac{d\psi_2}{dx}\right)_{x=-a} \ ; \ \left(\frac{d\psi_2}{dx}\right)_{x=a} = \left(\frac{d\psi_3}{dx}\right)_{x=a}$$

$\therefore$ $$Ae^{-\alpha a} = 2B\,Cos(-\beta a) = 2B\cos\beta a \qquad [10]$$

And $$2B\cos\beta a = Ge^{-\alpha a} = Ae^{-\alpha a} \ [\because A = G]$$

$$Ae^{-\alpha a} = 2B\cos\beta a \qquad [11]$$

And $$A\alpha e^{-\alpha a} = -2B\,\beta Sin(-\beta a) = 2B\,\beta Sin\beta \qquad [12]$$

$$-2B\,\beta Sin\beta a = -\alpha Ge^{-\alpha a} = -A\alpha e^{-\alpha a} \ [\because A = G]$$

Or, $$A\alpha e^{-\alpha a} = 2B\,\beta Sin\beta a \qquad [13]$$

From equation [13]

$$Ae^{-\alpha a} = \frac{\beta}{\alpha} 2B\sin\beta a \qquad [14]$$

From equations [14] & [11], we have

$$\frac{\beta}{\alpha} 2B\sin\beta a = 2B\cos\beta a \quad Or, \tan\beta a = \frac{\alpha}{\beta} = \left(\frac{2m}{\hbar^2} W_0 \times \frac{\hbar^2}{2m} \frac{1}{V_0 - W_0}\right)^{\frac{1}{2}} = \left(\frac{W_0}{V_0 - W_0}\right)^{\frac{1}{2}}$$

5.8 Attractive Square well Potential

Consider an attractive square well potential as shown in figure. Let the potential between $x = a\ and\ x = -a\ be\ V_0$ and the potential is zero elsewhere I.e.,

$$V(x) = 0\,;\ for\ |x| > a\ \ and\ V(x) = -V_0\,;\ \ for\ |x| < a$$

In which V_0 is the depth of the well & it is a positive number. Let a stream of electrons be directed from the left. According to Classical Mechanics, no electron is reflected from the edges but due to wave nature some electrons be reflected from the sharp edges at $x = a$ & $x = -a$. As a result, there will be reflected & transmitted waves [17.26].

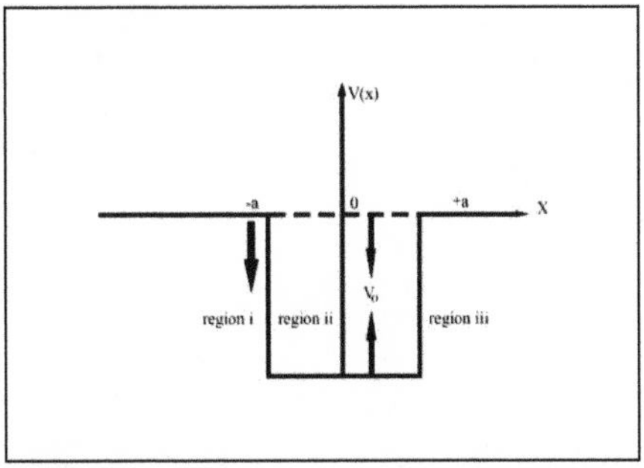

Fig.[5.8]: Shows an Attractive Square well Potential

CASE - A: The Schrödinger wave equation for the first region, is where $V = 0$.

$$\frac{d^2\psi_1}{dx^2} + \frac{2m}{\hbar^2}E\psi_1 = 0\ [\because \mathrm{V} = 0] \tag{1}$$

$$\frac{d^2\psi_1}{dx^2} + \frac{{p_1}^2}{\hbar^2}\psi_1 = 0 \tag{2}$$

Where $$p_1 = (2mE)^{\frac{1}{2}} \tag{3}$$

The solution of equation [2] is

$$\psi_1 = Ae^{ip_1x/\hbar} + Be^{-ip_1x/\hbar} \tag{4}$$

Where A & B are constants of integration.

CASE - B: The Schrödinger equation for the second region, where $V = -V_0$

$$\frac{d^2\psi_2}{dx^2} + \frac{2m}{\hbar^2}(E+V_0)\psi_2 = 0 \tag{5}$$

$$\frac{d^2\psi_2}{dx^2} + \frac{{p_2}^2}{\hbar^2}\psi_2 = 0 \tag{6}$$

Where $$p_2 = [2m(E+V_0)]^{\frac{1}{2}} \tag{7}$$

The solution of equation [6] is

$$\psi_2 = Ce^{ip_2x/\hbar} + De^{-ip_2x/\hbar} \qquad [8]$$

Where C & D are constants of integration.

CASE -C: The Schrödinger equation for the third region where $V = 0$.

$$\frac{d^2\psi_3}{dx^2} + \frac{2m}{\hbar^2}E\psi_3 = 0 \qquad [9]$$

$$\frac{d^2\psi_3}{dx^2} + \frac{{p_1}^2}{\hbar^2}\psi_3 = 0 \qquad [10]$$

The solution of equation [10] is

$$\psi_3 = Fe^{ip_1x/\hbar} + Ge^{-ip_1x/\hbar} \qquad [11]$$

Where F & G are constants of integration.

In the third region there is no reflected wave; the factor $Ge^{-ip_1x/\hbar}$ is discarded.

Hence $$\psi_3 = Fe^{ip_1x/\hbar} \qquad [12]$$

Where F is a constant of integration.

Now, the Values of these constants can be obtained by applying the following boundary conditions:

$$(\psi_1)_{x=-a} = (\psi_2)_{x=-a}; \ and \quad (\frac{d\psi_1}{dx})_{x=-a} = (\frac{d\psi_2}{dx})_{x=-a}$$

And $$(\psi_2)_{x=a} = (\psi_2)_{x=a}; \ \& \ (\frac{d\psi_2}{dx})_{x=a} = (\frac{d\psi_3}{dx})_{x=a}$$

Applying these conditions, we have

$$Ae^{-ip_1a/\hbar} + Be^{ip_1a/\hbar} = Ce^{-ip_2a/\hbar} + De^{ip_2a/\hbar} \qquad [13]$$

$$\left(\frac{ip_1}{\hbar}\right)Ae^{-ip_1a/\hbar} - \left(\frac{ip_1}{\hbar}\right)Be^{ip_1a/\hbar} = \left(\frac{ip_2}{\hbar}\right)Ce^{-ip_2a/\hbar} - \left(\frac{ip_2}{\hbar}\right)De^{ip_2a/\hbar} \qquad [14]$$

And $$Ce^{ip_2a/\hbar} + De^{-ip_2a/\hbar} = Fe^{ip_1a/\hbar} \qquad [15]$$

$$\left(\frac{ip_2}{\hbar}\right)Ce^{-ip_2a/\hbar} - \left(\frac{ip_2}{\hbar}\right)De^{-ip_2a/\hbar} = \left(\frac{ip_1}{\hbar}\right)Fe^{ip_1a/\hbar} \qquad [16]$$

From equation [14], we rewrite

$$Ae^{-ip_1a/\hbar} - Be^{ip_1a/\hbar} = \frac{P_2}{P_1} Ce^{-ip_2a/\hbar} - \frac{P_2}{P_1} De^{ip_2a/\hbar} \quad [17]$$

Now adding equations [13] & [17]

$$2Ae^{-ip_1a/\hbar} = \left(1+\frac{P_2}{P_1}\right) Ce^{-ip_2a/\hbar} + \left(1-\frac{P_2}{P_1}\right) De^{ip_2a/\hbar}$$

$$A = \frac{1}{2}\left[\left(1+\frac{P_2}{P_1}\right) Ce^{i(p_1-p_2)a/\hbar} + \left(1-\frac{P_2}{P_1}\right) De^{i(p_1+p_2)a/\hbar}\right] \quad [18]$$

From equation [16] we rewrite

$$Ce^{ip_2a/\hbar} - De^{-ip_2a/\hbar} = \frac{P_1}{P_2} Fe^{ip_1a/\hbar} \quad [19]$$

Now adding equations [15] & [19]

$$2Ce^{ip_2a/\hbar} = \left(1+\frac{P_1}{P_2}\right) Fe^{ip_1a/\hbar} \quad [20]$$

$$\therefore \quad C = \frac{1}{2}\left[\left(1+\frac{P_1}{P_2}\right) Fe^{i(p_1-p_2)a/\hbar}\right] \quad [21]$$

$$D = \frac{1}{2}\left[1-\frac{P_1}{P_2}\right] Fe^{i(p_1+p_2)a/\hbar} \quad [22]$$

Therefore equation [18] become,

$$A = \frac{1}{4}\left(1+\frac{P_2}{P_1}\right)\left(1+\frac{P_1}{P_2}\right) Fe^{2i(p_1-p_2)a/\hbar} + \frac{1}{4}\left(1-\frac{P_2}{P_1}\right)\left(1-\frac{P_1}{P_2}\right) Fe^{2i(p_1+p_2)a/\hbar}$$

$$A = \frac{F}{4}e^{2ip_1a/\hbar}\left[\left(1+\frac{P_2}{P_1}+\frac{P_1}{P_2}+1\right)e^{-2ip_2a/\hbar} + \left(1-\frac{P_2}{P_1}-\frac{P_1}{P_2}+1\right)e^{2ip_2a/\hbar}\right]$$

$$= \frac{F}{4}e^{2ip_1a/\hbar}\left[2e^{-2ip_2a/\hbar} + \left(\frac{P_2}{P_1}+\frac{P_1}{P_2}\right)e^{-2ip_2a/\hbar} + 2e^{2ip_2a/\hbar} - \left(\frac{P_2}{P_1}+\frac{P_1}{P_2}\right)e^{2ip_2a/\hbar}\right]$$

$$= \frac{F}{4}e^{2ip_1a/\hbar}\left[2Cos\left(\frac{2p_2a}{\hbar}\right) - 2iSin\left(\frac{2p_2a}{\hbar}\right) + \left(\frac{P_2}{P_1}+\frac{P_1}{P_2}\right)Cos\left(\frac{2p_2a}{\hbar}\right) - i\left(\frac{P_2}{P_1}+\frac{P_1}{P_2}\right)Sin\left(\frac{2p_2a}{\hbar}\right) + 2Cos\left(\frac{2p_2a}{\hbar}\right) + 2iSin\left(\frac{2p_2a}{\hbar}\right) - \left(\frac{P_2}{P_1}+\frac{P_1}{P_2}\right)Cos\left(\frac{2p_2a}{\hbar}\right) - i\left(\frac{P_2}{P_1}+\frac{P_1}{P_2}\right)Sin\left(\frac{2p_2a}{\hbar}\right)\right]$$

$$= \frac{F}{4}e^{2ip_1a/\hbar}\left[4\,Cos\left(\frac{2p_2a}{\hbar}\right) - 2i\left(\frac{P_2}{P_1}+\frac{P_1}{P_2}\right)Sin\left(\frac{2p_2a}{\hbar}\right)\right]$$

$$= Fe^{2ip_1a/\hbar}\left[Cos\left(\frac{2p_2a}{\hbar}\right) - \frac{i}{2}\left(\frac{P_2}{P_1}+\frac{P_1}{P_2}\right)Sin\left(\frac{2p_2a}{\hbar}\right)\right] \quad [23]$$

If N_i be the incident flux or current density then $N_I = (A \times A)v = |A|^2 v$ [24]

Similarly, the transmitted flux is $N_T = (F^{\times}F)v = |F|^2 v$ [25]

Hence the transmission coefficient

$$T = \frac{\text{Transmission flux}}{Incident\ flux} = \frac{|F|^2 v}{|A|^2 v} = \frac{|F|^2}{|A|^2}$$

$$T = \frac{|F|^2}{\left\|\left[Fe^{2ip_1a/\hbar}\left\{Cos\left(\frac{2p_2a}{\hbar}\right)-\frac{i}{2}\left(\frac{P_2}{P_1}+\frac{P_1}{P_2}\right)Sin\left(\frac{2p_2a}{\hbar}\right)\right\}\right]\right\|^2}$$

$$= \frac{1}{\left[Cos^2\left(\frac{2p_2a}{\hbar}\right)-\frac{1}{4}\left(\frac{P_2}{P_1}+\frac{P_1}{P_2}\right)^2 Sin^2\left(\frac{2p_2a}{\hbar}\right)\right]}$$

$$= \frac{1}{\left[1-Sin^2\left(\frac{2p_2a}{\hbar}\right)+\frac{1}{4}\left(\frac{P_2}{P_1}+\frac{P_1}{P_2}\right)^2 Sin^2\left(\frac{2p_2a}{\hbar}\right)\right]}$$

$$= \frac{1}{\left[1+\frac{1}{4}\left\{\left(\frac{P_2}{P_1}+\frac{P_1}{P_2}\right)^2-4\right\}Sin^2\left(\frac{2p_2a}{\hbar}\right)\right]}$$

$$T = \frac{1}{1+\frac{1}{4}\left[\left\{\left(\frac{P_1}{P_2}-\frac{P_2}{P_1}\right)^2-4\right\}Sin^2\left(\frac{2p_2a}{\hbar}\right)\right]} \quad [26]$$

We have noted that:

[1] ***When*** $\boldsymbol{P_1 = P_2}$ then $T = 1$; that means there is no potential well at all. But if $P_1 \neq P_2$ then the transmissvity is less than unity $I.e.,\ T < 1$. It means that some reflection has taken place. Even though when $P_1 \neq P_2$ the transmissivity $T = 1$

[2] ***When*** $\boldsymbol{Sin^2\left(\frac{2p_2a}{\hbar}\right) = 0}$ or, $\frac{2p_2a}{\hbar} = n\pi$ or, $P_2 = \frac{n\pi\hbar}{2a}$ or, $= \frac{2p_2a}{\pi\hbar}$; Where n is an integer.

The number of half-wavelengths contained in the well region

$$= \frac{2a}{\frac{1}{2}\lambda} = \frac{2a}{\frac{1}{2}\frac{h}{P_2}} = \frac{2aP_2}{h} = \frac{4aP_2}{2\pi\hbar} = \frac{2ap_2}{\pi\hbar} = n$$

Therefore, the well region contains an integer number of half- wavelengths

Exercise

1. Define potential barrier and Tunnel effect.
2. Explain the Quantum theory for single step potential barrier.
3. Show that the total number of particles is conserved for single step potential barrier.
4. Prove that the sum of the reflection coefficient & Transmission coefficient is unity in Quantum theory for single step potential barrier.
5. Find the energy levels for one dimensional square-well potential of finite depth or finite potential well.
6. Define the Bound State. Explain the Penetration of a potential barrier: TUNNEL EFECT
7. Find Energy levels of a particle enclosed with one dimensional rigid wall with infinite potential well

$$V(x) = 0\,;\quad for -a < x < a \;\; or,\; |x| < a$$
$$\text{And}\quad V(x) = \infty\,;\quad for\;\; -a > x > a\;\; or,\; |x| > a$$

8. Find the energy levels and corresponding normalized Eigen function of a particle in one dimensional potential-well of the form

$$V(x) = \infty;\; for\;\; x < 0 \;\;\&\;\; for\;\; x > 0$$
$$\text{And}\quad V(x) = 0\;\;;\; for\;\; 0 < x < a$$

9. A particle of mass m is confined within one dimensional potential well defined by

$$V = -V_0\,;\;\; \text{for } |x| < a$$
$$V = \quad 0\;\;;\;\; \text{for } |x| > a$$

The total energy of the particle is $E = -W$. Where W is positive. Assuming that the wave function of the particle in the lowest state(W_0) is an even function of x. Show that

$$\tan \beta a = \left(\frac{W_0}{V_0 - W_0}\right)^{\frac{1}{2}};\ \text{Where}\ \beta^2 = \frac{8\pi^2 m}{\hbar^2}(V_0 - W_0)$$

10. For an attractive square well potential as shown in figure.

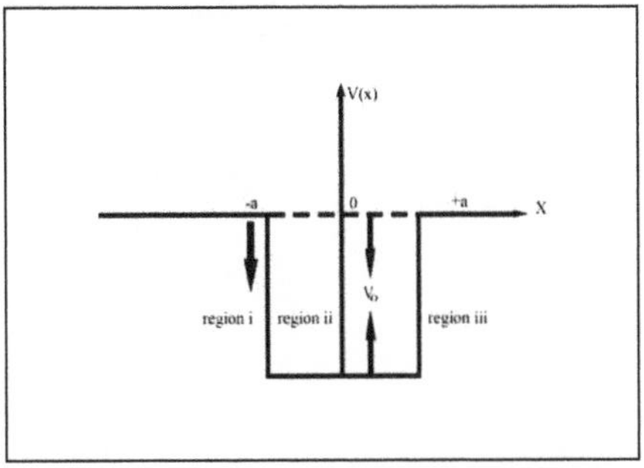

The potential between $x = a \; and \; x = -a \;\; be \; V_0$ and the potential is zero elsewhere I.e.,

$$V(x) = 0\,;\; for\; |x| > a \;\;\; and\;\; V(x) = \; -V_0\,;\;\; for\; |x| < a$$

In which V_0 is the depth of the well & it is a positive number. Show that when

(a) Transmission is unity
(b) Transmission is less than unity
(c) The well region contains an integer number of half- wavelengths
(d) Transmission is unity; what does it mean
(e) Transmission is less than unity; what does it mean

11. Show that in one dimensional problems, the energy spectrum of the bound states is always non- degenerate I.e., no two linearly independent Eigen functions have the same energy Eigen values.

12. Find Eigen functions and the energy spectrum of a particle in the potential well given by

$$V(x) = 0 \quad \text{if } |x| < a$$
$$= \infty \quad \text{if } |x| > a$$

13. Find the energies of the bound states of a particle in the symmetric potential well,

$$V(x) = -V_0 \quad \text{if } |x| < a$$
$$= 0 \quad \text{if } |x| > a$$

14. Find the energies of the bound states of a particle in the potential well given by

$$V(x) = \infty \quad \text{if } x < 0$$
$$= -V_0 \quad \text{if } 0 < x < a$$
$$= 0 \quad \text{if } x > a$$

15. Find the energy values of a particle in the asymmetric potential well is given by

$$V(x) = V_2 \quad \text{for } x < 0$$
$$= 0 \quad \text{for } 0 < x < a$$
$$= V_1 \quad \text{for } x > a$$

16. Consider a step potential barrier,

$$V(x) = 0 \quad \text{for } x < 0$$
$$= V_0 \quad \text{for } 0 < x < d$$
$$= 2V_0 \quad \text{for } x > d$$

17. Find the lowest energy state and the wave function for that state in potential well

$$V(x) = \infty \quad \text{if } x < 0$$
$$= 0 \quad \text{if } 0 < x < a$$
$$= V_0 \quad if \ x > a$$

18. Solve the Schrödinger equation for

a. $V(x) = \frac{V_0}{cos^2 ax}$ b. $V(x) = -\frac{V_0}{1+e^{x/a}}$ c. $V(x) = -kx$

19. For the potential barrier,

$$V(x) = 0 \quad \text{for } x < 0$$
$$= V_0 \quad \text{for } 0 < x < a$$
$$= 2V_0 \quad \text{for } x > a$$

20. Work out the one dimensional motion of a particle cross a potential barrier and calculate the transmission and reflection coefficients.

Chapter 6 Quantum Mechanics of Linear Harmonic Oscillator

6.1 Bound States and Free States

If one dimensional motion of a particle is assumed to take place with zero potential energy over a fixed distance and if the potential energy is assumed to become infinite at the extremities of the distance, it is described as a particle in a one dimensional box. This is the simplest example of all motions in bound states. *When the motion of a particle is confined to a limited region such that the particle moves back and forth in the region, the particle is said to be in a bound state* [5,7]. But in practice such motion is not possible. However, the Schrödinger equation will first be applied to study the motion of a particle in a one dimensional box. Because the study will show how quantum numbers, discrete values of energy and zero point energy arise. *The Simple Harmonic Oscillator (S.H.O) is a bound state problem and has discreet energy level.*

If one dimensional motion of a particle is assumed to take place with constant potential energy, [For convenience, the constant potential energy is taken to be zero] it can be that the particle has definite value of total energy and definite value of momentum. But the position of the particle is completely unknown. So, *when the particle is not subjected to any external forces and the particle moves in a region in which its potential energy is constant, it is said to be a free particle and the system is termed as Free states.*

6.2 Linear Harmonic Oscillator [L.H.O.]

A harmonic oscillator is a physical system that, when displaced from equilibrium, experiences a restoring force proportional to the displacement. A harmonic oscillator that is displaced and then let go will oscillate sinusoidally. Examples from classical physics are a mass attached to a spring and a simple pendulum swinging through a small angle. When a particle oscillates about its mean position along a straight line under the action of a force which is directed towards

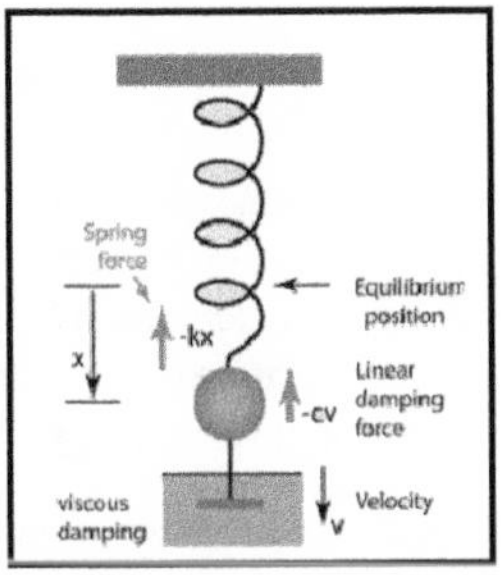

Fig.[6.1]: Shows a Linear Harmonic Oscillator [L.H.O]

the mean position and this particle is said proportional to the displacement at any instant from this position, the motion o to be simple harmonic and the Oscillating particle is called a *simple harmonic oscillator [S.H.O] or linear harmonic oscillator [L.H.O].* In classical mechanics, the lowest energy state of a **harmonic oscillator** occurs when it is at rest in its equilibrium position. In a general manner, harmonic oscillation is a first approximation for a imaginating a vibrating system by a physicist. In other words, all friction forces in a vibrating system are to be neglected, before it can be assimilated to a harmonic oscillator. In classical mechanics, the solution of the Newton's equation for such a system is a sinusoidal (harmonic) function which represents in fact an eternal oscillation without damping.

6.3 Harmonic motion and Simple Harmonic Motion and difference between them

Harmonic motion would just be any motion that is periodic. For example, motion following a kind of square wave would be harmonic. The simple pendulum undergoes Harmonic Motion.

In mechanics and physics, simple harmonic motion is a special type of periodic motion or oscillation motion where the restoring force is directly proportional to the displacement and acts in the direction opposite to that of displacement. Simple harmonic motion is motion that is specifically sinusoidal; that is, it can be described by a sine wave. The reason simple harmonic motion is "simple" is because any kind of harmonic motion can be constructed as superposition of SHMs of different frequencies, via the Fourier transform. Simple harmonic motion has important properties, for example, the period of oscillation does not depend on the amplitude of the motion and lots of systems do undergo simple harmonic motion even if sometimes it is an approximation.

A good example of the difference between Harmonic Motion and Simple Harmonic Motion is the simple pendulum. The simple pendulum undergoes harmonic motion, however for small angles of oscillation it does approximately undergo simple harmonic motion.

6.4 Quantum Mechanically Harmonic Oscillator

A quantum mechanically Harmonic Oscillator vibrates, with zero-point energy, even in its ground state [7]. This result plays an important role in the theory of metals. In quantum mechanics, when a quadratic potential (representing a harmonic oscillation) is introduced in the Hamiltonian of the Schrödinger's equation a set of discrete energy levels with associate wave functions are obtained. Importantly, the ground state (lowest energy level) has a non-null energy $\frac{1}{2}h\upsilon\left[=\frac{1}{2}2\pi\hbar\frac{1}{T}=\frac{2\pi h}{2}.\frac{\omega}{2\pi}=\frac{1}{2}\hbar\omega\right]$ (where h is the Planck's constant and υ stands for vibrational frequency) with a Gaussian function as its wave function. More interestingly, the successive energy states of a harmonic oscillator are separated by a unique energy interval $(h\upsilon)$

$\left[\hbar\omega = \frac{h}{2\pi}\cdot\frac{2\pi}{T} = h\upsilon\right]$ as Max Planck had predicted it smartly several years before the publication of the Schrödinger's equation.

A diatomic molecule vibrates somewhat like two masses on a spring with a potential energy that depends upon the square of the displacement from equilibrium. But the energy levels are quantized at equally spaced values.

$$E_n = (n + \tfrac{1}{2})\hbar\omega \qquad n = 0,1,2,3 \ldots$$
$$\omega = 2\pi(\text{frequency})$$
$$\hbar = \text{Planck's constant}/2\pi$$

The energy levels of the quantum harmonic oscillator are and for a diatomic molecule the natural frequency is of the form

$$\omega = \sqrt{\frac{k}{m_r}} \qquad \begin{array}{l} k - \text{bond force constant} \\ m_r = \text{reduced mass} \end{array}$$

Where the reduced mass is given by

$$m_r = \frac{m_1 m_2}{m_1 + m_2}$$

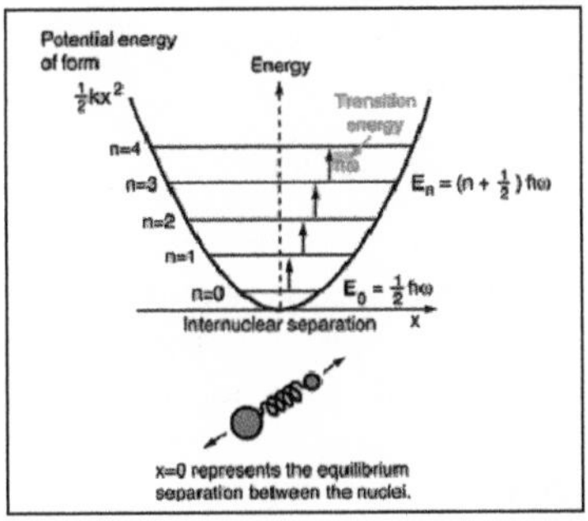

Fig.[6.4]: Sows Quantum Mechanically Harmonic Oscillator

This form of the frequency is the same as that for the classical simple harmonic_oscillator. *The most surprising difference for the quantum case is the so-called "zero-point vibration" of the n =0 ground state. This implies that molecules are not completely at rest, even at absolute zero temperature.* The quantum harmonic oscillator has implications far beyond the simple diatomic molecule. It is the foundation for the understanding of complex modes of vibration in larger molecules, the motion of atoms in a solid lattice, the theory of heat capacity, etc. In real systems, energy spacing are equal only for the lowest levels where the potential is a good approximation of the "mass on a spring" type harmonic potential. *The harmonic terms which appear in the potential for a diatomic molecule are useful for mapping the detailed potential of such systems.*

6.5 Simple Harmonic Oscillator is an interesting problem

- The inter particle forces of the Simple Harmonic Oscillator (S.H.O.) are linear functions of the relative displacement. So the motion is bounded & Schrödinger wave equation is exactly solvable
- The Simple Harmonic Oscillator (S.H.O.) is a bound state problem and has discreet energy level
- The energy levels of the linear harmonic oscillator are non-degenerate
- The successive energy levels are equally spaced
- The separation between two adjacent energy levels being $\hbar\omega$
- More interestingly, the successive energy states of a harmonic oscillator are separated by a unique energy interval hυ $\left[\hbar\omega = \frac{h}{2\pi}\cdot\frac{2\pi}{T} = h\upsilon\right]$ as Max Planck had predicted it smartly several years before the publication of the Schrödinger's equation
- At temperature above 0^0k, the atoms in a Crystal are temporarily displaced from their normal positions in the structure due to absorption of thermal energy. Consequently inter atomic forces obeying Hook's low acts on the displaced atoms. Under the action of such restoring forces each atom vibrates about its normal position [which is the correct position in the ideal structure].Thus the vibrations of each atom are similar to those of a Simple Harmonic Oscillator (S.H.O.)

6.6 Degenerate and Non-Degenerate system

If the particle having the same energy in an excited state will have several different stationary states or different wave functions. i.e. there are a number of independent quantum states of a system, each belonging to the same energy level. Such states and energy-levels are said to be degenerate. If there are n linearly independent wave functions $\psi_1, \psi_2, \psi_3 \ldots\ldots\ldots\ldots\ldots\ldots\ldots\ldots \psi_n$ belonging to the same energy state then the energy level is said to be n-fold degenerate. It can be easily shown that any linearly combination of the degenerate wave functions $= C_1\psi_1 + C_2\psi_2 + C_3\psi_3 + \ldots\ldots\ldots\ldots\ldots\ldots\ldots\ldots C_n\psi_n$; is also an Eigen function belonging to the same energy Eigen value.

If there are a number of independent quantum states of a system, each belonging to the different energy level. Such states and energy-levels are said to be non-degenerate.

In a cubical box, the energy depends on the sum of the squares of the Quantum numbers i.e.

$$E_{n_x,n_y,n_z} = \frac{\pi^2\hbar^2}{2m}\left[\frac{n_x^2}{a^2} + \frac{n_y^2}{b^2} + \frac{n_z^2}{c^2}\right] \qquad [1]$$

Where $n_x = 1, 2, 3, \ldots\ldots\ldots$

$n_y = 1, 2, 3, \ldots\ldots\ldots$

$n_z = 1, 2, 3, \ldots\ldots\ldots$

The lowest possible energy i.e. the energy in the ground state occurs when $n_x = n_y = n_z = 1$ and it depends on the values of a, b and c. If the particle is confined in a cubical box in which a = b = c = L. In this case the energy of the particle in the ground can be obtained from equation [1]

$$E_{111} = \frac{3\pi^2\hbar^2}{2mL^2} \qquad [2]$$

No other state will have this energy and this state has only one wave function. Therefore, the ground state and the energy-levels are said to be non-degenerate.

6.7 Number of energy- levels with the corresponding quantum numbers and the degree of degeneracy

In a cubical box, the energy depends on the sum of the squares of the Quantum numbers I.e.,

$$E_{n_x,n_y,n_z} = \frac{\pi^2\hbar^2}{2m}\left[\frac{n_x^2}{a^2} + \frac{n_y^2}{b^2} + \frac{n_z^2}{c^2}\right] \qquad [1]$$

Where $n_x = 1,2,3, \ldots\ldots\ldots$

$n_y = 1,2,3, \ldots\ldots\ldots$

$n_z = 1,2,3, \ldots\ldots\ldots$

The above gives the Eigen values of the energy of the particle. These values are called the energy-levels of the particle. Consequently, the particle having the same energy in an excited state will have several different stationary states or different wave functions. i.e., there are a number of independent quantum states of a system, each belonging to the same energy level. Such states and energy-levels are said to be degenerate. If there are n linearly independent wave functions Ψ_1, Ψ_2, Ψ_3, Ψ_4...Ψ_n belonging to the same energy state then the energy level is said to be n-fold degenerate[5].

It can be easily shown that any linearly combination of the degenerate wave functions $= C_1\psi_1 + C_2\psi_2 + C_3\psi_3 + \ldots\ldots\ldots\ldots C_n\psi_n$; is also an eigen function belonging to the same energy Eigen value. For example for the first excited state, the values of the quantum numbers are:

n_x	n_y	n_z
2	1	1
1	2	1
1	1	2

Thus there are three different Eigen functions and hence three different stationary states. Each state has the same energy $\frac{6\pi^2\hbar^2}{2mL^2}$. Therefore, the first excited state is said to be triply degenerate or three-fold degenerate. The number of independent wave functions for the stationary states of an energy-level is called degeneracy of the level [17].

For a particle in a cubical box a number of energy levels with the corresponding quantum numbers and the degree of degeneracy are given in the following table:

Energy Levels	Quantum Numbers (n_x, n_y, n_z)	Degree of Degeneracy
$\frac{3\pi^2\hbar^2}{2ML^2}$	(111)	Non-degenerate
$\frac{6\pi^2\hbar^2}{2mL^2}$	(211), (121), (112)	Three-fold degenerate
$\frac{9\pi^2\hbar^2}{2mL^2}$	(221), (212), (122)	Three-fold degenerate
$\frac{11\pi^2\hbar^2}{2mL^2}$	(311), (131), (113)	Three-fold degenerate
$\frac{12\pi^2\hbar^2}{2mL^2}$	(222)	Non-degenerate
$\frac{14\pi^2\hbar^2}{2mL^2}$	(123), (132), (213) (231), (312), (321)	Six- fold degenerate
$\frac{17\pi^2\hbar^2}{2mL^2}$	(322), (232), (223)	Three-fold degenerate
$\frac{18\pi^2\hbar^2}{2mL^2}$	(441), (141), (114)	Three-fold degenerate
$\frac{19\pi^2\hbar^2}{2mL^2}$	(331), (313), (133)	Three-fold degenerate
$\frac{21\pi^2\hbar^2}{2mL^2}$	(421), (412), (241) (214), (124), (142)	Six-fold degenerate

6.8 Energy Eigen value of one dimensional Linear Harmonic Oscillator

For Linear Harmonic Oscillator [L.H.O], the time independent Schrödinger equation along the

X- direction

$$-\frac{\hbar^2}{2m}\nabla^2 + V(x)\psi = E\psi \qquad [1]$$

$$\frac{d^2\psi}{dx^2} + \frac{2m}{\hbar^2}[E - V(x)]\psi = 0 \qquad [2]$$

Where, $V(x) = \frac{1}{2}\kappa x^2 = \frac{1}{2}m\omega^2 x^2$ [3]

$$\frac{d^2\psi}{dx^2} + \frac{2m}{\hbar^2}\left[E - \frac{1}{2}m\omega^2 x^2\right]\psi = 0 \qquad [4]$$

$$\frac{d^2\psi}{dx^2} + [\frac{2m}{\hbar^2}E - \frac{m^2\omega^2 x^2}{\hbar^2}]\psi = 0 \qquad [5]$$

To simplify the equation [5] we introduce a dimensionless independent quantity which is related to x is given by

$$y = \sqrt{\frac{m\omega}{\hbar}}\,x \quad \text{or,} \quad x^2 = \frac{\hbar y^2}{m\omega}$$

Now $\frac{d\psi}{dx} = \frac{d\psi}{dy}\cdot\frac{dy}{dx} = \frac{d\psi}{dy}\frac{d}{dx}\left[\sqrt{\frac{m\omega}{\hbar}}x\right] = \frac{d\psi}{dy}\left[\sqrt{\frac{m\omega}{\hbar}}\right]$

$$\frac{d^2\psi}{dx^2} = \frac{d}{dy}\frac{dy}{dx}\frac{d\psi}{dy}\left[\sqrt{\frac{m\omega}{\hbar}}\right] = \frac{d^2\psi}{dy^2}\frac{dy}{dx}\left[\sqrt{\frac{m\omega}{\hbar}}\right]$$

$$\frac{d^2\psi}{dx^2} = \frac{d^2\psi}{dy^2}\frac{dy}{dx}\left[\sqrt{\frac{m\omega}{\hbar}}\right] = \frac{d^2\psi}{dy^2}\sqrt{\frac{m\omega}{\hbar}}\sqrt{\frac{m\omega}{\hbar}} = \frac{d^2\psi}{dy^2}\frac{m\omega}{\hbar}$$

Hence equation [5] becomes

$$\frac{m\omega}{\hbar}\frac{d^2\psi}{dy^2} + [\frac{2m}{\hbar^2}E - \frac{m^2\omega^2}{\hbar^2}\frac{\hbar y^2}{m\omega}]\psi = 0 \qquad [6]$$

$$\frac{m\omega}{\hbar}\frac{d^2\psi}{dy^2} + \frac{m\omega}{\hbar}[\frac{2E}{\hbar\omega} - y^2]\psi = 0$$

$$\frac{d^2\psi}{dy^2} + [\frac{2E}{\hbar\omega} - y^2]\psi = 0$$

$$\frac{d^2\psi}{dy^2} + [\lambda - y^2]\psi = 0 \qquad [7]$$

Where $\lambda = \frac{2E}{\hbar\omega}$ [8]

Equation [7] is extremely a simplified form. But it is not easy to solve. For large value of y I.e., when $y^2 >> \lambda$; then equation [7] becomes

$$\frac{d^2\psi}{dy^2} - y^2\psi = 0 \qquad [9]$$

For large value of y; the approximate solution of equation [9]

$$\psi = e^{-\frac{y^2}{2}} + e^{+\frac{y^2}{2}} \qquad [10]$$

When y→ ∞ , ψ→ 0. So we reject the term $e^{+\frac{y^2}{2}}$

$$\therefore \qquad \psi = e^{-\frac{y^2}{2}} \qquad [11]$$

$$\psi = e^{-\infty} + e^{+\infty}$$
$$= \frac{1}{e^{\infty}} + \infty$$
$$= \frac{1}{\infty} + \infty$$
$$= 0 + \infty$$
$$= \infty$$

But if we substitute $\psi = e^{-\frac{y^2}{2}}$ in equation [9]; the equation [9] is not Satisfied. It is satisfied only if the equation [9] is in the form

$$\frac{d^2\psi}{dy^2} - (y^2 - 1)\psi = 0 \qquad [12]$$

Let us try to justify $\psi = e^{-\frac{y^2}{2}}$ is the solution of equation [12] and it can be shown in the following way:

$$[\psi = e^{-\frac{y^2}{2}}$$

$$\frac{d\psi}{dy} = \left(-\frac{1}{2}\right) e^{-\frac{y^2}{2}}.2y = ye^{-\frac{y^2}{2}}$$

$$\frac{d^2\psi}{dy^2} = -y\left(-\frac{1}{2}\right) e^{-\frac{y^2}{2}}.2y - e^{-\frac{y^2}{2}}.1 = y^2\psi - \psi = (y^2 - 1)\psi]$$

$$\frac{d^2\psi}{dy^2} - (y^2 - 1)\psi = 0]$$

For large value of y; equation [12] reduces to equation [9]. Hence for large value of y, the accurate solution must be of the form

$$\psi = \mathrm{H}(y)\ e^{-\frac{y^2}{2}} \qquad [13]$$

Where, H(y) is finite polynomial in y.

$$\frac{d\psi}{dy} = e^{-\frac{y^2}{2}}\frac{dH}{dy} + H\left(-\frac{1}{2}\right)2ye^{-\frac{y^2}{2}} = \left(\frac{dH}{dy} - Hy\right)e^{-\frac{y^2}{2}}$$

$$\frac{d^2\psi}{dy^2} = \left(\frac{d^2H}{dy^2} - H - y\frac{dH}{dy}\right)e^{-\frac{y^2}{2}} + \left(\frac{dH}{dy} - Hy\right)\left(-\frac{1}{2}\right)2ye^{-\frac{y^2}{2}}$$

$$\frac{d^2\psi}{dy^2} = \left(\frac{d^2H}{dy^2} - H - y\frac{dH}{dy} - y\frac{dH}{dy} + Hy^2\right)e^{-\frac{y^2}{2}}$$

$$\frac{d^2\psi}{dy^2} = \left(\frac{d^2H}{dy^2} - H - 2y\frac{dH}{dy} + Hy^2\right)e^{-\frac{y^2}{2}}$$

Therefore equation [7] becomes

$$\left(\frac{d^2H}{dy^2} - 2y\frac{dH}{dy} - H + Hy^2\right)e^{-\frac{y^2}{2}} + (\lambda - y^2)\text{H}e^{-\frac{y^2}{2}} = 0$$

$$\frac{d^2H}{dy^2} - 2y\frac{dH}{dy} - H(\lambda - 1) = 0 \quad \text{Since } e^{-\frac{y^2}{2}} \neq 0 \tag{14}$$

Equation [14] represents Hermite differential equation. The solution of equation [14] will be the form

$$H(y) = A_0 + A_1y + A_2y^2 + A_3y^3 + \cdots\cdots\cdots\cdots\cdots\cdots\cdots\cdots\cdots\cdots \tag{15a}$$

$$H(y) = \textstyle\sum_{n=0}^{\infty} A_n y^n \tag{15b}$$

Differentiating [15] with respect to y

$$\frac{dH}{dy} = A_1 + 2A_2y + 3A_3y^2 + \cdots\cdots\cdots\cdots\cdots\cdots\cdots\cdots\cdots\cdots \tag{16}$$

Multiplying both sides by (−2y)

$$-2y\frac{dH}{dy} = -2y\,[A_1 + 2A_2y + 3A_3y^2 + \cdots\cdots\cdots\cdots\cdots\cdots\cdots\cdots\,]$$

$$= \textstyle\sum_{n=0}^{\infty} -2nA_n y^n \tag{17}$$

$$\frac{d^2H}{dy^2} = 2A_2 + 6A_3y + \cdots\cdots\cdots\cdots\cdots\cdots\cdots\cdots\cdots\cdots\cdots\cdots\cdots\cdots$$

$$= \textstyle\sum_{n=0}^{\infty}(n+2)(n+1)A_{n+2}y^n \tag{18}$$

Therefore equation [14] becomes

$$\textstyle\sum_{n=0}^{\infty}(n+2)(n+1)A_{n+2}y^n + \sum_{n=0}^{\infty} -2nA_n y^n + (\lambda - 1)\sum_{n=0}^{\infty} A_n y^n = 0$$

Or, $\sum_{n=0}^{\infty}[(n+2)(n+1)A_{n+2} - 2nA_n + (\lambda - 1)A_n\,]y^n = 0$

Or, $\sum_{n=0}^{\infty}[(n+2)(n+1)A_{n+2} - (2n + 1 - \lambda)A_n\,]y^n = 0$

The coefficient of each power of y must be vanished separately. Hence we have

$$(n+2)(n+1)A_{n+2} = (2n + 1 - \lambda)A_n$$

$$A_{n+2} = \frac{(2n+1-\lambda)}{(n+2)(n+1)} A_n \qquad [19]$$

Equation [19] represents the recurrence formula connecting the coefficients A_{n+2} and A_n.

When $= 0, 2, 4, 6$ ……………………………………………; Then from equation [19], we get

$$A_2 = \frac{1-\lambda}{2!} A_0$$

$$A_4 = \frac{5-\lambda}{3.4} A_2 = \frac{5-\lambda}{3.4} . \frac{1-\lambda}{2} A_0 = \frac{(1-\lambda)(5-\lambda)}{4!} A_0$$

$$A_6 = \frac{(1-\lambda)(5-\lambda)(9-\lambda)}{6!} A_0$$

When $= 1, 3, 5, 7$ ……………………………………………; Then from equation [19],

$$A_3 = \frac{3-\lambda}{6} A_1 = \frac{3-\lambda}{3!} A_1$$

$$A_5 = \frac{5-\lambda}{20} A_3 = \frac{(3-\lambda)(7-\lambda)}{5!} A_1$$

Substituting the coefficients in equation [15a]

$$H(y) = \left[1 + \frac{1-\lambda}{2!} y^2 + \frac{(1-\lambda)(5-\lambda)}{4!} y^4 + \cdots\right] A_0 + \left[y + \frac{3-\lambda}{3!} y^3 + \frac{(3-\lambda)(7-\lambda)}{5!} y^5 + \cdots\right] A_1 \qquad [20]$$

The above equation [20] shows that the polynomial $H(y)$ is the sum of two infinite series. This means that if $H(y)$ does not terminate for some value of n, the wave function $\psi = \mathrm{H}(y)\ e^{-\frac{y^2}{2}}$ will become infinite as y becomes infinite.

In order to avoid this difficulty, we cut-off this series solution after A_n I.e., we assume that all the terms after A_n is zero. Hence equation [19] becomes

$$\frac{(2n+1-\lambda)}{(n+2)(n+1)} A_n = 0 \qquad [21]$$

Or, $\quad (2n + 1 - \lambda) = 0 \quad$ [Since $A_n \neq 0$]

Or, $\quad \lambda = 2n + 1 \qquad [22]$

Finally from equation [8], we get

$$\lambda = \frac{2E}{\hbar\omega} \quad or, 2n + 1 = \frac{2E_n}{\hbar\omega}$$

$$2E_n = (2n + 1)\hbar\omega = 2\left(n + \frac{1}{2}\right)\hbar\omega$$

$$\boxed{E_n = \left(n + \frac{1}{2}\right)\hbar\omega} \qquad [23]$$

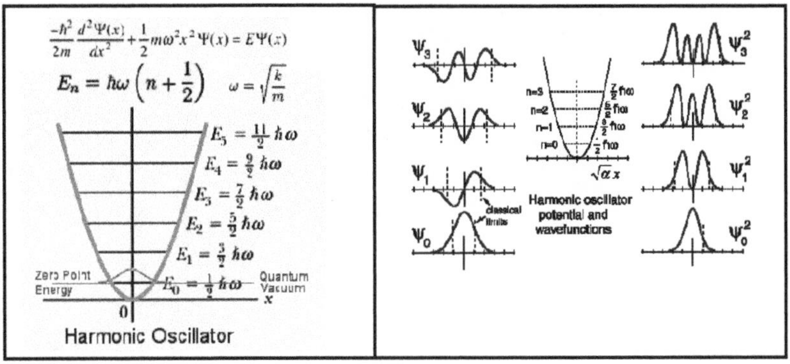

Fig.[6.6]: Shows Energy Eigen value and wave functions of one dimensional Linear Harmonic Oscillator[L.H.O.]

Where $n = 0, 1, 2, 3, 4$ ………………………………………………….

Equation [23] represents the energy Eigen values of a Linear Harmonic Oscillator

- When $n = 0$ in equation [23] we get

$$E_0 = \frac{1}{2}\hbar\omega \qquad [24]$$

Equation [23] represents the ground state energy or the zero-point vibrational energy of the linear harmonic oscillator [L.H.O.]

6.9 Non-Degeneracy of Linear Harmonic Oscillator

The non-degeneracy of a system can be explained uniquely from the energy Eigen value of a Linear Harmonic Oscillator. The energy Eigen value equation of a linear harmonic oscillator is

$$E_n = \left(n + \frac{1}{2}\right)\hbar\omega \qquad [1]$$

Where n is an integer. [$n = 0, 1, 2, 3,$ …………………………………………….]

Therefore, the ground state energy

$$E_0 = \frac{1}{2}\hbar\omega \qquad [2]$$

In a ground state configuration, all of the electrons are in as low an energy level as it is possible for them to be, when an electron absorbs energy, it occupies a higher energy orbital, and is said to be in an excited state. The first, second, third and so on excited states of L.H.O are

$$E_1 = \frac{3}{2}\hbar\omega \ , \ E_2 = \frac{5}{2}\hbar\omega \ , E_3 = \frac{7}{2}\hbar\omega, \dots\dots\dots\dots\dots, E_n = \left(n+\frac{1}{2}\right)\hbar\omega \qquad [3]$$

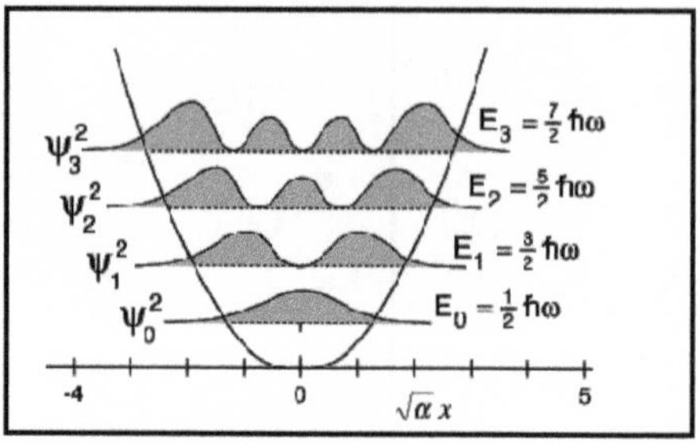

Fig.[6.7]: Shows non-degeneracy of Linear Harmonic Oscillator [L.H.O.]

So that the energy levels of the linear harmonic oscillator are discrete. I.e., *there are a number of independent quantum states, each belonging to the different energy level. Such states and energy-levels are said to be non-degenerate.* Also note that

- The successive energy levels are equally spaced
- The separation between two adjacent energy levels being $\hbar\omega$
- More interestingly, the successive energy states of a harmonic oscillator are separated by a unique energy interval hυ $\left[\hbar\omega = \frac{h}{2\pi}.\frac{2\pi}{T} = h\upsilon\right]$ as Max Planck had predicted it smartly several years before the publication of the Schrödinger's equation

6.10 Zero point energy and its physical significance

A quantum mechanically Harmonic Oscillator vibrates, with zero-point energy, even in its ground state. This result plays an important role in the theory of metals. *The most surprising difference for the quantum case is the so-called "zero-point vibration" of the n =0 ground state. This implies that molecules are not completely at rest, even at absolute zero temperature.* The energy Eigen values of a Linear Harmonic Oscillator is

$$\boxed{E_n = \left(n+\frac{1}{2}\right)\hbar\omega} \qquad [1]$$

- ❖ When $n = 0$ in equation [1] we get

$$\boxed{E_0 = \frac{1}{2}\hbar\omega} \qquad [2]$$

Equation [2] represents the ground state energy or the zero-point vibrational energy of the Linear Harmonic Oscillator [L.H.O]. *It means that a harmonic oscillator in equilibrium with its surroundings would approach an energy* $E = E_0$*;But not as* $E = 0$*; as the temperature approaches* 0^0K

6.11 An alternate approach of ground state energy of Linear Harmonic Oscillator

Consider a Linear Harmonic Oscillator [L.H.O.] whose total energy is

$$E = \frac{P_x^2}{2m} + \frac{1}{2}\, m\omega^2 x^2 \qquad [1]$$

We assume that the particle is confined to a region ~ a. I.e., x~ Δx~ a.

From the uncertainty relation

$$\Delta x \Delta p_x \geq \frac{\hbar}{2} \qquad [2]$$

$$\Delta p_x \geq \frac{\hbar}{2a} \qquad [3]$$

And the momentum is given by

$$p_x \sim \Delta p_x \sim \frac{\hbar}{2a} \qquad [4]$$

Thus the total energy will be

$$E = \frac{\hbar^2}{8ma^2} + \frac{1}{2}\, m\omega^2 a^2 \qquad [5]$$

The lowest state energy of the Linear Harmonic Oscillator [L.H.O] will corresponds to the minimum value of E for which

$$\frac{dE}{da} = 0 \qquad [6]$$

Or, $$-\frac{2\hbar^2}{8ma^3} + \frac{1}{2}\, 2m\omega^2 a = 0 \qquad [7]$$

Or, $$\frac{\hbar^2}{4ma^3} = m\omega^2 a \qquad [8]$$

Or, $$a^4 = \frac{\hbar^2}{4m^2\omega^2} \quad Or, a^2 = \frac{\hbar}{2m\omega} \quad Or, a = \sqrt{\frac{\hbar}{2m\omega}} \qquad [9]$$

From equation [5], the lowest state energy of the Linear Harmonic Oscillator [L.H.O] is

Or, $$E_0 = \frac{\hbar^2}{8m} . \frac{2m\omega}{\hbar} + \frac{1}{2}\, m\omega^2 . \frac{\hbar}{2m\omega} = \frac{1}{4}\, \hbar\omega + \frac{1}{4}\, \hbar\omega = \frac{1}{2}\, \hbar\omega \qquad [10]$$

Equation [10] represents the ground state energy or the zero-point vibrational energy of the Linear Harmonic Oscillator [L.H.O]. It means that a harmonic oscillator in equilibrium with its surroundings would approach an energy $E = E_0$; but not as $= 0$; as the temperature approaches 0^0K.

6.12 Ground state energy of Linear Harmonic Oscillator with relevant potential energy

The Hamiltonian operator for the Linear Harmonic Oscillator [L.H.O] is

$$\widehat{H} = -\frac{\hbar^2}{2m}\frac{\partial^2}{\partial x^2} + \frac{1}{2}m\omega^2 x^2 \qquad [1]$$

The time independent Schrödinger equation

$$\left(-\frac{\hbar^2}{2m}\frac{\partial^2}{\partial x^2} + \frac{1}{2}m\omega^2 x^2\right)\psi(x) = E\psi(x) \qquad [2]$$

For the ground state, the time independent Schrödinger equation can be written as

$$\left(-\frac{\hbar^2}{2m}\frac{\partial^2}{\partial x^2} + \frac{1}{2}m\omega^2 x^2\right)\psi_0 = E_0\psi_0 \qquad [3]$$

The Eigen function of the Hamiltonian operator for the ground state of the Linear Harmonic Oscillator [L.H.O] is

$$\psi_0 = \left(\frac{\alpha}{\pi}\right)^{\frac{1}{4}} e^{-\alpha x^2} \qquad [4]$$

Where $$\alpha = \frac{m\omega}{\hbar} \qquad [5]$$

Now, $$\frac{\partial\psi_0}{\partial x} = \left(\frac{\alpha}{\pi}\right)^{\frac{1}{4}}\left(-\frac{\alpha}{2}\right)2xe^{-\alpha x^2} = -\alpha x\psi_0$$

And $$\frac{\partial^2\psi_0}{\partial x^2} = -\alpha x\frac{\partial\psi_0}{\partial x} - \alpha\psi_0 = -\alpha x(-\alpha x\psi_0) - \alpha\psi_0 = \alpha^2 x^2\psi_0 - \alpha\psi_0 \qquad [6]$$

Multiplying both sides by $\left(-\frac{\hbar^2}{2m}\right)$, we have

$$-\frac{\hbar^2}{2m}\frac{\partial^2\psi_0}{\partial x^2} = -\frac{\hbar^2}{2m}\alpha^2 x^2\psi_0 + \frac{\hbar^2}{2m}\alpha\psi_0$$

$$= -\frac{\hbar^2}{2m}\frac{m^2\omega^2}{\hbar^2}x^2\psi_0 + \frac{\hbar^2}{2m}\frac{m\omega}{\hbar}\psi_0$$

$$= -\frac{1}{2}m\omega^2 x^2\psi_0 + \frac{1}{2}\hbar\omega\psi_0$$

Or, $$-\frac{\hbar^2}{2m}\frac{\partial^2\psi_0}{\partial x^2}+\frac{1}{2}m\omega^2x^2\psi_0=\frac{1}{2}\hbar\omega\psi_0$$

Or, $$\left[-\frac{\hbar^2}{2m}\frac{\partial^2}{\partial x^2}+\frac{1}{2}m\omega^2x^2\right]\psi_0=\frac{1}{2}\hbar\omega\psi_0 \qquad [7]$$

Now comparing equation [3] and [7] we have

$$\boxed{E_0=\frac{1}{2}\hbar\omega} \qquad [8]$$

6.13 Matrix form of energy Eigen value of a Linear Harmonic Oscillator

The Hamiltonian operator for a Linear Harmonic Oscillator in one dimension

$$H=\frac{p_x^2}{2m}+\frac{1}{2}m\omega^2x^2 \qquad [1]$$

The Eigen value equation

$$H|E>=E|E> \qquad [2]$$

Let us consider two operators

$$a=\frac{1}{(2m\hbar\omega)^{\frac{1}{2}}}(m\omega x+ip_x) \qquad [3]$$

$$a^\dagger=\frac{1}{(2m\hbar\omega)^{\frac{1}{2}}}(m\omega x-ip_x) \qquad [4]$$

Since x and p_x are Hermitian operator so a and $a^\dagger$ will conjugate with respect to each other I.e.,

$$(a^\dagger)^\dagger=a \qquad [5]$$

Now, $[a,a^\dagger]=aa^\dagger-a^\dagger a$

$$[a,a^\dagger]=\frac{1}{2m\hbar\omega}\{(m\omega x+ip_x)(m\omega x-ip_x)-(m\omega x-ip_x)(m\omega x+ip_x)\}$$

$$=\frac{1}{2m\hbar\omega}\{m^2\omega^2x^2+p_x^2-im\omega xp_x+im\omega p_xx-m^2\omega^2x^2-im\omega xp_x+im\omega p_xx-p_x^2\}$$

$$=\frac{1}{2m\hbar\omega}\{-2im\omega xp_x+2im\omega p_xx\}=\frac{1}{2m\hbar\omega}\{-2im\omega(xp_x-p_xx)\}=\frac{1}{2m\hbar\omega}\{-2im\omega[x,p_x]\}$$

$$[a,a^\dagger]=\frac{1}{2m\hbar\omega}(-2im\omega)i\hbar \qquad \because [x,p_x]=i\hbar$$

$$[a,a^\dagger]=\frac{1}{2m\hbar\omega}(2m\hbar\omega)=1 \qquad \boxed{[a,a^\dagger]=1} \qquad [6]$$

Again $$aa^\dagger=\frac{1}{2m\hbar\omega}\{(m\omega x+ip_x)(m\omega x-ip_x)\}$$

$$= \frac{1}{2m\hbar\omega}\{m^2\omega^2x^2 + im\omega x p_x - im\omega p_x x + p_x^2\}$$

$$= \frac{1}{\hbar\omega}\left\{\frac{p_x^2}{2m} + \frac{1}{2}m\omega^2x^2 - \frac{i\omega}{2}(xp_x - p_x x)\right\}$$

$$= \frac{1}{\hbar\omega}\left\{\frac{p_x^2}{2m} + \frac{1}{2}m\omega^2x^2 - \frac{i\omega}{2}(x, p_x)\right\} = \frac{1}{\hbar\omega}\left\{\mathrm{H} - \frac{i\omega}{2}i\hbar\right\}$$

$$aa^\dagger = \frac{1}{\hbar\omega}\left\{\mathrm{H} + \frac{1}{2}\hbar\omega\right\} \Rightarrow \mathrm{H} = \hbar\omega\left(aa^\dagger - \frac{1}{2}\right)$$

Similarly, $\mathrm{H} = \hbar\omega\left(a^\dagger a + \frac{1}{2}\right) = \hbar\omega\left(\mathrm{N} + \frac{1}{2}\right)$ $[\because \mathrm{N} = a^\dagger a]$

Now, $[\mathrm{H}, a^\dagger] = \left[\hbar\omega\left(a^\dagger a + \frac{1}{2}\right), a^\dagger\right]$

$$= \hbar\omega\left\{[a^\dagger a] + \left[\frac{1}{2}, a^\dagger\right]\right\}$$

$$= \hbar\omega\{a^\dagger[a, a^\dagger] + [a^\dagger, a^\dagger]a\}$$

$$= \hbar\omega\{a^\dagger . 1 + 0\}$$

$$\Rightarrow [\mathrm{H}, a^\dagger] = \hbar\omega a^\dagger$$

$$\Rightarrow [\mathrm{H}, a^\dagger]\,|\mathrm{E}> = \hbar\omega a^\dagger|\,\mathrm{E}>$$

$$\Rightarrow \mathrm{H}a^\dagger|\mathrm{E}> - a^\dagger\mathrm{H}|\mathrm{E}> = \hbar\omega a^\dagger|\mathrm{E}>$$

$\Rightarrow \mathrm{H}a^\dagger|\mathrm{E}> - a^\dagger\mathrm{H}|\mathrm{E}> = \hbar\omega a^\dagger|\mathrm{E}>1;$ [Using equation (1)]

$$\Rightarrow \mathrm{H}a^\dagger|\mathrm{E}> = (\mathrm{E} + \hbar\omega)a^\dagger|\mathrm{E}> \quad [7]$$

Similarly, $\mathrm{H}a|\mathrm{E}> = (\mathrm{E} - \hbar\omega)a|\mathrm{E}>$ [8]

From equation [7] and [8], it is clear that the Hamiltonian operator H has the Eigen vectors $a^\dagger|E>$ and $a|E>$ and the corresponding Eigen values are $(\mathrm{E} + \hbar\omega)$ and $(\mathrm{E} - \hbar\omega)$ respectively. The value of E increasing and decreasing by $a^\dagger$ and a; So $a^\dagger$ and a are called the raising and lowering operators. Let the lowest value of E is E_0 and its Eigen state is $|\mathrm{E}_0>$; so

$$\mathrm{H}|\mathrm{E}_0> = \mathrm{E}_0|\mathrm{E}_0> \quad \because a|\mathrm{E}_0> = 0$$

$$\Rightarrow a^\dagger a|\mathrm{E}_0> = 0 \Rightarrow \left(\frac{\mathrm{H}}{\hbar\omega} - \frac{1}{2}\right)|\mathrm{E}_0> = 0 \quad \left[\because a^\dagger a = \frac{1}{\hbar\omega}\left(\mathrm{H} - \frac{1}{2}\hbar\omega\right)\right]$$

$$\Rightarrow \frac{\mathrm{E}_0}{\hbar\omega} = \frac{1}{2} \quad [\because |\mathrm{E}_0> \neq 0]$$

$$\Rightarrow \mathrm{E}_0 = \frac{1}{2}\hbar\omega \quad [9]$$

Equation [9] represents the ground state energy of a Linear Harmonic Oscillator.

Again, $Ha^{\dagger}|E_0> = (E_0 + \hbar\omega)a^{\dagger}|E_0>$

$$\Rightarrow Ha^{\dagger}|E_0> = \left(\frac{1}{2}\hbar\omega + \hbar\omega\right)a^{\dagger}|E_0>$$

$$\Rightarrow H|E_1> = \left(1 + \frac{1}{2}\right)\hbar\omega|E_1>$$

Similarly, $Ha^{\dagger}a^{\dagger}|E_0> = \left(2 + \frac{1}{2}\right)\hbar\omega a^{\dagger}a^{\dagger}|E_0>$

$$\Rightarrow H|E_2> = \left(2 + \frac{1}{2}\right)\hbar\omega|E_2>$$

$$\therefore H(a^{\dagger})^n|E_0> = \left(n + \frac{1}{2}\right)\hbar\omega(a^{\dagger})^n|E_0>$$

$$\Rightarrow H|E_n> = \left(n + \frac{1}{2}\right)\hbar\omega|E_n>$$

$$\boxed{\therefore E_n = \left(n + \frac{1}{2}\right)\hbar\omega} \quad [10]$$

Equation [10] gives the energy Eigen values of a Linear Harmonic Oscillator by the matrix form.

Exercise

1. Explain Bound States and Free States.
2. Define Linear Harmonic Oscillator [L.H.O.]. What is the difference between Harmonic motion and Simple Harmonic Motion?
3. Give a brief description on Quantum Mechanically Harmonic Oscillator.
4. Why Simple Harmonic Oscillator is an interesting problem?
5. State and explain the Degenerate and Non-Degenerate system.
6. Find the Number of energy- levels with the corresponding quantum numbers and the degree of degeneracy.
7. Find the energy Eigen value of one dimensional Linear Harmonic Oscillator.
8. Explain the non-degeneracy of Linear Harmonic Oscillator.
9. What is Zero point energy of Linear Harmonic Oscillator? Explain its physical significance.
10. Calculate the ground state energy of Linear Harmonic Oscillator by an alternate approach of Heisenberg space and time relation.
11. Calculate the ground state energy of Linear Harmonic Oscillator with relevant potential energy

$$\psi_0 = \left(\frac{\alpha}{\pi}\right)^{\frac{1}{4}} e^{-\alpha x^2} \quad \text{Where} \quad \alpha = \frac{m\omega}{\hbar}$$

12. Determine the Matrix form of energy Eigen value of a Linear Harmonic Oscillator.

Chapter 7 Atomic Orbitals of Hydrogen Atom

7.1 Atomic Spectra

When gaseous hydrogen in a glass tube is excited by a 5000-volt electrical discharge, four lines are observed in the visible part of the emission spectrum: red at 656 nm, blue-green at 486 nm, blue violet at 434 nm and violet at 410 nm.

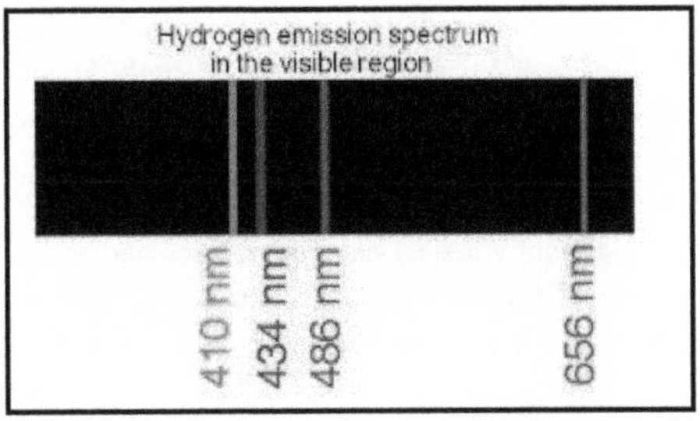

Fig.[7.1] Shows visible spectrum of atomic hydrogen.

Other series of lines have been observed in the ultraviolet and infrared regions. Rydberg (1890) found that all the lines of the atomic hydrogen spectrum could be fitted to a single formula

$$\frac{1}{\lambda} = \mathcal{R}\left(\frac{1}{n_1^2} - \frac{1}{n_2^2}\right), n_1 = 1, 2, 3, \dots\dots\dots\dots, \qquad n_2 > n_1 \qquad [1]$$

where $\mathcal{R}$, known as the Rydberg constant, has the value 109,677 cm^{-1} for hydrogen. The reciprocal of wavelength, in units of cm^{-1}, is in general use by spectroscopists. This unit is also designated *wave numbers*, since it represents the number of wavelengths per cm. The Balmer series of spectral lines in the visible region, shown in Fig.7.1 correspond to the values $n_1 = 2$, $n_2 = 3$, 4, 5, and 6. The lines with $n_1 = 1$ in the ultraviolet make up the Lyman series. The line with $n_2 = 2$, designated the Lyman alpha, has the longest wavelength (lowest wave number) in this series, with $1/\lambda = 82.258$ cm $\lambda = 121.57$ nm.

Other atomic species have line spectra, which can be used as a "fingerprint" to identify the element. However, no atom other than hydrogen has a simple relation analogous to (1) for its spectral frequencies. Bohr in1913 proposed that all atomic spectral lines arise from transitions between discrete energy levels, giving a photon such that

$$\Delta E = hv = \frac{hC}{\lambda} \qquad [2]$$

This is called the Bohr frequency condition. We now understand that the atomic transition energy ΔE is equal to the energy of a photon, as proposed earlier by Planck and Einstein.

7.2The Bohr Atom

The nuclear model proposed by Rutherford in 1911 pictures the atom as a heavy, positively-charged nucleus, around which much lighter, negatively-charged electrons circulate, much like planets in the Solar system. This model is however completely untenable from the standpoint of classical electromagnetic theory, for an accelerating electron (circular motion represents an acceleration) should radiate away its energy. In fact, a hydrogen atom should exist for no longer than 5×10^{-11}sec, time enough for the electron's death spiral into the nucleus. This is one of the worst quantitative predictions in the history of physics. It has been called the Hindenberg disaster on an atomic level. (Recall that the Hindenberg, a hydrogen-filled dirigible, crashed and burned in a famous disaster in 1937.)

Bohr sought to avoid an atomic catastrophe by proposing that certain orbits of the electron around the nucleus could be exempted from classical electrodynamics and remain stable. The Bohr model was quantitatively successful for the hydrogen atom, as we shall now show. We recall that the attraction between two opposite charges, such as the electron and proton, is given by Coulomb's law

$$F = -\frac{e^2}{r^2} \text{ [gaussian units] and } F = -\frac{e^2}{4\pi\epsilon_0 r^2} \text{ [SI units]} \quad [1]$$

We prefer to use the gaussian system in applications to atomic phenomena. Since the Coulomb attraction is a central force (dependent only on r), the potential energy is related by

$$F = -\frac{dV(r)}{dr} \quad [2]$$

We find therefore, for the mutual potential energy of a proton and electron

$$V(r) = -\frac{e^2}{r} \quad [3]$$

Bohr considered an electron in a circular orbit of radius r around the proton. To remain in this orbit, the electron must be experiencing a centripetal acceleration

$$a = -\frac{v^2}{r} \quad [4]$$

Where v is the speed of the electron.

Using [1] and [3] in Newton's second law, we find

$$\frac{e^2}{r^2} = \frac{mv^2}{r} \quad [5]$$

Where m is the mass of the electron.

For simplicity, we assume that the proton mass is infinite [actually $m_p \approx 1836m_e$] so that the proton's position remains fixed.

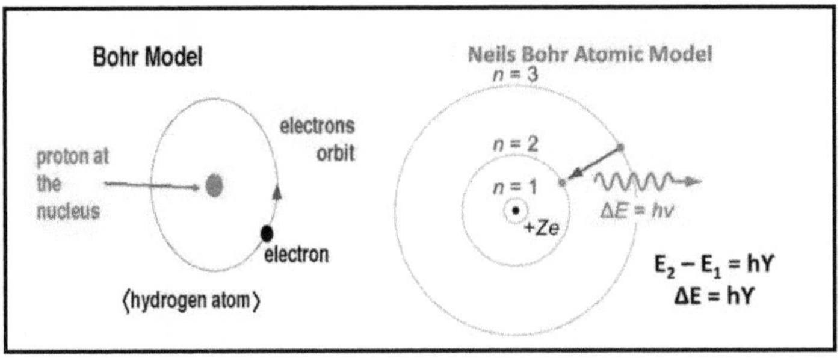

Fig.[7.2]: Shows Neils Bohr Atomic Model

We will later correct for this approximation by introducing reduced mass. The energy of the hydrogen atom is the sum of the kinetic and potential energies:

$$E = T + V = \frac{1}{2}mv^2 - \frac{e^2}{r} \qquad [6]$$

Using equation [5], we see that

$$T = -\frac{1}{2}V \quad \text{and } E = \frac{1}{2}V = -T \qquad [7]$$

This is the form of the virial theorem for a force law varying as r^{-2}. Note that the energy of a bound atom is negative, since it is lower than the energy of the separated electron and proton, which is taken to be zero. For further progress, we need some restriction on the possible values of r or v. This is where we can introduce the quantization of angular momentum $\bar{L} = \bar{r} \times \bar{p}$. Since $\bar{p}$ is perpendicular to $\bar{r}$ we can write simply

$$L = rp = mvr \qquad [8]$$

Using [7], we find also that

$$r = \frac{L^2}{me^2} \qquad [9]$$

We introducing angular momentum quantization, writing

$$L = n\hbar, \qquad n = 1, 2, 3, \dots \dots \dots \dots \dots \dots \dots \qquad [10]$$

excluding $n = 0$, since the electron would then not be in a circular orbit. The allowed orbital radii are then given by

$$r_n = n^2 a_0 \qquad [11]$$

Where $$a_0 = \frac{\hbar^2}{me^2} = 5.29 \times 10^{-11}\text{m} = 0.529\text{Å} \qquad [12]$$

which is known as the Bohr radius. The corresponding energy is

$$E_n = -\frac{e^2}{2a_0n^2} = -\frac{me^4}{2\hbar^2n^2}, \qquad n = 1, 2, 3, \ldots \qquad [13]$$

Rydberg's formula [1] can now be deduced from the Bohr model. We have

$$\frac{hc}{\lambda} = E_{n_2} - E_{n_1} = \frac{2\pi^2 me^4}{h^2}\left(\frac{1}{n_1^2} - \frac{1}{n_2^2}\right) \qquad [14]$$

And the Rydbeg constant can be identified as $\mathcal{R} = \frac{2\pi^2 me^4}{h^3c} \approx 109{,}737\ cm^{-1}$ [15]

The slight discrepancy with the experimental value for hydrogen [109677] is due to the finite proton mass. This will be corrected later. The Bohr model can be readily extended to hydrogen like ions, systems in which a single electron orbits a nucleus of arbitrary atomic number Z. Thus Z = 1 for hydrogen, Z = 2 for He^+, Z = 3 for Li^{++}, and so on. The Coulombs potential [3]

$$V(r) = -\frac{Ze^2}{r} \qquad [16]$$

The radius of the orbit [11] becomes

$$r_n = \frac{n^2a_0}{Z} \qquad [17]$$

And the energy becomes

$$E_n = -\frac{Z^2e^2}{2a_0n^2} \qquad [18]$$

de-Broglie's proposal that electrons can have wavelike properties was actually inspired by the Bohr atomic model

$$L = rp = n\hbar = \frac{nh}{2\pi} \qquad [19]$$

We find

$$2\pi r = \frac{nh}{p} = n\lambda \qquad [20]$$

Therefore, each allowed orbit traces out an integral number of de- Broglie wave lengths. Wilson [1915] and Sommerfeld [1916] generalized Bohr's formula for the allowed orbits to

$$\oint p\,dr = nh \qquad [21]$$

7.3 Bohr- Sommerfeld Orbit

The Sommerfeld-Wilson quantum conditions $\oint pdr = nh$ reduce to Bohr's results for circular orbits, but allow, in addition, elliptical orbits along which the momentum p is variable. According to Kepler's first law of planetary motion, the orbits of planets are ellipses with the Sun at one focus. Fig.[7.3] shows the generalization of the Bohr theory for hydrogen, including the elliptical orbits. The lowest energy state is $n = 1$ still a circular orbit. But $n = 2$ allows an elliptical orbit in addition to the circular one; $n = 3$ has three possible orbits, and so on. The energy still depends on n alone, so that the elliptical orbits represent degenerate states. Atomic spectroscopy shows in fact that energy levels with $n > 1$ consist of multiple states, as implied by the splitting of atomic lines by an electric field (Stark effect) or a magnetic field (Zeeman effect). Some of these generalized orbits are drawn schematically in the following figure.

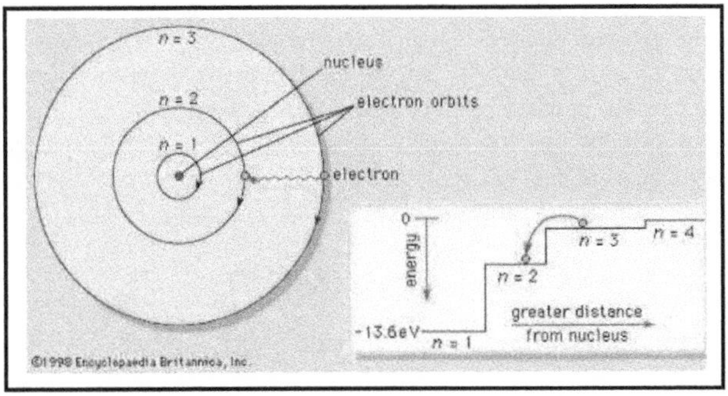

Fig.[7.3]: Bohr-Sommerfeld orbits for n= 1; 2; 3

The Bohr model was an important first step in the historical development of quantum mechanics. It introduced the quantization of atomic energy levels and gave quantitative agreement with the atomic hydrogen spectrum. With the Sommerfeld-Wilson generalization, it accounted as well for the degeneracy of hydrogen energy levels. Although the Bohr model was able to sidestep the atomic "Hindenberg disaster," it cannot avoid what we might call the "Heisenberg disaster." By this we mean that the assumption of well-defined electronic orbits around a nucleus is completely contrary to the basic premises of quantum mechanics. Another flaw in the Bohr picture is that the angular momenta are all too large by one unit, for example, the ground state actually has zero orbital angular momentum (rather than h- bar)

7.4 Orbitals

Before we go into great detail about the quantum numbers, it is important to note that when we say location, I mean probable location. There is really no way to know *exactly* where an electron is at a given time; they are very elusive. But it is possible to determine which specific three-dimensional region it is probably in. These three-dimensional boundaries where an electron is most likely found are called an atomic orbital. By solving the Schrödinger equation [HΨ=EΨ, Since H is a Hamiltonian operator] we obtain a set of mathematical equations, called wave function [Ψ], which describes the probability of finding electrons at certain energy levels within an atom. *A wave function for an electron in an atom is called an atomic orbital*; this atomic orbital describes a region of space in which there is a high probability of finding the electron.

7.5 Atomic Orbitals

The general solution for $R_{nl}(r)$ has a rather complicated form which we give without proof:

$$R_{nl}(r) = N_{nl}\rho^{l}L_{n+l}^{2l+1}(\rho)e^{-\rho/2} \qquad where \equiv \frac{2Zr}{n} \qquad [1]$$

Here L_{n+l}^{2l+1} is an associated Laguerre polynomial and N_{nl} is the normalization constant. The angular momentum quantum number l is by convention designated by a code: s for $l= 0$, p for $l=1$, d for $l= 2$, f for $l=3$, g for $l=4$, and so on. The first four letters come from an old classification scheme for atomic spectral lines: *sharp, principal, diffuse and fundamental.* Although these designations have long since outlived their original significance, they remain in general use. The solutions of the hydrogenic Schrodinger equation in spherical polar coordinates can now be written in full

$$\psi_{nlm}(r,\theta,\varphi) = R_{nl}(r)Y_{lm}(\theta,\varphi)$$

$$n = 1,2 \ldots.. \qquad l = 0,\ 1, \ldots\ldots\ldots\ldots.., n-1 \qquad m = 0,\ \pm1,\ \pm2 \ldots\ldots\ldots\ldots.., \pm l \qquad [2]$$

where Y_{lm} are the spherical harmonics. Table below enumerates all the hydrogenic functions we will actually need. These are in the following table- 1:

$\psi_{1s} = \frac{1}{\sqrt{\pi}} e^{-r}$	ψ_{3p_x}, ψ_{3p_y} analogous
$\psi_{2s} = \frac{1}{2\sqrt{2\pi}}\left(1-\frac{r}{2}\right)e^{-r/2}$	$\psi_{3d_{z^2}} = \frac{1}{81\sqrt{6\pi}}(3z^2-r^2)e^{-r/3}$
$\psi_{2p_z} = \frac{1}{4\sqrt{2\pi}} z\, e^{-r/2}$	$\psi_{3d_{zx}} = \frac{\sqrt{2}}{81\sqrt{\pi}} zx\, e^{-r/3}$
ψ_{2p_x}, ψ_{2p_y} analogous	$\psi_{3d_{yz}}, \psi_{3d_{xy}}$ analogous
$\psi_{3s} = \frac{1}{81\sqrt{3\pi}}(27-18r+2r^2)e^{-r/3}$	$\psi_{3d_{x^2-y^2}} = \frac{1}{81\sqrt{\pi}}(x^2-y^2)e^{-r/3}$
$\psi_{3p_z} = \frac{\sqrt{2}}{81\sqrt{\pi}}(6-r)z\, e^{-r/3}$	

Table shows the real hydrogenic functions in atomic units

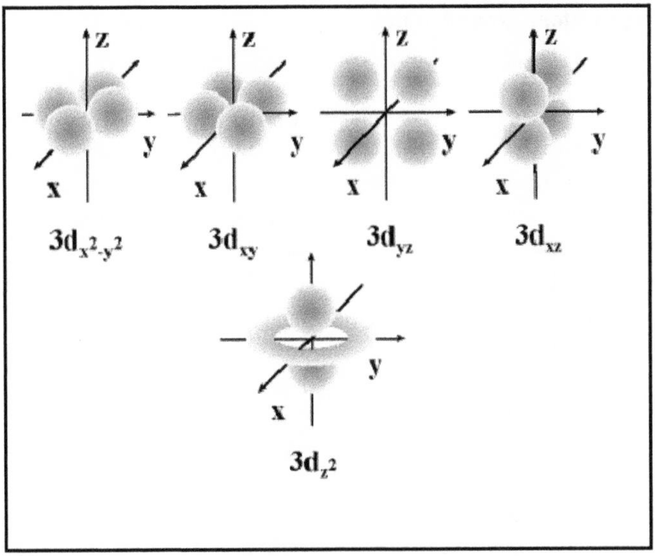

Fig.[7.5(a)]: Types of atomic orbitals of Hydrogen atom

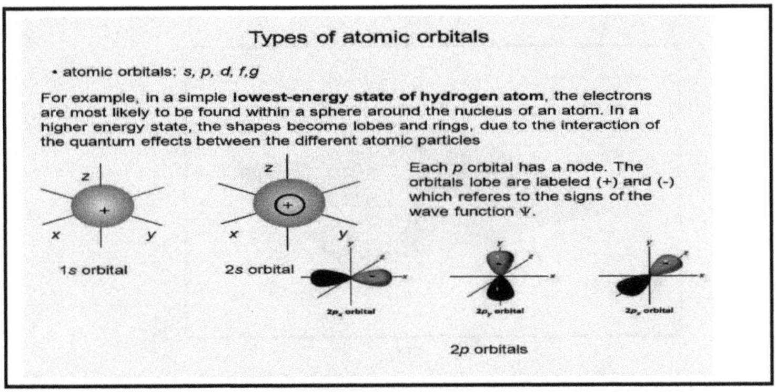

Fig.[7.5(b)]: Types of atomic orbitals of Hydrogen atom

7.6 Hydrogen atom ground state: The ground state wave function of hydrogen atom

$$\psi_{100} = \frac{1}{\sqrt{\pi a_0^3}}\, e^{-\frac{r}{a_0}} \qquad [1]$$

There are a number of different ways of representing hydrogen atom wave functions graphically. We will illustrate some of these for the 1s ground state. In atomic units equation becomes

$$\psi_{1s}(r) = \frac{1}{\sqrt{\pi}}\, e^{-r} \qquad [2]$$

is a decreasing exponential function of a single variable r and is simply plotted below

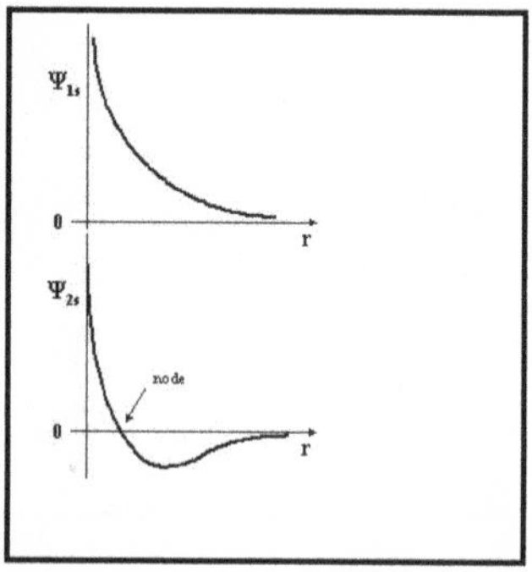

Fig.[7.6.1]: Shows the wave functions for 1s and 2s orbital for atomic hydrogen. The 2s-function [scaled by a factor of 2] has a node at r = 2 Bohr.

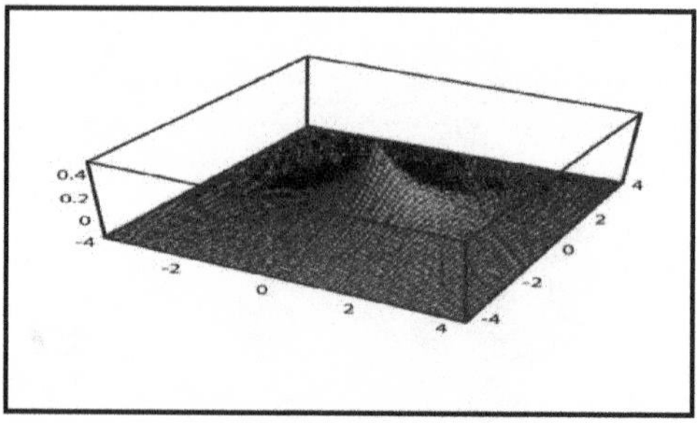

Fig.[7.6.2]: Shows the Contour map of 1s orbital in the x, y plane

Fig.[7.6.2] gives a somewhat more pictorial representation, a three dimensional contour plot of $\psi_{1s}(r)$ as a function of x and y in the x, y plane.

According to Born's interpretation of the wave function, the probability per unit volume of finding the electron at the point[r, θ, ϕ] is equal to the square of the normalized wave function

$$\rho_{1s}(r) = |\psi_{1s}(r)|^2 = \frac{1}{\pi}e^{-2r} \quad [3]$$

This is represented in Fig.[7.6.3] by a scatter plot describing a possible sequence of observations of the electron position. Although results of individual measurements are not predictable, a statistical pattern does emerge after a sufficiently large number of measurements.

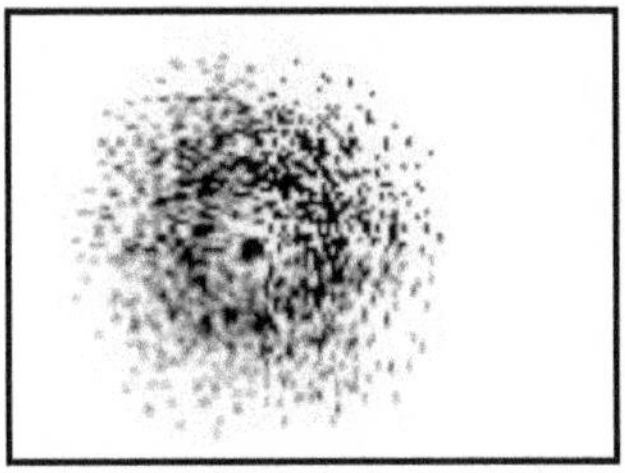

Fig.[7.6.3]: Shows the Scatter plot of electron position measurements in hydrogen 1s orbital

The probabilty density is normalized such that

$$\int_0^{\infty} \rho_{1s}(r)\, 4\pi r^2 dr = 1 \quad [4]$$

In some ways $\rho(r)$ does not provide the best description of the electron distribution, since the region around $r = 0$, where the wave function has its largest values, is a relatively small fraction of the volume accessible to the electron. Larger radii r represent larger physical regions since, in spherical polar coordinates, a value of r is associated with a shell of volume $4\pi r^2 dr$. A more significant measure is therefore the radial distribution function [RDF].

$$D_{1s}(r) = 4\pi r^2\, |\psi_{1s}(r)|^2 \quad [5]$$

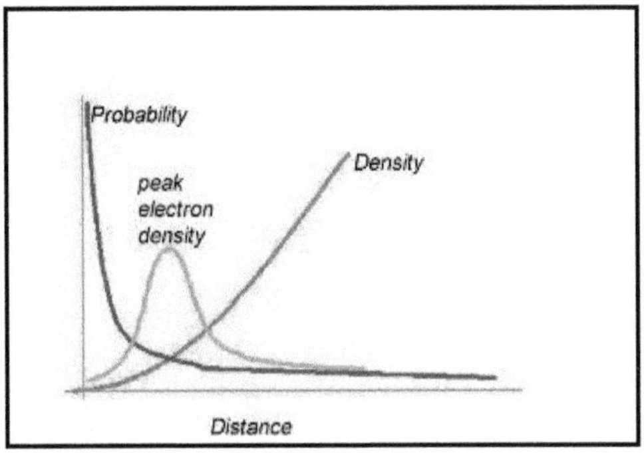

Fig.[7.6.4]: Shows Density $\rho(r)$ and RDF $D(r)$ for hydrogen 1s orbital

Equation [5] which represents the probability density within the entire shell of radius r, normalized such that

$$\int_0^{\infty} D_{1s}(r)dr = 1 \qquad [6]$$

The functions $\rho_{1s}(r)$ and $D_{1s}(r)$ are both shown in Fig.[7.6.4]. Remarkably, the 1s RDF has its maximum at $= a_0$, equal to the radius of the first Bohr orbit.

7.7 p- orbitals and d- orbitals

The lowest-energy solutions deviating from spherical symmetry are the 2p- orbitals. Using equation [1] and [2] from the previous section 7.5

$$R_{nl}(r) = N_{nl}\rho^l L_{n+l}^{2l+1}(\rho)e^{-\rho/2} \qquad where \equiv \frac{2Zr}{n}$$

$$\psi_{nlm}(r,\theta,\varphi) = R_{nl}(r)Y_{lm}(\theta,\varphi)$$

$n = 1,2 \ldots..$ $\qquad l = 0,\ 1, \ldots\ldots\ldots\ldots.., n-1$ $\qquad m = 0,\ \pm1,\ \pm2 \ldots\ldots\ldots\ldots.., \pm l$
and the $l = 1$ spherical harmonics, we find three degenerate Eigen functions:

$$\psi_{210}(r,\theta,\varphi) = \frac{1}{4\sqrt{2\pi}}\, re^{-r/2}cos\theta \qquad [1]$$

And

$$\psi_{21\pm1}(r,\theta,\varphi) = \mp\frac{1}{4\sqrt{2\pi}}\, re^{-r/2}sin\theta e^{\pm i\varphi} \qquad [2]$$

The function ψ_{210} is real and contains the factor $rcos\theta$, which is equal to the cartesian variable z. In chemical applications, this is designated as a $2p_z$ orbitals:

$$\psi_{2p_z} = \frac{1}{4\sqrt{2\pi}}\, ze^{-r/2} \qquad [3]$$

A contour plot is shown in Fig.[7.7.1] Note that that this function is cylidrically symmetrical about the z axis with a node in the *x, y* plane. The $\psi_{21\pm1}$ are complex functions and not as easy to represent graphically.

Making use of the Euler formulas for *sin* and *cosine*

$$cos\varphi = \frac{e^{i\varphi}+e^{-i\varphi}}{2} \qquad \text{And } sin\varphi = \frac{e^{i\varphi}-e^{-i\varphi}}{2i} \qquad [4]$$

and noting that the combinations $sin\theta cos\varphi$ and $sin\theta sin\varphi$ corresponds the cartesian variables *x* and *y*, respectively, we can define the alternative *2p* orbitals

$$\psi_{2p_x} = \frac{1}{\sqrt{2}}(\psi_{21-1} - \psi_{211}) = \frac{1}{4\sqrt{2\pi}}\, x\, e^{-r/2} \qquad [5]$$

and

$$\psi_{2p_y} = -\frac{i}{\sqrt{2}}(\psi_{21-1} + \psi_{211}) = \frac{1}{4\sqrt{2\pi}}\, y e^{-r/2} \qquad [6]$$

Clearly, these have the same shape as the $2p_z$- orbital, but are oriented along the x- and y- axes respectively. The three fold degeneracy of the p- orbitals is very clearly shown by the geometric equivalence the functions $2p_x$, $2p_y$ and $2p_z$, which is not obvious for the spherical harmonics. The functions listed in the table- 1 are, in fact, the real forms for all atomic orbitals, which are more useful in chemical applications. All higher p- orbitals have analogous functional forms *xf(r), y f(r) and z f(r)* and are likewise 3-fold degenerate.

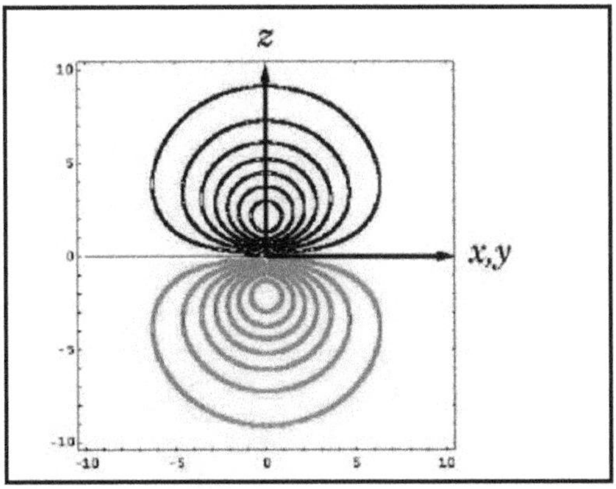

Fig.[7.7.1]: Shows Contour plot of $2p_Z$ orbital. Negative values are shown in red in the lower portion. Scale units in Bohrs.

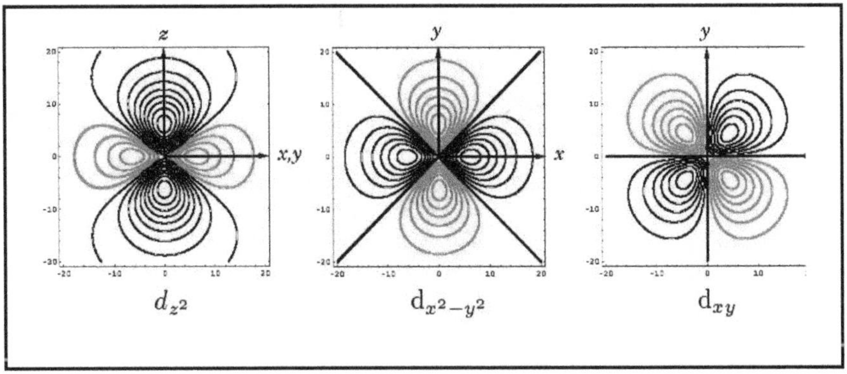

Fig.[7.7.2]: Shows Contour plots of 3d orbitals

The orbital ψ_{320} is like ψ_{210} , is a real function. It is known in chemistry as the d_{z^2} – orbital and can be expressed as a cartesian factor times a function of r.

$$\psi_{3d_{z^2}} = \psi_{320} = (3z^2 - r^2)f(r) \qquad [7]$$

A contour plot is shown in Fig.[7.7.2]. This function is also cylindrically symmetric about the z-axis with two angular nodes-the conical surfaces with $3z^2 - r^2 = 0$. The remaining four 3d orbitals are complex functions containing the spherical harmonics $Y_{2\pm1}$ *and* $Y_{2\pm2}$ are not shown in Fig.[7.7.2]. We can again construct real functions from linear combinations, the result being four geometrically equivalent "four-leaf clover" functions with two perpendicular planar nodes. These orbitals are designed $d_{x^2-y^2}$, d_{xy}, d_{zx} and d_{yz}. Two of them are shown in figure [7.7.2] The d_{z^2} orbital has a different shape. However, it can be expressed in terms of two non- standard d-orbitals $d_{z^2-x^2}$ *and* $d_{y^2-z^2}$. The latter functions, along with $d_{x^2-y^2}$ add to zero and thus constitute a linearly *dependent set.* Two combinations of these three functions can be chosen as independent Eigen functions.

7.8 Radial Distribution Function [RDF]

The electron charge distribution in an orbital $\psi_{nlm}(r)$ is given by

$$\rho(r) = |\psi_{nlm}(r)|^2 \qquad [1]$$

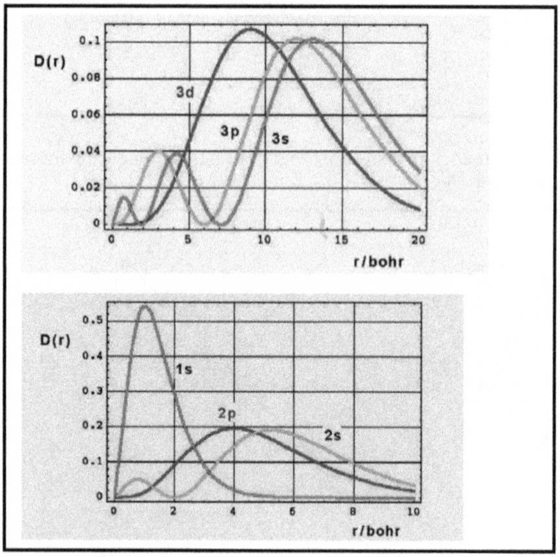

Fig.[7.8]: Some radial distribution function[RDF]

which for the s- orbitals is a function of r alone. The radial distribution function can be defined, even for orbitals containing angular dependence, by

$$D_{nl}(r) = r^2|R_{nl}(r)|^2 \quad [2]$$

This represents the electron density in a shell of radius r, including all values of the angular variables θ, ϕ. Fig.[7.8] shows plots of the radial distribution function[RDF] for the first few hydrogen orbitals.

A more significant measure is therefore the radial distribution function[RDF].

$$D_{1s}(r) = 4\pi r^2|\psi_{1s}(r)|^2 \quad [3]$$

The probabilty density is normalized such that

$$\int_0^\infty \rho_{1s}(r)4\pi r^2 dr = 1 \quad [4]$$

Equation [3] which represents the probabilty density within the entire shell of radius r, normalized such that

$$\int_0^\infty D_{1s}(r)dr = 1 \quad [5]$$

7.9 Significance of Quantum Numbers

A quantum number describes a specific aspect of an electron. Quantum numbers often describe specifically the energy levels of electrons in atoms, but other possibilities include angular momentum, spin, etc., we have four ways of defining the properties of an electron or four quantum numbers. *The first three quantum numbers [n, l, m_l,] describe the size, shape, and orientation of the atomic orbitals in space.* The fourth [m_s] specifies how many electrons can occupy that orbitals. This is an additional quantum number which does not follow the Schrödinger wave equation but is introduced to account for electron spin.

7.10 Quantum Numbers

Quantum numbers describe values of conserved quantities in the dynamics of a quantum system. In the case of electrons, the quantum numbers can be defined as *"the sets of numerical values which give acceptable solutions to the Schrödinger wave equation for the hydrogen atom".* An important aspect of quantum mechanics is the quantization of the observable quantities, since quantum numbers are discrete sets of integers or half-integers, although they could approach infinity in some cases. This distinguishes quantum mechanics from classical mechanics where the values that characterize the system such as mass, charge, or momentum, range continuously. Quantum numbers often describe specifically the energy levels of electrons in atoms, but other possibilities include angular momentum, spin, etc. Any quantum system can have one or more quantum numbers; it is thus difficult to list all possible quantum numbers. *Quantum numbers are derived from the mathematical solutions of Schrödinger's Wave Equation for the hydrogen atom.*

7.11 Principal Quantum Number[n]

We find Principal Quantum Number[n] by solving the differential equation for $R(r)$

$$\frac{1}{r^2}\frac{\partial}{\partial r}\left(r^2\frac{\partial R}{\partial r}\right)+\left[\frac{2m}{\hbar^2}\left(\frac{e^2}{4\pi\varepsilon_0 r}+E\right)-\frac{l(l+1)}{r^2}\right]R(r)=0.$$

It can be solved only for energies E which satisfy the same condition as we found on the energies for the Bohr atom

$$E_n = -\frac{me^4}{32\pi^2\epsilon_0^2\ \hbar^2}\cdot\frac{1}{n^2} = \frac{E_1}{n^2},\quad n = 1, 2, 3, \ldots..$$

This $n = 1, 2, 3, \ldots..$ is known as the *principal quantum number.*

The first quantum number that describes an electron is called the principal quantum number. It is often symbolized by the letter *n*. This number tells us the energy level or size of an orbital. The principal quantum number [n] can have only positive non-zero integral values (n = 1, 2,3,4...).*This means that in an atom, the electron can have only certain energies.* The principal quantum number also determines the mean distance of the electron from the nucleus (**size**). Greater the value of n farther is the electron from the nucleus. All orbitals that have the same value of n are said to be in the same shell [level]

Each principal shell can accommodate a maximum of $2n^2$ electrons.

- n = 1, number of electrons: 2
- n = 2, number of electrons: 8
- n = 3, number of electrons: 18
- n = 4, number of electrons: 32

The higher the number, the larger the region is. So let's take the electron configuration for hydrogen atom with $n = 1$ the electron is in its ground state; if the electron is in the $n = 2$ orbital, it is in an excited state. The total number of orbitals for a given n value is n^2**.**

7.12 Angular momentum [secondary, azimuthal] Quantum number

We find orbital quantum number by solving the differential equation for $\Theta(\theta)$

$$\frac{1}{\sin\theta}\frac{\partial}{\partial\theta}(\sin\theta\frac{\partial\Theta(\theta)}{\partial\theta}) + [l(l+1) - \frac{m_l^2}{\sin^2\theta}]\,\Theta(\theta) = 0$$

It involves the term $\quad l(l+1) - \dfrac{m_l^2}{\sin^2\theta}$

It turns out from the above differential equations that the equation for $\Theta(\theta)$ can be solved only if l is an integer greater than or equal to the absolute value of m_l.

Now we have found another quantum number, the *orbital quantum number*, and the requirement on l can be restated as

$$\mathrm{m}_l = 0,\ \pm 1, \pm 2, \pm 3, \dots\dots\dots\dots, \pm l$$

The azimuthal or orbital angular momentum quantum number, l is related to the geometrical shape of the orbital. The value of l may be zero or a positive integer less than or equal to n–1 (The values of l are integers that depend on the value of the principal quantum number, n). For any given value of n, the possible range of values for l goes from 0 to n -1. That is $1 = 0, 1, 2, \dots\dots\dots\dots, (n-1)$

- If n = 1, there's only one possible value of l; that is, 0. If n = 2, there are two values for l: 0 and 1. If n = 3, l has three values: 0, 1, and 2
- These values of l have letter designations: *s, p, d, and f.* So if l = 0 we have *s* orbital; if l = 1, the orbital is *p;* if l = 2, the orbital is *d, and if* l = 3, the orbital is *f, if* l = 4, the orbital is *g*, *if* l = 5, the orbital is *h* and so on. The first four letters come from an old classification scheme for atomic spectral lines: *sharp, principal, diffuse and fundamental.*

l	0	1	2	3	4	5	...
Letter	*s*	*p*	*d*	*f*	*g*	*h*	...

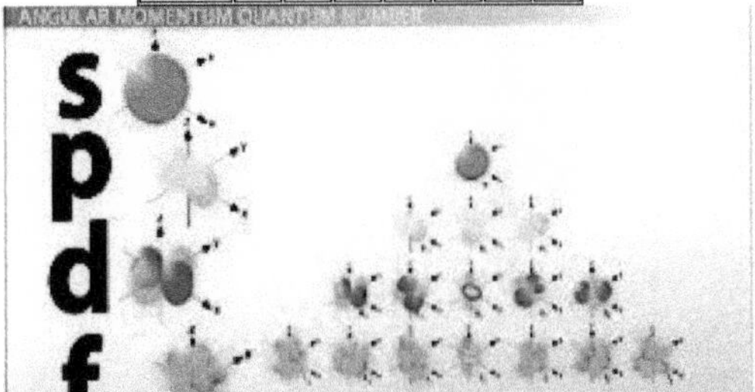

Fig.[7. 13]The angular momentum quantum number refers to the shapes of orbitals

- Since this is the shape quantum number, use those letter designations to help us remember the electron density shapes. 's' is spherical; 'p' looks like a peanut; 'd' reminds me of a daisy, and 'f' is generally harder to predict.

7.13 Magnetic Quantum Number [m_l]

We find Magnetic Spin Quantum Number by solving the differential equation for ϕ.

$$\frac{\partial^2\Phi(\phi)}{\partial\phi^2} + m_l^{\,2}\Phi(\phi) = 0$$

The above equation has solutions which are Sines and Cosines. We write the general solution

$$\Phi(\phi) = Ae^{im_l\varphi}$$

We will get the constant by normalization. Now, since ϕ and [$\phi+2\pi$] represent a single point in space, we must have

$$\Phi(\phi) = Ae^{im_l\varphi} = Ae^{im_l(\phi+2\pi)}. \textit{ This happens only for } \quad \mathrm{m}_l = 0, \ \pm1, \pm2, \pm3, ...$$

For reasons which are not yet obvious, $\mathrm{m}_l = [\pm1, \pm2, \pm3, ...]$ is called the *magnetic quantum number*. This describes the orientation of the orbital in space, or the direction along the axes in which it faces. Within a subshell, the value of this depends on the value of l. For a certain value of l, there are $(2l + 1)$ integral values, or $m_l = -l, \ldots 0, \ldots +l$. Thus the *s* subshell has only one orbtal, the *p* subshell has three orbitals, and so on.

7.14 Magnetic Spin Quantum number

The quantum number, m_s, describes the spin of the electron [Electrons behave in some respects

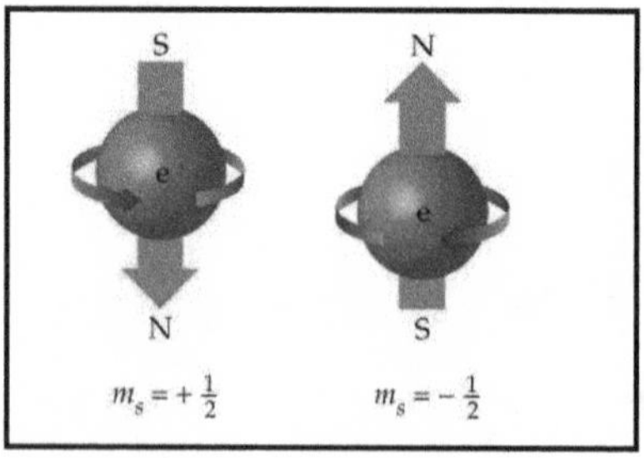

Fig.[7. 14]The Spin angular momentum quantum number refers to the shapes of orbitals

as if they were tiny charges spheres spinning around an axis.] whether it is clockwise or anticlockwise. The quantum number does not arise while solving SWE. The clockwise and anticlockwise direction of electron spin has arbitrarily been assigned the values as +½ and –½ respectively.

The Pauli Exclusion Principle states that no two electrons in the same atom can have identical values for all four of their quantum numbers. This means that in an atom with more than one electron, no two electrons can have the same set of 4 quantum numbers. What this means uniquely? *This means that no more than two electrons can occupy the same orbital, and that two electrons in the same orbital must have* ***opposite spins.*** Electromagnetic theory states that a spinning charge generates a magnetic field. Because an electron spins, it creates a magnetic field, which can be oriented in one of two directions. For two electrons in the same orbital, the spin must be opposite to each other; the spins are said to be paired. These substances are not attracted to magnets and are said to be diamagnetic. Atoms with more electrons that spin in one direction than another contain unpaired electrons. These substances are weakly attracted and are said to be paramagnetic.

7.15 Table of Allowed Quantum Numbers[32]

n	l	m_l	**Number of orbitals**	**Orbital Name**	**Number of electrons**
1	0	0	1	1*s*	2
2	0	0	1	2*s*	2
	1	-1, 0, +1	3	2*p*	6
3	0	0	1	3*s*	2
	1	-1, 0, +1	3	3*p*	6
	2	-2, -1, 0, +1, +2	5	3*d*	10
4	0	0	1	4*s*	2
	1	-1, 0, +1	3	4*p*	6
	2	-2, -1, 0, +1, +2	5	4*d*	10
	3	-3, -2, -1, 0, +1, +2, +3	7	4*f*	14

Exercise

1. Explain atomic Spectra.
2. Give an idea about the Bohr model. Show that Wilson [1915] and Sommerfeld [1916] generalized Bohr's formula for the allowed orbits to $\oint pdr = nh.$
3. Write some comments on Bohr- Sommerfeld Orbit.
4. Write a short note on:

(a) Orbitals
(b) Atomic Orbitals
(c) p- orbitals and d- orbitals
(d) Radial Distribution Function [RDF]

5. Define Quantum Numbers. What is the Significance of Quantum Numbers?
6. Explain the following Quantum Numbers

(a) Principal Quantum Number[n]
(b) Angular momentum [secondary, azimuthal] Quantum number
(c) Magnetic Quantum Number [m_l]
(d) Magnetic Spin Quantum number

7. Determine the value of the longest wavelength of the Paschen series.
8. Derive an express for the time period of an electron in an n-th Bohr orbit.
9. Using the Bohr theory, quantise a hydrogen like atom for the case of elliptic orbits. Show that the period of electron remains the same as for circular orbits.
10. Give an account of Bohr theory of hydrogen spectrum and show how Rydberg constant was obtained in terms of fundamental constants.

Chapter 8 Quantum mechanics of Hydrogen like atoms

8.1 Hydrogen atom is an interesting problem

The hydrogen atom is a real physical system that can be treated exactly by Quantum Mechanics. That is, Quantum Mechanics works very well for describing the hydrogen atom.

- The hydrogen atom is a system consists of a proton & a single electron bound by the electrostatic force of attraction. So it is a two body system. *The Schrödinger wave equation for this system can be solved exactly.*
- This electrostatic force is radially direct. It is a central force problem, the angular momentum is conserved & the problem is spherically symmetric.
- The hydrogen atom is the most abundant & the simplest atom in the Universe.

8.2 Two body system problem with central force interaction

We consider a hydrogen atom for two body system. The mass of the electron and the mass of the proton are m_1 and m_2 respectively. Both are at the head of the vector $\bar{r}_1$and $\bar{r}_2$ of a Laboratory system. These $\bar{r}_1$ and $\bar{r}_2$ are called Laboratory coordinates. Therefore, *a coordinate system in which the bombarded particles or target is initially at rest is known as Laboratory frame or system*[5.24].

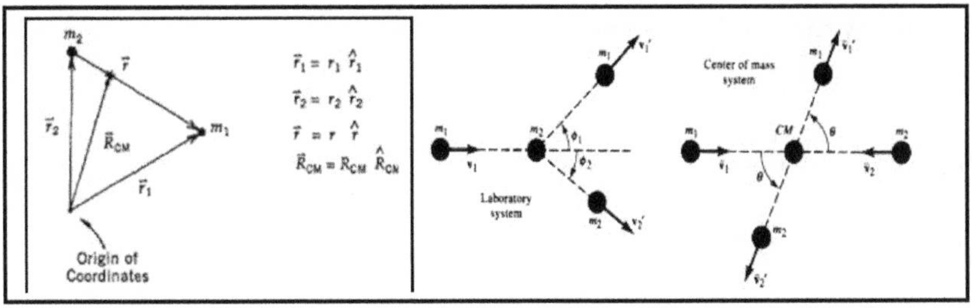

Fig.[8.2]: Shows a comparison of Laboratory and centre of mass coordinates.

Furthermore, the two masses m_1 and m_2 can also be located with respect to the centre of mass by radii vector $\acute{r}_1$ and $\acute{r}_2$ [The location of m_1 and m_2 are $\acute{r}_1$ and $\acute{r}_2$ respectively with respect to Center of mass [CM] system].These are called centre of mass coordinates. Therefore, *a coordinate system in which the centre of mass of two colliding particles is initially and always at rest is known as centre of mass frame or system.* Center of mass [CM] is the head of the vector $\bar{R}$ of the Laboratory system. $\bar{r} = \bar{r}_1 - \bar{r}_2 = \acute{r}_1 - \acute{r}_2$ is the position of the electron with respect to the proton. For these two particles with respect to Center of mass [CM] system, the classical Hamiltonian function H_{fCM} can be written as

$$H_{fCM} = \frac{1}{2}m_1\dot{\bar{r}}_1^2 + \frac{1}{2}m_2\dot{\bar{r}}_2^2 - \frac{e^2}{4\pi\varepsilon_0|\bar{r}_1 - \bar{r}_2|} \quad [1]$$

$$H_{fCM} = \frac{1}{2}m_1\dot{\bar{r}}_1^2 + \frac{1}{2}m_2\dot{\bar{r}}_2^2 - \frac{e^2}{4\pi\varepsilon_0 r} = \frac{1}{2}m_1\left(\frac{m_2}{m_1+m_2}\dot{\bar{r}}\right)^2 + \frac{1}{2}m_2\left(-\frac{m_1}{m_1+m_2}\dot{\bar{r}}\right)^2 - \frac{e^2}{4\pi\varepsilon_0 r}$$

Using $\dot{\bar{r}}_1 = \frac{m_2}{m_1+m_2}\dot{\bar{r}}$ and $\dot{\bar{r}}_2 = -\frac{m_1}{m_1+m_2}\dot{\bar{r}}$

$$H_{fCM} = \frac{1}{2}\frac{m_1 m_2}{(m_1+m_2)^2}(m_1+m_2)\dot{\bar{r}}^2 - \frac{e^2}{4\pi\varepsilon_0 r} = \frac{1}{2}\frac{m_1 m_2}{(m_1+m_2)}\dot{\bar{r}}^2 - \frac{e^2}{4\pi\varepsilon_0 r}$$

$$H_{fCM} = \frac{1}{2}\mu\dot{\bar{r}}^2 - \frac{e^2}{4\pi\varepsilon_0 r} \quad [2]$$

Where $\mu = \frac{m_1 m_2}{(m_1+m_2)}$ is called the reduced mass. Hamiltonian operator corresponding to the above classical Hamiltonian function H_{fCM} can be obtained by replacing the momentum $\mu\dot{\bar{r}}$ in equation [2] by the operator $\frac{\hbar}{i}\bar{\nabla}$; where $\bar{\nabla}$ corresponds to the coordinates r, θ *and* φ. Therefore, the Hamiltonian operator is

$$H_{op} = -\frac{\hbar^2}{2\mu}\nabla^2 - \frac{e^2}{4\pi\varepsilon_0 r} \quad [3]$$

8.3 Energy Eigen value equation of hydrogen atom

From three dimensional Schrödinger equation

$$\nabla^2\psi(\bar{r}) + \frac{2m}{\hbar^2}[E - V(\bar{r})]\psi(\bar{r}) = 0 \quad [1]$$

In spherical polar coordinate, the Laplacian operator ∇^2 can be expressed as

$$\nabla^2 = \frac{1}{r^2}\frac{\partial}{\partial r}(r^2\frac{\partial}{\partial r}) + \frac{1}{r^2\sin\theta}\frac{\partial}{\partial\theta}(\sin\theta\frac{\partial}{\partial\theta}) + \frac{1}{r^2\sin\theta}\frac{\partial^2}{\partial\phi^2}$$

$$\nabla^2\psi = \frac{1}{r^2}\frac{\partial}{\partial r}(r^2\frac{\partial\psi}{\partial r}) + \frac{1}{r^2\sin\theta}\frac{\partial}{\partial\theta}(\sin\theta\frac{\partial\psi}{\partial\theta}) + \frac{1}{r^2\sin\theta}\frac{\partial^2\psi}{\partial\phi^2} \quad [2]$$

$$= [\frac{1}{r^2}\frac{\partial}{\partial r}(r^2\frac{\partial}{\partial r}) + 2r\frac{\partial}{\partial r}) + \frac{1}{r^2\sin\theta}\frac{\partial}{\partial\theta}(\sin\theta\frac{\partial}{\partial\theta}) + \frac{1}{r^2\sin\theta}\frac{\partial^2}{\partial\varphi^2}]\psi$$

$$= \frac{\partial^2\psi}{\partial r^2} + \frac{2}{r}\frac{\partial\psi}{\partial r} + \frac{1}{r^2\sin\theta}\frac{\partial}{\partial\theta}(\sin\theta\frac{\partial\psi}{\partial\theta}) + \frac{1}{r^2\sin^2\theta}\frac{\partial^2\psi}{\partial\varphi^2}$$

Therefore, equation [1] becomes

$$\frac{\partial^2\psi}{\partial r^2}+\frac{2}{r}\frac{\partial\psi}{\partial r}+\frac{1}{r^2\sin\theta}\frac{\partial}{\partial\theta}(\sin\theta\frac{\partial\psi}{\partial\theta})+\frac{1}{r^2\sin^2\theta}\frac{\partial^2\psi}{\partial\varphi^2}+\frac{2m}{\hbar^2}[E-V(\vec{r})]\psi=0 \qquad [3]$$

Let the trial equation of equation [3]

$$\psi_{nlm}(r,\theta,\phi)=R(r)\Theta(\theta)\Phi(\phi)=R(r)Y_{lm}(\theta,\Phi) \qquad [4]$$

- The function $R(r)$[radial wave function] describes how the Eigen function $\psi_{nlm}(r,\theta,\phi)$ varies along the line joining the electron and the proton, for given values of θ *and* ϕ
- The function $\Theta(\theta)$[zenith part] describes how the Eigen function $\psi_{nlm}(r,\theta,\phi)$ varies with the zenith angle θ along a meridian on a sphere of radius r centered at the proton, for given values of r *and* ϕ
- The function $\Phi(\phi)$[azimuthal wavefunction] describes how the Eigen function $\psi_{nlm}(r,\theta,\phi)$ varies with the azimuthal angle ϕ on a sphere of radius r centered at the proton, for given values of r *and* θ

The form of the Hamiltonian operator will be

$$\hat{H}=-\frac{\hbar^2}{2m}\nabla^2-\frac{e^2}{r} \qquad [5]$$

$$\hat{H}=-\frac{\hbar^2}{2m}[\frac{1}{r^2}\frac{\partial}{\partial r}(r^2\frac{\partial}{\partial r})+\frac{1}{r^2\sin\theta}\frac{\partial}{\partial\theta}(\sin\theta\frac{\partial}{\partial\theta})+\frac{1}{r^2\sin^2\theta}\frac{\partial^2}{\partial\phi^2}]-\frac{e^2}{r}$$

As the energy occurs in the radial equation, so we consider the radial part of the Laplacian operator

$$\nabla^2=\frac{\partial^2}{\partial r^2}+\frac{2}{r}\frac{\partial}{\partial r} \qquad [6]$$

From equation [3], the radial part of the Schrödinger wave equation

$$\frac{\partial^2 u(r)}{\partial r^2}+\left[\frac{2m}{\hbar^2}[E-V(r)]-\frac{\lambda}{r^2}\right]u(r) \qquad [7]$$

Let $\psi=u(r)=rR \quad Or,\ \frac{\partial u(r)}{\partial r}=r\frac{\partial R}{\partial r}+R\frac{\partial r}{\partial r}=r\frac{\partial R}{\partial r}+R$

$$Or,\ \frac{\partial^2 u(r)}{\partial r^2}=\frac{\partial R}{\partial r}+r\frac{\partial^2 R}{dr^2}+\frac{\partial R}{\partial r}=r\frac{\partial^2 R}{dr^2}+2\frac{\partial R}{\partial r}$$

Therefore equation [7] becomes

$$r\frac{\partial^2 R}{dr^2}+2\frac{\partial R}{\partial r}+\left[\frac{2m}{\hbar^2}[E-V(r)]-\frac{\lambda}{r^2}\right]rR=0 \qquad [8]$$

Or,
$$\frac{\partial^2 R}{dr^2}+\frac{2}{r}\frac{\partial R}{\partial r}+\left[\frac{2m}{\hbar^2}[E-V(r)]-\frac{\lambda}{r^2}\right]R=0$$

Or,
$$\frac{\partial^2 R}{dr^2}+\frac{2}{r}\frac{\partial R}{\partial r}+\left[\frac{2m}{\hbar^2}[E-V(r)]-\frac{l(l+1)}{r^2}\right]R=0 \qquad [9]$$

Or,
$$\frac{\partial^2 R}{dr^2}+\frac{2}{r}\frac{\partial R}{\partial r}+\left[\left[\frac{2m}{\hbar^2}E-\frac{2m}{\hbar^2}V(r)\right]-\frac{l(l+1)}{r^2}\right]R=0$$

Or,
$$\frac{\partial^2 R}{dr^2}+\frac{2}{r}\frac{\partial R}{\partial r}+\left[[K^2-U(r)]-\frac{l(l+1)}{r^2}\right]R=0 \qquad [10]$$

Where $K^2=\frac{2m}{\hbar^2}E \quad And\ U(r)=\frac{2m}{\hbar^2}V(r)$ [11]

and l is the orbital quantum number.

There are two sets of solution of equation [9].The first set is applicable when $E<0$, corresponds to bound states. The second set is applicable to unbound states. For hydrogen atom, we solve the equation [9] for $E<0$. Now rearranging [9]

$$\frac{\partial^2 R}{dr^2}+\frac{2}{r}\frac{\partial R}{\partial r}+\left[\frac{2mE}{\hbar^2}+\frac{2me^2}{4\pi\epsilon_0\hbar^2}\frac{1}{r}-\frac{l(l+1)}{r^2}\right]R=0 \qquad [12]$$

Now we introduce dimensionless independent variables

$$\rho=\alpha r \qquad Or, \alpha=\frac{\rho}{r} \qquad [13]$$

Here $\langle\rho^2\rangle=\langle x^2\rangle+\langle y^2\rangle$ is the mean square of the perpendicular distance of the electron from the field axis through the nucleus[19]. But $\langle r^2\rangle=\langle x^2\rangle+\langle y^2\rangle+\langle z^2\rangle$ is the mean square of the electrons from the nucleus. For a spherically symmetrical distribution we have $\langle x^2\rangle=\langle y^2\rangle=\langle z^2\rangle$. So that $\langle\rho^2\rangle=2\langle x^2\rangle\ and\ \langle r^2\rangle=3\langle x^2\rangle \ \ or, \langle x^2\rangle=\frac{1}{2}\langle\rho^2\rangle \quad \therefore \langle r^2\rangle=\frac{3}{2}\langle\rho^2\rangle$

$$\frac{\partial R}{\partial r}=\frac{\partial R}{\partial\rho}\frac{\partial\rho}{\partial r}=\frac{\partial R}{\partial\rho}\alpha \quad and \quad \frac{\partial^2 R}{\partial r^2}=\alpha^2\frac{\partial^2 R}{\partial\rho^2}$$

Hence equation [12] becomes

$$\alpha^2\frac{\partial^2 R}{\partial\rho^2}+\frac{2}{\rho}\alpha^2\frac{\partial R}{\partial\rho}+\left[\frac{2mE}{\hbar^2}+\frac{2me^2}{4\pi\epsilon_0\hbar^2}\frac{\alpha}{\rho}-\frac{l(l+1)}{\rho^2}\alpha^2\right]R=0$$

$$\frac{\partial^2 R}{\partial\rho^2}+\frac{2}{\rho}\frac{\partial R}{\partial\rho}+\left[\frac{2mE}{\hbar^2}\frac{1}{\alpha^2}+\frac{2me^2}{4\pi\epsilon_0\hbar^2}\frac{1}{\alpha\rho}-\frac{l(l+1)}{\rho^2}\right]R=0 \qquad [14]$$

Substituting $\frac{2mE}{\hbar^2}\frac{1}{\alpha^2}=-\frac{1}{4}$ [15]

And $$\frac{2me^2}{4\pi\epsilon_0\,\hbar^2} = \lambda\alpha \tag{16}$$

Hence equation [14] becomes

$$\frac{\partial^2 R}{\partial\rho^2} + \frac{2}{\rho}\frac{\partial R}{\partial\rho} + \left[-\frac{1}{4} + \frac{\lambda}{\rho} - \frac{l(l+1)}{\rho^2}\right]R = 0 \tag{17}$$

Now the exact solution

$$R(\rho) = F(\rho)e^{-\frac{\rho}{2}} \tag{18}$$

Where $F(\rho)$ is a finite polynomial in ρ

$$\frac{\partial R}{\partial\rho} = \frac{\partial F}{\partial\rho}e^{-\frac{\rho}{2}} - \frac{1}{2}F(\rho)e^{-\frac{\rho}{2}} = e^{-\frac{\rho}{2}}\left[\frac{\partial F}{\partial\rho} - \frac{1}{2}F\right]$$

$$\frac{\partial^2 R}{\partial\rho^2} = \left[\frac{\partial^2 F}{\partial\rho^2} - \frac{\partial F}{\partial\rho} + \frac{1}{4}F\right]e^{-\frac{\rho}{2}}$$

∴ Equation [17] becomes

$$\left[\frac{\partial^2 F}{\partial\rho^2} - \frac{\partial F}{\partial\rho} + \frac{1}{4}F + \frac{2}{\rho}\left(\frac{\partial F}{\partial\rho} - \frac{1}{2}F\right) + \left(\frac{\lambda}{\rho} - \frac{1}{4} - \frac{l(l+1)}{\rho^2}\right)F\right]e^{-\frac{\rho}{2}} = 0$$

$$\frac{\partial^2 F}{\partial\rho^2} + \frac{\partial F}{\partial\rho}\left(\frac{2}{\rho} - 1\right) + \frac{1}{4}F - \frac{1}{4}F - \frac{F}{\rho} + \left(\frac{\lambda}{\rho} - \frac{l(l+1)}{\rho^2}\right)F = 0$$

$$\frac{\partial^2 F}{\partial\rho^2} + \frac{\partial F}{\partial\rho}\left(\frac{2}{\rho} - 1\right) + \left(\frac{\lambda}{\rho} - \frac{1}{\rho} - \frac{l(l+1)}{\rho^2}\right)F = 0$$

$$\frac{\partial^2 F}{\partial\rho^2} + \frac{\partial F}{\partial\rho}\left(\frac{2}{\rho} - 1\right) + \left(\frac{\lambda-1}{\rho} - \frac{l(l+1)}{\rho^2}\right)F = 0 \tag{19}$$

Let the solution of equation [18]

$$F(\rho) = \sum_{k=0}^{\infty} A_k\,\rho^{s+k} \tag{20}$$

$$\sum_{k=0}^{\infty}(s+k)(s+k-1)A_k\,\rho^{s+k-2} + \left(\frac{2}{\rho} - 1\right)\sum_{k=0}^{\infty}(s+k)\,A_k\,\rho^{s+k-1} + \left(\frac{\lambda-1}{\rho} - \frac{l(l+1)}{\rho^2}\right)\sum_{k=0}^{\infty} A_k\,\rho^{s+k}$$

$$\sum_{k=0}^{\infty}(s+k)(s+k-1)A_k\,\rho^{s+k-2} + \sum_{k=0}^{\infty}2(s+k)\,A_k\,\rho^{s+k-2} - \sum_{k=0}^{\infty}(s+k)\,A_k\,\rho^{s+k-1}$$
$$+ \sum_{k=0}^{\infty}(\lambda-1)\,A_k\,\rho^{s+k-1} - \sum_{k=0}^{\infty}l(l+1)\,A_k\,\rho^{s+k-2} = 0$$

$$\sum_{k=0}^{\infty}[(s+k)(s+k-1) + 2(s+k) - l(l+1)]\,A_k\,\rho^{s+k-2}$$
$$-\sum_{k=0}^{\infty}[(s+k) - (\lambda-1)]\,A_k\,\rho^{s+k-1} = 0$$

$$\sum_{k=0}^{\infty}[(s+k)(s+k-1) - l(l+1)]\,A_k\,\rho^{s+k-2} - \sum_{k=0}^{\infty}[(s+k+1-\lambda)]\,A_k\,\rho^{s+k-1} = 0$$

To put $k = 0$; and equating the lowest power of ρ to zero.

$$[s(s+1)-l(l+1)]\, A_0 = 0$$

$$[s^2 + s - l^2 - l]\, A_0 = 0$$

$$[(s-l)(s+l+1)]\, A_0 = 0$$

$$[(s-l)(s+l+1) = 0 \quad ; \qquad \text{Since } A_0 \neq 0$$

It means that $\quad s = l \quad or, \quad s = -(l+1).$

The value $s = -(l+1)$ does not satisfying the condition of well- behaved function at the origin and only accepted value of $s = l$. It has been shown in the following box.

But $F(\rho)$diverge for $\rho = 0$; when $s = -(l+1)$
$R(\rho) = F(\rho)e^{-\frac{\rho}{2}}$; $F(\rho)$is finite polynomial in ρ
$R(\rho)$ will be finite when $\rho = 0 \qquad \therefore e^{-0} = 1$
$F(\rho) = A_k\, \rho^{s+k} = \rho^l \rho^k = \rho^0 \rho^0 = 1 \quad \therefore F(\rho) = 1$
$F(\rho) = \rho^{-(l+1)} = \frac{1}{\rho} = \infty$

Again, equating the coefficient of ρ^{s+k-1} to zero.

$$[(s+k+1)(s+k+2) - l(l+1)]\, A_{k+1} - (s+k+1-\lambda)A_k = 0$$

$$A_{k+1} = \frac{s+k+1-\lambda}{(s+k+1)(s+k+2)-l(l+1)}\, A_k \qquad [21]$$

$$= \frac{l+k+1-\lambda}{(l+k+1)(l+k+2)-l(l+1)}\, A_k \qquad [22]$$

Equation [22] is the recurrence formula connecting the coefficients A_{k+1} and A_k. If $F(\rho)$ does not terminate for some value of ρ the wave function $R(\rho) = F(\rho)e^{-\frac{\rho}{2}}$ will become infinite. In order to avoid this difficulty, the function $F(\rho)$terminates for some values of k. i.e. we assume that all the terms after A_k is zero. Therefore equation [22] becomes

$$\frac{l+k+1-\lambda}{(l+k+1)(l+k+2)-l(l+1)}\, A_k = 0 \qquad [23]$$

Since $A_k \neq 0$; then equation [23] becomes

$$l + k + 1 - \lambda = 0 \qquad or,\ \lambda = l + k + 1 \qquad [24]$$

For energy of an atomic state of hydrogen like atom defined by the Principal quantum number. So replacing λ by n

$$\boxed{n = l + k + 1} \qquad [25]$$

Where n = Principal quantum number [$n = 1, 2, 3, \ldots\ldots\ldots\ldots\ldots\ldots\ldots\ldots..$]

k = Radial quantum number

l = Orbital quantum number

Therefore, the Eigen values of hydrogen atom can be calculated from equation [16]

$$\lambda = \frac{2me^2}{4\pi\epsilon_0 \hbar^2}\frac{1}{\alpha} \qquad \text{or, } n = \frac{2me^2}{4\pi\epsilon_0 \hbar^2}\frac{1}{\alpha} \qquad \text{or, } n = \frac{2me^2}{4\pi\epsilon_0 \hbar^2}\left[-\sqrt{\frac{\hbar^2}{8mE}}\right] \text{; using equation [15]}$$

Or, $$n^2 = \left(\frac{2me^2}{4\pi\epsilon_0 \hbar^2}\right)^2 \frac{\hbar^2}{8mE} \quad or \quad E_n = -\frac{me^4}{32\pi^2\epsilon_0^2\ \hbar^2}\cdot\frac{1}{n^2}, \qquad [26]$$

Equation [26] represents the energy Eigen value or energy level of hydrogen atom.

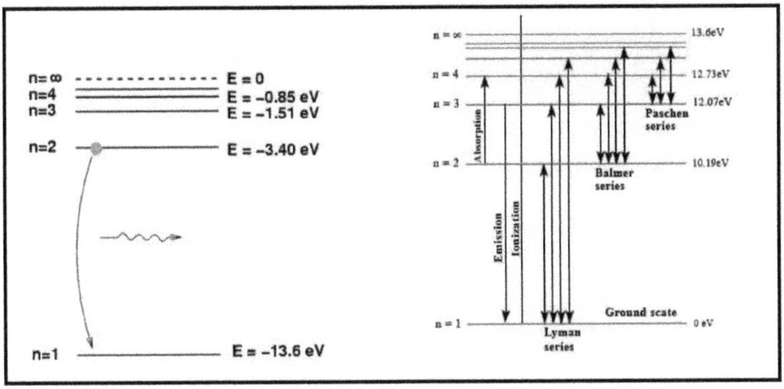

Fig.[8.3]: Energy levels of atomic hydrogen

8.4 Ground state energy of hydrogen atom

The energy Eigen value or energy level of hydrogen atom

$$E_n = -\frac{me^4}{32\pi^2\epsilon_0^2\ \hbar^2}\cdot\frac{1}{n^2} \qquad [1]$$

When $= 1$; the above equation becomes

$$E_1 = -\frac{me^4}{32\pi^2\epsilon_0^2\ \hbar^2} = -\frac{9.1\times10^{-31}\,kg\times\left(1.6\times10^{-19}coul\right)^4}{32\times9.87\times\left(8.9\times10^{-12}\,\frac{coul^2}{newton-m^2}\right)^2\times(1.054\times10^{-34}joule-second)^2}$$

$$= -\frac{9.1\times(1.6)^4\times10^{-107}}{32\times9.87\times(8.9)^2\times(1.054\times)^2\times10^{-92}}\ Newton-meter = -\frac{9.1\times(1.6)^4\times10^{-15}}{32\times9.87\times(8.9)^2\times(1.054\times)^2}\ joule$$

$$= -\frac{9.1\times(1.6)^4\times10^{-15}\times1.6\times10^{-19}}{32\times9.87\times(8.9)^2\times(1.054\times)^2}\ electron-volt[eV] = -13.6\ electron-volt[eV]$$

Electron volts (eV) are a convenient unit for atomic energies. One eV is defined as the energy an electron gains when accelerated across a potential difference of 1 volt. The ground state energy of hydrogen atom -13.6eV.

$\therefore$ *The ground state energy of hydrogen atom is* $-13.6\ electron-volt[eV]$

In a ground state configuration, all of the electrons are in as low an energy level as it is possible for them to be. When an electron absorbs energy, it occupies a higher energy orbital, and is said to be in an excited state. The first, second, third and so on excited states of Hydrogen atom are $E_2 = -3.40eV\ when\ n = 2$, $E_3 = -1.51eV\ when\ n = 2$, $E_4 = -0.085eV\ when\ n = 3$, $E = 0, when\ n = \infty$. The excited states of Hydrogen atom are as shown in figure [8.3].

- Calculation for SI unit of the ground state of hydrogen atom

$$\frac{kg\times coul^4}{\left(\frac{coul^2}{newton-m^2}\right)^2 (joule-second)^2} = \frac{kg\times coul^4\times(nt-m^2)\times(nt-m^2)}{(joule-second)\times coul^4\times\ (joule-second)}$$

$$= \frac{kg\times(nt-m)\times(nt-m)\times m^2}{(joule-second)\times\ (joule-second)} = \frac{kg\times joule\times joule\times m^2}{(joule-second)\times\ (joule-second)} = \frac{kg-m^2}{Sec^2}$$

$$= (Kg-m/Sec^2).m = nt-m = joule$$

8.5 Spherical harmonics $Y_{\ell,m}(\theta,\phi)$ is calculated for the values of $l = 3$ and $m = 2$

The Common, normalized Eigen function of $L^2\ and\ L_z\ denote\ by\ Y_{\ell,m}(\theta,\phi)$. Such that $L^2Y_{\ell,m}(\theta,\phi) = \ell(\ell+1)\hbar^2Y_{\ell,m}(\theta,\phi)$ and $L^2Y_{\ell,m}(\theta,\phi) = m\hbar Y_{\ell,m}(\theta,\phi)$ are called Spherical harmonics. Here $\ell = 0,1,2,3,\ldots\ and\ m = 0,\pm1,\pm2,\pm3,\ldots,\pm\ell$. The Spherical harmonics are given by

$$Y_{\ell,m}(\theta,\phi) = \Theta_{\ell,m}(\theta)\Phi_m(\phi)\ ;\ where\ \Phi_m(\phi) = \frac{1}{\sqrt{2\pi}}e^{im\phi} \qquad (normalised)$$

And $\Theta_{\ell,m}(\theta) = \left[\frac{2\ell+1}{2}\frac{(\ell-|m|)!}{(\ell+|m|)!}\right]^{1/2} P_\ell^{|m|}(\cos\theta)$ (normalised) giving

$$Y_{\ell,m}(\theta,\phi) = \left[\frac{2\ell+1}{4\pi}\frac{(\ell-|m|)!}{(\ell+|m|)!}\right]^{1/2} P_\ell^{|m|}(\cos\theta)e^{i|m|\phi} \quad ; \quad \text{For} \quad m \geq 0$$

$$= \left[\frac{2\ell+1}{4\pi}\frac{(\ell-|m|)!}{(\ell+|m|)!}\right]^{1/2} P_\ell^{|m|}(\cos\theta)e^{-i|m|\phi} \quad ; \quad \text{For} \quad m < 0$$

$\therefore Y_{\ell,m}(\theta,\phi) = Y_{\ell,-m}^{\times}(\theta,\phi)$ We get $|Y_{\ell,m}(\theta,\phi)|^2 = \frac{1}{2\pi}|\Theta_{\ell,m}(\theta)|^2$, independent of ϕ.

The orthonormality of $Y_{\ell,m}(\theta,\phi)$ is given by

$(Y_{\ell',m'}, Y_{\ell,m}) = \int_0^{\pi}\int_0^{2\pi} Y_{\ell',m'}^{\times} Y_{\ell,m} \sin\theta\, d\theta\, d\phi = \delta_{\ell\ell'}\delta_{mm'}$; $\sin\theta\, d\theta\, d\phi$ comes from θ and ϕ parts of volume element $(dr)(rd\theta)(r\sin\theta d\phi) = (r^2dr)(\sin\theta d\theta)(d\phi)$

For the values of $l = 3$ and $m = 2$, the Spherical harmonics $Y_{\ell,m}(\theta,\phi)$ or $Y_{3,2}(\theta,\phi)$ is calculated as follows:

$$Y_{\ell,m}(\theta,\phi) = \left[\frac{2\ell+1}{4\pi}\frac{(\ell-|m|)!}{(\ell+|m|)!}\right]^{1/2} P_\ell^{|m|}(\cos\theta)e^{i|m|\phi} \quad \text{for} \quad m \geq 0$$

Gives $Y_{3,2}(\theta,\phi) = \left[\frac{2\times3+1}{4\pi}\frac{(3-2)!}{(3+2)!}\right]^{1/2} P_3^{|2|}(\cos\theta)e^{i2\phi} = \left[\frac{7}{4\pi\times5!}\right]^{\frac{1}{2}} 15z(1-z^2)e^{i2\phi}$

$$= \left[\frac{105}{32\pi}\right]^{1/2} \cos\theta\, \sin^2\theta e^{i2\phi}. \quad where\ z = \cos\theta$$

8.6 Ground state wave function of hydrogen atom

For energy of an atomic state of hydrogen like atom is defined by the Principal quantum number.

$$\boxed{n = l + k + 1} \qquad [1]$$

Where n = Principal quantum number

k = Radial quantum number

And l = Orbital quantum number

When $n = 1$; That is $l = 0$ *and* $k = 0$

Let us take $\psi_{nlm}(r,\theta,\phi) = R_{nl}(r)Q_{lm}(\theta)Q_m(\phi) = R_{nl}(r)Y_{lm}(\theta,\Phi)$

$$\psi_{100} = a_0 e^{-\frac{\rho}{2}} \frac{1}{\sqrt{4\pi}} \quad [2]$$

Where $R_{nl}(r) = a_0 e^{-\frac{\rho}{2}}$ [3]

And $Y_{lm}(\theta,\Phi) = \frac{1}{\sqrt{4\pi}}$ [4]

$$[\frac{1}{\sin\theta}\frac{\partial}{\partial\theta}(\sin\theta\frac{\partial Y}{\partial\theta}) + \frac{1}{\sin^2\theta}\cdot\frac{\partial^2 Y}{\partial\varphi^2} + \lambda Y_{lm}(\theta,\phi) = 0]$$

$$Y_{lm}(\theta,\varphi) = (-1)^m \left[\frac{2l+1}{4\pi}\frac{(l-m)!}{(l+m)!}\right]^{1/2} P_l^m(cos\theta)e^{im\varphi}\ ; for\ m \geq 0$$

$$Y_{00}(\theta,\phi) = (-1)^0 \frac{1}{\sqrt{4\pi}} P_0(\cos\theta)e^0 = \frac{1}{\sqrt{4\pi}} \quad [\therefore P_0(\cos\theta)]$$

Now we introduce dimensionless independent variables

$$\rho = \alpha r \qquad Or, \alpha = \frac{\rho}{r} \quad [5]$$

Here $\langle\rho^2\rangle = \langle x^2\rangle + \langle y^2\rangle$ is the mean square of the perpendicular distance of the electron from the field axis through the nucleus[19]. But $\langle r^2\rangle = \langle x^2\rangle + \langle y^2\rangle + \langle z^2\rangle$ is the mean square of the electrons from the nucleus. For a spherically symmetrical distribution we have $\langle x^2\rangle = \langle y^2\rangle = \langle z^2\rangle$.

So that $\langle\rho^2\rangle = 2\langle x^2\rangle\ and\ \langle r^2\rangle = 3\langle x^2\rangle$

Or, $\langle x^2\rangle = \frac{1}{2}\langle\rho^2\rangle \qquad \therefore \langle r^2\rangle = \frac{3}{2}\langle\rho^2\rangle$ [6]

From the normalization condition in spherical polar coordinates, we have

$$\int_{r=0}^{\infty}\int_{\theta=0}^{\pi}\int_{\varphi=0}^{2\pi}(\psi_{100})^\times\,\psi_{100}\,d\tau = 1$$

$$\int_{r=0}^{\infty}\int_{\theta=0}^{\pi}\int_{\varphi=0}^{2\pi}\left(a_0 e^{-\frac{\rho}{2}}\frac{1}{\sqrt{4\pi}}\right)^\times a_0 e^{-\frac{\rho}{2}}\frac{1}{\sqrt{4\pi}} r^2 \sin\theta d\theta d\varphi dr = 1$$

$$\frac{a_0^2}{4\pi}\int_{\varphi=0}^{2\pi} d\varphi \int_{\theta=0}^{\pi} \sin\theta\, d\theta \int_{r=0}^{\infty} e^{-\rho} r^2 dr = 1$$

$$\frac{a_0^2}{4\pi} 2\pi . 2 \int_{r=0}^{\infty} e^{-\rho} r^2 dr = 1$$

$$a_0^2 \int_{r=0}^{\infty} e^{-\rho}\, r^2 dr \;= 1$$

$$a_0^2 \int_{r=0}^{\infty} e^{-\rho} \left(\frac{\rho}{\alpha}\right)^2 \frac{d\rho}{\alpha} = 1 \qquad \boxed{\rho = \alpha r,\;\; d\rho = \alpha dr\,, dr = \frac{d\rho}{\alpha}}$$

$$\frac{a_0^2}{\alpha^3} \int_{r=0}^{\infty} \rho^2\, e^{-\rho}\, d\rho = 1 \qquad [7]$$

From the definition of Gamma function

$$\int_{r=0}^{\infty} e^{-x} x^n\, dx = \Gamma(n+1) = n! \qquad [8]$$

Therefore equation [7] becomes $\frac{a_0^2}{\alpha^3}\, 2! = 1$ *or,* $a_0 = \left(\frac{\alpha^3}{2}\right)^{\frac{1}{2}}$. From equation [2]

$$\psi_{100} = \frac{1}{\sqrt{4\pi}} \left(\frac{\alpha^3}{2}\right)^{\frac{1}{2}} e^{-\frac{\alpha r}{2}} \qquad [9]$$

The energy Eigen value or energy level of hydrogen atom

$$E_n = -\frac{me^4}{32\pi^2\epsilon_0^2\;\hbar^2} \cdot \frac{1}{n^2}\;;\quad |E_n| = \frac{me^4}{32\pi^2\epsilon_0^2\;\hbar^2} \cdot \frac{1}{n^2}$$

Also we have, $\frac{2mE}{\hbar^2}\frac{1}{\alpha^2} = -\frac{1}{4}$ $\quad Or, \alpha^2 = \frac{8m}{\hbar^2}|E_n| = \frac{8m}{\hbar^2}\frac{me^4}{32\pi^2\epsilon_0^2\;\hbar^2} \cdot \frac{1}{n^2}\quad or, \alpha = \frac{me^2}{2\pi\varepsilon_0\hbar^2} \cdot \frac{1}{n^2}$

When = 1 ; the above equation becomes $\alpha = \frac{me^2}{2\pi\varepsilon_0\hbar^2} \cdot \frac{1}{n^2} = \frac{me^2}{2\pi\varepsilon_0\hbar^2} = \frac{2}{\frac{4\pi\varepsilon_0\hbar^2}{me^2}} = \frac{2}{a_0}$

Where $a_0 = \frac{4\pi\varepsilon_0\hbar^2}{me^2}$ is the first Bohr radius. Again from equation [5]

$$\rho = \alpha r = \frac{2}{a_0} r \qquad [10]$$

Finally equation [9] becomes

$$\psi_{100} \;= \frac{\alpha^{\frac{3}{2}}}{\sqrt{2}}\frac{1}{\sqrt{4\pi}} e^{-\frac{\alpha r}{2}} = \frac{1}{\sqrt{2}}\left(\frac{2}{a_0}\right)^{\frac{3}{2}} \frac{1}{\sqrt{4\pi}} e^{-\frac{2r}{2a_0}} = \frac{1}{\sqrt{\pi}}\left(\frac{1}{a_0}\right)^{\frac{3}{2}} e^{-\frac{r}{a_0}} \qquad \boxed{\psi_{100} = \frac{1}{\sqrt{\pi a_0^3}}\; e^{-\frac{r}{a_0}}} \qquad [11]$$

Equation [12] represents the ground state wave function of hydrogen atom.

8.7 Degeneracy of the hydrogen atom ground state

The ground state wave function of hydrogen atom $\quad \psi_{100} = \frac{1}{\sqrt{\pi a_0^3}}\; e^{-\frac{r}{a_0}}$ [1]

There are a number of different ways of representing hydrogen atom wave functions graphically. We will illustrate some of these for the 1s ground state. In atomic units equation becomes

$$\psi_{1s}(r) = \frac{1}{\sqrt{\pi}}\; e^{-r} \qquad [2]$$

is a decreasing exponential function of a single variable r and is simply plotted below

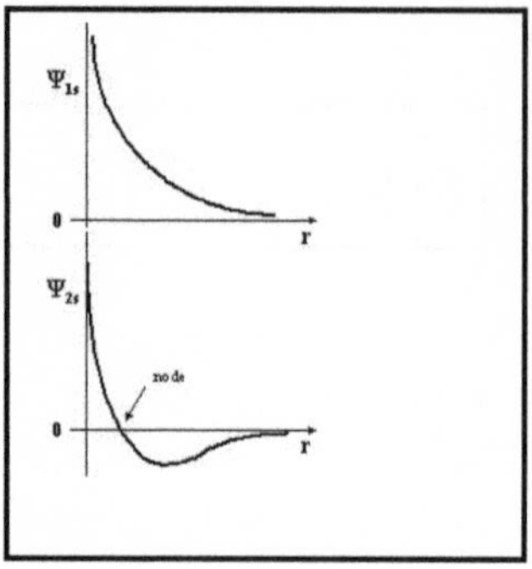

Fig.[8.7.1]: Shows the wave functions for 1s and 2s orbital for atomic hydrogen. The 2s-function [scaled by a factor of 2] has a node at r = 2 Bohr

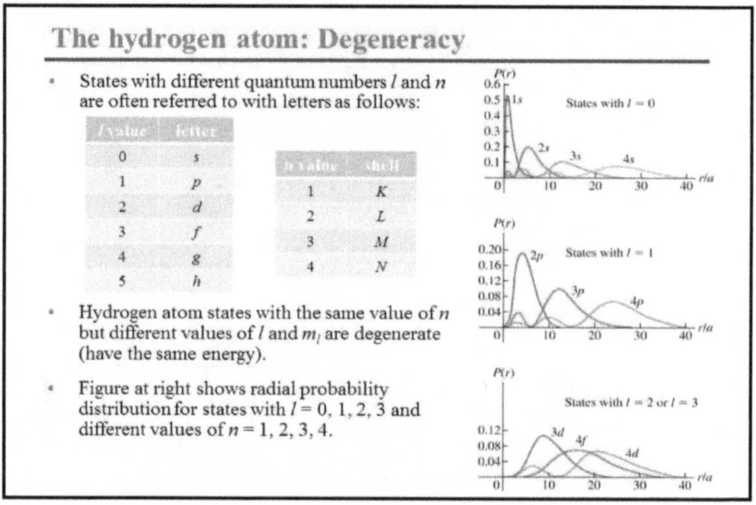

Fig.[8.7.2]: Shows the degeneracy of Hydrogen atom

Now we want to present a clear idea about the degeneracy of Hydrogen atom

If the particle having the same energy in an excited state will have several different stationary states or different wave functions. I.e., there are a number of independent quantum states of a system, each belonging to the same energy level. Such states and energy-levels are said to be degenerate. If there are n linearly independent wave functions $\psi_1, \psi_2, \psi_3 \dots\dots\dots\dots\dots\dots\dots\dots \psi_n$ belonging to the same energy state then the energy level is said to be n-fold degenerate. It can be easily shown that any linearly combination of the degenerate wave functions $= C_1\psi_1 + C_2\psi_2 +$

$C_3\psi_3 +$ ……………………… $C_n\psi_n$; is also an Eigen function belonging to the same energy Eigen value.

8.8 Table for Normalized Polar and radial Wave Function

n	l	m	$Qm(\theta)$	$R_{nl}(r)$
1	0	0	$\frac{1}{\sqrt{2}}$	$2(\frac{1}{a_0})^{\frac{3}{2}} e^{-r/} a_0$
2	0	0	$\frac{1}{\sqrt{2}}$	$\frac{1}{2\sqrt{2}}(\frac{1}{a_0})^{\frac{3}{2}} (2-\frac{r}{a_0} e)^{-r/2a_0}$
2	1	0	$\frac{\sqrt{6}}{2}\cos\theta$	$\frac{1}{2\sqrt{6}}(\frac{1}{a_0})^{\frac{3}{2}} \frac{r}{a_0} e^{-r/2a_0}$
2	1	±1	$\frac{\sqrt{6}}{2}\cos\theta$	$\frac{1}{2\sqrt{6}}(\frac{1}{a_0})^{\frac{3}{2}} \frac{r}{a_0} e^{-r/2a_0}$
3	0	0	$\frac{\sqrt{1}}{2}$	$\frac{2}{81\sqrt{3}}(\frac{1}{a_0})$ $(27\text{-}18\frac{r}{a_0} + 2\frac{r^2}{a_0^2})$ $e^{-r/3a_0}$
3	0	0	$\frac{\sqrt{6}}{2}\cos\theta$	$\frac{4}{81\sqrt{6}}(\frac{1}{a_0})^{\frac{3}{2}}(6-\frac{r}{a_0})\frac{r}{a_0} e^{-r/3a_0}$
3	1	±1	$\frac{\sqrt{3}}{2}\sin\theta$	$\frac{4}{81\sqrt{6}}(\frac{1}{a_0})^{\frac{3}{2}}(6-\frac{r}{a_0})\frac{r}{a_0} e^{-r/3a_0}$
3	1	0	$\frac{\sqrt{10}}{4}(3cos^2\theta - 1)$	$\frac{4}{81\sqrt{30}}(\frac{1}{a_0})\frac{3}{2}\frac{r^2}{a_0^2} e^{-r/3a_0}$
3	2	±1	$\frac{\sqrt{15}}{2}\sin\theta\ cos\theta$	$\frac{4}{81\sqrt{30}}(\frac{1}{a_0})\frac{3}{2}\frac{r^2}{a_0^2} e^{-r/3a_0}$
3	2	±2	$\frac{\sqrt{15}}{4}\sin^2\theta$	$\frac{4}{81\sqrt{30}}(\frac{1}{a_0})\frac{3}{2}\frac{r^2}{a_0^2} e^{-r/3a_0}$

8.9 Table for the Normalized Complete wave functions for the hydrogen atom

The complete wave function for the hydrogen atom is written as[26]

$$\psi_{nlm}(r,\theta,\phi) = R_{nl}(r)Q_{lm}(\theta)Q_m(\phi)$$

n	l	m	$\psi_{nlm}(r,\theta,\phi)$
1	0	0	$\frac{1}{\sqrt{\pi}}\left(\frac{1}{a_0}\right)^{\frac{3}{2}}.\, e^{-r/a_0}$
2	0	0	$\frac{1}{4\sqrt{2\pi}}\left(\frac{1}{a_0}\right)^{\frac{3}{2}}.\left(2-\frac{r}{a_0}\right).\, e^{-r/2a_0}$
2	1	0	$\frac{1}{4\sqrt{2\pi}}\left(\frac{1}{a_0}\right)^{\frac{3}{2}}.\frac{r}{a_0}.\, e^{-r/2a_0}.\cos\theta$
2	1	±1	$\frac{1}{8\sqrt{2\pi}}\left(\frac{1}{a_0}\right)^{\frac{3}{2}}.\frac{r}{a_0}.\, e^{-r/2a_0}.\sin\theta\ e^{\pm i\phi}.$
3	0	0	$\frac{1}{81\sqrt{3\pi}}\left(\frac{1}{a_0}\right)^{\frac{3}{2}}\left(27\text{-}18\frac{r}{a_0}+2\frac{r^2}{a_0^2}\right) e^{-r/3a_0}$
3	1	0	$\frac{1}{81}\sqrt{\frac{2}{\pi}}\left(\frac{1}{a_0}\right)^{\frac{3}{2}}\left(6-\frac{r}{a_0}\right).\frac{r}{a_0}.\, e^{-r/3a_0}\cos\theta$
3	1	±1	$\frac{1}{81\sqrt{1\pi}}\left(\frac{1}{a_0}\right)^{\frac{3}{2}}\left(6-\frac{r^2}{a_0}\right).\frac{r}{a_0}.\, e^{-\frac{r}{3a_0}}\sin\theta.\, e^{\pm i\phi}$
3	2	0	$\frac{1}{81\sqrt{6\pi}}\left(\frac{1}{a_0}\right)^{\frac{3}{2}}.\frac{r^2}{a_0}.\, e^{-r/3a_0}.(3\cos^2\theta-1)$
3	2	±1	$\frac{1}{81\sqrt{\pi}}\left(\frac{1}{a_0}\right)^{\frac{3}{2}}.\frac{r^2}{a_0}.\, e^{-r/3a_0}.\sin\theta.\cos\theta.\, e^{\pm i\phi}$
3	2		$\frac{1}{162\sqrt{\pi}}\left(\frac{1}{a_0}\right)^{\frac{3}{2}}.\ \frac{r^2}{a_0}.\, e^{-r/3a.}\ \sin^2\theta.\, e^{\pm 2i\phi}$

Where $a_0 = \frac{4\pi\varepsilon_0\hbar^2}{m_0 e^2} = 0.53\text{Å}$ = radius of the first Bohr orbit.

8.10 Volume element in terms of spherical polar coordinates

The volume element $dV = d\tau = d^3r = dxdydz$ can be determined in terms of spherical polar coordinates. The relation between Cartesian coordinates (x, y, z) and spherical polar coordinates

$$x = r\sin\theta\cos\varphi,\ \ y = r\sin\theta\sin\varphi \quad z = r\cos\theta \qquad [1]$$

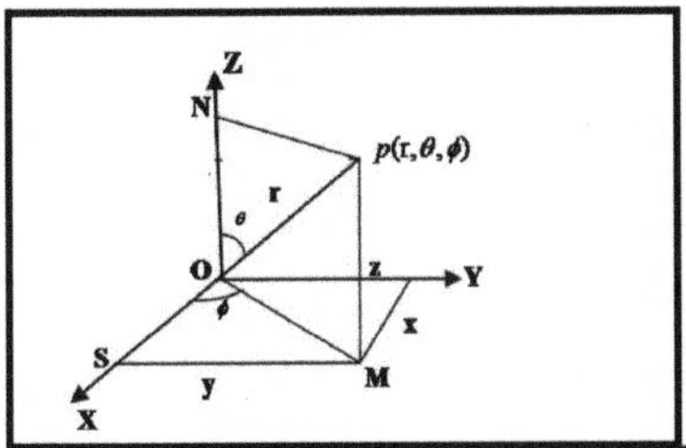

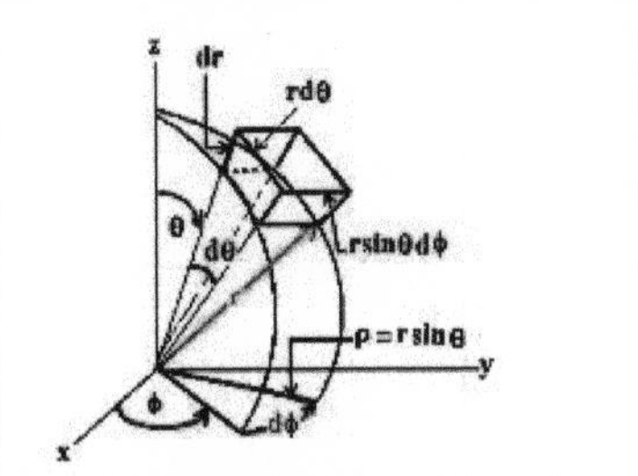

Volume element = $r^2 dr \sin\theta d\theta\, d\phi$

$$Cos\theta = \frac{ON}{OP} \quad or, \quad ON = OPCos\theta$$

$$or, Z = r\ Cos\theta$$

$$Sin\theta = \frac{MS}{OM}\ \ \&\ \ Sin\theta = \frac{PN}{OP}$$

$$MS = OMSin\phi \qquad PN = OPSin\ \theta$$

$$y = PNSin\varphi = OP\ Sin\theta Sin\phi = r\ Sin\theta\ Sin\ \phi$$

$$y = r\ Sin\theta\ Sin\ \phi$$

$$Cos\,\phi = \frac{OS}{OM} \quad or, \quad OS = OMCos\phi$$

$$x = PNCos\varphi = OPSin\theta Cos\varphi$$

$$x = r\sin\theta\ \ Cos\phi$$

$$\therefore \qquad dxdydz = J\left[\frac{x,y,z}{r,\theta,\varphi}\right] drd\theta d\varphi \qquad [2]$$

$$= \frac{\partial(x,y,z)}{\partial(r,\theta,\varphi)}\, drd\theta d\varphi \qquad [3]$$

$$= \begin{vmatrix} \frac{\partial x}{\partial r} & \frac{\partial x}{\partial \theta} & \frac{\partial x}{\partial \varphi} \\ \frac{\partial y}{\partial r} & \frac{\partial y}{\partial \theta} & \frac{\partial y}{\partial \varphi} \\ \frac{\partial z}{\partial r} & \frac{\partial z}{\partial \theta} & \frac{\partial z}{\partial \varphi} \end{vmatrix} drd\theta d\varphi$$

$$= \begin{vmatrix} \sin\theta\cos\varphi & r\cos\theta\cos\varphi & -\sin\theta\sin\varphi \\ \sin\theta\sin\varphi & r\cos\theta\sin\varphi & r\sin\theta\cos\varphi \\ \cos\theta & -r\sin\theta & 0 \end{vmatrix} drd\theta d\varphi$$

$$= \ r^2\sin\theta \begin{vmatrix} \sin\theta\cos\varphi & \cos\theta\cos\varphi & -\sin\varphi \\ \sin\theta\sin\varphi & \cos\theta\sin\varphi & \cos\varphi \\ \cos\theta & -\sin\theta & 0 \end{vmatrix} drd\theta d\varphi$$

$$= r^2 \sin\theta[\sin\theta \cos\varphi\,(0 + \sin\theta \cos\varphi) + \cos\theta \cos\varphi\,(\cos\theta \cos\varphi - 0) - \sin\varphi\,(-sin^2\theta \sin\varphi) - cos^2\theta \sin\varphi]\, drd\theta d\varphi$$

$$= r^2 \sin\theta[sin^2\theta cos^2\varphi + cos^2\theta cos^2\varphi + sin^2\theta sin^2\varphi + cos^2\theta sin^2\varphi]\, drd\theta d\varphi$$

$$= r^2 \sin\theta[(sin^2\theta + cos^2\theta)cos^2\varphi + +sin^2\varphi((sin^2\theta + cos^2\theta]\, drd\theta d\varphi$$

$$= r^2 \sin\theta[(sin^2\theta + cos^2\theta) + (sin^2\varphi + cos^2\varphi)]\, drd\theta d\varphi$$

$$= r^2 \sin\theta\, drd\theta d\varphi \qquad [4]$$

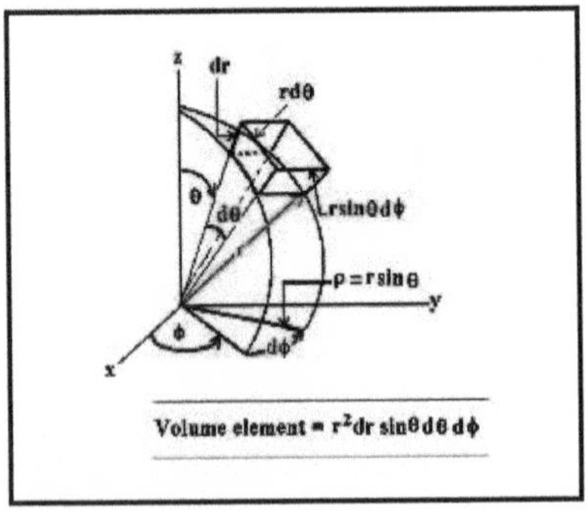

Fig.[8.10]: Shows Volume element in terms of spherical polar coordinates

Equation [4] represents the volume element in terms of spherical polar coordinates [20,21].

8.11 Del or Nabla [∇] and Laplacian operator [∇²] in spherical polar coordinates

The relation between Cartesian coordinates (x, y, z) and spherical polar coordinates(r, θ, φ)

$$x = r\sin\theta\cos\varphi,\ \ y = r\sin\theta\sin\varphi,\ \ z = r\cos\theta$$

The position vector in three dimensions

$$\mathbf{r} = \mathbf{i}x + \mathbf{j}y + \mathbf{k}z = \mathbf{i}\, r\sin\theta\cos\varphi + \mathbf{j}\, r\sin\theta\sin\varphi + \mathbf{k}\, r\cos\theta$$

In that case, the unit vector

$$\mathbf{u_r} = \frac{\partial \mathbf{r}/\partial r}{|\partial \mathbf{r}/\partial r|},\ \mathbf{u_\theta} = \frac{\partial \mathbf{r}/\partial \theta}{|\partial \mathbf{r}/\partial \theta|},\ \mathbf{u_\varphi} = \frac{\partial \mathbf{r}/\partial \varphi}{|\partial \mathbf{r}/\partial \varphi|}$$

We know, Del or Nabla can be written in curvilinear coordinates

$$\nabla = \frac{u_1}{h_1}\frac{\partial}{\partial q_1} + \frac{u_2}{h_2}\frac{\partial}{\partial q_2} + \frac{u_3}{h_3}\frac{\partial}{\partial q_3}$$

In Cartesian coordinates $(q_1, q_2, q_3) = (x, y, z)$; $\mathbf{u_1} = \mathbf{i}, \mathbf{u_2} = \mathbf{j}, \mathbf{u_3} = \mathbf{k}$ **and** $h_1 = h_2 = h_3$

So, Del or Nabla can be written in Cartesian coordinates

$$\boxed{\nabla = \mathbf{i}\frac{\partial}{\partial x} + \mathbf{j}\frac{\partial}{\partial y} + \mathbf{k}\frac{\partial}{\partial z}} \qquad [1]$$

In spherical polar coordinates

$(q_1, q_2, q_3) = (r, \theta, \varphi)$; $\mathbf{u_1} = \mathbf{u_r}$, $\mathbf{u_2} = \mathbf{u_\theta}$, $\mathbf{u_3} = \mathbf{u_\varphi}$; $h_1 = 1$, $h_2 = r$, $h_3 = r\sin\theta$

Therefore, equation [1] can be written in the form

$$\boxed{\nabla = \mathbf{u_r}\frac{\partial}{\partial r} + \mathbf{u_\theta}\frac{1}{r}\frac{\partial}{\partial \theta} + \mathbf{u_\varphi}\frac{1}{r\sin\theta}\frac{\partial}{\partial \varphi}}$$

Where $\mathbf{u_r} = \mathbf{i}\,\sin\theta\cos\varphi + \mathbf{j}\,\sin\theta\sin\varphi + \mathbf{k}\,\cos\theta$

$\mathbf{u_\theta} = \mathbf{i}\,\cos\theta\cos\varphi + \mathbf{j}\,\cos\theta\sin\varphi - \mathbf{k}\sin\theta$

$\mathbf{u_\varphi} = -\mathbf{i}\sin\varphi + \mathbf{j}\cos\varphi$

So, $\mathbf{u_r}.\mathbf{u_r} = \mathbf{u_\theta}.\mathbf{u_\theta} = \mathbf{u_\varphi}.\mathbf{u_\varphi} = 1$ and $\mathbf{u_r}.\mathbf{u_\theta} = \mathbf{u_\theta}.\mathbf{u_\varphi} = \mathbf{u_\varphi}.\mathbf{u_r} = 0$

$\therefore$ $\mathbf{u_r}\sin\theta + \mathbf{u_\theta}\cos\theta = \mathbf{i}\,\cos\varphi + \mathbf{j}\,\sin\varphi$

In that case,

$$\frac{\partial \mathbf{u_r}}{\partial r} = 0, \qquad \frac{\partial \mathbf{u_\theta}}{\partial r} = 0, \qquad \frac{\partial \mathbf{u_\varphi}}{\partial r} = 0$$

$$\frac{\partial \mathbf{u_r}}{\partial \theta} = \mathbf{i}\,\cos\theta\cos\varphi + \mathbf{j}\,\cos\theta\sin\varphi - \mathbf{k}\sin\theta = \mathbf{u_\theta}$$

$$\frac{\partial \mathbf{u_\theta}}{\partial \theta} = -[\mathbf{i}\,\sin\theta\cos\varphi + \mathbf{j}\,\sin\theta\sin\varphi + \mathbf{k}\,\cos\theta] = -\mathbf{u_r}$$

$$\frac{\partial \mathbf{u_\theta}}{\partial \varphi} = -(\mathbf{i}\cos\varphi + \mathbf{j}\sin\varphi) = -\mathbf{u_r}\sin\theta + \mathbf{u_\theta}\cos\theta$$

Now $\nabla^2 = \nabla.\nabla = \left(\mathbf{u_r}\frac{\partial}{\partial r} + \mathbf{u_\theta}\frac{1}{r}\frac{\partial}{\partial \theta} + \mathbf{u_\varphi}\frac{1}{r\sin\theta}\frac{\partial}{\partial \varphi}\right).\left(\mathbf{u_r}\frac{\partial}{\partial r} + \mathbf{u_\theta}\frac{1}{r}\frac{\partial}{\partial \theta} + \mathbf{u_\varphi}\frac{1}{r\sin\theta}\frac{\partial}{\partial \varphi}\right)$

$$= \frac{\partial^2}{\partial r^2} + 0 + 0 + \mathbf{u_\theta}\frac{1}{r}\frac{\partial \mathbf{u_r}}{\partial \theta}\frac{\partial}{\partial r} + \mathbf{u_\theta}\frac{1}{r}\frac{\partial \mathbf{u_\theta}}{\partial \theta}\frac{1}{r}\frac{\partial}{\partial \theta} + \frac{1}{r^2}\frac{\partial^2}{\partial \theta^2} + 0 + \mathbf{u_\varphi}\frac{1}{r\sin\theta}\frac{\partial \mathbf{u_r}}{\partial \varphi}.\frac{\partial}{\partial r} + \mathbf{u_\varphi}\frac{1}{r\sin\theta}\frac{\partial \mathbf{u_\theta}}{\partial \varphi}\frac{1}{r}\frac{\partial}{\partial \theta}$$

$$+\mathbf{u_\varphi}\frac{1}{r\sin\theta}\frac{\partial \mathbf{u_\varphi}}{\partial \varphi}\frac{1}{r\sin\theta}\frac{\partial}{\partial \varphi} + \frac{1}{r^2}\frac{1}{\sin^2\theta}\frac{\partial^2}{\partial \varphi^2}$$

$$\nabla^2 = \frac{\partial^2}{\partial r^2} + \mathbf{u}_\theta \frac{1}{r}\frac{\partial}{\partial r}\mathbf{u}_\theta + \mathbf{u}_\theta \frac{1}{r}(-\mathbf{u}_r)\frac{1}{r}\frac{\partial}{\partial \theta} + \frac{1}{r^2}\frac{\partial^2}{\partial \theta^2} + \mathbf{u}_\varphi \frac{1}{r\sin\theta}\mathbf{u}_\varphi \sin\theta . \frac{\partial}{\partial r} + \mathbf{u}_\varphi \frac{1}{r\sin\theta}\mathbf{u}_\varphi \cos\theta . \frac{1}{r}\frac{\partial}{\partial \theta}$$

$$+\mathbf{u}_\varphi \frac{1}{r\sin\theta}(-\mathbf{u}_r \sin\theta - \mathbf{u}_\theta \cos\theta)\frac{1}{r\sin\theta}\frac{\partial}{\partial \varphi} + \frac{1}{r^2}\frac{1}{\sin^2\theta}\frac{\partial^2}{\partial \varphi^2}$$

$$\nabla^2 = \frac{\partial^2}{\partial r^2} + \frac{2}{r}\frac{\partial}{\partial r} + \frac{1}{r^2}\frac{\partial^2}{\partial \theta^2} + \frac{\cos\theta}{r^2 \sin\theta}\frac{\partial}{\partial \theta} + \frac{1}{r^2}\frac{1}{\sin^2\theta}\frac{\partial^2}{\partial \varphi^2}$$

$$\boxed{\nabla^2 = \frac{1}{r^2}\frac{\partial}{\partial r}\left(r^2\frac{\partial}{\partial r}\right) + \frac{1}{r^2}\frac{1}{\sin\theta}\frac{\partial}{\partial \theta}\left(\sin\theta\frac{\partial}{\partial \theta}\right) + \frac{1}{r^2}\frac{1}{\sin^2\theta}\frac{\partial^2}{\partial \varphi^2}} \quad [3]$$

8.12 Average value $\langle r \rangle$ and root mean square value $\langle r^2 \rangle^{\frac{1}{2}}$ in hydrogen atom by using the ground state wave function

[a] The average or expectation value of $\langle r \rangle$

The average or expectation value of r

$$\langle r \rangle = \int \psi^{\times} r \psi d\tau \quad [1]$$

Here the ground state wave function of hydrogen atom

$$\psi_{100} = \frac{1}{\sqrt{\pi a_0^3}} e^{-\frac{r}{a_0}} \quad [2]$$

And its complex conjugate $\psi_{100}^{\times} = \frac{1}{\sqrt{\pi a_0^3}} e^{-\frac{r}{a_0}}$

$$\langle r \rangle = \int_{r=0}^{\infty}\int_{\theta=0}^{\pi}\int_{\varphi=0}^{2\pi} \frac{1}{\sqrt{\pi a_0^3}} e^{-\frac{r}{a_0}} r \frac{1}{\sqrt{\pi a_0^3}} e^{-\frac{r}{a_0}} r^2 \sin\theta d\theta d\varphi dr$$

$$\langle r \rangle = \frac{1}{\pi a_0^3} 4\pi \int_{r=0}^{\infty} r^3 e^{-\frac{2r}{a_0}} dr$$

From the standard integral

$$\int_0^{\infty} x^n e^{-\alpha x} dx = \frac{n!}{\alpha^{n+1}} \quad [3]$$

$$\langle r \rangle = \frac{4\pi}{\pi a_0^3}\frac{3!}{\left(\frac{2}{a_0}\right)^{3+1}} = \frac{4\pi}{\pi a_0^3}.\frac{3.2.1}{\left(\frac{2}{a_0}\right)^4} = \frac{4\pi}{\pi a_0^3}.\frac{6a_0^4}{16}, \quad [4]$$

$$\boxed{\langle r \rangle = \frac{3}{2}a_0}$$

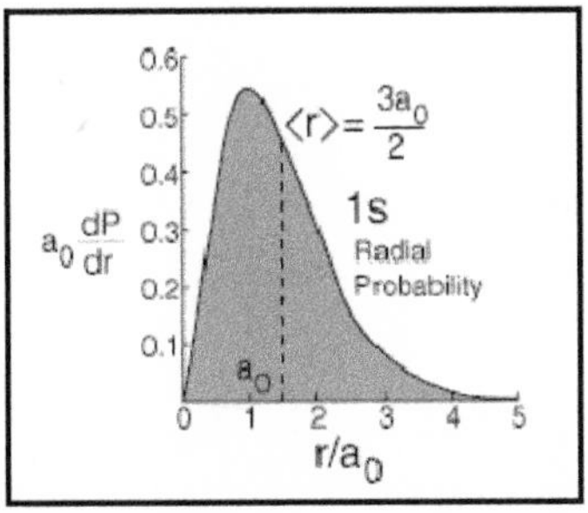

Fig.[8. 11]: Shows Average value $\langle r\rangle$

[b] Root mean square value of $\langle r^2\rangle^{\frac{1}{2}}$

The average or expectation value of r

$$\langle r\rangle = \int \psi^{\times} r\psi d\tau \qquad [1]$$

Here the ground state wave function of hydrogen atom

$$\psi_{100} = \frac{1}{\sqrt{\pi a_0^3}}\ e^{-\frac{r}{a_0}} \qquad [2]$$

And its complex conjugate $\psi_{100}^{\times} = \frac{1}{\sqrt{\pi a_0^3}}\ e^{-\frac{r}{a_0}}$ [3]

$$\therefore \qquad \langle r^2\rangle = \int_{r=0}^{\infty}\int_{\theta=0}^{\pi}\int_{\varphi=0}^{2\pi} \frac{1}{\sqrt{\pi a_0^3}}\ e^{-\frac{r}{a_0}}\ r^2\ \frac{1}{\sqrt{\pi a_0^3}}\ e^{-\frac{r}{a_0}}\quad r^2 \sin\theta d\theta d\varphi dr$$

$$\langle r^2\rangle = \frac{1}{\pi a_0^3}\int_0^{\theta} \sin\theta \int_0^{2\pi} d\varphi \int_0^{\infty} r^4 e^{-\frac{2r}{a_0}}\ dr = \frac{1}{\pi a_0^3}\ 2.2\pi \int_0^{\infty} r^4 e^{-\frac{2r}{a_0}}\ dr$$

$$\langle r^2\rangle = \frac{4}{a_0^3}\int_0^{\infty} r^4 e^{-\frac{2r}{a_0}}\ dr \qquad [4]$$

From the standard integral

$$\int_{r=0}^{\infty} r^n\ e^{-\alpha r}\ dr = \frac{n!}{\alpha^{n+1}}$$

$$\langle r^2\rangle = \frac{4}{a_0^3}\cdot\frac{4!}{\left(\frac{2}{a_0}\right)^5} = \frac{4}{a_0^3}\cdot\frac{4.3.2.1}{\frac{32}{a_0^5}} = \frac{4}{a_0^3}\cdot\ \frac{24}{32}\ a_0^5 = 3a_0^2$$

$$\sqrt{\langle r^2\rangle} = \sqrt{3a_0^2} \qquad Or, \langle r^2\rangle^{\frac{1}{2}} = \sqrt{3}\, a_0$$

8.13 Average value of $\langle\frac{1}{r}\rangle$ in the hydrogen atom

The average or expectation value of $\langle\frac{1}{r}\rangle$

$$\langle\frac{1}{r}\rangle = \int \psi^{\times}(\frac{1}{r})\psi d\tau \qquad [1]$$

Here the ground state wave function of hydrogen atom

$$\psi_{100} = \frac{1}{\sqrt{\pi a_0^3}} e^{-\frac{r}{a_0}} \qquad [2]$$

And its complex conjugate $\psi_{100}^{\times} = \frac{1}{\sqrt{\pi a_0^3}} e^{-\frac{r}{a_0}}$ [3]

$$\therefore \quad \langle\frac{1}{r}\rangle = \int_{r=0}^{\infty}\int_{\theta=0}^{\pi}\int_{\varphi=0}^{2\pi} \frac{1}{\sqrt{\pi a_0^3}} e^{-\frac{r}{a_0}} \left(\frac{1}{r}\right) \frac{1}{\sqrt{\pi a_0^3}} e^{-\frac{r}{a_0}} \quad r^2 \sin\theta d\theta d\varphi dr$$

$$\langle\frac{1}{r}\rangle = \frac{4}{a_0^3}\int_0^{\infty} r\ e^{-\frac{2r}{a_0}}\ dr \qquad [4]$$

From the standard integral

$$\int_0^{\infty} x^n\, e^{-ax}\, dx = \frac{n!}{a^{n+1}} \qquad [5]$$

$$\langle\frac{1}{r}\rangle = \frac{4}{a_0^3}. \quad \frac{1!}{\left(\frac{2}{a_0}\right)^2} = \frac{4}{a_0^3}.\frac{a_0^2}{4} \qquad [6]$$

$$\boxed{\langle\frac{1}{r}\rangle = \frac{1}{a_0}}$$

Exercise

1. Why Hydrogen atom is an interesting problem?
2. Explain two body system problems with central force interaction.
3. Show that the energy Eigen value equation or energy level of hydrogen atom

$$E_n = -\frac{me^4}{32\pi^2\epsilon_0^2\ \hbar^2} \cdot \frac{1}{n^2}$$

4. Prove that the ground state energy of hydrogen atom

$$E_0 = -13.6\, electron - volt[eV]$$

5. Show that the ground state wave function of hydrogen atom

$$\psi_{100} = \frac{1}{\sqrt{\pi a_0^3}}\ e^{-\frac{r}{a_0}}$$

6. Calculate the Spherical harmonics $Y_{\ell,m}(\theta,\phi)$ for the values of = 3 and m = 2 .
7. Give a concrete feature of degeneracy of the hydrogen atom ground state.
8. Calculate the Volume element $dV = d\tau = d^3r = dxdydz$ in terms of spherical polar coordinates.
9. Determine the mathematical expression of Del or Nabla [∇] in Cartesian coordinates and Laplacian operator[∇^2] in spherical polar coordinates.
10. Calculate the Average value $\langle r \rangle$ and root mean square value $\langle r^2 \rangle^{\frac{1}{2}}$ in hydrogen atom by using the ground state wave function.
11. Calculate the Average value of $\langle \frac{1}{r} \rangle$ in the hydrogen atom.

Chapter 9 Orbital angular momentum in Quantum mechanics

9.1 Orbital angular momentum, its components and their quantum mechanical equivalents

In classical mechanics, the orbital angular momentum of a particle relative to the centre of rotation is a vector quantity $\vec{L}$. It is given by the following equation[5,30]

$$\bar{L} = \bar{r} \times \bar{p} \qquad [1]$$

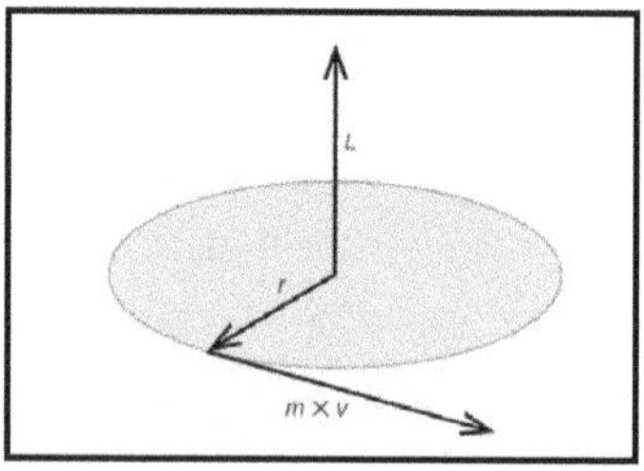

Fig.[9.1]: Shows the orbital angular momentum

Where $\vec{r}$ is the position of the particle with respect to the origin. $\vec{P}$ is the linear momentum of the particle. The cross product ($\vec{r} \times \vec{p}$) is defined to mean that the vector $\vec{L}$ is perpendicular to the plane formed by $\vec{r}$ & $\vec{p}$.

Now $\vec{r} = \hat{i}\,x + \hat{j}\,y + \hat{k}\,z$, $\vec{P} = \hat{i}P_x + \hat{j}P_y + \hat{k}P_z$ and $\vec{L} = \hat{i}L_x + \hat{j}L_y + \hat{k}L_z$

Therefore [1] becomes

$$\vec{L} = \begin{vmatrix} \hat{i} & \hat{j} & \hat{k} \\ x & y & z \\ P_x & P_y & P_z \end{vmatrix}$$

Now the components of the orbital angular momentum $\vec{L}$

$$\left.\begin{aligned} L_x &= yp_z - zp_y \\ L_y &= zp_x - xp_z \\ L_z &= xp_y - yp_x \end{aligned}\right\}$$

In order to discuss the orbital angular momentum in quantum mechanics, the associate operators p_x, p_y & p_z are replaced by their quantum mechanical equivalents

$-i\hbar\frac{\partial}{\partial x}$, $-i\hbar\frac{\partial}{\partial y}$ & $-i\hbar\frac{\partial}{\partial z}$ *respectively I.e.,* $p_x = -i\hbar\frac{\partial}{\partial x}$ $p_y = -i\hbar\frac{\partial}{\partial y}$ *and* $p_z = -i\hbar\frac{\partial}{\partial z}$

In terms of quantum mechanical equivalents, the components of the orbital angular momentum

$$\hat{L}_x = -i\hbar\left(y\frac{\partial}{\partial z} - z\frac{\partial}{\partial y}\right),\quad \hat{L}_y = -i\hbar\left(z\frac{\partial}{\partial x} - x\frac{\partial}{\partial z}\right) \quad \& \quad \hat{L}_z = -i\hbar\left(x\frac{\partial}{\partial y} - y\frac{\partial}{\partial x}\right)$$

9.2 Basic relation between Position and Linear momentum

The position and linear momentum for one dimension is defined as $x = x \text{ and } p_x = -i\hbar\frac{\partial}{\partial x}$. We want to establish the commutation relation between position and linear momentum, I.e., $[x, p_x]$.

Therefore, $[x, p_x] = (xp_x - p_x x) = -i\hbar\left(x\frac{\partial}{\partial x} - \frac{\partial}{\partial x}x\right)$

$$[x, p_x]f = -i\hbar\left(x\frac{\partial}{\partial x} - \frac{\partial}{\partial x}x\right)f = -i\hbar\left(x\frac{\partial f}{\partial x} - x\frac{\partial f}{\partial x} - f\frac{\partial x}{\partial x}\right) = -i\hbar(-f) = i\hbar f$$

$$[x, p_x] = i\hbar \qquad [1]$$

From the above equation, there is an uncertainty between position and linear momentum.

9.3 Commutation rule of Orbital angular momentum

Now, we want to establish the following commutation relations of orbital angular momentum.

1. $[L_x, L_y]$ 2. $[L_y, L_z]$ 3. $[L_z, L_x]$ 4. $\vec{L}\times\vec{L} = i\hbar\,\vec{L}$

5. $[L^2, L_x]$ 6. $[L^2, L_x]$ 7. $[L^2, L_x]$

1. $[L_x, L_y] = (L_xL_y - L_yL_x)$

$$= \left\{-i\hbar\left(y\frac{\partial}{\partial z} - z\frac{\partial}{\partial y}\right) - i\hbar\left(z\frac{\partial}{\partial x} - x\frac{\partial}{\partial y}\right)\right\} - \left\{-i\hbar\left(z\frac{\partial}{\partial x} - x\frac{\partial}{\partial z}\right) - i\hbar\left(y\frac{\partial}{\partial z} - z\frac{\partial}{\partial y}\right)\right\}$$

$$= (-i\hbar)(-i\hbar)\left(y\frac{\partial}{\partial x} - x\frac{\partial}{\partial y}\right)$$

$$= -(i\hbar)(i\hbar)\left(x\frac{\partial}{\partial y} - y\frac{\partial}{\partial x}\right)$$

$$= i\hbar\, L_z$$

Similarly, we can prove that

$$[L_y, L_z] = i\hbar\, L_x$$

$$\& \quad [L_z, L_x] = i\hbar\, L_y$$

4. The components of the orbital angular momentum $\vec{L}$

$\vec{L} = \hat{i}L_x + \hat{j}L_y + \hat{k}L_z$ *Where* $L_x = yp_z - zp_y$, $L_y = zp_x - xp_z$, $L_z = xp_y - yp_x$

The cross product or vector product of orbital angular momentum operator in three dimensions can be written by the determinant

$$\vec{L}\times\vec{L} = \begin{vmatrix} \hat{i} & \hat{j} & \hat{k} \\ L_x & L_y & L_z \\ L_x & L_y & L_z \end{vmatrix}$$

$$= \hat{i}\,(L_yL_z - L_zL_y) + \hat{j}\,(L_zL_x - L_xL_z) + \hat{k}\,(L_xL_y - L_yL_x)$$

$$= \hat{i}\,[L_y\,,L_z] + \hat{j}\,[L_z\,,L_x] + \hat{k}\,[L_x\,,L_y]$$

$$= \hat{i}\,(i\hbar\,L_x) + \hat{j}\,(i\hbar\,L_y) + \hat{k}\,(i\hbar\,L_z) = i\hbar\,(\hat{i}\,L_x + \hat{j}\,L_y + \hat{k}\,L_z) = i\hbar\,\vec{L}$$

Finally $\vec{L}\times\vec{L} = i\hbar\,\vec{L}$

5. Commutation relation between the square of the total orbital angular momentum and its first component:

$$L^2 = L_x^2 + L_y^2 + L_z^2 \text{ And } L_x = yp_z - zp_y$$

$$\therefore [L^2, L_x] = [L_x^2 + L_y^2 + L_z^2, L_x]$$

$$= [L_x^2, L_x] + [L_y^2, L_x] + [L_z^2, L_x] = [L_xL_x, L_x] + [L_yL_y, L_x] + [L_zL_z, L_x]$$

$$= (L_xL_xL_x - L_xL_xL_x) + (L_yL_yL_x - L_xL_yL_y) + (L_zL_zL_x - L_xL_zL_z)$$

$$= L_yL_yL_x - L_xL_yL_y + L_zL_zL_x - L_xL_zL_z$$

Now adding and subtracting the terms $L_yL_xL_y$and $L_zL_xL_z$

$$= L_yL_yL_x - L_xL_yL_y + L_zL_zL_x - L_xL_zL_z + L_yL_xL_y - L_yL_xL_y + L_zL_xL_z - L_zL_xL_z$$

$$= (L_yL_yL_x - L_xL_yL_y + L_yL_xL_y - L_yL_xL_y) + (L_zL_zL_x - L_xL_zL_z + L_zL_xL_z - L_zL_xL_z)$$

$$= L_y(\,L_yL_x - L_xL_y) + (L_yL_x - L_xL_y)L_y + L_z(L_zL_x - L_xL_z) + (L_zL_x - L_xL_z)L_z$$

$$= L_y[\,L_y, L_x] + [L_y, L_x]L_y + L_z[L_z, L_x] + [L_z, L_x]L_z$$

$$= L_y(-i\hbar L_z) + (-i\hbar L_z)L_y + L_z\left(i\hbar L_y\right) + \left(i\hbar L_y\right)L_z$$

$$= -i\hbar L_yL_z - i\hbar L_zL_y + i\hbar L_zL_y + i\hbar L_yL_z = 0$$

Finally, $\left[L^2, L_x\right] = 0$; *Similarly, we can prove that* $\left[L^2, L_y\right] = 0$ *and* $\left[L^2, L_z\right] = 0$

6. **An alternate approach** of the commutation relation between the square of the total orbital angular momentum and its first component:

$$L^2 = L_x^2 + L_y^2 + L_z^2 \text{ And } L_x = yp_z - zp_y$$

$$[L^2, L_x] = [L_x^2 + L_y^2 + L_z^2, L_x]$$
$$= [L_x^2, L_x] + [L_y^2, L_x] + [L_z^2, L_x]$$
$$= [L_xL_x, L_x] + [L_yL_y, L_x] + [L_zL_z, L_x]$$
$$= L_x[L_x, L_x] + [L_x, L_x]L_x + L_y[L_y, L_x] + [L_y, L_x]L_y + L_z[L_z, L_x] + [L_z, L_x]L_z$$

$$\boxed{[ab, c] = a[b, c] + [a, c]b}$$

$$= 0 + 0 - i\hbar L_yL_z - i\hbar L_zL_y + i\hbar L_zL_y + i\hbar L_yL_z$$

$[L^2, L_x] = 0$; *Similarly, we can prove that* $\left[L^2, L_y\right] = 0$ *and* $\left[L^2, L_z\right] = 0$

9.4 Ladder operator [Raising and Lowering operator] of orbital angular momentum and their some commutation rules

Let us define Raising and Lowering operator of orbital angular momentum

$$L_+ = L_x + iL_y \ And \ L_- = L_x - iL_y$$

$$(L_+)^\dagger = (L_x + iL_y)^\dagger = L_x - iL_y = L_-$$

$$And \quad (L_-)^\dagger = (L_x - iL_y)^\dagger = L_x + iL_y = L_+$$

$$(L_+)^\dagger = L_- \ and \ (L_-)^\dagger = L_+$$

Some commutation rules:

1. $[L_z, L_\pm] = \hbar L_\pm$ 2. $[L_+, L_-] = 2\hbar L_z$ 3. $[L^2, L_\pm] = 0$ 4. $[L_x, L_+] = -\hbar L_z$ 5. $[L_y, L_+] = -i\hbar L_z$

6. $[L_z, L_+] = \hbar L_+$ 7. $[L_x, L_-] = \hbar L_z$ 8. $[L_y, L_-] = -i\hbar L_z$ 9. $[L_z, L_-] = -\hbar L_-$

1. $L.H.S = [L_z, L_\pm] = (L_zL_\pm - L_\pm L_z) = L_z(L_x \pm iL_y) - (L_x \pm iL_y)L_z$

$$= L_zL_x \pm iL_zL_y - L_xL_z \mp iL_yL_z$$
$$= (L_zL_x - L_xL_z) \pm i(L_zL_y - L_yL_z) = [L_z, L_x] \pm i[L_z, L_y]$$
$$= i\hbar L_y \pm i(-i\hbar L_x) = i\hbar L_y \pm \hbar L_x = \hbar(iL_y \pm L_x) = \hbar L_\pm$$

2. $L.H.S = [L_+, L_-] = (L_+L_- - L_-L_+) = (L_x + iL_y)(L_x - iL_y) - (L_x - iL_y)(L_x + iL_y)$

$$= L_xL_x - iL_xL_y + iL_yL_x + L_yL_y - L_xL_x - iL_xL_y + iL_yL_x - L_yL_y$$
$$= 2iL_yL_x - 2iL_xL_y = -2iL_xL_y + 2iL_yL_x$$
$$= -2i(L_xL_y - L_yL_x) = -2i[L_x, L_y] = -2i(i\hbar L_z) = 2\hbar L_z$$

3. $L.H.S = [L^2, L_\pm] = [L_x^2 + L_y^2 + L_z^2, L_\pm] = [L_x^2, L_\pm] + [L_y^2, L_\pm] + [L_z^2, L_\pm]$

$$= [L_xL_x, L_\pm] + [L_yL_y, L_\pm] + [L_zL_z, L_\pm]$$

Now using the formula $\boxed{[ab,c]=a[b,c]+[a,c]b}$

$= L_x[L_x, L_\pm] + [L_x, L_\pm]L_x + L_y[L_y, L_\pm] + [L_y, L_\pm]L_y + L_z[L_z, L_\pm] + [L_z, L_\pm]L_z$

$= L_x(L_xL_\pm - L_\pm L_x) + (L_xL_\pm - L_\pm L_x)L_x + L_y(L_yL_\pm - L_\pm L_y) + (L_yL_\pm - L_\pm L_y)L_y$

$+L_z(L_zL_\pm - L_\pm L_z) + (L_zL_\pm - L_\pm L_z)L_z$ [1]

Consider the terms $L_x(L_xL_\pm - L_\pm L_x) + (L_xL_\pm - L_\pm L_x)L_x$ from the above equation,

Therefore,

$L_x(L_xL_\pm - L_\pm L_x) + (L_xL_\pm - L_\pm L_x)L_x$

$= L_x\{L_x(L_x \pm iL_y) - (L_x \pm iL_y)L_x\} + \{L_x(L_x \pm iL_y) - (L_x \pm iL_y)L_x\}L_x$

$= L_x\left\{L_xL_x \pm iL_xL_y - L_xL_x \mp iL_yL_x\right\} + \left\{L_xL_x \pm iL_xL_y - L_xL_x \mp iL_yL_x\right\}L_x$

$= L_x\left(\pm iL_xL_y \mp iL_yL_x\right) + \left(\pm iL_xL_y \mp iL_yL_x\right)L_x$

$= \pm iL_xL_xL_y \mp iL_xL_yL_x \pm iL_xL_yL_x \mp iL_yL_xL_x$

$= \pm iL_x(L_xL_y - L_yL_x) \pm i(L_xL_y - L_yL_x)L_x$

$= \pm iL_x[L_x, L_y] \pm i[L_x, L_y]L_x$

$= \pm iL_x(i\hbar L_z) \pm i(i\hbar L_z)L_x$

$= \pm \hbar L_zL_x \mp \hbar L_zL_x = 0$

Similarly, we can show that

$L_y(L_yL_\pm - L_\pm L_y) + (L_yL_\pm - L_\pm L_y)L_y \; and \; L_z(L_zL_\pm - L_\pm L_z) + (L_zL_\pm - L_\pm L_z)L_z = 0$

Finally, $[L^2, L_\pm] = 0$

$4. L.H.S = [L_x, L_+] = (L_xL_+ - L_+L_x) = L_x(L_x + iL_y) - (L_x + iL_y)L_x$

$= (L_xL_x + +iL_xL_y - L_xL_x - iL_yL_x) = (iL_xL_y - iL_yL_x) = i[L_x, L_y] = i(i\hbar L_z) = -\hbar L_z$

$5. L.H.S = [L_y, L_+] = [L_y, (L_x + iL_y)] = [L_y, L_x] + i[L_y, L_y] = -i\hbar L_z$

$6. L.H.S = [L_z, L_+] = [L_z, L_+] = [L_z, (L_x + iL_y)] = [L_z, L_x] + i[L_z, L_y] = i\hbar L_y + i(-i\hbar L_x)$

$= \hbar L_x + i\hbar L_y = \hbar(L_x + iL_y) = \hbar L_+$

$7. L.H.S = [L_x, L_-] = [L_x, (L_x - iL_y)] = [L_x, L_x] - i[L_x, L_y] = -i(i\hbar L_z) = \hbar L_z$

$8. L.H.S = [L_y, L_-] = [L_y, (L_x - iL_y)] = [L_y, L_x] - i[L_y, L_y] = -i\hbar L_z$

$9. L.H.S = [L_z, L_-] = [L_z, (L_x - iL_y)] = [L_z, L_x] - i[L_z, L_y] = i\hbar L_y - i(-i\hbar L_x)$

$= -\hbar L_x + i\hbar L_y = -\hbar(L_x - iL_y) = -\hbar L_-$

9.5 Commutation rule of Orbital angular momentum with Linear momentum and position

$$[L_x, p_x] = 0\,,\ [L_x, p_z] = -i\hbar p_y\ ,\ [L_x, p_y] = i\hbar p_z\ ,$$
$$[L_x, y] = i\hbar z\,,\ [L_z, x] = i\hbar y\ ,[L_x, x] = 0$$

1. $L.H.S = [L_x, p_x]$

$$= L_x p_x - p_x L_x$$
$$= (yp_z - zp_y)p_x - p_x(yp_z - zp_y)$$
$$= (-i\hbar)^2\left\{y\frac{\partial}{\partial z} - z\frac{\partial}{\partial y}\right\}\frac{\partial}{\partial x} - (-i\hbar)^2\left\{\frac{\partial}{\partial x}\left(y\frac{\partial}{\partial z} - z\frac{\partial}{\partial y}\right)\frac{\partial}{\partial x}\right\}$$
$$= (-i\hbar)^2\left\{y\frac{\partial}{\partial z}\frac{\partial}{\partial x} - z\frac{\partial}{\partial y}\frac{\partial}{\partial x} - \frac{\partial y}{\partial x}\frac{\partial}{\partial z} - y\frac{\partial}{\partial x}\frac{\partial}{\partial z} + z\frac{\partial}{\partial y}\frac{\partial z}{\partial x}\right\}$$
$$= (-i\hbar)^2\left\{\frac{\partial}{\partial y}\frac{\partial z}{\partial x} - \frac{\partial y}{\partial x}\frac{\partial}{\partial z}\right\} = 0$$

2. $L.H.S = [L_z, p_x]$

$$= (L_z p_x - p_x L_z)$$
$$= (xp_y - yp_x)p_x - p_x(xp_y - yp_x)$$
$$= (-i\hbar)^2\left\{\left(x\frac{\partial}{\partial y} - y\frac{\partial}{\partial x}\right)\frac{\partial}{\partial x} - \frac{\partial}{\partial x}\left(x\frac{\partial}{\partial y} - y\frac{\partial}{\partial x}\right)\right\}$$
$$= (-i\hbar)^2\left\{x\frac{\partial}{\partial y}\frac{\partial}{\partial x} - y\frac{\partial}{\partial x}\frac{\partial}{\partial x} - \frac{\partial x}{\partial x}\frac{\partial}{\partial y} - x\frac{\partial}{\partial x}\frac{\partial}{\partial y} + \frac{\partial y}{\partial x}\frac{\partial}{\partial x} + y\frac{\partial}{\partial x}\frac{\partial}{\partial x}\right\}$$
$$= (-i\hbar)^2\frac{\partial}{\partial y} = i\hbar\left(-i\hbar\frac{\partial}{\partial y}\right)$$

$\therefore [L_z, P_x] = i\hbar P_y$

Similarly, we can prove that $[L_x, p_z] = -i\hbar p_y\ ,\ [L_x, p_y] = i\hbar p_z$

4. $[L_x, x] = (yp_z - zp_y)x - x(yp_z - zp_y) = -i\hbar\left\{\left(y\frac{\partial}{\partial z} - z\frac{\partial}{\partial y}\right)x - x\left(y\frac{\partial}{\partial z} - z\frac{\partial}{\partial y}\right)\right\}$

$$[L_x, x]f = -i\hbar\left\{xy\frac{\partial f}{\partial z} + yf\frac{\partial x}{\partial z} - zx\frac{\partial f}{\partial y} - zf\frac{\partial x}{\partial y} - xy\frac{\partial f}{\partial z} + xz\frac{\partial f}{\partial y}\right\}$$
$$= -i\hbar\left(yf\frac{\partial x}{\partial z} - zf\frac{\partial x}{\partial y}\right) \qquad \therefore\ [L_x, x] = 0$$

Similarly, we can prove that $[L_x, y] = i\hbar z\,,\ [L_z, x] = i\hbar y$

9.6 Relation between Cartesian coordinates and Spherical polar coordinates

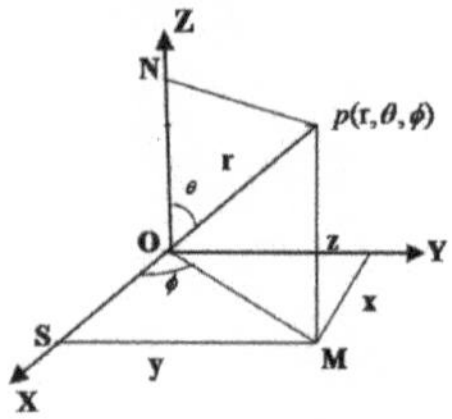

$Cos\theta = \frac{ON}{OP}$ or, $ON = OPCos\theta$ or, $Z = r\,Cos\theta$, $Sin\phi = \frac{MS}{OM}$ and $Sin\theta = \frac{PN}{OP}$

$MS = OMSin\varphi$ and $PN = OPSin\theta$ or, $y = PNSin\varphi = OP\ Sin\theta Sin\phi = r\ Sin\theta\ Sin\ \phi$

$Cos\varphi = \frac{OS}{OM}$ or, $OS = OMCos\varphi$ or, $x = PNCos\varphi = OPSin\theta Cos\varphi$ or, $x = r\sin\theta Cos\varphi$

The relation between Cartesian coordinates (x, y, z) and spherical polar coordinates

$$x = r\sin\theta\cos\varphi\,, y = r\sin\theta\sin\varphi\,, z = r\cos\theta\ and\ r = \sqrt{x^2 + y^2 + z^2}$$

9.7 Components of orbital angular momentum operators in spherical polar coordinates

In classical mechanics, the orbital angular momentum of a particle relative to the centre of rotation is a vector quantity $\vec{L}$. It is given by the following equation

$$\bar{L} = \bar{r} \times \bar{p} \quad [1]$$

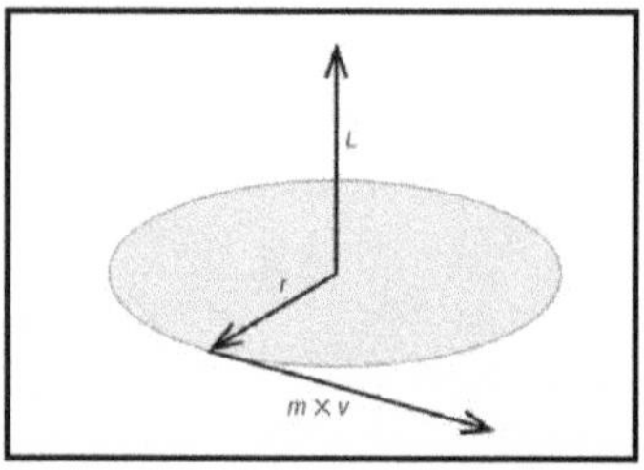

Fig.[9.7.1]: Shows the orbital angular momentum

Where $\vec{r}$ is the position of the particle with respect to the origin. $\vec{P}$ is the linear momentum of the particle. The cross product $(\vec{r} \times \vec{p})$ is defined to mean that the vector $\vec{L}$ is perpendicular to the plane formed by $\vec{r}$ & $\vec{p}$.

Now $\vec{r}=\hat{i}\,x+\hat{j}\,y+\hat{k}\,z$, $\vec{P}=\hat{i}P_x+\hat{j}P_y+\hat{k}P_z$ and $\vec{L}=\hat{i}L_x+\hat{j}L_y+\hat{k}L_z$

Therefore [1] becomes

$$\vec{L}=\begin{vmatrix}\hat{i} & \hat{j} & \hat{k}\\ x & y & z\\ P_x & P_y & P_z\end{vmatrix}$$

Now the components of the orbital angular momentum $\vec{L}$

$$L_x = yp_z - zp_y$$
$$L_y = zp_x - xp_z$$
$$L_z = xp_y - yp_x$$

In order to discuss the orbital angular momentum in quantum mechanics, the associate operators p_x, p_y & p_z are replaced by their quantum mechanical equivalents[3,9]

$$-i\hbar\frac{\partial}{\partial x}\;,\; -i\hbar\frac{\partial}{\partial y}\;\&\; -i\hbar\frac{\partial}{\partial z}\;\; respectively\; I.e.,\; p_x=-i\hbar\frac{\partial}{\partial x}\;\; p_y=-i\hbar\frac{\partial}{\partial y}\;\; and\;\; p_z=-i\hbar\frac{\partial}{\partial z} \qquad [2]$$

In terms of quantum mechanical equivalents, the components of the orbital angular momentum

$$\hat{L}_x=-i\hbar\left(y\frac{\partial}{\partial z}-z\frac{\partial}{\partial y}\right),\; \hat{L}_y=-i\hbar\left(z\frac{\partial}{\partial x}-x\frac{\partial}{\partial z}\right)\;\&\; \hat{L}_z=-i\hbar\left(x\frac{\partial}{\partial y}-y\frac{\partial}{\partial x}\right) \qquad [3]$$

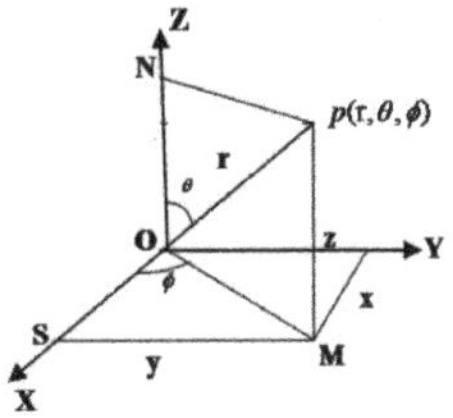

Fig.[9.7.2] Shows the Spherical polar coordinates

$$Cos\theta=\frac{ON}{OP}\;\; or,\;\; ON=OPCos\theta\;\; or,\;\; Z=r\,Cos\theta\;,\;\; Sin\phi=\frac{MS}{OM}\;\; and\;\; Sin\theta=\frac{PN}{OP}$$

$$MS = OMSin\varphi \; and \; PN = OPSin\theta \quad or,\, y = PNSin\varphi = OP\; Sin\theta Sin\phi = r\; Sin\theta\; Sin\;\phi$$

$$Cos\varphi=\frac{OS}{OM}\; or, OS = OMCos\varphi \quad or,\quad x = PNCos\varphi = OPSin\theta Cos\varphi \;\; or,\quad x = r\sin\theta Cos\varphi$$

From spherical polar coordinate x, y and z can be written as

$$x = r\sin\theta\cos\varphi\,, y = r\sin\theta\sin\varphi\,, z = r\cos\theta \; and\; r=\sqrt{x^2+y^2+z^2} \qquad [4]$$

Now, using calculus the operators $\frac{\partial}{\partial x},\frac{\partial}{\partial y}$ and $\frac{\partial}{\partial z}$ can be written in the followings:

$$\left.\begin{aligned} \frac{\partial}{\partial x} &= \frac{\partial r}{\partial x}\frac{\partial}{\partial r} + \frac{\partial \theta}{\partial x}\frac{\partial}{\partial \theta} + \frac{\partial \varphi}{\partial x}\frac{\partial}{\partial \varphi} \\ \frac{\partial}{\partial y} &= \frac{\partial r}{\partial y}\frac{\partial}{\partial r} + \frac{\partial \theta}{\partial y}\frac{\partial}{\partial \theta} + \frac{\partial \varphi}{\partial y}\frac{\partial}{\partial \varphi} \\ \frac{\partial}{\partial z} &= \frac{\partial r}{\partial z}\frac{\partial}{\partial r} + \frac{\partial \theta}{\partial z}\frac{\partial}{\partial \theta} + \frac{\partial \varphi}{\partial z}\frac{\partial}{\partial \varphi} \end{aligned}\right\} \quad [5]$$

$$\therefore \frac{\partial r}{\partial x} = \frac{\partial}{\partial x}(\sqrt{x^2+y^2+z^2}) = \frac{\partial}{\partial x}(x^2+y^2+z^2)^{\frac{1}{2}}$$

$$or, \frac{\partial r}{\partial x} = \frac{1}{2}.\frac{1}{(\sqrt{x^2+y^2+z^2})}.2x = \frac{x}{r} = \frac{r\sin\theta\cos\varphi}{r}$$

$$or, \quad \frac{\partial r}{\partial x} = \sin\theta\cos\varphi$$

$$similarly, \ \frac{\partial r}{\partial y} = \frac{y}{r} = \frac{r\sin\theta\sin\varphi}{r} = \sin\theta\sin\varphi$$

$$and \quad \frac{\partial r}{\partial z} = \frac{z}{r} = \frac{r\cos\theta}{r} = \cos\theta$$

$$Again \quad cos\theta = \frac{z}{r} = \frac{z}{\sqrt{x^2+y^2+z^2}}$$

$$or, -Sin\,\theta\frac{\partial \theta}{\partial z} = \frac{1}{\sqrt{x^2+y^2+z^2}} + z(-\frac{1}{2})\frac{2z}{\left(\sqrt{x^2+y^2+z^2}\right)^3} = \frac{1}{r} - \frac{z^2}{r^3} = \frac{1}{r} - \frac{r^2\cos^2\theta}{r^3}$$

$$= \frac{1}{r}(1-\cos^2\theta) = \frac{\sin^2\theta}{r}$$

$$or, \quad \frac{\partial \theta}{\partial z} = -\frac{\sin\theta}{r}$$

$$Similarly, \ \frac{\partial \theta}{\partial x} = \frac{\cos\theta\cos\varphi}{r}$$

$$and \quad \frac{\partial \theta}{\partial y} = \frac{\cos\theta\sin\varphi}{r}$$

$$Finally \quad \frac{\partial \theta}{\partial x} = \frac{\cos\theta\cos\varphi}{r}, \frac{\partial \theta}{\partial y} = \frac{\cos\theta\sin\varphi}{r} \ and \ \frac{\partial \theta}{\partial z} = -\frac{\sin\theta}{r} \quad [6]$$

$$\text{And, } \tan\varphi = \frac{y}{x} \ or, \ \sec^2\varphi\frac{\partial \varphi}{\partial x} = -\frac{y}{x^2} = -\frac{r\sin\theta\sin\varphi}{r^2\sin^2\theta\cos^2\varphi}$$

$$Or, \ \sec^2\varphi\frac{\partial \varphi}{\partial x} = -\frac{\sin\varphi}{r\sin\theta}\frac{1}{\cos^2\varphi}$$

$$\frac{\partial \varphi}{\partial x} = -\frac{\sin\varphi}{r\sin\theta}$$

Similarly, $\frac{\partial\varphi}{\partial y}=\frac{\cos\varphi}{r\sin\theta}$

And $\frac{\partial\varphi}{\partial z}=0$

Finally, $\frac{\partial\varphi}{\partial x}=-\frac{\sin\varphi}{r\sin\theta},\frac{\partial\varphi}{\partial y}=\frac{\cos\varphi}{r\sin\theta}$ *and* $\frac{\partial\varphi}{\partial z}=0$ [7]

From equation [5] we have

$$\frac{\partial}{\partial x}=\sin\theta\cos\varphi\frac{\partial}{\partial r}+\frac{\cos\theta\cos\varphi}{r}\frac{\partial}{\partial\theta}-\frac{\sin\varphi}{r\sin\theta}\frac{\partial}{\partial\varphi}$$

$$\frac{\partial}{\partial y}=\sin\theta\sin\varphi\frac{\partial}{\partial r}+\frac{\cos\theta\sin\varphi}{r}\frac{\partial}{\partial\theta}+\frac{\cos\varphi}{r\sin\theta}\frac{\partial}{\partial\varphi}$$

$$\frac{\partial}{\partial z}=\cos\theta\frac{\partial}{\partial r}-\frac{\sin\theta}{r}\frac{\partial}{\partial\theta}$$

Therefore, equation [3] becomes

$$\hat{L}_x=-i\hbar\left[y\frac{\partial}{\partial z}-z\frac{\partial}{\partial y}\right]$$

$$=-i\hbar\left[r\sin\theta\sin\varphi\left(\cos\theta\frac{\partial}{\partial r}-\frac{\sin\theta}{r}\frac{\partial}{\partial\theta}\right)-r\cos\theta\left(\sin\theta\sin\varphi\frac{\partial}{\partial r}+\frac{\cos\theta\sin\varphi}{r}\frac{\partial}{\partial\theta}+\frac{\cos\varphi}{r\sin\theta}\frac{\partial}{\partial\varphi}\right)\right]$$

$$=-i\hbar[-(\sin^2\theta+\cos^2\theta)\sin\varphi\frac{\partial}{\partial\theta}-\cos\varphi\cot\theta\frac{\partial}{\partial\varphi}]=i\hbar[\sin\varphi\frac{\partial}{\partial\theta}+\cot\theta\cos\varphi\frac{\partial}{\partial\varphi}]$$

Similarly, $\hat{L}_y=i\hbar[-\cos\varphi\frac{\partial}{\partial\theta}+\cot\theta\sin\varphi\frac{\partial}{\partial\varphi}]$ *And* $\hat{L}_z=-i\hbar\frac{\partial}{\partial\theta}$

Finally,

$$\hat{L}_x=i\hbar[\sin\varphi\frac{\partial}{\partial\theta}+\cot\theta\cos\varphi\frac{\partial}{\partial\varphi}]$$

$$\hat{L}_y=i\hbar[-\cos\varphi\frac{\partial}{\partial\theta}+\cot\theta\sin\varphi\frac{\partial}{\partial\varphi}] \qquad [9]$$

$$\&\ \hat{L}_z=-i\hbar\frac{\partial}{\partial\varphi}$$

Therefore,

$\hat{L}^2=\hat{L}_x^2+\hat{L}_y^2+\hat{L}_z^{\,2}$ $or, \hat{L}^2=\hat{L}_x\hat{L}_x+\hat{L}_y\hat{L}_y+\hat{L}_z\hat{L}_z$ [10]

Now using equation [9] in equation [10], we can obtain of operator L^2 as follows:

$\hat{L}^2=\hat{L}_x^2+\hat{L}_y^2+\hat{L}_z^{\,2}$ $or, \hat{L}^2=\hat{L}_x\hat{L}_x+\hat{L}_y\hat{L}_y+\hat{L}_z\hat{L}_z$

$$= \frac{-\hbar}{i}\left(\sin\varphi\frac{\partial}{\partial\theta}+\cot\theta\cos\varphi\frac{\partial}{\partial\varphi}\right)\frac{-\hbar}{i}\left(\sin\varphi\frac{\partial}{\partial\theta}+\cot\theta\cos\varphi\frac{\partial}{\partial\varphi}\right)$$

$$+\frac{-\hbar}{i}\left(-\cos\varphi\frac{\partial}{\partial\theta}+\cot\theta\sin\varphi\frac{\partial}{\partial\varphi}\right)\frac{-\hbar}{i}\left(-\cos\varphi\frac{\partial}{\partial\theta}+\cot\theta\sin\varphi\frac{\partial}{\partial\varphi}\right)+\left(\frac{\hbar}{i}\frac{\partial}{\partial\varphi}\right)\left(\frac{\hbar}{i}\frac{\partial}{\partial\varphi}\right)$$

$$= -\hbar^2\left[\sin\varphi\frac{\partial}{\partial\theta}\left(\sin\varphi\frac{\partial}{\partial\theta}\right)+\cot\theta\cos\varphi\frac{\partial}{\partial\varphi}\left(\sin\varphi\frac{\partial}{\partial\theta}\right)\right.$$

$$\left.+\sin\varphi\frac{\partial}{\partial\theta}\left(\cot\theta\cos\varphi\frac{\partial}{\partial\varphi}\right)+\cot\theta\cos\varphi\frac{\partial}{\partial\varphi}\left(\cot\theta\cos\varphi\frac{\partial}{\partial\varphi}\right)\right]$$

$$-\hbar^2\left[-\cos\varphi\frac{\partial}{\partial\theta}\left(-\cos\varphi\frac{\partial}{\partial\theta}\right)+\cot\theta\sin\varphi\frac{\partial}{\partial\varphi}\left(-\cos\varphi\frac{\partial}{\partial\theta}\right)\right]$$

$$-\hbar^2\left[\frac{\partial^2}{\partial\phi^2}-\cos\varphi\frac{\partial}{\partial\theta}\left(\cot\theta\sin\varphi\frac{\partial}{\partial\varphi}\right)+\cot\theta\sin\varphi\frac{\partial}{\partial\varphi}\left(\cot\theta\sin\varphi\frac{\partial}{\partial\varphi}\right)\right]$$

$$= -\hbar^2\left[\sin^2\varphi\frac{\partial^2}{\partial\theta^2}+\cot\theta\cos\varphi\cos\varphi\frac{\partial}{\partial\theta}+\cot\theta\cos\varphi\sin\varphi\frac{\partial^2}{\partial\varphi\partial\theta}\right]$$

$$-\hbar^2\left[-\sin\varphi\cos ec^2\theta\cos\varphi\frac{\partial}{\partial\varphi}+\sin\varphi\cot\theta\cos\varphi\frac{\partial^2}{\partial\theta\partial\varphi}-\cot\theta\cos\varphi\cot\theta\sin\varphi\frac{\partial}{\partial\varphi}\right]$$

$$-\hbar^2\left[+\cot\theta\cos\varphi\cot\theta\cos\varphi\frac{\partial^2}{\partial\phi^2}+\cos^2\varphi\frac{\partial^2}{\partial\theta^2}+\cot\theta\sin\varphi\sin\varphi\frac{\partial}{\partial\theta}\right]$$

$$-\hbar^2\left[-\cot\theta\sin\varphi\cos\varphi\frac{\partial^2}{\partial\varphi\partial\theta}+\cos\varphi\cos ec^2\theta\sin\varphi\frac{\partial}{\partial\varphi}-\cos\varphi\cot\theta\sin\varphi\frac{\partial^2}{\partial\theta\partial\varphi}\right]$$

$$-\hbar^2\left[+\cot\theta\sin\varphi\cot\theta\cos\varphi\frac{\partial}{\partial\varphi}+\cot\theta\sin\varphi\cot\theta\sin\varphi\frac{\partial^2}{\partial\phi^2}\right]-\hbar^2\frac{\partial^2}{\partial\phi^2}$$

$$= -\hbar^2\left[\sin^2\varphi\frac{\partial^2}{\partial\theta^2}+\cot\theta\cos^2\varphi\frac{\partial}{\partial\theta}+\cot\theta\cos\varphi\sin\varphi\frac{\partial^2}{\partial\varphi\partial\theta}\right]$$

$$-\hbar^2\left[-\sin\varphi\cos\varphi\frac{\partial}{\partial\varphi}-\sin\varphi\cos\varphi\cot^2\theta\frac{\partial}{\partial\varphi}+\sin\varphi\cos\varphi\cot\theta\frac{\partial^2}{\partial\theta\partial\varphi}\right]$$

$$-\hbar^2\left[-\cot^2\theta\sin\varphi\cos\varphi\frac{\partial}{\partial\varphi}+\cot^2\theta\cos^2\varphi\frac{\partial^2}{\partial\varphi^2}+\cos^2\varphi\frac{\partial^2}{\partial\theta^2}\right]$$

$$-\hbar^2\left[+\cot\theta\sin^2\varphi\frac{\partial}{\partial\theta}-\cot\theta\sin\varphi\cos\varphi\frac{\partial^2}{\partial\varphi\partial\theta}+\cos\varphi\sin\varphi\frac{\partial}{\partial\varphi}\right]$$

$$-\hbar^2\left[+\cos\varphi\cot^2\theta\sin\varphi\frac{\partial}{\partial\varphi}-\cos\varphi\cot\theta\sin\varphi\frac{\partial^2}{\partial\theta\partial\varphi}+\cot^2\theta\sin\varphi\cos\varphi\frac{\partial}{\partial\varphi}+\cot^2\theta\sin^2\varphi\frac{\partial^2}{\partial\varphi^2}\right]$$

$$-\hbar^2\frac{\partial^2}{\partial\varphi^2}$$

$$=-\hbar^2\left[\frac{\partial^2}{\partial\theta^2}+\cot\theta\frac{\partial}{\partial\theta}+\cot^2\theta\frac{\partial^2}{\partial\varphi^2}\right]-\hbar^2\frac{\partial^2}{\partial\varphi^2}$$

$$=-\hbar^2\left[\frac{\partial^2}{\partial\theta^2}+\cot\theta\frac{\partial}{\partial\theta}+\operatorname{cosec}^2\theta\frac{\partial^2}{\partial\varphi^2}\right]$$

$$L^2=-\hbar^2\left[\frac{1}{\sin\theta}\frac{\partial}{\partial\theta}\left(\sin\theta\frac{\partial}{\partial\theta}\right)+\frac{1}{\sin^2\theta}\frac{\partial^2}{\partial\varphi^2}\right] \quad [11]$$

Equation [11] represents the expression for x, y and z components of the square of the total orbital angular momentum operators in spherical polar coordinates.

9.8 Common Eigen functions [spherical harmonics] of both L^2 and L_Z

In classical mechanics, the orbital angular momentum of a particle relative to the centre of rotation is a vector quantity $\vec{L}$. It is given by the following equation

$$\bar{L}=\bar{r}\times\bar{p}$$

Or, $\bar{L}=\bar{r}\times\frac{\hbar}{i}\vec{\nabla} \quad [\because \bar{p}=-i\hbar\vec{\nabla}]$

Or, $\bar{L}\Psi=\frac{\hbar}{i}(\bar{r}\times\bar{\nabla})\Psi$ *Or,* $\bar{L}.\bar{L}\Psi=\left(\frac{\hbar}{i}(\bar{r}\times\vec{\nabla})\right).\left(\frac{\hbar}{i}(\bar{r}\times\vec{\nabla}\Psi)\right)$

Or, $L^2\Psi=\left(\frac{\hbar}{i}\right)^2\left[\left(\bar{r}\times\vec{\nabla}\right).(\bar{r}\times\bar{\nabla}\Psi)\right]\Rightarrow L^2\Psi=-\hbar^2\vec{r}.\vec{\nabla}\times(r\times\vec{\nabla}\Psi)$ [1]

[By interchanging the dot & cross product]

Again know that

$$\vec{\nabla}\times(\bar{F}\times\bar{G})=(\bar{\nabla}.\bar{G})\bar{F}-(\bar{\nabla}.\bar{F})\vec{G}+(\bar{G}.\bar{\nabla})\bar{F}-\left(\vec{F}.\vec{\nabla}\right)\vec{G}$$

So, $\vec{\nabla}\times(\bar{r}\times\vec{\nabla}\Psi)=(\bar{\nabla}.\bar{\nabla}\Psi)\bar{r}-(\bar{\nabla}.\bar{r})\vec{\nabla}\Psi+(\bar{\nabla}\Psi.\bar{\nabla})\bar{r}-\left(\vec{r}.\vec{\nabla}\right)\vec{\nabla}\Psi$ [2]

Therefore equation [1] becomes

$$L^2\Psi = -\hbar^2\vec{r}.[\nabla^2\Psi\vec{r} - (\bar{\nabla}.\bar{r})\bar{\nabla}\Psi + (\bar{\nabla}\Psi.\bar{\nabla})\bar{r} - (\bar{r}.\bar{\nabla})\bar{\nabla}\Psi]$$

$$= -\hbar^2 r^2\nabla^2\Psi - \hbar^2[r.(\bar{\nabla}\Psi.\bar{\nabla})\bar{r} - \vec{r}.(\bar{\nabla}.\bar{r})\bar{\nabla}\Psi - \vec{r}.\left(\vec{r}.\vec{\nabla}\right)\vec{\nabla}\Psi] \qquad [3]$$

Now Consider the following terms from the above equation [3]

$\bar{r}.(\bar{\nabla}\Psi.\bar{\nabla})\,\vec{r}$:

$$= \bar{r}.[(\hat{i}\frac{\partial\Psi}{\partial x} + \hat{j}\frac{\partial\Psi}{\partial y} + \hat{k}\frac{\partial\Psi}{\partial z}).(\hat{i}\frac{\partial}{\partial x} + \hat{j}\frac{\partial}{\partial y} + \hat{k}\frac{\partial}{\partial z})](\hat{i}x + \hat{j}y + \hat{k}z)$$

$$= (\hat{i}x + \hat{j}y + \hat{k}z).[\frac{\partial\Psi}{\partial x}\frac{\partial}{\partial x} + \frac{\partial\Psi}{\partial y}\frac{\partial}{\partial y} + \frac{\partial\Psi}{\partial z}\frac{\partial}{\partial z}](\hat{i}x + \hat{j}y + \hat{k}z)$$

$$= (\hat{i}x + \hat{j}y + \hat{k}z).(\hat{i}\frac{\partial\Psi}{\partial x} + \hat{j}\frac{\partial\Psi}{\partial y} + \hat{k}\frac{\partial\Psi}{\partial z}) = x\frac{\partial\Psi}{\partial x} + y\frac{\partial\Psi}{\partial y} + z\frac{\partial\Psi}{\partial z} = r\frac{\partial\Psi}{\partial r}$$

$\bar{r}.(\bar{\nabla}.\bar{r})\bar{\nabla}\Psi$:

$$= \vec{r}.[(\hat{i}\frac{\partial}{\partial x} + \hat{j}\frac{\partial}{\partial y} + \hat{k}\frac{\partial}{\partial z}).(\hat{i}x + \hat{j}y + \hat{k}z)\vec{\nabla}\Psi = \vec{r}.\,3\bar{\nabla}\Psi$$

$$= 3\left[(\hat{i}x + \hat{j}y + \hat{k}z).\left(\hat{i}\frac{\partial\Psi}{\partial x} + \hat{j}\frac{\partial\Psi}{\partial y} + \hat{k}\frac{\partial\Psi}{\partial z}\right)\right] = 3\left[x\frac{\partial\Psi}{\partial x} + y\frac{\partial\Psi}{\partial y} + z\frac{\partial\Psi}{\partial z}\right] = 3r\frac{\partial\Psi}{\partial r}$$

And $\vec{r}.\left(\vec{r}.\vec{\nabla}\right)\vec{\nabla}\Psi = r^2\frac{\partial^2\Psi}{\partial r^2}$

Hence equation [3] becomes

$$L^2\Psi \quad = -\hbar^2 r^2\nabla^2\Psi - \hbar^2[r\frac{\partial\Psi}{\partial r} - 3r\frac{\partial\Psi}{\partial r} - r^2\frac{\partial^2\Psi}{\partial r^2}]$$

$$L^2\Psi \quad = -\hbar^2 r^2\nabla^2\Psi - \hbar^2[-2r\frac{\partial\Psi}{\partial r} - r^2\frac{\partial^2\Psi}{\partial r^2}] \Rightarrow L^2\psi = -\hbar^2 r^2\nabla^2\Psi + \hbar^2\frac{\partial}{\partial r}(r^2\frac{\partial\Psi}{\partial r})$$

$$L^2\Psi \quad \Rightarrow \hbar^2 r^2[-\nabla^2\Psi + \frac{1}{r^2}\frac{\partial}{\partial r}(r^2\frac{\partial\Psi}{\partial r})]$$

$$\therefore \quad \frac{1}{\hbar^2 r^2}L^2\Psi = -\nabla^2\Psi + \frac{1}{r^2}\frac{\partial}{\partial r}(r^2\frac{\partial\Psi}{\partial r}) \qquad [4]$$

From spherical polar coordinate $\nabla^2\Psi$ can be expressed as

$$\nabla^2\Psi = \frac{1}{r^2}\frac{\partial}{\partial r}(r^2\frac{\partial^2\Psi}{\partial r}) + \frac{1}{r^2\sin\theta}\frac{\partial}{\partial\theta}(\sin\theta\frac{\partial\Psi}{\partial\theta}) + \frac{1}{r^2\sin^2\theta}\frac{\partial^2\Psi}{\partial\Phi^2}$$

Hence equation [4] becomes

$$\frac{1}{\hbar^2 r^2}L^2\Psi = -\frac{1}{r^2}\frac{\partial}{\partial r}(r^2\frac{\partial\Psi}{\partial r}) - \frac{1}{r^2\sin\theta}\frac{\partial}{\partial\theta}(\sin\theta\frac{\partial\Psi}{\partial\theta}) - \frac{1}{r^2\sin^2\theta}.\frac{\partial^2\Psi}{\partial\Phi^2} + \frac{1}{r^2}\frac{\partial}{\partial r}(r^2\frac{\partial\Psi}{\partial r})$$

$$Or, \frac{1}{\hbar^2 r^2} L^2\Psi = -[\frac{1}{r^2 \sin\theta}\frac{\partial}{\partial\theta}(\sin\theta\frac{\partial\Psi}{\partial\theta}) + \frac{1}{r^2\sin\theta}\frac{\partial^2\Psi}{\partial\Phi^2}]$$

$$Or, L^2\Psi = -\hbar^2[\frac{1}{\sin\theta}\frac{\partial}{\partial\theta}(\sin\theta\frac{\partial\Psi}{\partial\theta}) + \frac{1}{\sin^2\theta}\frac{\partial^2\Psi}{\partial\Phi^2}] \quad [5]$$

Equation [5] represents just the angular part of the Laplacian operator. If we denote the Eigen function of this operator by the angular function then Eigen value equation can be taken

$$L^2 Y_l^m(\theta,\varphi) = -\hbar^2[\frac{1}{\sin\theta}\frac{\partial}{\partial\theta}(\sin\theta\frac{\partial}{\partial\theta}) + \frac{1}{\sin^2\theta}\frac{\partial^2}{\partial\varphi^2}]Y_l^m(\theta,\varphi) \quad \left[Denoting \quad \Psi = Y_l^m(\theta,\varphi)\right]$$

$$L^2 Y_l^m(\theta,\varphi) = -\hbar^2[-l(l+1)]Y_l^m(\theta,\varphi) \quad [6]$$

$$Where \quad \frac{1}{\sin\theta}\frac{\partial}{\partial\theta}(\sin\theta\frac{\partial}{\partial\theta}) + \frac{1}{\sin^2\theta}\frac{\partial^2}{\partial\Phi^2} = -l(l+1) \quad [7]$$

Therefore, $$\boxed{L^2 Y_l^m(\theta,\varphi) = l(l+1)\hbar^2 Y_l^m(\theta,\varphi) \quad [8]}$$

Where $l(l+1)\hbar^2$ [with l = 0, 1, 2, ------------- (n-1)] represents the Eigen value of L^2 & $Y_l^m(\theta,\phi)$ being the corresponding Eigen function.

In spherical polar coordinates, the z- component of the orbital angular momentum operator is

$$L_z = -i\hbar\frac{\partial}{\partial\Phi}$$

$$Or, \quad L_z\ \Psi(\theta,\varphi) = -i\hbar\frac{\partial}{\partial\Phi}\Psi(\theta,\varphi) \quad [9]$$

Where, $\Psi(\theta,\varphi)$ is the angular part of the wave function and can be written by the separation of variables. I.e.,

$$\Psi(\theta,\varphi) = \Theta(\theta)\Phi(\varphi) = Y_l^m(\theta,\varphi) \quad [10]$$

Therefore, equation [10] becomes

$$Or, L_z\ \Theta(\theta)\Phi(\varphi) = -i\hbar\frac{\partial}{\partial\Phi}\Theta(\theta)\Phi(\varphi) \quad [11]$$

$$Or, L_z Y_l^m(\theta,\varphi) = -i\hbar\frac{\partial}{\partial\varphi}Y_l^m(\theta,\varphi)$$

$$Let \quad \Phi(\varphi) = A\exp(im_l\varphi)\ or, \frac{\partial\Phi(\varphi)}{\partial\varphi} = Aim_l\exp(im_l\varphi) = im_l\ \Phi(\varphi)$$

From equation [11]

$$L_z\ \Theta(\theta)\Phi(\varphi) = -i\hbar\ \Theta(\theta)\,\mathrm{i}m_l\ \Phi(\varphi) = -i^2\hbar m_l\ \Theta(\theta)\,\Phi(\varphi) == m_l\,\hbar\,Y_l^m(\theta,\varphi)$$

$$\boxed{L_z\,Y_l^m(\theta,\varphi) = m_l\,\hbar\,Y_l^m(\theta,\varphi)} \qquad [12]$$

From equation [12] where $m_l\hbar$ [with $l = 0,\ 1,\ 2,\ldots\ldots(n-1)$] represents the Eigen value of L_z & $Y_l^m(\theta,\phi)$ being the corresponding Eigen function. *Thus the spherical Harmonics $Y_l^m(\theta,\phi)$ are the simultaneous Eigen functions of the operators L^2 and L_Z belonging to the Eigen values $l(l+1)\hbar^2$ and $m_l\hbar$ respectively.*

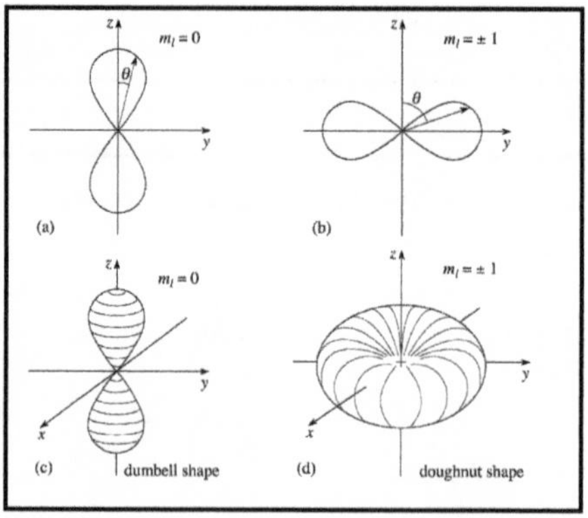

Fig.[9.8.1]: Shows different orbitals of magnetic quantum number

The Eigen values of L^2 are $\lambda = l(l+1)\hbar^2$ Where $l = 0, 1, 2, 3, \ldots\ldots(n-1)$. Since $\langle L_z^2\rangle$ cannot be greater than L^2, we have $m^2\hbar^2 \le l(l+1)\hbar^2$ or $m^2 \le l(l+1)$. $\therefore |m| < |l|$. For $m = l+1$, $m^2 = l^2 + 2l + 1 > l(l+1)$ which violates the condition $m^2 \le l(l+1)$. $\therefore$ $-l \le m \le +l$. $\therefore$ $m = -l, -l+1, -l+2, \ldots, 0, \ldots, l-2, l-1, +l$. There are $(2l+1)$ number of values of m. Since $\Theta(\theta)$ depends on both m and l, $\Theta(\theta)$ is labelled as Θ_l^m. The Eigen functions of L^2 and L_Z are denoted by $\Psi_{l,m}(\theta,\phi) = \Theta_l^m(\cos\theta)\,e^{im\phi}$.

For a particular Eigen function $\Psi_{l,m}$; We have particular values of l *and* m. Thus L^2 and L_Z have particular values. If we know the particular values of L^2 and L_Z, we can deduce the angle between the vector $\vec{L}$ and z-axis. Because the values of l and m are discrete, $\vec{L}$ can assume certain allowed orientations. For a given value of l, there will be $(2l + 1)$ allowed directions of $\vec{L}$, as m ranges over$(2l + 1)$ integer values from $-l\ to + l$.

The maximum value of $|L_Z| = |\ l|\hbar < \sqrt{l(l+1)}\hbar$. $\therefore$ $\vec{L}$ cannot point along z axis. For $l = 1$,

$m = (-1, 0, +1)$ $\therefore L = \sqrt{l(l+1)}\hbar = \sqrt{1(1+1)}\hbar = \sqrt{2}\hbar = 1.4\hbar$. $L_Z = -\hbar, 0, +\hbar$.

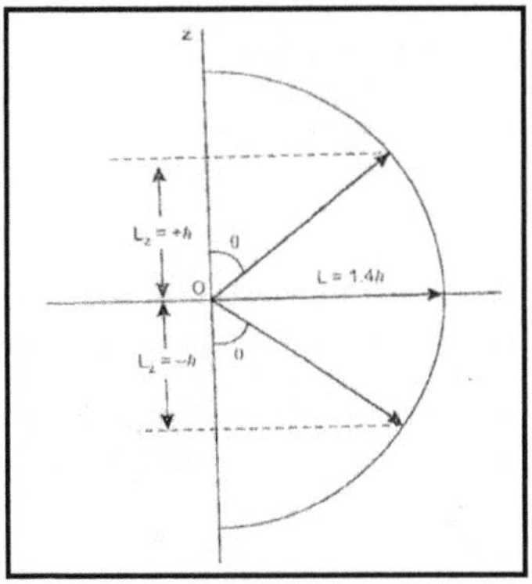

Fig.[9.8.2]: Shows there are 3 possible orientations of $\vec{L}$

Here $\theta = cos^{-1}\left(\frac{\hbar}{1.4\hbar}\right) = cos^{-1}\left(\frac{1}{\sqrt{2}}\right) = 45^0$

for $l = 2$, $m = -2, -1, 0, +1, +2$. $\therefore L = \sqrt{l\,(l+1)}\hbar = \hbar\sqrt{2(2+1)} = \sqrt{6}\,\hbar = 2.5\,\hbar$

$L_Z = -2\hbar, -\hbar, 0, \hbar, 2\hbar$.

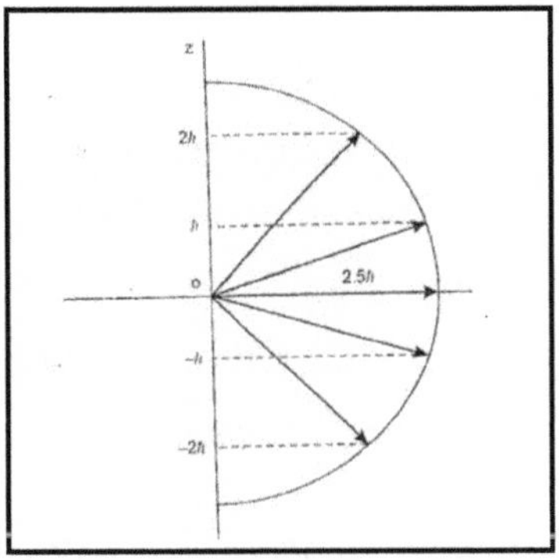

Fig.[9.8.3]: Shows there are 5 possible orientations of $\vec{L}$

Thus there are 5 possible orientations of $\vec{L}$ as shown in the Fig [9.8.3]. In macroscopic systems, L is so large that l is a very large number. The number of allowed values of m is also very large. Thus the billions of allowed orientations are indistinguishable from a continuum; and we can say that there is no restriction on the orientation of $\vec{L}$. But in atomic systems, *I.e.*, L is small and l is also small and hence the number of allowed orientations is also handful; so this angle quantization (called space quantization) is discernible and leads to important effects[7,8].

9.9 Series solution of Differential equation

In this section, we will get introduced to the method of obtaining series solution as well as polynomial solution of differential equation. We are interested the series solution of differential equation to solve

- The Eigen value equation of simple harmonic oscillator
- The Eigen value equation of L^2 ; where L is orbital angular momentum

Linear differential equations with constant coefficients have solution of the type e^{mx}. If the coefficients are functions of the variable, the solution is an infinite series such as

$$y = \sum_{n=0}^{\infty} a_n x^n$$

Let us consider the equation

$$a_0(x)\frac{d^2y}{dx^2} + a_1(x)\frac{dy}{dx} + a_2(x)y = 0 \qquad [1]$$

Or,
$$\frac{d^2y}{dx^2} + Q_1(x)\frac{dy}{dx} + Q_2(x)y = 0$$

Where
$$Q_1(x) = \frac{a_1(x)}{a_0(x)} \ and \ Q_2(x) = \frac{a_2(x)}{a_0(x)}$$

If $Q_1$and Q_2 are analytic at $x = x_0$, $i.e.$, if $a_0(x = x_0) \neq 0, x_0$ is called an ordinary point of equation [1]; others wise x_0 is called a singular point of equation [1], If x_0 is an ordinary point of equation [1], $y = \sum_{n=0}^{\infty} a_n(x - x_0)^n$ is a solution of equation [1]. If x_0 is a regular singular point of the differential equation, a solution is

$$y = |x - x_0|^r \sum_{n=0}^{\infty} a_n(x - x_0)^n = \sum_{n=0}^{\infty} a_n(x - x_0)^{n+r}$$

Where, r is a constant, called exponent of the differential equation. r is determined from an equation called indicial equation.

By regular singular point, we mean $(x - x_0)Q_1 \ and \ (x - x_0)^2Q_2$ are analytic at $x = x_0$. If $(x - x_0)Q_1$ or $(x - x_0)^2Q_2$ or both are not analytic at $x = x_0$; x_0 is called irregular singular point. Now we will obtain series solution of the differential equation.

$$(1-z^2)\frac{d^2y}{dz^2} - 2z\frac{dy}{dz} + \lambda y = 0 \qquad [2]$$

Where z is the variable, λ is a constant and $y = f(z)$ is a solution. This is a linear differential equation with variable coefficients. It can be written as

$$\frac{d^2y}{dz^2} - \frac{2z}{1-z^2}\frac{dy}{dz} + \frac{\lambda}{1-z^2}y = 0$$

$z = 0$ is an ordinary point, $i.e., \frac{2z}{1-z^2} \neq \infty$ $at\ z = 0$ and $\frac{\lambda}{1-z^2} \neq \infty$ $at\ z = 0.$

$\therefore$ $Trial\ solution\ is\ y = \sum_{\gamma=0}^{\infty} a_\gamma (z-0)^\gamma = \sum_{\gamma=0}^{\infty} a_\gamma z^\gamma.$

$$\frac{dy}{dz} = \sum_{\gamma=1}^{\infty} a_\gamma\, \gamma z^{\gamma-1}\ and\ \frac{d^2y}{dz^2} = \sum_{\gamma=2}^{\infty} a_\gamma \gamma(\gamma-1) z^{\gamma-2}$$; From equation [2]

$$(1-z^2)\sum_{\gamma=2}^{\infty} a_\gamma\, \gamma(\gamma-1)z^{\gamma-2} - 2z\sum_{\gamma=1}^{\infty} a_\gamma\, \gamma z^{\gamma-1} + \lambda \sum_{\gamma=0}^{\infty} a_\gamma\, z^\gamma = 0$$

$$\sum_{\gamma=2}^{\infty} a_\gamma\, \gamma(\gamma-1)z^{\gamma-2} - \sum_{\gamma=2}^{\infty} a_\gamma\, \gamma(\gamma-1)z^{\gamma} - 2\sum_{\gamma=1}^{\infty} a_\gamma \gamma z^\gamma + \lambda\sum_{\gamma=0}^{\infty} a_\gamma z^\gamma = 0$$

$$\sum_{\gamma=0}^{\infty} a_{\gamma+2}(\gamma+2)(\gamma+1)z^\gamma - \sum_{\gamma=2}^{\infty} a_\gamma\, \gamma(\gamma-1)z^\gamma - 2\sum_{\gamma=1}^{\infty} a_\gamma \gamma z^\gamma + \lambda\sum_{\gamma=0}^{\infty} a_\gamma z^\gamma = 0 \qquad [3]$$

Since equation [3] holds for all values of z, the coefficient of each power of z in equation [3] must be zero. The coefficient of z^γ is zero.

$$\therefore \quad a_{\gamma+2}(\gamma+2)(\gamma+1) - a_\gamma \gamma(\gamma-1) - 2a_\gamma \gamma + \lambda a_\gamma = 0$$

$$a_{\gamma+2} = \frac{\gamma(\gamma-1)+2\gamma-\lambda}{(\gamma+2)(\gamma+1)}\, a_\gamma$$

$$\therefore \quad a_{\gamma+2} = \frac{\gamma(\gamma+1)-\lambda}{(\gamma+2)(\gamma+1)}\, a_\gamma \qquad [4]$$

This is called recurrence relation or recursion relation which defines the solution of equation [2]. From equation [4], we can obtain

$$a_2 = -\frac{\lambda}{2}a_0, (\gamma = 0);\ \ a_4 = \frac{2.3-\lambda}{3.4}a_2, (\gamma = 2);\ \ = -\frac{2.3-\lambda}{3.4}.\frac{\lambda}{2}a_0$$

$$a_6 = \frac{4.5-\lambda}{5.6}a_4, (\gamma = 4) = -\frac{4.5-\lambda}{5.6}.\frac{2.3-\lambda}{3.4}.\frac{\lambda}{2}a_0$$

………………………………………………………………………………………………

And

$$a_3 = \frac{1.2-\lambda}{2.3}a_1, (\gamma = 1);\ \ a_5 = \frac{3.4-\lambda}{4.5}a_3, (\gamma = 3) = \frac{3.4-\lambda}{4.5}.\frac{1.2-\lambda}{2.3}a_1$$

$$a_7 = \frac{5.6-\lambda}{6.7} a_5, (\gamma = 5) = \frac{5.6-\lambda}{6.7} . \frac{3.4-\lambda}{4.5} . \frac{1.2-\lambda}{2.3} a_1$$

……………………………………………………………………………………………………

Therefore, $y = \sum_{\gamma=0}^{\infty} a_\gamma z^\gamma$

$$= a_0 + a_1 z + a_2 z^2 + a_3 z^3 + a_4 z^4 + a_5 z^5 + a_6 z^6 + \dots\dots\dots\infty$$

$$= a_0 + a_2 z^2 + a_4 z^4 + a_6 z^6 + \dots\dots\infty + a_1 z + a_3 z^3 + +a_5 z^5 + \dots\dots\infty$$

$$= a_0 \left(1 - \frac{\lambda}{2} z^2 - \frac{2.3-\lambda}{3.4} . \frac{\lambda}{2} z^4 - \frac{4.5-\lambda}{5.6} . \frac{2.3-\lambda}{3.4} . \frac{\lambda}{2} z^6 - \dots\dots\dots\infty\right)$$

$$+ a_1 \left(z + \frac{1.2-\lambda}{2.3} z^3 + \frac{3.4-\lambda}{4.5} . \frac{1.2-\lambda}{2.3} z^5 + \dots\dots\dots\infty\right)$$

$$= a_0 y_1 + a_1 y_2 \qquad [5]$$

In equation [5] a_0 and a_1 are arbitrary constants. The solution of equation [5] is a linear combination of two series in z, one containing only even powers of z and the other containing only odd powers of z.

From equation [4], we find that, if $\lambda = \ell\,(\ell + 1)\,;\ where\ \ell = 0, 1, 2, 3, \dots\dots\dots$

$$a_{\gamma+2} = \frac{\gamma(\gamma+1)-\ell(\ell+1)}{(\gamma+2)(\gamma+1)} a_\gamma$$

$$a_{\ell+2} = \frac{\ell(\ell+1)-\ell(\ell+1)}{(l+2)(\ell+1)} a_\ell = 0$$

$$a_{\ell+4} = 0\ ; \quad a_{\ell+6} = 0\ ;\ a_{\ell+8} = 0\ ; \dots\dots\dots$$

- If ℓ is even, the even series terminates at the term in z^ℓ. We must set $a_1 = 0$ to eliminate the infinite odd series to obtain finite solution
- If ℓ is odd, the odd series terminates at the term in z^ℓ. So we must set $a_1 = 0$ to eliminate the infinite even series to obtain finite solution

Thus $\lambda = \ell(\ell + 1)$ is required to obtain finite solution of equation [2]; Here $\ell = 0,1,2,3, \dots\dots,$

The finite solutions are polynomials which are uniquely defined apart from an arbitrary constant $a_0\ or\ a_1$ which are chosen so that $P_\ell(1) = 1$ by convention and the differential equation [2] becomes

$$(1 - z^2) \frac{d^2 y}{dz^2} - 2z \frac{dy}{dz} + \ell(\ell + 1)\, y = 0$$; called Legendre differential equation.

The polynomial solution are denoted as $P_\ell(z)$. From equation [5]

$$P_\ell(z) = a_0\left(1 - \frac{\ell(\ell+1)}{2}z^2 - \frac{2.3-\ell(\ell+1)}{3.4}.\frac{\ell(\ell+1)}{2}z^4 - \frac{4.5-\ell(\ell+1)}{5.6}.\frac{2.3-\ell(\ell+1)}{3.4}.\frac{\ell(\ell+1)}{2}z^6 - \dots\dots\infty\right)$$

$$+a_1\left(z + \frac{1.2-\ell(\ell+1)}{2.3}z^3 + \frac{3.4-\ell(\ell+1)}{4.5}.\frac{1.2-\ell(\ell+1)}{2.3}z^5 + \dots\dots\infty\right)$$

Using this equation, we will now obtain expressions for $P_0(z), P_1(z), P_2(z), P_3(z), P_4(z), P_5(z)$ to learn method of obtaining polynomial solutions.

$P_0(z) = a_0,\quad taking\ a_1 = 0\ ;\ P_0(z) = 1,\qquad taking\ P_\ell(1) = P_0(1) = 1$

$P_1(z) = a_1 z,\quad taking\ a_0 = 0$

$= z,\qquad taking\ P_\ell(1) = P_1(1) = 1$

$P_2(z) = a_0 - a_0\frac{2.3}{2}z^2,\quad taking\ a_1 = 0$

$= a_0(1-3z^2)$

$= -\frac{1}{2}(1-3z^2)$

$= \frac{1}{2}(3z^2-1)$

$taking\ P_\ell(1) = P_2(1) = 1,$
$P_2(1) = a_0(1-3) = -2a_0 = 1$
$\therefore a_0 = -\frac{1}{2}$

$P_3(z) = a_1\left(z + \frac{1.2-3.4}{2.3}z^3\right), taking\ a_0 = 0$

$= a_1\left(z + \frac{-5}{3}z^3\right) = -\frac{3}{2}\left(z - \frac{5}{3}z^3\right)$

$= \left(\frac{5}{2}z^3 - \frac{3}{2}z\right) = \frac{1}{2}(5z^3-3z)$

$taking\ P_\ell(1) = P_3(1) = 1$
$a_1\left(1-\frac{5}{3}\right) = 1$
or, $a_1\frac{-2}{3} = 1$
$\therefore a_1 = -\frac{3}{2}$

$P_4(z) = a_0\left(1 - \frac{4.5}{2}z^2 - \frac{2.3-4.5}{3.4}.\frac{4.5}{2}z^4\right)\ taking\ a_1 = 0$

$= a_0\left(1 - 10z^2 + \frac{7}{6}.10z^4\right)$

$= a_0\left(1 - 10z^2\frac{35}{3}z^4\right)$

$= \frac{3}{8}\left(1 - 10z^2 + \frac{35}{3}z^4\right)$

$= \frac{1}{8}(35z^4 - 30z^2 + 3)$

$taking\ P_\ell(1) = P_4(1) = 1$
$a_0\left(1 - 10 + \frac{35}{2}\right) = 1$
or, $a_0\left(\frac{-27+35}{3}\right) = 1$
or, $a_0\frac{8}{3} = 1\ \therefore a_0 = \frac{3}{8}$

$$P_5(z) = a_1\left(z + \frac{1.2-5.6}{2.3}z^3 + \frac{3.4-5.6}{4.5}.\frac{1.2-536}{2.3}z^4\right) \; ; taking\ a_0 = 0$$

$$= a_1\left(z + \frac{-28}{6}z^3 + \frac{-18}{20}.\frac{-28}{6}z^5\right)$$

$$= a_1\left(z - \frac{14}{3}z^3 + \frac{21}{5}z^5\right)$$

$$= \frac{15}{8}\left(z - \frac{14}{3}z^3 + \frac{21}{5}z^5\right)$$

$$= \frac{1}{8}(15z - 70z^3 + 63z^5)$$

$$= \frac{1}{8}(63z^5 - 70z^3 + 15z)$$

taking $P_\ell(1) = P_5(1) = 1$

$$a_0\left(1 - \frac{14}{3} + \frac{21}{5}\right) = 1$$

or, $a_1 \frac{15-70+63}{15} = 1$

or, $a_1 \frac{8}{15} = 1 \;\therefore a_1 = \frac{15}{8}$

9.10 Normalized solution of Associated Legendre equation and first few associated Legendre functions without normalization

As far as the state $\Psi_{\ell,m}(\theta,\phi)$ is concerned, there is no restriction on the azimuthal angle of $\vec{L}$, about z-axis, that is, we cannot say where on the XY-plane the projection of $\vec{L}$ with be.

In atomic systems, associated with an orbital angular momentum $\vec{L}$, there is a magnetic dipole moment. Introduction of a magnetic field along z causes the energy of the atom to change by certain allowed values due to the orientational potential energy of the magnetic moment for the allowed orientations. Thus space quantization manifests itself through discrete changes in atomic levels in a magnetic field. Hence the quantum number *m* is called magnetic *quantum number.* From the *associated Legendre equation*[24]

$$(1-z^2)\frac{d^2\Theta}{dz^2} - 2z\frac{d\Theta}{dz} + \left(\ell(\ell+1) - \frac{m^2}{1-z^2}\right)\Theta = 0 \qquad [1]$$

For, $m = 0$ equation [1] reduces to what is called Lengendre differential equation having polynomial solution called Legendre polynomial denoted by $P_\ell(z)$. The first few Legendre polynomials are (see Section 9.9 ; series solution of differential equation).

$P_0(z) = 1$; $P_1(z) = z$; $P_2(z) = \frac{1}{2}(3z^2 - 1)$; $P_3(z) = \frac{1}{2}(5z^3 - 3z)$

$P_4(z) = \frac{1}{8}(35z^4 - 30z^2 + 3)$; $P_5(z) = \frac{1}{8}(63z^5 - 70z^3 + 15z)$

Legendre polynomials are given by Rodrigue's formula

$P_\ell(z) = \frac{1}{2^\ell \ell!}\frac{d^\ell}{dz^\ell}(z^2-1)^\ell$ which satisfies Legendre equation and the convention $P_\ell(1) = 1$.

For *m* not necessarily equal to zero, solutions of equation [1] are given by

$P_\ell^{|m|}(z) = (1-z^2)^{\frac{|m|}{2}} \frac{d^{|m|}}{dz^{|m|}} P_\ell(z) = (1-z^2)^{\frac{|m|}{2}} G(z)$; $where \quad G(z) = \frac{d^{|m|}}{dz^{|m|}} P_\ell(z)$ and $|m| = 0, 1, 2, 3, \ldots$. $P_\ell^{|m|}(z)$ are called assocated Lagaendre function.

For $|m| > l$, $P_\ell^{|m|}(z) = 0. Thus\ |m| < l.$ $\therefore m = 0, \pm 1, \pm 2, \pm 3, \ldots, \pm \ell$. Since equation [1] is independent of the sing of m, the solution $P_\ell^{|m|}(z)$ can be characterised by ℓ and $|m|$. $P_\ell^{|m|}(z)$ is the product of $(1-z^2)^{|m|/2}$ and a polynomial $G(z) = \frac{d^{|m|}}{dz^{|m|}} P_\ell(z)$ of degree $\ell - |m|$, since $P_\ell(z)$ is a polynomial of degree ℓ.

$$P_\ell^{|m|}(-z) = (-1)^{\ell-|m|} P_\ell^{|m|}(z), \int_{-1}^{+1} P_\ell^{|m|}(z) P_{\ell'}^{|m|}(z) dz = \frac{2}{2\ell+1} \frac{(\ell+|m|)!}{(\ell-|m|)!} \delta_{\ell\ell'}$$

Here $dz = d(\cos\theta) = -\sin\theta\, d\theta$. Hence $P_\ell^{|m|}(z)$ given above are not normalised (to unity).

$\therefore$ Normalized solutions of equation [1] are

$$\Theta_{\ell,m}(\theta) = \left[\frac{2\ell+1}{2} \frac{(\ell-|m|)!}{(\ell+|m|)!}\right]^{1/2} P_\ell^{|m|}(cos\theta)$$

The orthonormality of $\Theta_{\ell,m\,(\theta)}$'s given by $\left(\Theta_{\ell,m}(\theta), \Theta_{\ell',m}(\theta)\right) = \boldsymbol{\delta_{\ell\ell'}}$ is to be written in integral from like $\int_0^\pi \Theta_{\ell,m}^\times(\theta)\, \Theta_{\ell',m}(\theta) \sin\theta d\theta = \boldsymbol{\delta_{\ell\ell'}}$. $\sin\theta\ d\theta\ comes\ from\ \theta$ Part of volume element $(dr)(rd\theta)(r\sin\theta d\phi) = (r^2 dr)(\sin\theta d\theta)(d\phi)$.

The lst few associated Lagendre functions without normalisation are

$$P_1^1(z) = (1-z^2)^{1/2},\ P_2^1(z) = 3(1-z^2)^{1/2}\, z,\ P_2^2(z) = 3(1-z^2)$$

$$P_3^1(z) = \tfrac{3}{2}(1-z^2)^{1/2}(5z^2-1),\ P_3^2(z) = 15z(1-z^2),\ P_3^3(z) = 15(1-z^2)^{3/2}$$

Calculations of $P_3^1(z)$ and $P_3^2(z)$ are shown below, as two examples.

We have $P_3(z) = \frac{1}{2}(5z^3 - 3z). P_\ell^{|m|}(z) = (1-z^2)^{|m|/2} \frac{d^{|m|}}{dz^{|m|}} P_l(z)$ gives

$$P_3^1(z) = (1-z^2)^{\frac{1}{2}} \frac{d}{dz} P_3(z) = (1-z^2)^{1/2} \frac{d}{dz}\left[\frac{1}{2}(5z^3-3z)\right]$$

$$= \frac{1}{2}(1-z^2)^{\frac{1}{2}}(15z^2-3) = \frac{3}{2}(1-z^2)^{\frac{1}{2}}(5z^2-1)$$

Again, $P_\ell^{|m|}(z) = (1-z^2)^{|m|/2} \frac{d^{|m|}}{dz^{|m|}} P_l(z)$ gives

$$P_3^2(z) = (1 - z^2)^{2/2} \frac{d^2}{dz^2} P_3(z) = (1 - z^2) \frac{d^2}{dz^2} \left[\frac{1}{2} (5z^3 - 3z) \right]$$

$$= \frac{1}{2}(1 - z^2) \frac{d}{dz}(15z^2 - 3) = \frac{1}{2}(1 - z^2)(30z) = 15z(1 - z^2)$$

9.11 Schrödinger equation can be separated into radial, angular and azimuthal parts for spherical harmonics

Spherically symmetric potentials are those in which the potential energy of the particle does not depend upon θ & ϕ. It is only the function of position r. Coulomb potential, Gravitation potential, Yukawa potential & Gaussian potential are spherically symmetric potentials[10,11]. From Schrödinger equation

$$\left[\frac{-\hbar^2}{2m} \nabla^2 + V(\overline{r}) \right] \psi(\overline{r}) = E\psi(\overline{r})$$

Or, $$\nabla^2 \psi(\overline{r}) + \frac{2m}{\hbar^2}[E - V(\overline{r})]\, \psi(\overline{r}) = 0 \qquad [1]$$

It is generally impossible to obtain analytical solution of (3-D) Schrödinger equation; unless it can be separated into total differential equation in each of the three space coordinates. Now in terms of spherical polar coordinates $\nabla^2 \psi$ can be expressed as

$$\nabla^2 \psi = \frac{1}{r^2} \frac{\partial}{\partial r}(r^2 \frac{\partial \psi}{\partial r}) + \frac{1}{r^2 \sin\theta} \frac{\partial}{\partial \theta}(\sin\theta \frac{\partial \psi}{\partial \theta}) + \frac{1}{r^2 \sin\theta}$$

$$\nabla^2 \psi = [\frac{1}{r^2} \frac{\partial}{\partial r}(r^2 \frac{\partial \psi}{\partial r}) + \frac{1}{r^2 \sin\theta} \frac{\partial}{\partial \theta}(\sin\theta \frac{\partial \psi}{\partial \theta}) + \frac{1}{r^2 \sin\theta} \frac{\partial^2}{\partial \theta}$$

$$= [\frac{1}{r^2} \frac{\partial}{\partial r}(r^2 \frac{\partial \psi}{\partial r}) + 2r \frac{\partial}{\partial r}) + \frac{1}{r^2 \sin\theta} \frac{\partial}{\partial \theta}(\sin\theta \frac{\partial}{\partial \theta}) + \frac{1}{r^2 \sin\theta} \frac{\partial^2}{\partial \varphi^2}]\psi$$

$$= \frac{\partial^2}{\partial r^2} + \frac{2}{1} \frac{\partial}{\partial r} + \frac{1}{r^2 \sin\theta} \frac{\partial}{\partial \theta}(\sin\theta \frac{\partial}{\partial \theta}) + \frac{1}{r^2 \sin\theta} \frac{\partial^2}{\partial \varphi^2}$$

$$= \frac{\partial^2 \psi}{\partial r^2} + \frac{2}{1} \frac{\partial \psi}{\partial r} + \frac{1}{r^2 \sin\theta} \frac{\partial}{\partial \theta}(\sin\theta \frac{\partial \psi}{\partial \theta}) + \frac{1}{r^2 \sin^2\theta} \frac{\partial^2}{\partial \varphi^2}$$

$$= \frac{1}{r} \frac{\partial^2}{\partial r^2}(r\psi) + \frac{1}{r^2}[\frac{1}{\sin^2\theta} \frac{\partial^2}{\partial \varphi^2}(\sin\theta \frac{\partial \psi}{\partial \theta}) + \frac{1}{\sin^2\theta} \frac{\partial^2 \psi}{\partial \varphi^2}] \qquad [2]$$

Now consider the term

$$\frac{\partial}{\partial r}(r\psi) = r\frac{\partial \psi}{\partial r} + \psi$$

Or, $$\frac{\partial^2}{\partial r^2}(r\psi) = \frac{\partial \psi}{\partial r} + \frac{\partial \psi}{\partial r} + r\frac{\partial^2 \psi}{\partial r^2}$$

$$= 2\frac{\partial \psi}{\partial r} + r\frac{\partial^2 \psi}{\partial r^2} = r(\frac{\partial^2 \psi}{\partial r^2} + \frac{2}{r} + \frac{\partial \psi}{\partial r})$$

$$\Rightarrow \frac{1}{r}\frac{\partial^2}{\partial r^2}(r\psi) = \frac{\partial^2 \psi}{\partial r^2} + \frac{2}{r} + \frac{\partial \psi}{\partial r} \qquad [3]$$

$$Note \quad \Lambda = \frac{1}{\sin\theta}\frac{\partial}{\partial\theta}\left(\sin\frac{\partial}{\partial\theta}\right) + \frac{1}{\sin^2\theta}\frac{\partial^2}{\partial\Phi^2}$$

Therefore equation [2] becomes

$$\nabla^2\Psi = \frac{1}{r}\frac{\partial^2}{\partial r^2}(r\Psi) + \frac{1}{r^2}\Lambda\Psi \qquad [4]$$

Finally equation [1] becomes

$$\frac{1}{r}\frac{\partial^2}{\partial r^2}(r\Psi) + \frac{1}{r^2}\Lambda\Psi + \frac{2m}{\hbar^2}[E - V(r)] \qquad [5]$$

Let the solution of equation [5]

$$\Psi(r,\theta,\phi) = R(r)\Theta(\theta)\Phi(\phi) = R(\mathrm{r})Y_l^m(\theta,\phi)$$

Where R(r) is independent of polar & azimuthal angles.

$$\therefore \frac{1}{r}\frac{\partial^2}{\partial r^2}\left(rR(r)Y_l^m(\theta,\varphi)\right) + \frac{1}{r^2}\Lambda R(r)Y_l^m(\theta,\phi) + \frac{2m}{\hbar^2}[E - V(r)]R(r)Y_l^m(\theta,\phi) = 0$$

$$Or, Y(\theta,\phi)\frac{1}{r}\frac{\partial^2}{\partial r^2}\left(rR(r)\right) + \frac{2m}{\hbar^2}[E - V(r)]R(r)Y_l^m(\theta,\phi) = -\frac{1}{r^2}\Lambda R(r)Y_l^m(\theta,\phi)$$

$$Or, \frac{1}{rR(r)}\frac{\partial^2}{\partial r^2}(rR(r)) + \frac{2m}{\hbar^2}[E - v(r)] = -\frac{\Lambda Y_l^m(\theta,\phi)}{Y_l^m(\theta,\phi)}$$

$$Or, r^2[\frac{1}{rR(r)}\frac{\partial^2}{\partial r^2}\left(rR(r)\right) + \frac{2m}{\hbar^2}[E - V(r)] = -\frac{\Lambda Y_l^m(\theta,\phi)}{Y_l^m(\theta,\phi)} \qquad [6]$$

In the above equation the L.H.S member depends on the variable r while the R.H.S member is independent of r. Consequently, this equation can be satisfied identically only if each side is a constant λ I.e.,

$$r^2[\frac{1}{rR(r)}\frac{\partial^2}{\partial r^2}\left(rR(r)\right) + \frac{2m}{\hbar^2}[E - V(r)] = \lambda \qquad [7]$$

$$\& \quad -\frac{\Lambda Y_l^m(\theta,\phi)}{Y_l^m(\theta,\phi)} = \lambda \qquad [8]$$

The above equation [8] is the angular equation .we define a new radial function

$$u(r) = rR(r) \qquad [9]$$

Therefore equation [7] becomes

$$r^2[\frac{1}{u(r)}\frac{\partial^2 u(r)}{\partial r^2}+\frac{2m}{\hbar^2}(E-V(r)]=\lambda$$

$$Or,\ \frac{r^2}{u(r)}[\frac{\partial^2 u(r)}{\partial r^2}+\frac{2m}{\hbar^2}(E-V(r))u(\mathrm{r})]=\lambda$$

$$Or,\ \frac{\partial^2 u(r)}{\partial r^2}+\frac{2m}{\hbar^2}[E-V(r)]u(\mathrm{r})]=\lambda\frac{u(r)}{r^2}$$

$$Or,\frac{\partial^2 u(r)}{\partial r^2}+\left[\frac{2m}{\hbar^2}(E-V(r)-\frac{\lambda}{r^2}\right]u(r)=0 \qquad [10]$$

The above equation is the radial equation. Again from equation [8]

$$-\frac{\Lambda Y_l^m(\theta,\phi)}{Y_l^m(\theta,\phi)}=\lambda\ or,\ \Lambda Y(\theta,\varphi)+\lambda Y_l^m(\theta,\phi)=0 \qquad [11]$$

Now we use the separation of variable to solve the equation [11]. Assume the separation form to be

$$Y_l^m(\theta,\phi)=\Theta(\theta)\Phi(\phi) \qquad [12]$$

$$\therefore [\frac{1}{\sin\theta}\frac{\partial}{\partial\theta}(\sin\theta\frac{\partial}{\partial\theta})+\frac{1}{\sin^2\theta}\frac{\partial^2}{\partial\phi^2}]\ \Theta(\theta)\Phi(\phi)=-\lambda\Theta(\theta)\Phi(\phi)$$

$$Or,\Phi(\varphi)[\frac{1}{\sin\theta}\frac{\partial}{\partial\theta}(\sin\theta\frac{\partial}{\partial\theta})\ \Theta(\theta)]+\frac{\Theta(\theta)}{\sin^2\theta}\frac{\partial^2\Phi(\phi}{\partial\phi^2}=-\lambda\Theta(\theta)\Phi(\phi)$$

$$Or,\Phi(\phi)[\frac{1}{\sin\theta}\frac{\partial}{\partial\theta}(\sin\theta\frac{\partial}{\partial\theta})+\lambda]\ \Theta(\theta)=-\frac{\Theta(\theta)}{\sin^2\theta}\frac{\partial^2\Phi(\phi}{\partial\phi^2}$$

$$Or,\frac{\sin^2\theta}{\Theta(\theta)}[\frac{1}{\sin\theta}\frac{\partial}{\partial\theta}(\sin\theta\frac{\partial}{\partial\theta})+\lambda]\Theta(\theta)=--\frac{1}{\Phi(\phi)}\frac{\partial^2\Phi(\phi)}{\partial\phi^2}$$

In the above equation the L.H.S is independent of $\Phi(\phi)$ while the R.H.S is independent of $\Theta(\theta)$ consequently this equation can be satisfied identically if each side is a constant m^2. I.e.,

$$\frac{\sin^2\theta}{\Theta(\theta)}[\frac{1}{\sin\theta}\frac{\partial}{\partial\theta}(\sin\theta\frac{\partial}{\partial\theta})+\lambda]\Theta(\theta)=m^2$$

$$Or,\frac{1}{\sin\theta}\frac{\partial}{\partial\theta}(\sin\theta\frac{\partial\Theta(\theta)}{\partial\theta})+(\lambda-\frac{m^2}{\sin^2\theta})]\ \Theta(\theta)=0 \qquad [13]$$

This is the differential equation for $\Theta(\theta)$ coordinate.

And $$-\frac{1}{\Phi(\phi)}\frac{\partial^2\Phi(\phi)}{\partial\phi^2}=m^2 \qquad Or,\frac{\partial^2\Phi(\phi)}{\partial\phi^2}+m^2\Phi(\phi)=0$$

This is the differential equation for $\Phi(\phi)$ coordinate.

Exercise

1. Define Orbital angular momentum. Calculate its components and their quantum mechanical equivalents.

2. Establish the basic relation between Position and Linear momentum.

3. Find the following commutation rule of orbital angular momentum:

$$a.\ [L_x, L_y] \quad b.\ [L_y, L_z] \quad c.\ [L_z, L_x] \quad d.\ \vec{L}\times\vec{L} = i\hbar\,\vec{L}$$
$$e.\ [L^2, L_x] \quad f.\ [L^2, L_x] \quad g.\ [L^2, L_x]$$

4. Define Ladder operator [Raising and Lowering operator] of orbital angular momentum and develop their following some commutation rules:

$$a.\,[L_z, L_\pm] = \hbar L_\pm \quad b.\,[L_+, L_-] = 2\hbar L_z \quad c.\,[L^2, L_\pm] = 0 \quad d.\,[L_x, L_+] = -\hbar L_z$$
$$e.\,[L_y, L_+] = -i\hbar L_z \quad f.\,[L_z, L_+] = \hbar L_+ \quad g.\,[L_x, L_-] = \hbar L_z \quad h.\,[L_y, L_-] = -i\hbar L_z$$
$$i.\,[L_z, L_-] = -\hbar L_-$$

5. Establish the following Commutation rule of Orbital angular momentum with Linear momentum and position.

$$a.\,[L_x, p_x] = 0 \quad b.\,[L_x, p_z] = -i\hbar p_y \quad c.\,[L_x, p_y] = i\hbar p_z \quad d.\,[L_x, p_z] = -i\hbar p_y$$
$$e.\,[L_x, p_y] = i\hbar p_z \quad f.\,[L_x, y] = i\hbar z \quad g.\,[L_z, x] = i\hbar y \quad h.\,[L_x, x] = 0$$

6. Find the relation between Cartesian coordinate and Spherical polar coordinate?

7. Calculate the expression for x, y and z components of the square of the total orbital angular momentum operators in spherical polar coordinates.

8. Show that the spherical Harmonics $Y_l^m(\theta,\phi)$ are the simultaneous Eigen functions of the operators L^2 and L_Z belonging to the Eigen values $l(l+1)\hbar^2$ and $m_l\hbar$ respectively.

9. Find the series solution of differential equation.

10. Find the first few associated Legendre functions without normalization from normalized solution of Associated Legendre equation.

11. Show that Schrödinger equation can be separated into radial, angular and azimuthal parts for spherical harmonics.

12. Show that the ground state of a particle in a spherically symmetric potential must have zero orbital angular momentum.

Chapter 10 Generalized Angular momentum in Quantum Mechanics

10.1 Generalized angular momentum operator

We define generalized angular momentum operator[5]

$$\vec{J} = [J_x , J_y , J_z] \qquad [1]$$

The generalized angular momentum operator or total angular momentum $\vec{J}$ can be written as the sum of the orbital & spin angular momentum I.e.,

$$\vec{J} = \vec{L} + \vec{S}$$

So, $\vec{J}$ have the components: $J_x = L_x + S_x \quad J_y = L_y + S_y \quad \& \; J_z = L_z + S_z$

The components of the orbital angular momentum operators satisfy the following relations:

1.$\left[L_x , L_y\right]$ 2.$\left[L_y , L_z\right]$ 3.$[L_z , L_x]$ 4.$\vec{L} \times \vec{L} = i\hbar\, \vec{L}$ 5.$\left[L^2 , L_x\right]$ 6.$\left[L^2 , L_y\right]$ 7.$\left[L^2 , L_z\right]$

Again, the components of the spin angular momentum operators satisfy the following commutation relations:

$$\left[S_x , S_y\right] = i\hbar\, S_z \quad \left[S_y , S_z\right] = i\hbar\, S_x \quad \& \quad [S_z , S_x] = i\hbar\, S_y$$

But spin angular momentum $\vec{S}\,[S_x , S_y , S_z]$ & orbital angular momentum $\vec{L}[L_x , L_y , L_z]$ are with different degree of freedom [they have common Eigen functions]. So, that they commute to with each other.

$$\left[\vec{L} , \vec{S}\right] = 0$$

Since the generalized angular momentum operator $\vec{J} = [J_x , J_y , J_z]$ is Hermition operator. Hence the components of the generalized angular momentum operators satisfy the following commutation relations:

$$\left[J_x , J_y\right] = i\hbar J_z , \left[J_y , J_z\right] = i\hbar J_x , [J_z , J_x] = i\hbar J_y , \vec{J} \times \vec{J} = i\hbar\, \vec{J} , \left[J^2 , J_x\right] = \left[J^2 , J_x\right] = \left[J^2 , J_z\right] = 0$$

10.2 Commutation rules of generalized angular momentum operators

Now, we shall establish the following commutation relations of generalized angular momentum Operators.

$$\left[J_x , J_y\right] = i\hbar J_z , \quad \left[J_y , J_z\right] = i\hbar J_x , \quad [J_z , J_x] = i\hbar J_y , \vec{J} \times \vec{J} = i\hbar\, \vec{J}$$

$$\left[J^2 , J_x\right] = 0 \quad \left[J^2 , J_x\right] = 0 \quad \left[J^2 , J_z\right] = 0$$

- **Consider the first relation:**

$$\left[J_x, J_y\right] = i\hbar J_z$$

Let us define total angular momentum of $\overrightarrow{J_1}$ and $\overrightarrow{J_2}$ as vector sum:

$$\vec{J} = \overrightarrow{J_1} + \overrightarrow{J_2}$$

This means $\quad J_x = J_{1x} + J_{2x},\ J_y = J_{1y} + J_{2y},\ \ J_z = J_{1z} + J_{2z}$

We need to verify if $\vec{J}$ qualifies as an angular momentum in Quantum Mechanics I.e., we need to show that

$$[J_x, J_y] = i\hbar J_z,\ [J_y, J_z] = i\hbar J_x,\ [J_z, J_x] = i\hbar J_y$$

So that $\vec{J}$ qualifies to be an angular momentum (vector) in Quantum Mechanics.

$$[J_y, J_z]\Psi = J_y J_z \Psi - J_z J_y \Psi$$

$$= (J_{1y} + J_{2y})(J_{1z} + J_{2z})\Psi - (J_{1z} + J_{2z})(J_{1y} + J_{2y})\Psi$$

$$= J_{1y}(J_{1z} + J_{2z})\Psi + J_{2y}(J_{1z} + J_{2z})\Psi - J_{1z}(J_{1y} + J_{2y})\Psi - J_{2z}(J_{1y} + J_{2y})\Psi$$

$$= J_{1y}J_{1z}\Psi + J_{1y}J_{2z}\Psi + J_{2y}J_{1z}\Psi + J_{2y}J_{2z}\Psi - J_{1z}J_{1y}\Psi - J_{1z}J_{2y}\Psi - J_{2z}J_{1y}\Psi - J_{2z}J_{2y}\Psi$$

Because operator of any component of $\overrightarrow{J_1}$ commutes with operator of any component of $\overrightarrow{J_2}$.

$$[J_y, J_z]\Psi = (J_{1y}J_{1z} - J_{1z}J_{1y})\Psi + (J_{2y}J_{2z} - J_{2z}J_{2y})\Psi$$

$$= [J_{1y}, J_{1z}]\Psi + [J_{2y}, J_{2z}]\Psi$$

$$= i\hbar J_{1x}\Psi + i\hbar J_{2x}\Psi$$

$$= i\hbar(J_{1x} + J_{2x})\Psi = i\hbar J_x \Psi$$

Thus $\quad [J_y, J_z] = i\hbar J_x$; Similarly, $[J_y, J_z] = i\hbar J_x,\ [J_z, J_x] = i\hbar J_y$

- **Consider the relation:**

$$\vec{J} \times \vec{J} = i\hbar\, \vec{J}$$

The components of the generalized angular momentum $\vec{J}$

$$\vec{J} = \hat{i}J_x + \hat{j}J_y + \hat{k}J_z \ and \ J_x = L_x + S_x,\ J_y = L_y + S_y \ \&\ J_z = L_z + S_z$$

$$But \quad L_x = yp_z - zp_y,\ L_y = zp_x - xp_z \ and \quad L_z = xp_y - yp_x$$

$$\therefore \vec{J} \times \vec{J} = \begin{vmatrix} \hat{i} & \hat{j} & \hat{k} \\ J_x & J_y & J_z \\ J_x & J_y & J_z \end{vmatrix}$$

$$= \hat{i}(J_y J_z - J_z J_y) + \hat{j}\,(J_z J_x - J_x J_z) + \hat{k}\,(J_x J_y - J_y J_x)$$

$$= \hat{i}\,[J_y\,, J_z] + \hat{j}\,[J_z\,, J_x] + \hat{k}\,[J_x\,, J_y]$$
$$= \hat{i}\,(i\hbar\; J_x) + \hat{j}\,(i\hbar\; J_y) + \hat{k}\,(i\hbar\; J_z)$$
$$= i\hbar(\,\hat{i}\; J_x + \hat{j}\; J_y + \hat{k}\; J_z)$$
$$= i\hbar\; \vec{J}$$

Finally $\vec{J} \times \vec{J} = i\hbar\vec{J}$

- **Consider the relation:**

$$\left[J^2, \mathrm{J}_x\right] = 0$$

The square of the generalized angular momentum and its components are

$J^2 = J_x^2 + J_y^2 + J_z^2$ And $J_x = L_x + S_x$; where $L_x = yp_z - zp_y$

$$\therefore [J^2, J_x] = [J_x^2 + J_y^2 + J_z^2, J_x]$$
$$= [J_x^2, J_x] + [J_y^2, J_x] + [J_z^2, J_x]$$
$$= [J_xJ_x, J_x] + [J_yJ_y, J_x] + [J_zJ_z, J_x]$$
$$= (J_xJ_xJ_x - J_xJ_xJ_x) + (J_yJ_yJ_x - J_xJ_yJ_y) + (J_zJ_zJ_x - J_xJ_zJ_z)$$
$$= J_yJ_yJ_x - J_xJ_yJ_y + J_zJ_zJ_x - J_xJ_zJ_z$$

Now adding and subtracting the terms $J_yJ_xJ_y$ and $J_zJ_xJ_z$

$$= J_yJ_yJ_x - J_xJ_yJ_y + J_zJ_zJ_x - J_xJ_zJ_z + J_yJ_xJ_y - J_yJ_xJ_y + J_zJ_xJ_z - J_zJ_xJ_z$$
$$= (J_yJ_yJ_x - J_xJ_yJ_y + J_yJ_xJ_y - J_yJ_xJ_y) + (J_zJ_zJ_x - J_xJ_zJ_z + J_zJ_xJ_z - J_zJ_xJ_z)$$
$$= J_y(J_yJ_x - J_xJ_y) + (J_yJ_x - J_xJ_y)J_y + J_z(J_zJ_x - J_xJ_z) + (J_zJ_x - J_xJ_z)J_z$$
$$= J_y[J_y, J_x] + [J_y, J_x]J_y + J_z[J_z, J_x] + [J_z, J_x]J_z$$
$$= J_y(-i\hbar J_z) + (-i\hbar J_z)J_y + J_z(i\hbar J_y) + (i\hbar J_y)J_z$$
$$= -i\hbar J_yJ_z - i\hbar J_zJ_y + i\hbar J_zJ_y + i\hbar J_yJ_z$$
$$= 0$$

Finally, $\left[J^2, \mathrm{J}_x\right] = 0$; *Similarly, we can prove that* $\left[J^2, \mathrm{J}_y\right] = 0$ *and* $\left[J^2, \mathrm{J}_z\right] = 0$

- **An alternate approach** of the square of the generalized angular momentum and its components:

  $J^2 = J_x^2 + J_y^2 + J_z^2$ And $J_x = L_x + S_x$; Where $L_x = yp_z - zp_y$

$$[J^2, J_x] = [J_x^2 + J_y^2 + J_z^2, J_x]$$
$$= [J_x^2, J_x] + [J_y^2, J_x] + [J_z^2, J_x]$$
$$= [J_x J_x, J_x] + [J_y J_y, J_x] + [J_z J_z, J_x]$$
$$= J_x[J_x, J_x] + [J_x, J_x]J_x + J_y[J_y, J_x] + [J_y, J_x]J_y + J_z[J_z, J_x] + [J_z, J_x]J_z$$

$$\boxed{[ab, c] = a[b, c] + [a, c]b}$$

$$= 0 + 0 - i\hbar J_y J_z - i\hbar J J_y + i\hbar J_z J_y + i\hbar J_y J_z$$

$[J^2, J_x] = 0$; *Similarly, we can prove that* $[J^2, J_y] = 0$ *and* $[J^2, J_z] = 0$

10.3 Raising and lowering operators of generalized angular momentum

Suppose, Φ_k be the angular momentum components Eigen function & let $J_z \Phi_k = K \Phi_k$. The $J_+ \Phi_k = \Phi$ is also an J_z Eigen state belonging to the Eigen value $(K + \hbar)$

$$\therefore \quad J_z \Phi_k = K \Phi_k$$

$$or, \quad J_+(J_z \Phi_k) = J_+(K \Phi_k) \qquad [1]$$

[*Operating by* J_+ *on the both sides*]

We know that $[J_+, J_z] = -\hbar J_+$

$$or, \quad J_+ J_z - J_z J_+ = -\hbar J_+$$

$$J_+ J_z = J_z J_+ - \hbar J_+$$

Therefore,

$$J_+(J_z \Phi_k) = J_+(K \Phi_k)$$

$$\text{or,} \quad (J_z J_+ - \hbar J_+)\,\Phi_k = J_+(K \Phi_k)$$

$$or, \quad J_z J_+ \Phi_k = J_+(K \Phi_k) + \hbar J_+ \Phi_k$$

$$or, \quad J_z J_+ \Phi_k = J_+ \Phi_k (K + \hbar) \qquad [2]$$

Where J_+ is the angular momentum raising operator .Thus J_+ increase the value of m by 1 in the spherical harmonics & that is why J_+ is called the raising operator[7].

$$J_z \Phi_K = (K + \hbar)\Phi; \; \textit{where } (K + \hbar) \textit{ is an eigen value of } J_z \qquad [3]$$

❖ Lowering operator

Suppose Φ_k be the angular momentum components Eigen function & let $J_z \Phi_k = K\Phi_k$. The $J_+ \Phi_k = \Phi$ is also an J_z Eigen state belonging to the Eigen value $(K - \hbar)$.

$$\therefore \quad J_z \Phi_k = K \Phi_k$$

$or, \quad J_-(J_z\Phi_k) = J_-(K\Phi_k)$

[Operating by J_- on the both sides]

or, $\quad J_-J_z\Phi_k = K\,J_-\Phi_k$ [1]

We know that $[J_-, J_z] = -\hbar J_-$

$or, \qquad J_-J_z - J_zJ_- = -\hbar J_-$

$or, \qquad J_-J_z = J_zJ_- + \hbar J_-$

$\therefore \quad (J_zJ_- + \hbar J_-)\,\Phi_k = KJ_-\Phi_k$

$\Rightarrow (J_zJ_- + \hbar J_-)\,\Phi_k = KJ_-\Phi_k$

$\Rightarrow J_zJ_-\Phi_k + \hbar J_-\Phi_k = KJ_-\Phi_k$

$\Rightarrow J_zJ_-\Phi_k = J_-\Phi_k(K - \hbar)$ [2]

Where J_- is the angular momentum lowering operator .Thus J_- decrease the value of m by 1 in the spherical harmonics & that is why $J_-$is called the lowering operator $J_z\Phi_k = (K-\hbar)\Phi_k$

$J_z\Phi = (K-\hbar)\Phi$ [3]

Where $(K-\hbar)$ *is an eigen value of* J_z

10.4 Some properties of ladder operators $J_+$and J_-

Three properties of ladder operators J_+ and J_- and can verify in the following way:

- J_+ and J_- are adjoint of each other, i.e., $J_+^\dagger = J_-$ and $J_-^\dagger = J_+$.
- $J_+\Psi_{j,m} = \sqrt{j(j+1) - m(m+1)}\hbar\Psi_{j,m+1}$ showing that $\Psi_{j,m}$ which is common eigenfunction of J^2 and J_z is not an eigenfunction of J_+ .
- $J_-\Psi_{j,m} = \sqrt{j(j+1) - m(m-1)}\hbar\Psi_{j,m-1}$showing that $\Psi_{j,m}$ which is common eigenfunction of J^2 and J_z is not an eigenfunction of J_- .

First properties: J_+ and $J_-$are adjoint to each other. So from the defination of adjoint

$(\Psi_1, J_+\Psi_2) = (J_-\Psi_1, \Psi_2)$, I.e., $J_+^\dagger = J_-$

And $\quad (\Psi_1, J_-\Psi_2) = (J_+\Psi_1, \Psi_2)$, I.e., $J_-^\dagger = J_+$ can be proved in the following:

$$(\Psi_1, J_+\Psi_2) = \left(\Psi_1, (J_x + iJ_y)\Psi_2\right)$$

$$= (\Psi_1, J_x\Psi_2) + \left(\Psi_1, iJ_y\Psi_2\right)$$

$$= (\Psi_1, J_x\Psi_2) + i\left(\Psi_1, J_y\Psi_2\right)$$

$$= (J_x\Psi_1, \Psi_2) + i\left(J_y\Psi_1, \Psi_2\right)$$ [$\because J_x$and $J_y$are Hermitian.]

$$= (J_x\Psi_1, \Psi_2) + i^*(J_y\Psi_1, \Psi_2)$$
$$= \left((J_x + i^*J_y)\Psi_1, \Psi_2\right)$$
$$= \left((J_x - iJ_y)\Psi_1, \Psi_2\right) = (J_-\Psi_1, \Psi_2)$$

$$\therefore (\Psi_1, J_+\Psi_2) = (J_-\Psi_1, \Psi_2) \quad \text{i.e., } J_+^\dagger = J_-$$

Again,

$$(\Psi_1, J_-\Psi_2) = \left(\Psi_1, (J_x - iJ_y)\Psi_2\right)$$
$$= (\Psi_1, J_x\Psi_2) + \left(\Psi_1, -iJ_y\Psi_2\right)$$
$$= (\Psi_1, J_x\Psi_2) - i\left(\Psi_1, J_y\Psi_2\right)$$
$$= (J_x\Psi_1, \Psi_2) - i\left(J_y\Psi_1, \Psi_2\right) \because J_x \text{ and } J_y \text{ are Hermitian.}$$
$$= (J_x\Psi_1, \Psi_2) + \left(iJ_y\Psi_1, \Psi_2\right)$$
$$= \left(J_x\Psi_1 + iJ_y\Psi_1, \Psi_2\right) = \left((J_x + iJ_y)\Psi_1, \Psi_2\right) = (J_+\Psi_1, \Psi_2)$$

$$\therefore (\Psi_1, J_-\Psi_2) = (J_+\Psi_1, \Psi_2) \quad \text{i.e., } J_-^\dagger = J_+$$

Second properties: $J_+\Psi_{j,m} = \sqrt{j(j+1) - m(m+1)}\hbar\Psi_{j,m+1}$

Let $\Psi_{j,m}$ are common orthonormal Eigen functions of both J^2 and J_z.

$\therefore$ $J^2\Psi_{j,m} = j(j+1)\hbar^2\Psi_{j,m}$

And $J_z\Psi_{j,m} = m\hbar\Psi_{j,m}$

$J_+\Psi_{j,m}$ is the Eigen function of J_z with Eigen value $m\hbar + \hbar = (m+1)\hbar$

I.e., $J_z\left(J_+\Psi_{j,m}\right) = (m+1)\hbar\left(J_+\Psi_{j,m}\right)$

Again, $\Psi_{j,m+1}$ is also the Eigen function of J_z with the same Eigen value $(m+1)\hbar$

I.e., $J_z\Psi_{j,m+1} = (m+1)\hbar\Psi_{j,m+1}$

In non-degenerate case, $J_+\Psi_{j,m}$ and $\Psi_{j,m+1}$ are same or can differ at best by a constant multiple, I.e., $J_+\Psi_{j,m}$ and $\Psi_{j,m+1}$ are linearly dependent.

Let $J_+\Psi_{j,m} = N_+\Psi_{j,m+1}$

$$\therefore \left(J_+\Psi_{j,m}, J_+\Psi_{j,m}\right) = \left(N_+\Psi_{j,m+1}, N_+\Psi_{j,m+1}\right)$$

$= N_+^* N_+ (\Psi_{j,m+1}, \Psi_{j,m+1}) = |N_+|^2$ [$\because$ $\Psi_{j,m+1}$ is taken to be normalized.]

$(J_+\Psi_{j,m}, J_+\Psi_{j,m}) = |N_+|^2$

$\therefore |N_+|^2 = (J_+\Psi_{j,m}, J_+\Psi_{j,m})$

$= (\Psi_{j,m}, J_- J_+ \Psi_{j,m})$ [$\therefore J_+^\dagger = J_-$]

$= (\Psi_{j,m} (J^2 - J_z^2 - \hbar J_z)\Psi_{j,m})$

$= (\Psi_{j,m}, (j(j+1)\hbar^2 - m^2\hbar^2 - \hbar(m\hbar))\Psi_{j,m}) = (j(j+1)\hbar^2 - m^2\hbar^2 - m\hbar^2)(\Psi_{j,m}, \Psi_{j,m})$

$= j(j+1)\hbar^2 - m^2\hbar^2 - m\hbar^2 = j(j+1)\hbar^2 - m(m+1)\hbar^2$[$\because$ $\Psi_{j,m}$ is taken to be normalized.]

$\therefore |N_+|^2 = [j(j+1) - m(m+1)]\hbar^2 \therefore N_+ = \sqrt{j(j+1) - m(m+1)}\hbar e^{i\delta}$; where δ is real.

$\therefore N_+ = \sqrt{j(j+1) - m(m+1)}\hbar$ taking = 0 .

$\therefore J_+\Psi_{j,m} = \sqrt{j(j+1) - m(m+1)}\hbar\Psi_{j,m+1}$

$\therefore$ $\Psi_{j,m}$ is not an Eigen function of J_+. This is because $\Psi_{j,m}$ is an Eigen function of J_z and that J_+ and J_z do not commute.

Third properties: $J_-\Psi_{j,m} = \sqrt{j(j+1) - m(m-1)}\hbar\Psi_{j,m-1}$

$J_-\Psi_{j,m}$ and $\Psi_{j,m-1}$ are eigenfunctions of J_z with same eigenvalue $m\hbar - \hbar = (m-1)\hbar$

I.e., $J_z(J_-\Psi_{j,m}) = (m\hbar - \hbar)(J_-\Psi_{j,m})$ and $J_z\Psi_{j,m-1} = (m-1)\hbar\Psi_{j,m-1}$

$\therefore$ In non-degenerate case, $J_-\Psi_{j,m}$ and $\Psi_{j,m-1}$are same or can differ at best by a multiplicative constant, i.e., they are linearly dependent. Let

$J_-\Psi_{j,m} = N_-\Psi_{j,m-1}$

$\therefore (J_-\Psi_{j,m}, J_-\Psi_{j,m}) = (N_-\Psi_{j,m-1}, N_-\Psi_{j,m-1}) = N_-^* N_- (\Psi_{j,m-1}, \Psi_{j,m-1}) = |N_-|^2$

[$\because$ $\Psi_{j,m-1}$ is taken to be normalized.]

$\therefore |N_-|^2 = (J_-\Psi_{j,m}, J_-\Psi_{j,m})$

$= (\Psi_{j,m}, J_+ J_- \Psi_{j,m})$ $\therefore J_-^\dagger = J_+$

$= (\Psi_{j,m}, (J^2 - J_z^2 + \hbar J_z)\Psi_{j,m}) = (\Psi_{j,m}, (j(j+1)\hbar^2 - m^2\hbar^2 + \hbar(m\hbar))\Psi_{j,m})$

$= (j(j+1)\hbar^2 - m^2\hbar^2 + m\hbar^2)(\Psi_{j,m}, \Psi_{j,m}) = j(j+1)\hbar^2 - m^2\hbar^2 + m\hbar^2$

$= j(j+1)\hbar^2 - m(m-1)\hbar^2$

$\therefore |N_-|^2 = [j(j+1) - m(m-1)]\hbar^2 \therefore N_- = \sqrt{j(j+1) - m(m-1)}\hbar e^{i\delta}$; where δ is real.

$\therefore \quad N_- = \sqrt{j(j+1) - m(m-1)}\hbar \qquad$ taking $= 0$.

$\therefore J_-\Psi_{j,m} = \sqrt{j(j+1) - m(m-1)}\hbar\Psi_{j,m-1}$ $\therefore \Psi_{j,m}$ is not an Eigen function of J_z. This is because $\Psi_{j,m}$ is an eigenfunction of J_z and that J_- and J_z do not commute.

10.5 Commutation relations of Ladder operators [raising and lowering operators] and the product of raising and lowering operators

1. $[J_+, J_-] = 2\hbar J_z$ 2. $[J_-, J_+] = \hbar J_-$ 3. $[J_+, J_z] = -\hbar J_+$ 4. $[J_z, J_\pm] = \hbar J_\pm$
5. $[J^2, J_-] = 0$

1. Let us define the raising and lowering operators are $J_+ = J_x + iJ_y$ *and* $J_- = J_x - iJ_y$

$$
\begin{aligned}
L.H.S &= [J_+, J_-] \\
&= (J_+ J_- - J_- J_+) \\
&= (J_x + iJ_y)(J_x - iJ_y) - (J_x - iJ_y)(J_x + iJ_y) \\
&= (J_x^2 - iJ_xJ_y + iJ_yJ_x + J_y^2) - (J_x^2 + iJ_xJ_y - iJ_yJ_x + J_y^2) \\
&= J_x^2 - iJ_xJ_y + iJ_yJ_x + J_y^2 - J_x^2 - iJ_xJ_y + iJ_yJ_x - J_y^2 \\
&= 2iJ_yJ_x - 2iJ_xJ_y \\
&= -2i\,(J_xJ_y - J_yJ_x) \\
&= -2i[J_x, J_y] \\
&= -2i\,(i\hbar\, J_z) \\
&= 2\hbar\, J_z
\end{aligned}
$$

$$\therefore [J_+, J_-] = 2\hbar\, J_z \quad (proved)$$

2. $[J_+, J_z] = -\hbar J_+$

$$L.H.S = [J_+, J_z] = J_+ J_z - J_z J_+ = (J_x + iJ_y)J_z - J_z(J_x + iJ_y)$$
$$= J_x J_z + iJ_yJ_z - J_zJ_x - iJ_zJ_y = (J_x J_z - J_zJ_x) + i(J_yJ_z - J_zJ_y)$$
$$= [J_x, J_z] + i[J_y, J_z] = -i\hbar J_y + i(i\hbar J_x) = -i\hbar J_y - \hbar J_x = -\hbar(J_x + iJ_y) = -\hbar J_+$$

$$\therefore [J_+, J_z] = -\hbar J_+ \qquad (proved)$$

3. $[J_-, J_z] = \hbar J_-$

$$\text{L.H.S} = [J_-, J_z] = (J_- J_z - J_z J_-) = (J_x - iJ_y) J_z - J_z (J_x - iJ_y)$$
$$= J_x J_z - iJ_y J_z - J_z J_x + iJ_z J_y = (J_x J_z - J_z J_x) - i(J_y J_z - J_z J_y)$$
$$= [J_x, J_z] - i[J_y, J_z] = -i\hbar J_y - i(i\hbar J_x) = -i\hbar J_y + \hbar J_x = \hbar(J_x - iJ_y) = \hbar J_-$$

$\therefore \quad [J_-, J_z] = \hbar J_- \qquad$ (proved)

4. $[J_z, J_\pm] = \hbar J_\pm$

We know, the raising operator $J_+ = J_x + iJ_y$ & the lowering operator $J_- = J_x - iJ_y$

$$L.H.S = [J_z, J_\pm] = (J_z J_\pm - J_\pm J_z) = J_z (J_x \pm iJ_y) - (J_x \pm iJ_y) J_z$$
$$= J_z J_x \pm iJ_z J_y - J_x J_z \mp iJ_y J_z = (J_z J_x - J_x J_z) \pm i(J_z J_y - J_y J_z)$$
$$= [J_z, J_x] \pm i[J_z, J_y] = i\hbar J_y \pm i(-i\hbar J_x)$$
$$= i\hbar J_y \pm \hbar J_x = \hbar(J_y \pm iJ_x) = \hbar J_\pm = J_\pm \qquad [\because \hbar = 1]$$

5. $[J^2, J_-] = 0$

Product of raising and lowering operator

$$J_+ J_- = (J_x + iJ_y)(J_x - iJ_y) = J_x^2 - iJ_x J_y + iJ_y J_x + J_y^2 = J_x^2 + J_y^2 - i(J_x J_y - J_y J_x)$$
$$= J_x^2 + J_y^2 - i[J_x, J_y] = J_x^2 + J_y^2 + J_z^2 - J_z^2 + \hbar J_Z \quad Or,\ J_+ J_- = J^2 - J_z^2 + \hbar J_Z$$

Similarly, $J_- J_+ = J_x^2 + J_y^2 + i(J_x J_y - J_y J_x) = J^2 - J_z^2 - \hbar J_Z$

Now adding the product

$$J_+ J_- + J_- J_+ = J_x^2 + J_y^2 - i(J_x J_y - J_y J_x) + J_x^2 + J_y^2 + i(J_x J_y - J_y J_x)$$
$$2(J_x^2 + J_y^2) = J_+ J_- + J_- J_+$$
$$J_x^2 + J_y^2 = \frac{1}{2}(J_+ J_- + J_- J_+)$$

But we know

$$J^2 = J_x^2 + J_y^2 + J_z^2 = \tfrac{1}{2}(J_+ J_- + J_- J_+) + J_z^2$$

$$[J^2, J_+] = \left[\frac{1}{2}(J_+ J_- + J_- J_+) + J_z^2, J_+\right] = \frac{1}{2}[J_+ J_-, J_+] + \frac{1}{2}[J_- J_+, J_+] + [J_z^2, J_+]$$
$$= \tfrac{1}{2}[J_+, J_+]J_- + \tfrac{1}{2}J_+[J_-, J_+] + \tfrac{1}{2}[J_-, J_+]J_+ + \tfrac{1}{2}J_-[J_+, J_+] + [J_z, J_+]J_z + J_z[J_z, J_+]$$
$$= \tfrac{1}{2}J_+(-2\hbar J_z) + \tfrac{1}{2}(-2\hbar J_z)J_+ + (\hbar J_+)J_z + J_z(\hbar J_+) \quad ;\text{Since} \quad [J_+, J_+] = 0$$
$$= -\hbar J_+ J_z - \hbar J_z J_+ + \hbar J_+ J_z + \hbar J_z J_+ = 0$$

$[J^2,J_+] = 0$; Similarly $[J^2,J_-] = 0$

10.6 Ladder method

We know that $J_z\Phi_k = K\Phi_k$

$$\Rightarrow J_- J_z \Phi_k = K J_- \Phi_k [operating\ by\ J_-] \qquad [1]$$

$$Again \quad [J_-, J_z] = \hbar J_- \Rightarrow J_- J_z - J_z J_- = \hbar J_- \Rightarrow J_- J_z = J_z J_- + \hbar J_- \qquad [2]$$

Therefore [1] becomes

$$(J_z J_- + \hbar J_-)\Phi_k = KJ_-\Phi_k \Rightarrow J_z J_- \Phi_k = KJ_-\Phi_k - \hbar J_-\Phi_k \Rightarrow J_z J_- \Phi_k = J_-\Phi_k (K-\hbar) \qquad [3]$$

Where J_- is the angular momentum lowering operator

$$J_z\Phi_k = (K-\hbar)\Phi_k \Rightarrow J_z\Phi = (K-\hbar)\Phi \qquad [4] \quad [\because J_-\Phi_k = \Phi]$$

Again operating J_- on the both sides of equation [4]

$$J_- J_z \Phi = J_-(K-\hbar)\Phi \Rightarrow J_- J_z J_- \Phi_k = J_-(K-\hbar) J_- \Phi_k$$

$$\Rightarrow (J_z J_- + \hbar J_-) J_- \Phi_k = (K-\hbar) J_-^2 \Phi_k \Rightarrow J_z J_-^2 \Phi_k + \hbar J_-^2 \Phi_k = (K-\hbar) J_-^2 \Phi_k$$

$$\Rightarrow J_z J_-^2 \Phi_k = (K-2\hbar) J_-^2 \Phi_k$$

If we operate J_- *in* f times, we get

$$J_z(J_-, J_- J_-, J_- J_- J_-, \ldots\ldots\ f\ times)\Phi_k = (K - f\hbar) J_-^f \Phi_k$$

$$i.e \quad J_z J_-^f \Phi_k = (K - f\hbar) J_-^f \Phi_k \qquad [5]$$

Again,

$$J_z\Phi_k = K\Phi_k \Rightarrow J_+ J_z \Phi_k = K J_+ \Phi_k [operating\ by\ J_+] \qquad [6]$$

Again, we know that

$$[J_+, J_z] = -\hbar J_+ \Rightarrow J_+ J_z - J_z J_+ = -J_+\hbar \Rightarrow J_+ J_z = J_z J_+ - J_+\hbar \qquad [7]$$

From equation [6]

$$J_+ J_z \Phi_k = K J_+ \Phi_k \Rightarrow (J_z J_+ - J_+\hbar)\Phi_k = KJ_+\Phi_k \Rightarrow J_z J_+ \Phi_k = (K+\hbar) J_+ \Phi_k$$

Where J_+ is the angular momentum raising operator

$$\Rightarrow J_+ J_z J_+ \Phi_k = J_+(K+\hbar) J_+ \Phi_k \qquad [Operating\ by\ J_+\ on\ the\ both\ sides]$$

$$\Rightarrow J_+ J_z J_+ \Phi_k = (K+\hbar) J_+^2 \Phi_k$$

$$\Rightarrow (J_z J_+ - J_+\hbar) J_+ \Phi_k = (K+\hbar) J_+^2 \Phi_k \Rightarrow J_z J_+^2 \Phi_k - \hbar J_+^2 \Phi_k = (K+\hbar) J_+^2 \Phi_k$$

$$\Rightarrow J_z J_+^2 \Phi_k = (K+2\hbar) J_+^2 \Phi_k \qquad [8]$$

If we operate J_+ in f times we get finally,

$$J_z J_+^f \Phi_k = (K + f\hbar) J_+^f \Phi_k \qquad [9]$$

$$\& \quad J_z J_-^f \Phi_k = (K - f\hbar) J_-^f \Phi_k \qquad [5]$$

If we operate J_- or J_+ in f times on the Eigen function Φ_k ,we get the Eigen values $(K - f\hbar) \& (K + f\hbar)$. So this method is called Ladder method.

10.7 Introduction to addition of generalized angular momentum

Angular momenta, being vectors, could be added vectorially. Thus, for a system consisting of two subsystems with angular momenta $\vec{J_1}$ and $\vec{J_2}$,we can define a resultant or total angular momentum $\vec{J}$

$$\vec{J} = \vec{J_1} + \vec{J_2} = \vec{J_2} + \vec{J_1} \qquad [1]$$

The resultant or total generalized angular momentum $\vec{J} = \vec{J_1} + \vec{J_2}$ of the two given vectors $\vec{J_1}$ and $\vec{J_2}$ depends on relative orientation of $\vec{J_1}$ and $\vec{J_2}$ as shown in Figures [a] and [b]. We can draw triangles with J_1, J_2 and J as its sides, and thereby get allowed relative orientations of the three vectors.

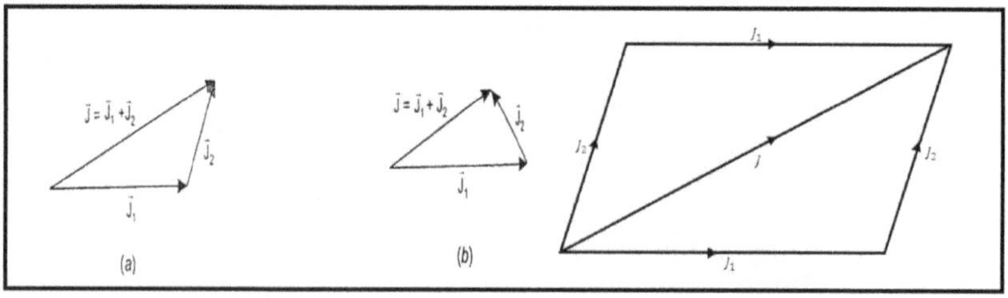

Fig.[10.7]: Shows addition of two angular momentum

10.8 Addition of two generalized angular momenta

Let us consider $\vec{J_1}$ and $\vec{J_2}$ be two generalized angular momenta. For a system consisting of two subsystems with angular momenta $\vec{J_1}$ and $\vec{J_2}$,we can define a resultant or total angular momentum $\vec{J}$

$$\vec{J} = \vec{J_1} + \vec{J_2} = \vec{J_2} + \vec{J_1}$$

The addition of the two generalized angular momenta $\vec{J_1}$ and $\vec{J_2}$ and the components of $\vec{J_1}$ satisfy the following commutation relations I.e.,

$$[J_{1x}, J_{1y}] = i\hbar J_{1z}\,,\ [J_{1y}, J_{1z}] = i\hbar J_{1x}\,,\ [J_{1z}, J_{1x}] = i\hbar J_{1y}$$

$$[J_1^2, J_{1x}] = [J_1^2, J_{1y}] = [J_1^2, J_{1z}] = 0$$

The components of $\vec{J_2}$ also satisfy the commutation relations

$$[J_{2x}, J_{2y}] = i\hbar J_{2z}\,,\ [J_{2y}, J_{2z}] = i\hbar J_{2x}\,,\ [J_{2z}, J_{2x}] = i\hbar J_{2y}$$

$$[J_2^2, J_{2x}] = [J_2^2, J_{2y}] = [J_2^2, J_{2z}] = 0$$

In any case, $\vec{J_1}$ and $\vec{J_2}$ correspond to independent degrees of freedom (or coordinates). $\vec{J_1}$ and $\vec{J_2}$ are angular momenta of any type. They may be: [a] *orbital and spin angular momentum of a given particle [b] they may be orbital angular momenta of two different particles[c] they may be spin angular momentum of two different particles.* If generalized angular momenta $\vec{J_1}$ and $\vec{J_2}$ are orbital angular momenta of two different particles are given by

$$\bar{L}_1 = \vec{r_1} \times \vec{p_1} \text{ and } \bar{L}_2 = \vec{r_2} \times \vec{p_2}$$

Then $\hat{J}_{1x}$, $\hat{J}_{1y}$, $\hat{J}_{1z}$, $\hat{J}_1^2$ will contain x_1, y_1, z_1 while $\hat{J}_{2x}$, $\hat{J}_{2y}$, $\hat{J}_{2z}$, $\hat{J}_2^2$ will contain x_2, y_2, z_2. The 6 coordinates $x_1, y_1, z_1, x_2, y_2, z_2$ are independent. As such operator of any component of $\vec{J_1}$ commutes with operator of any component of $\vec{J_2}$.

$$J_{1x} = \frac{\hbar}{i}\left(y_1 \frac{\partial}{\partial z_1} - z_1 \frac{\partial}{\partial y_1}\right),$$

$$J_{2x} = \frac{\hbar}{i}\left(y_2 \frac{\partial}{\partial z_2} - z_2 \frac{\partial}{\partial y_2}\right)$$

Since $x_1, y_1, z_1, x_2, y_2, z_2$ are independent.

So, $J_{1x}J_{2x}\Psi = J_{2x}J_{1x}\Psi$ or, $[J_{1x}, J_{2x}] = 0$

Similarly, $[J_{1i}, J_{2j}] = 0$, $[J_1^2, J_{2j}] = [J_2^2, J_{1i}] = [J_1^2, J_2^2]$ with $i = x, y, z$ and $j = x, y, z$.

Let us define total angular momentum of $\vec{J_1}$ and $\vec{J_2}$ as vector sum of the two generalized angular momenta

$$\vec{J} = \vec{J_1} + \vec{J_2}$$

This means $J_x = J_{1x} + J_{2x}$, $J_y = J_{1y} + J_{2y}$, $J_z = J_{1z} + J_{2z}$

We need to show that

$$[J_x, J_y] = i\hbar J_z, \ [J_y, J_z] = i\hbar J_x, \ [J_z, J_x] = i\hbar J_y \text{ and } [J^2, J_x] = [J^2, J_y] = [J^2, J_z] = 0$$

So that $\vec{J}$ satisfies to be an angular momentum (vector) in Quantum Mechanics.

$$\begin{aligned}
[J_y, J_z]\Psi &= J_yJ_z\Psi - J_zJ_y\Psi \\
&= (J_{1y} + J_{2y})(J_{1z} + J_{2z})\Psi - (J_{1z} + J_{2z})(J_{1y} + J_{2y})\Psi \\
&= J_{1y}(J_{1z} + J_{2z})\Psi + J_{2y}(J_{1z} + J_{2z})\Psi - J_{1z}(J_{1y} + J_{2y})\Psi - J_{2z}(J_{1y} + J_{2y})\Psi \\
&= J_{1y}J_{1z}\Psi + J_{1y}J_{2z}\Psi + J_{2y}J_{1z}\Psi + J_{2y}J_{2z}\Psi - J_{1z}J_{1y}\Psi - J_{1z}J_{2y}\Psi - J_{2z}J_{1y}\Psi - J_{2z}J_{2y}\Psi
\end{aligned}$$

Because operator of any component of $\vec{J_1}$ commutes with operator of any component of $\vec{J_2}$.

$[J_y, J_z]\Psi = (J_{1y}J_{1z} - J_{1z}J_{1y})\Psi + (J_{2y}J_{2z} - J_{2z}J_{2y})\Psi$

$= [J_{1y}, J_{1z}]\Psi + [J_{2y}, J_{2z}]\Psi = i\hbar J_{1x}\Psi + i\hbar J_{2x}\Psi = i\hbar(J_{1x} + J_{2x})\Psi = i\hbar J_x\Psi$ Thus $[J_y, J_z] = i\hbar J_x$

Now, $[J^2, J_z]\Psi = [J^2, J_{1z}]\Psi + [J^2, J_{2z}]\Psi$

$$[J^2, J_{1z}]\Psi = [J_1^2 + J_2^2 + 2J_{1x}J_{2x} + 2J_{1y}J_{2y} + 2J_{1z}J_{2z}, J_{1z}]\Psi$$

$J^2 = \vec{J}.\vec{J} = (\vec{J_1} + \vec{J_2}).(\vec{J_1} + \vec{J_2}) = \vec{J_1}.\vec{J_1} + \vec{J_1}.\vec{J_2} + \vec{J_2}.\vec{J_1} + \vec{J_2}.\vec{J_2} = J_1^2 + J_2^2 + 2\vec{J_1}.\vec{J_2}$

Or, $J^2 = \vec{J}.\vec{J} = J_1^2 + J_2^2 + 2(J_{1x}J_{2x} + J_{1y}J_{2y} + J_{1z}J_{2z})$

Because $[J_{1i}, J_{2j}] = 0$ [with $i = x, y, z$ and $j = x, y, z$] I.e., operator of any component of $\vec{J_1}$ commutes with operator of any component of $\vec{J_2}$.

$$[J^2, J_{1z}]\Psi = [J_1^2, J_{1z}]\Psi + [J_2^2, J_{1z}]\Psi + 2[J_{1x}J_{2x}, J_{1z}]\Psi + 2[J_{1y}J_{2y}, J_{1z}]\Psi + [2J_{1z}J_{2z}, J_{1z}]\Psi$$

Because $[P + Q + R + S + T, U] = [P, U] + [Q, U] + [R, U] + [S, U] + [T, U]$

$$= 0 + 0 + 2J_{1x}[J_{2x}, J_{1z}]\Psi + 2[J_{1x}, J_{1z}]J_{2x}\Psi + 2J_{1y}[J_{2y}, J_{1z}]\Psi + 2[J_{1y}, J_{1z}]J_{2y}$$
$$+ 2J_{1z}[J_{2z}, J_{1z}]\Psi + [J_{1z}, J_{1z}]J_{2z}\Psi$$
$$= 2[J_{1x}, J_{1z}]J_{2x}\Psi + 2[J_{1y}, J_{1z}]J_{2y}\Psi$$
$$= 2(-i\hbar J_{1y})J_{2x}\Psi + 2i\hbar J_{1x}J_{2y}\Psi$$
$$= 2i\hbar(J_{1x}J_{2y} - J_{1y}J_{2x})\Psi \quad [1]$$

$$[J^2, J_{2z}]\Psi = [J_1^2 + J_2^2 + 2J_{1x}J_{2x} + 2J_{1y}J_{2y} + 2J_{1z}J_{2z}, J_{2z}]$$
$$= [J_1^2, J_{2z}]\Psi + [J_2^2, J_{2z}]\Psi + 2[J_{1x}J_{2x}, J_{2z}]\Psi + 2[J_{1y}J_{2y}, J_{2z}]\Psi + 2[J_{1z}J_{2z}, J_{2z}]$$
$$= 0 + 0 + 2[J_{1x}J_{2x}, J_{2z}]\Psi + 2[J_{1y}J_{2y}, J_{2z}]\Psi + 2[J_{1z}J_{2z}, J_{2z}]$$
$$= 2J_{1x}[J_{2x}, J_{2z}]\Psi + 2[J_{1x}, J_{2z}]J_{2x}\Psi + 2J_{1y}[J_{2y}, J_{2z}]\Psi + 2[J_{1y}, J_{2z}]J_{2y}\Psi$$
$$+ 2J_{1z}[J_{2z}, J_{2z}]\Psi + [J_{1z}, J_{2z}]J_{2z}\Psi$$
$$= 2J_{1x}(-i\hbar J_{2y})\Psi + 2J_{1y}(i\hbar J_{2x})\Psi = 2i\hbar(J_{1y}J_{2x} - J_{1x}J_{2y})\Psi \quad [2]$$

Adding equations [1] and [2], we get

$[J^2, J_{1z}]\Psi + [J^2, J_{2z}]\Psi = 2i\hbar(J_{1x}J_{2y} - J_{1y}J_{2x})\Psi + 2i\hbar(J_{1y}J_{2x} - J_{1x}J_{2y})\Psi$

Or, $[J^2, J_z]\Psi = 0$. Thus $[J^2, J_z] = 0$

Thus total angular momentum of $\vec{J_1}$ and $\vec{J_2}$ as vector sum of the two generalized angular momenta I.e., $\vec{J} = \vec{J_1} + \vec{J_2}$ *satisfies as an angular momentum in Quantum Mechanics.*

10.9 Clebsch–Gordan (CG) coefficients and its significance

The name derives from the German mathematicians Alfred Clebsch and Paul Gordan, who encountered an equivalent problem in invariant theory.

In physics, the **Clebsch–Gordan (CG) coefficients** are numbers that arise in angular momentum coupling in quantum mechanics. They appear as the expansion coefficients of total angular momentum Eigen states in an uncoupled tensor product basis. The addition of spins in quantum-mechanical terms can be read directly from this approach as spherical harmonics are Eigen functions of total angular momentum. From the formal definition of angular momentum, recursion relations for the Clebsch–Gordan coefficients can be found.

Clebsch-Gordan coefficients are the coefficients in the triple integration for rule for spherical harmonics. Clebsch-Gordan coefficients arise in quantum mechanical angular momentum calculations, as well as any other computation expressed in spherical harmonics (e.g., for rendering in computer graphics). In particular, they allow one to express the product of two spherical harmonic series as another spherical harmonic series. Unfortunately, although all coefficients are closed form and are not too hard to compute in specific cases (say if), there are no simple rules related the different coefficients together which makes analytic calculations with them extremely difficult (or impossible).

In more mathematical terms, the CG coefficients are used in representation theory, particularly of compact Lie groups, to perform the explicit direct sum decomposition of the tensor product of two irreducible representations (i.e., a reducible representation) into irreducible representations, in cases where the numbers and types of irreducible components are already known abstractly.

10.10 Common Eigen functions of J_1^2, J_2^2, J^2 and J_z

Number of allowed Ψ_1's is $(2j_1 + 1)$ for j_1. Number of allowed Ψ_2's is $(2j_2 + 1)$ for j_2. Number of allowed Ψ_1's and Ψ_2's are

$$\Psi = \Psi_1\Psi_2$$
$$= (2j_1 + 1)(2j_2 + 1)$$

We want to count number of allowed ϕ's. For a given particular value $j = j_p$,we have allowed

$$\phi\text{'s} = \phi(j_1, j_2, j_p, m)$$

where m can have $(2j_p + 1)$ number of values, namely,

$$-j_p,\ (-j_p + 1),\ (-j_p + 2), \dots\dots\dots\dots\dots,\ (j_p - 2),\ (j_p - 1),\ +j_p.$$

Thus, we have$(2j_p + 1)$ number of ϕ's for a given value of $j = j_p$. Values of j can be

$$(j_1 - j_2), (j_1 - j_2 + 1), (j_1 - j_2 + 2), \dots, (j_1 + j_2 - 2), (j_1 + j_2 - 1), (j_1 + j_2).$$

Hence grand total number of ϕ's is

$$N = \sum_{j=j_1-j_2}^{j_1+j_2}(2j + 1)$$
$$= [2(j_1 - j_2) + 1] + [2(j_1 - j_2 + 1) + 1] + [2(j_1 - j_2 + 2) + 1]$$
$$+[2(j_1 - j_2 + 3) + 1] + \cdots + [2(j_1 + j_2) + 1]$$

$$= [2(j_1 - j_2) + 1] + [2(j_1 - j_2 + 1) + 1] + [2(j_1 - j_2 + 2) + 1]$$

$$+[2(j_1 - j_2 + 3) + 1] + \cdots + [2(j_1 - j_2 + 2j_2) + 1] \quad [1]$$

We see that there are $2j_2 + 1$ number of terms on R.H.S. of equation [1]. Each term contains $[2(j_1 - j_2) + 1]$. Hence equation [1] gives

$$N = [2(j_1 - j_2) + 1](2j_2 + 1) + 0 + 2 + 4 + 6 + \cdots + 4j_2$$

$$= [2(j_1 - j_2) + 1](2j_2 + 1) + 2(1 + 2 + 3 + \cdots + 2j_2)$$

$$= [2(j_1 - j_2) + 1](2j_2 + 1) + 2\frac{2j_2(2j_2+1)}{2}$$

$$= (2j_2 + 1)[2(j_1 - j_2) + 1 + 2j_2] = (2j_1 + 1)(2j_2 + 1)$$

Total number of allowed ϕ's is$(2j_1 + 1)(2j_2 + 1)$ which is the same as total number of allowed Ψ's.

10.11Calculation of Clebsch-Gordan coefficients for $J_1 = 1$ and $J_2 = 1/2$

For a system consisting of two subsystems with angular momenta $\vec{J_1}$ and $\vec{J_2}$, we can define a resultant or total angular momentum $\vec{J}$

$$\vec{J} = \vec{J_1} + \vec{J_2} = \vec{J_2} + \vec{J_1}$$

Corresponding to the three angular momentum vectors $\vec{J_1}$, $\vec{J_2}$ and $\vec{J}$, we have the six Hermitian operators J_1^2, J_{1z},J_2^2,J_{2z} ,$J^2 and$ J_z. All these six Hermitian operators, however, do not commute among themselves [J_{1z} and J_{2z} do not commute with J^2]. But we can form two sets consisting of four operators each, which, together with the Hamiltonian, form complete obserbles for the system. These are:

[a] $J_1^2, J_2^2, J_{1z}, J_{2z}$

[b] J_1^2, J_2^2, J^2, J_z

Let us consider the case with $j_1 = 1$ and $j_2 = \frac{1}{2}$.

For $j_1 = 1,\ m_1 = -1, 0, +1$ and $For\ j_2 = \frac{1}{2},\ m_2 = -\frac{1}{2}, +\frac{1}{2}$

$\Psi(j_1, j_2, m_1, m_2)$ is the Common Eigen functions of generalized angular momentum and its components $J_1^2, J_2^2, J_{1z}, J_{2z}$ are given by

$$\Psi\left(1,\frac{1}{2},-1,-\frac{1}{2}\right), \Psi\left(1,\frac{1}{2},0,-\frac{1}{2}\right), \Psi\left(1,\frac{1}{2},+1,-\frac{1}{2}\right), \Psi\left(1,\frac{1}{2},-1,+\frac{1}{2}\right), \Psi\left(1,\frac{1}{2},0,+\frac{1}{2}\right)$$

$$\Psi\left(1,\frac{1}{2},+1,+\frac{1}{2}\right)$$

There are 6 Ψ's because total number of allowed Ψ's are

$$(2j_1+1)(2j_2+1) = (2\times 1+1)\left(2\times\tfrac{1}{2}+1\right) = 3\times 2 = 6.$$

We have $m_1 = -1, 0, +1,\ m_2 = -\frac{1}{2}, +\frac{1}{2};$ Thus $m = m_1 + m_2 = -\frac{3}{2}, -\frac{1}{2}, \frac{1}{2}, \frac{3}{2}.$

Allowed values of j are $(j_1 - j_2)$ to $(j_1 + j_2)$ are

$$(1-\tfrac{1}{2})\ to\ (1+\tfrac{1}{2}) \text{ Or, } (\tfrac{1}{2}\ to\ \tfrac{3}{2})$$

I.e., $j = \frac{1}{2}\ and\ \frac{3}{2}$ [in step of +1]

$For\ J = \frac{1}{2},\ m_1 = -\frac{1}{2}, +\frac{1}{2}\ and\ For\ J = \frac{3}{2},\ m_2 = -\frac{3}{2}, -\frac{1}{2}, +\frac{1}{2}, +\frac{3}{2}.$

$\phi(j_1, j_2, j, m)$ is the Common Eigen functions of generalized angular momentum and its components $J_1^2,\ J_2^2,\ J^2,\ J_z$ are given by

$$\phi\left(1, \tfrac{1}{2}, \tfrac{1}{2}, -\tfrac{1}{2}\right); \qquad \phi\left(1, \tfrac{1}{2}, \tfrac{1}{2}, +\tfrac{1}{2}\right); \qquad \phi\left(1, \tfrac{1}{2}, \tfrac{3}{2}, -\tfrac{3}{2}\right)$$

$$\phi\left(1, \tfrac{1}{2}, \tfrac{3}{2}, -\tfrac{1}{2}\right); \qquad \phi\left(1, \tfrac{1}{2}, \tfrac{3}{2}, +\tfrac{1}{2}\right); \qquad \phi\left(1, \tfrac{1}{2}, \tfrac{3}{2}, +\tfrac{3}{2}\right)$$

There are 6 ϕ's because total number of allowed Ψ's are

$(2j_1+1)(2j_2+1) = (2\times 1+1)\left(2\times\frac{1}{2}+1\right) = 3\times 2 = 6$; same as the number of Ψ's.

We have $m_1 = -\frac{1}{2}, +\frac{1}{2}$; $m_2 = -\frac{3}{2}, -\frac{1}{2}, +\frac{1}{2}, +\frac{3}{2}$

Thus $m = m_1 + m_2 = -\frac{3}{2}, -\frac{1}{2}, \frac{1}{2}, \frac{3}{2}.$

Any of the 6 ϕ's can be expressed as linear combination of the 6 Ψ's,

I.e., $$\phi(j_1, j_2, j, m) = \textstyle\sum_{m_1}\sum_{m_2} C^{j,m}_{m_1,m_2}\Psi(j_1, j_2, j, m_1, m_2) \qquad [1]$$

Where $C^{j,m}_{m_1,m_2}$ and the names, **Clebsch-Gordan, or C- coefficient.**

Since $m_1 + m_2 = m$, the Ψ's for which $m_1 + m_2 \neq m$ cannot be present in R.H.S. of equation [1] *I.e.*, the expansion coefficients involved are zero. Thus

$$\phi\left(1, \tfrac{1}{2}, \tfrac{1}{2}, -\tfrac{1}{2}\right) = C_1\Psi\left(1, \tfrac{1}{2}, 0, -\tfrac{1}{2}\right) + C_2\Psi\left(1, \tfrac{1}{2}, -1, +\tfrac{1}{2}\right) \qquad [2]$$

$$\phi\left(1, \tfrac{1}{2}, \tfrac{1}{2}, +\tfrac{1}{2}\right) = C_3\Psi\left(1, \tfrac{1}{2}, +1, -\tfrac{1}{2}\right) + C_4\Psi\left(1, \tfrac{1}{2}, 0, +\tfrac{1}{2}\right) \qquad [3]$$

$$\phi\left(1, \tfrac{1}{2}, \tfrac{3}{2}, -\tfrac{3}{2}\right) = \Psi\left(1, \tfrac{1}{2}, -1, -\tfrac{1}{2}\right) \qquad [4]$$

$$\phi\left(1, \tfrac{1}{2}, \tfrac{3}{2}, -\tfrac{1}{2}\right) = C_5\Psi\left(1, \tfrac{1}{2}, 0, -\tfrac{1}{2}\right) + C_6\Psi\left(1, \tfrac{1}{2}, -1, +\tfrac{1}{2}\right) \qquad [5]$$

$$\phi\left(1, \tfrac{1}{2}, \tfrac{3}{2}, +\tfrac{1}{2}\right) = C_7\Psi\left(1, \tfrac{1}{2}, +1, -\tfrac{1}{2}\right) + C_8\Psi\left(1, \tfrac{1}{2}, 0, +\tfrac{1}{2}\right) \qquad [6]$$

$\phi\left(1,\frac{1}{2},\frac{3}{2},+\frac{3}{2}\right)=\Psi\left(1,\frac{1}{2},+1,+\frac{1}{2}\right)$ [7]

There are 36 Clebsch-Gordan coefficients, 6 in each of equation (2) to (7). Only 10 are non-zero and hence have been written, 2 of them taken as 1 because of normalization, and 8 of them taken unknown, to be determined or calculated. The remaining 26 coefficients are zero because $m_1+m_2\neq m$ for the Ψ's involved[7].

We shall use

$$J_{\pm}\ \phi(j_1,j_2,j,m)=\sqrt{j(j+1)-m(m\pm 1)}\hbar\ \phi(j_1,j_2,j,m\pm 1)$$

$$J_{1\pm}\ \Psi(j_1,j_2,m_1,m_2)=\sqrt{j_1(j_1+1)-m_1(m_1\pm 1)}\hbar\ \Psi(j_1,j_2,m_1\pm 1,m_2)$$

$$J_{2\pm}\ \Psi(j_1,j_2,m_1,m_2)=\sqrt{j_1(j_1+1)-m_1(m_1\pm 1)}\hbar\ \Psi(j_1,j_2,j,m_1,m_2\pm 1)$$

From equation [7]

$$J_-\phi\left(j_{1=}1,j_2=\frac{1}{2},j=\frac{3}{2},m=+\frac{3}{2}\right)=J_-\Psi\left(j_{1=}1,j_2=\frac{1}{2},m_1=+1,m_2=+\frac{1}{2}\right)$$

Or, $J_-\phi\left(1,\frac{1}{2},\frac{3}{2},+\frac{3}{2}\right)=J_-\Psi\left(1,\frac{1}{2},+1,+\frac{1}{2}\right)$

Or, $J_-\phi\left(1,\frac{1}{2},\frac{3}{2},+\frac{3}{2}\right)=J_{1-}\Psi\left(1,\frac{1}{2},+1,+\frac{1}{2}\right)+J_{2-}\Psi\left(1,\frac{1}{2},+1,+\frac{1}{2}\right)$

Or, $\sqrt{\frac{3}{2}\left(\frac{3}{2}+1\right)-\frac{3}{2}\left(\frac{3}{2}-1\right)}\hbar\phi\left(1,\frac{1}{2},\frac{3}{2},\frac{3}{2}-1\right)=\sqrt{1(1+1)-1(1-1)}\hbar\Psi\left(1,\frac{1}{2},1-1,\frac{1}{2}\right)$

$+\sqrt{\frac{1}{2}\left(\frac{1}{2}+1\right)-\frac{1}{2}\left(\frac{1}{2}-1\right)}\hbar\Psi\left(1,\frac{1}{2},1,\frac{1}{2}-1\right)$

Or, $\sqrt{3}\hbar\phi\left(1,\frac{1}{2},\frac{3}{2},\frac{1}{2}\right)=\sqrt{2}\hbar\Psi\left(1,\frac{1}{2},0,\frac{1}{2}\right)+\hbar\Psi\left(1,\frac{1}{2},1,-\frac{1}{2}\right)$

$\phi\left(1,\frac{1}{2},\frac{3}{2},\frac{1}{2}\right)=\sqrt{\frac{3}{2}}\Psi\left(1,\frac{1}{2},0,\frac{1}{2}\right)+\frac{1}{\sqrt{3}}\Psi\left(1,\frac{1}{2},1,-\frac{1}{2}\right)$ [8]

Comparing equation [8] with equation [6], we get

$$C_7=\frac{1}{\sqrt{3}},\qquad C_8=\sqrt{\frac{2}{3}}$$

From equation [4]

$$J_+\phi\left(j_1=1,j_2=\frac{1}{2},j=\frac{3}{2},m=-\frac{3}{2}\right)=J_+\Psi\left(j_1=1,j_2=\frac{1}{2},m_1=-1,m_2=-\frac{1}{2}\right)$$

or, $J_+\phi\left(1,\frac{1}{2},\frac{3}{2},-\frac{3}{2}\right)=J_+\Psi\left(1,\frac{1}{2},-1,-\frac{1}{2}\right)$

or, $J_+\phi\left(1,\frac{1}{2},\frac{3}{2},-\frac{3}{2}\right)=J_{1+}\Psi\left(1,\frac{1}{2},-1,-\frac{1}{2}\right)+J_{2+}\Psi\left(1,\frac{1}{2},-1,-\frac{1}{2}\right)$

Or, $\sqrt{\frac{3}{2}\left(\frac{3}{2}+1\right)-\left(-\frac{3}{2}\right)\left(-\frac{3}{2}+1\right)}\hbar\phi\left(1,\frac{1}{2},\frac{3}{2},-\frac{3}{2}+1\right)$

$$=\sqrt{1(1+1)-(-1)(-1+1)}\hbar\Psi\left(1,\frac{1}{2},-1+1,-\frac{1}{2}\right)$$

$$+\sqrt{\frac{1}{2}\left(\frac{1}{2}+1\right)-\left(-\frac{1}{2}\right)\left(-\frac{1}{2}+1\right)}\hbar\Psi\left(1,\frac{1}{2},-1,-\frac{1}{2}+1\right)$$

Or, $\sqrt{3}\hbar\phi\left(1,\frac{1}{2},\frac{3}{2},-\frac{1}{2}\right)=\sqrt{2}\hbar\Psi\left(1,\frac{1}{2},0,-\frac{1}{2}\right)+\hbar\Psi\left(1,\frac{1}{2},-1,\frac{1}{2}\right)$

Or, $\phi\left(1,\frac{1}{2},\frac{3}{2},-\frac{1}{2}\right)=\sqrt{\frac{2}{3}}\Psi\left(1,\frac{1}{2},0,-\frac{1}{2}\right)+\frac{1}{\sqrt{3}}\Psi\left(1,\frac{1}{2},-1,\frac{1}{2}\right)$

Comparing equation [9] and equation, [5] we get

$$C_5=\sqrt{\frac{2}{3}},\quad C_6=\frac{1}{\sqrt{3}}$$

All the ϕ's are orthonormal. All the Ψ's are also orthonormal. Scalar product of any two ϕ's is zero. Taking scalar product of ϕ's of equation [2] and equation [5], we get

$$\left(\phi\left(1,\frac{1}{2},\frac{1}{2},-\frac{1}{2}\right),\phi\left(1,\frac{1}{2},\frac{3}{2},-\frac{1}{2}\right)\right)$$

$$=\left(C_1\Psi\left(1,\frac{1}{2},0,-\frac{1}{2}\right)+C_2\Psi\left(1,\frac{1}{2},-1,\frac{1}{2}\right),C_5\Psi\left(1,\frac{1}{2},0,-\frac{1}{2}\right)+C_6\Psi\left(1,\frac{1}{2},-1,\frac{1}{2}\right)\right)$$

Or, $0=\left(C_1\Psi\left(1,\frac{1}{2},0,-\frac{1}{2}\right),C_5\Psi\left(1,\frac{1}{2},0,-\frac{1}{2}\right)\right)+\left(C_2\Psi\left(1,\frac{1}{2},-1,\frac{1}{2}\right),+C_6\Psi\left(1,\frac{1}{2},-1,\frac{1}{2}\right)\right)$

$$+\left(C_1\Psi\left(1,\frac{1}{2},0,-\frac{1}{2}\right),C_6\Psi\left(1,\frac{1}{2},-1,\frac{1}{2}\right)\right)+\left(C_2\Psi\left(1,\frac{1}{2},-1,\frac{1}{2}\right),+C_5\Psi\left(1,\frac{1}{2},0,-\frac{1}{2}\right)\right)$$

Or, $0=C_1^*C_5\left(\Psi\left(1,\frac{1}{2},0,-\frac{1}{2}\right),\Psi\left(1,\frac{1}{2},0,-\frac{1}{2}\right)\right)+C_2^*C_6\left(\Psi\left(1,\frac{1}{2},-1,\frac{1}{2}\right),\Psi\left(1,\frac{1}{2},-1,\frac{1}{2}\right)\right)$

$$+C_1^*C_6\left(\Psi\left(1,\frac{1}{2},0,-\frac{1}{2}\right),\Psi\left(1,\frac{1}{2},-1,\frac{1}{2}\right)\right)+C_2^*C_5\left(\Psi\left(1,\frac{1}{2},-1,\frac{1}{2}\right),\Psi\left(1,\frac{1}{2},0,-\frac{1}{2}\right)\right)$$

Or, $0=C_1^*C_5(1)+C_2^*C_6(1)+C_1^*C_6(0)+C_2^*C_5(0)$

[$\because$ Ψ's are orthonormal]

Or, $C_1^*C_5+C_2^*C_6=0$ Or, $C_1^*\sqrt{\frac{2}{3}}+C_2^*\frac{1}{\sqrt{3}}=0$ Or, $C_1\sqrt{\frac{2}{3}}+C_2\frac{1}{\sqrt{3}}=0$ or, $C_2=-\sqrt{2}C_1$

Equation [2] gives

$$\phi\left(1,\frac{1}{2},\frac{1}{2},-\frac{1}{2}\right)=C_1\Psi\left(1,\frac{1}{2},0,-\frac{1}{2}\right)-\sqrt{2}C_1\Psi\left(1,\frac{1}{2},-1,+\frac{1}{2}\right)$$

Normalization of the ϕ requires

$$|C_1|^2+\left|-\sqrt{2}C_1\right|^2=1 \text{ because for } \phi=\textstyle\sum_i C_i\Psi_i \text{, we can write } \sum_i|C_i|^2=1$$

Or, $3|C_1|^2=1$ or, $|C_1|^2=\frac{1}{3}$ or, $C_1=\frac{1}{\sqrt{3}}e^{i\delta}$, where δ is real. $C_1=\frac{1}{\sqrt{3}}$ taking $\delta=0$.

Then $C_2 = -\sqrt{2}C_1$ gives $C_2 = -\sqrt{\frac{2}{3}}$ Thus $C_1 = \frac{1}{\sqrt{3}}$, $C_2 = -\sqrt{\frac{2}{3}}$

Taking scalar product of ϕ's of equations [B] and [E], we get

$$\left(\phi\left(1,\frac{1}{2},\frac{1}{2},+\frac{1}{2}\right),\phi\left(1,\frac{1}{2},\frac{3}{2},+\frac{1}{2}\right)\right)$$

$$= \left(C_3\Psi\left(1,\tfrac{1}{2},+1,-\tfrac{1}{2}\right) + C_4\Psi\left(1,\tfrac{1}{2},0,+\tfrac{1}{2}\right), C_7\Psi\left(1,\tfrac{1}{2},+1,-\tfrac{1}{2}\right) + C_8\Psi\left(1,\tfrac{1}{2},0,+\tfrac{1}{2}\right)\right)$$

Or, $0 = \left(C_3\Psi\left(1,\tfrac{1}{2},+1,-\tfrac{1}{2}\right), C_7\Psi\left(1,\tfrac{1}{2},+1,-\tfrac{1}{2}\right)\right) + \left(C_4\Psi\left(1,\tfrac{1}{2},0,+\tfrac{1}{2}\right), C_8\Psi\left(1,\tfrac{1}{2},0,+\tfrac{1}{2}\right)\right)$

$$+\left(C_3\Psi\left(1,\tfrac{1}{2},+1,-\tfrac{1}{2}\right), C_8\Psi\left(1,\tfrac{1}{2},0,+\tfrac{1}{2}\right)\right) + \left(C_4\Psi\left(1,\tfrac{1}{2},0,+\tfrac{1}{2}\right), C_7\Psi\left(1,\tfrac{1}{2},+1,-\tfrac{1}{2}\right)\right)$$

Or, $0 = C_3^*C_7(1) + C_4^*C_8(1) + C_3^*C_8(0) + C_4^*C_7(0)$

Or, $C_3^*C_7 + C_4^*C_8 = 0$

Or, $C_3^*\frac{1}{\sqrt{3}} + C_4^*\sqrt{\frac{2}{3}} = 0$

Or, $C_3 + \sqrt{2}C_4 = 0$ or, $C_4 = -\frac{C_3}{\sqrt{2}}$

Equation [3] gives

$$\phi\left(1,\tfrac{1}{2},\tfrac{1}{2},+\tfrac{1}{2}\right) = C_3\Psi\left(1,\tfrac{1}{2},+1,-\tfrac{1}{2}\right) - \frac{C_3}{\sqrt{2}}\Psi\left(1,\tfrac{1}{2},0,+\tfrac{1}{2}\right)$$

Normalization of this ϕ requires

$$|C_3|^2 + \left|-\frac{C_3}{\sqrt{2}}\right|^2 = 1 \text{ or, } C_3 = \sqrt{\frac{2}{3}}e^{i\delta}, \text{where } \delta \text{ is real} \quad \text{Or, } C_3 = \sqrt{\frac{2}{3}} \text{ taking } \delta = 0.$$

Then $C_4 = -\frac{C_3}{\sqrt{2}}$ gives $C_4 = -\frac{1}{\sqrt{3}}$ Thus $C_3 = +\sqrt{\frac{2}{3}}$, $C_4 = -\frac{1}{\sqrt{3}}$

We now write equation [2] to [7] in matrix form as:

$$\begin{pmatrix} \phi\left(1,\frac{1}{2},\frac{1}{2},-\frac{1}{2}\right) \\ \phi\left(1,\frac{1}{2},\frac{1}{2},+\frac{1}{2}\right) \\ \phi\left(1,\frac{1}{2},\frac{3}{2},-\frac{3}{2}\right) \\ \phi\left(1,\frac{1}{2},\frac{3}{2},-\frac{1}{2}\right) \\ \phi\left(1,\frac{1}{2},\frac{3}{2},+\frac{1}{2}\right) \\ \phi\left(1,\frac{1}{2},\frac{3}{2},+\frac{3}{2}\right) \end{pmatrix} = \begin{pmatrix} 0 & \frac{1}{\sqrt{3}} & 0 & -\sqrt{\frac{2}{3}} & 0 & 0 \\ 0 & 0 & \sqrt{\frac{2}{3}} & 0 & -\frac{1}{\sqrt{3}} & 0 \\ 1 & 0 & 0 & 0 & 0 & 0 \\ 0 & \sqrt{\frac{2}{3}} & 0 & \frac{1}{\sqrt{3}} & 0 & 0 \\ 0 & 0 & \frac{1}{\sqrt{3}} & 0 & \sqrt{\frac{2}{3}} & 0 \\ 0 & 0 & 0 & 0 & 0 & 1 \end{pmatrix} \begin{pmatrix} \Psi\left(1,\frac{1}{2},-1,-\frac{1}{2}\right) \\ \Psi\left(1,\frac{1}{2},0,-\frac{1}{2}\right) \\ \Psi\left(1,\frac{1}{2},+1,-\frac{1}{2}\right) \\ \Psi\left(1,\frac{1}{2},-1,+\frac{1}{2}\right) \\ \Psi\left(1,\frac{1}{2},0,+\frac{1}{2}\right) \\ \Psi\left(1,\frac{1}{2},+1,+\frac{1}{2}\right) \end{pmatrix}$$

We can see all the 36 Clebsch-Gordan coefficients in the square matrix.

10.12 Calculation of Clebsch-Gordan coefficients for $J_1 = 1/2$ and $J_2 = 1/2$

For a system consisting of two subsystems with angular momenta $\vec{J_1}$ and $\vec{J_2}$, we can define a resultant or total angular momentum $\vec{J}$

$$\vec{J} = \vec{J_1} + \vec{J_2} = \vec{J_2} + \vec{J_1}$$

Corresponding to the three angular momentum vectors $\vec{J_1}$, $\vec{J_2}$ and $\vec{J}$, we have the six Hermitian operators J_1^2, J_{1z}, J_2^2, J_{2z}, $J^2 and\ J_z$. All these six Hermitian operators, however, do not commute among themselves [J_{1z} and J_{2z} do not commute with J^2]. But we can form two sets consisting of four operators each, which, together with the Hamiltonian, form complete obserbles for the system. These are:

$$[a]\ J_1^2,\ J_2^2, J_{1z}, J_{2z} \qquad [b]\ J_1^2,\ J_2^2, J^2, J_z$$

Let us consider the case with $j_1 = \frac{1}{2}$ and $j_2 = \frac{1}{2}$.

For $j_1 = \frac{1}{2}$, $m_1 = -\frac{1}{2}, +\frac{1}{2}$. For $j_2 = \frac{1}{2}$, $m_2 = -\frac{1}{2}, +\frac{1}{2}$.

$\Psi(j_1, j_2, m_1, m_2)$ is the Common Eigen functions of generalized angular momentum and its components J_1^2, J_2^2, J_{1z}, J_{2z} are given by

$$\Psi\left(\frac{1}{2},\frac{1}{2},-\frac{1}{2},-\frac{1}{2}\right) \quad ; \quad \Psi\left(\frac{1}{2},\frac{1}{2},-\frac{1}{2},+\frac{1}{2}\right)$$

$$\Psi\left(\frac{1}{2},\frac{1}{2},+\frac{1}{2},-\frac{1}{2}\right) \quad ; \quad \Psi\left(\frac{1}{2},\frac{1}{2},+\frac{1}{2},+\frac{1}{2}\right)$$

There are 4 Ψ's because total number of allowed Ψ's are

$$(2j_1 + 1)(2j_2 + 1) = \left(2 \times \frac{1}{2} + 1\right)\left(2 \times \frac{1}{2} + 1\right) = 2 \times 2 = 4$$

We have $m_1 = -\frac{1}{2}, +\frac{1}{2}$, $m_2 = -\frac{1}{2}, +\frac{1}{2}$. Thus $m = m_1 + m_2 = -1, 0, +1$.

Allowed values of j are $(j_1 - j_2)$ to $(j_1 + j_2)$

I.e., $\left(\frac{1}{2} - \frac{1}{2}\right)$ to $\left(\frac{1}{2} + \frac{1}{2}\right)$ or, 0 to 1 I.e., $j = 0$ and 1. [in step of +1]

For $j = 0, m = 0$. For $= 1$, $m = -1, 0, +1$.

These values of m are the same as those already obtained using $m = m_1 + m_2$.

$\phi(j_1, j_2, j, m)$ is the Common Eigen functions of generalized angular momentum and its components J_1^2, J_2^2, J^2, J_z are given by

$$\phi\left(\frac{1}{2},\frac{1}{2},0,0\right) \quad ; \quad \phi\left(\frac{1}{2},\frac{1}{2},1,-1\right)$$

$$\phi\left(\frac{1}{2},\frac{1}{2},1,0\right) \quad ; \quad \phi\left(\frac{1}{2},\frac{1}{2},1,+1\right)$$

There are 4 ϕ's because total number of allowed ϕ's are

$(2j_1+1)(2j_2+1) = \left(2\times\frac{1}{2}+1\right)\left(2\times\frac{1}{2}+1\right) = 2\times 2 = 4$, same as the number of Ψ's.

Any of the 4ϕ's can be expressed as linear combination of the 4Ψ's, I.e.,

$$\phi(j_1, j_2, j, m) = \sum_{m_1}\sum_{m_2} C^{j,m}_{m_1,m_2}\Psi(j_1, j_2, j, m_1, m_2) \qquad [1]$$

Where $C^{j,m}_{m_1,m_2}$ and the names, **Clebsch-Gordan, or C- coefficient.**

Since $m_1 + m_2 = m$, the Ψ's for which $m_1 + m_2 \neq m$ cannot be present in RHS of equation [1], I.e., the expansion coefficients involved are zero.Thus

$$\phi\left(\frac{1}{2},\frac{1}{2},0,0\right) = C_1\Psi\left(\frac{1}{2},\frac{1}{2},-\frac{1}{2},+\frac{1}{2}\right) + C_2\Psi\left(\frac{1}{2},\frac{1}{2},+\frac{1}{2},-\frac{1}{2}\right) \qquad [A]$$

$$\phi\left(\frac{1}{2},\frac{1}{2},1,-1\right) = \Psi\left(\frac{1}{2},\frac{1}{2},-\frac{1}{2},-\frac{1}{2}\right) \qquad [B]$$

$$\phi\left(\frac{1}{2},\frac{1}{2},1,0\right) = C_3\Psi\left(\frac{1}{2},\frac{1}{2},-\frac{1}{2},+\frac{1}{2}\right) + C_4\Psi\left(\frac{1}{2},\frac{1}{2},+\frac{1}{2},-\frac{1}{2}\right) \qquad [C]$$

$$\phi\left(\frac{1}{2},\frac{1}{2},1,+1\right) = \Psi\left(\frac{1}{2},\frac{1}{2},+\frac{1}{2},+\frac{1}{2}\right) \qquad [D]$$

There are 16 Clebsch-Gordan coefficients, 4 in each of equation [A] to [D]; only 6 are non-zero and hence have been written, 2 of them taken as 1 because of normalization, and 4 of them taken unknown, to be determined or calculated. The remaining 10 coefficients are zero because $m_1 + m_2 \neq m$ for the Ψ's involved. We shall use

$$J_\pm\phi(j_1, j_2, j, m) = \sqrt{j(j+1) - m(m\pm 1)}\hbar\phi(j_1, j_2, j, m\pm 1)$$

$$J_{1\pm}\Psi(j_1, j_2, m_1, m_2) = \sqrt{j_1(j_1+1) - m_1(m_1\pm 1)}\hbar\Psi(j_1, j_2, m_1\pm 1, m_2)$$

$$J_{2\pm}\Psi(j_1, j_2, m_1, m_2) = \sqrt{j_2(j_2+1) - m_2(m_2\pm 1)}\hbar\Psi(j_1, j_2, j, m_1, m_2\pm 1)$$

From equation [D]

$$J_-\phi\left(j_{1=}\frac{1}{2}, j_2 = \frac{1}{2}, j = 1, m = +1\right) = J_-\Psi\left(j_{1=}\frac{1}{2}, j_2 = \frac{1}{2}, m_1 = +\frac{1}{2}, m_2 = +\frac{1}{2}\right)$$

Or, $$J_-\phi\left(\frac{1}{2},\frac{1}{2},1,+1\right) = J_-\Psi\left(\frac{1}{2},\frac{1}{2}+\frac{1}{2},+\frac{1}{2}\right)$$

Or, $$J_-\phi\left(\frac{1}{2},\frac{1}{2},1,+1\right) = J_{1-}\Psi\left(\frac{1}{2},\frac{1}{2},+\frac{1}{2},+\frac{1}{2}\right) + J_{2-}\Psi\left(\frac{1}{2},\frac{1}{2},+\frac{1}{2},+\frac{1}{2}\right)$$

$$\sqrt{1(1+1) - 1(1-1)}\hbar\phi\left(\frac{1}{2},\frac{1}{2},1,1-1\right) = \sqrt{\frac{1}{2}\left(\frac{1}{2}+1\right) - \frac{1}{2}\left(\frac{1}{2}-1\right)}\hbar\Psi\left(\frac{1}{2},\frac{1}{2},\frac{1}{2}-1,\frac{1}{2}\right)$$

$$+\sqrt{\frac{1}{2}\left(\frac{1}{2}+1\right) - \frac{1}{2}\left(\frac{1}{2}-1\right)}\hbar\Psi\left(\frac{1}{2},\frac{1}{2},\frac{1}{2},\frac{1}{2}-1\right)$$

Or, $\sqrt{2}\hbar\phi\left(\frac{1}{2},\frac{1}{2},1,0\right) = \hbar\Psi\left(\frac{1}{2},\frac{1}{2},-\frac{1}{2},+\frac{1}{2}\right) + \hbar\Psi\left(\frac{1}{2},\frac{1}{2},+\frac{1}{2},-\frac{1}{2}\right)$

Or, $\phi\left(\frac{1}{2},\frac{1}{2},1,0\right)=\frac{1}{\sqrt{2}}\Psi\left(\frac{1}{2},\frac{1}{2},-\frac{1}{2},+\frac{1}{2}\right)+\frac{1}{\sqrt{2}}\Psi\left(\frac{1}{2},\frac{1}{2},+\frac{1}{2},-\frac{1}{2}\right)$ [E]

Comparing equation [E] with equation [C], we get

$$C_3=\frac{1}{\sqrt{2}}\,,C_4=\frac{1}{\sqrt{2}}$$

All the ϕ's are orthonormal. All the Ψ's are also orthonormal. Scalar product of any two ϕ's is zero. Taking scalar product of ϕ's of equation [A] and equation [C], we get

$$\left(\phi\left(\frac{1}{2},\frac{1}{2},0,0\right),\phi\left(\frac{1}{2},\frac{1}{2},1,0\right)\right)$$

$$=\left(C_1\Psi\left(\frac{1}{2},\frac{1}{2},-\frac{1}{2},+\frac{1}{2}\right)+C_2\Psi\left(\frac{1}{2},\frac{1}{2},+\frac{1}{2},-\frac{1}{2}\right)+C_3\Psi\left(\frac{1}{2},\frac{1}{2},-\frac{1}{2},+\frac{1}{2}\right)+C_4\Psi\left(\frac{1}{2},\frac{1}{2},+\frac{1}{2},-\frac{1}{2}\right)\right)$$

$$0=\left(C_1\Psi\left(\frac{1}{2},\frac{1}{2},-\frac{1}{2},+\frac{1}{2}\right)+C_3\Psi\left(\frac{1}{2},\frac{1}{2},-\frac{1}{2},+\frac{1}{2}\right)\right)+\left(C_2\Psi\left(\frac{1}{2},\frac{1}{2},+\frac{1}{2},-\frac{1}{2}\right),C_4\Psi\left(\frac{1}{2},\frac{1}{2},+\frac{1}{2},-\frac{1}{2}\right)\right)$$

$$+\left(C_1\Psi\left(\frac{1}{2},\frac{1}{2},-\frac{1}{2},+\frac{1}{2}\right),C_4\Psi\left(\frac{1}{2},\frac{1}{2},+\frac{1}{2},-\frac{1}{2}\right)\right)+\left(C_2\Psi\left(\frac{1}{2},\frac{1}{2},+\frac{1}{2},-\frac{1}{2}\right),+C_3\Psi\left(\frac{1}{2},\frac{1}{2},-\frac{1}{2},+\frac{1}{2}\right)\right)$$

$$0=C_1^*C_3\left(\Psi\left(\frac{1}{2},\frac{1}{2},-\frac{1}{2},+\frac{1}{2}\right),\Psi\left(\frac{1}{2},\frac{1}{2},-\frac{1}{2},+\frac{1}{2}\right)\right)+C_2^*C_4\left(\Psi\left(\frac{1}{2},\frac{1}{2},+\frac{1}{2},-\frac{1}{2}\right),\Psi\left(\frac{1}{2},\frac{1}{2},+\frac{1}{2},-\frac{1}{2}\right)\right)$$

$$+C_1^*C_4\left(\Psi\left(\frac{1}{2},\frac{1}{2},-\frac{1}{2},+\frac{1}{2}\right),\Psi\left(\frac{1}{2},\frac{1}{2},+\frac{1}{2},-\frac{1}{2}\right)\right)+C_2^*C_3\left(\Psi\left(\frac{1}{2},\frac{1}{2},+\frac{1}{2},-\frac{1}{2}\right),\Psi\left(\frac{1}{2},\frac{1}{2},-\frac{1}{2},+\frac{1}{2}\right)\right)$$

$0=C_1^*C_3(1)+C_2^*C_4(1)+C_1^*C_4(0)+C_2^*C_3(0)$ $\because$ Ψ's are orthonormal

$C_1^*C_3+C_2^*C_4=0$ *Or*, $C_1^*\frac{1}{\sqrt{2}}+C_2^*\frac{1}{\sqrt{2}}=0$ Or, $C_1+C_2=0$ or, $C_2=-C_1$

Equation [A] gives

$$\phi\left(\frac{1}{2},\frac{1}{2},0,0\right)=C_1\Psi\left(\frac{1}{2},\frac{1}{2},-\frac{1}{2},+\frac{1}{2}\right)-C_1\Psi\left(\frac{1}{2},\frac{1}{2},+\frac{1}{2},-\frac{1}{2}\right)$$

Normalization of the ϕ requires

$$|C_1|^2+|-C_1|^2=1$$

because for $\phi=\sum_i C_i\Psi_i$, we can write $\sum_i |C_i|^2=1$

$$2|C_1|^2=1 \quad \text{or,}\ |C_1|^2=\frac{1}{2}$$

or, $C_1=\frac{1}{\sqrt{2}}e^{i\delta}$, where δ is real. or, $C_1=\frac{1}{\sqrt{2}}$ taking $\delta=0$.

Then $C_2=-C_1$ gives $C_2=-\frac{1}{\sqrt{2}}$ Thus $C_1=\frac{1}{\sqrt{2}}\,,C_2=\frac{1}{\sqrt{2}}$.

We now write equations [A] to [D] in matrix form as

$$\begin{pmatrix} \phi\begin{pmatrix}\frac{1}{2} & \frac{1}{2} & 0 & 0\end{pmatrix} \\ \phi\begin{pmatrix}\frac{1}{2} & \frac{1}{2} & 1 & -1\end{pmatrix} \\ \phi\begin{pmatrix}\frac{1}{2} & \frac{1}{2} & 1 & 0\end{pmatrix} \\ \phi\begin{pmatrix}\frac{1}{2} & \frac{1}{2} & 1 & +1\end{pmatrix} \end{pmatrix} = \begin{pmatrix} 0 & 1/\sqrt{2} & -1/\sqrt{2} & 0 \\ 1 & 0 & 0 & 0 \\ 0 & 1/\sqrt{2} & 1/\sqrt{2} & 0 \\ 0 & 0 & 0 & 1 \end{pmatrix} \begin{pmatrix} \Psi\left(\frac{1}{2},\frac{1}{2},-\frac{1}{2},-\frac{1}{2}\right) \\ \Psi\left(\frac{1}{2},\frac{1}{2},-\frac{1}{2},+\frac{1}{2}\right) \\ \Psi\left(\frac{1}{2},\frac{1}{2},+\frac{1}{2},-\frac{1}{2}\right) \\ \Psi\left(\frac{1}{2},\frac{1}{2},+\frac{1}{2},+\frac{1}{2}\right) \end{pmatrix}$$

We can see all the 16 Clebsch-Gordan coefficients in the square matrix.

10.13 The Selection Rules

For given values of j_1, j_2, m_1 and m_2, $C^{j_1 j_2 j}_{m_1 m_2 m}$ is non-zero only for certain allowed values of j and m. The rules specifying these allowed values are referred to as selection rules. These are:

Selection rule 1: $m = m_1 + m_2$;

Selection rule 2 : $|j_1 - j_2| \leq j \leq (j_1 + j_2)$, where, j varies by integer steps.

Selection rule 1: Follows from the relation $\hat{j}_z = \hat{j}_{1z} + \hat{j}_{2z}$.

For, $\hat{j}_z|jm> = (j_{1z} + j_{2z})|jm>$

$$= \sum_{m_1 m_2} C^{j_1 j_2 j}_{m_1 m_2\, m} (j_{1z} + j_{2z})|j_1 j_2 m_1 m_2>$$

$$= \sum_{m_1 m_2} (m_1 + m_2)\hbar\, C^{j_1 j_2 j}_{m_1 m_2 m}\, |j_1 j_2 m_1 m_2>$$

I.e., $\sum_{m_1 m_2} (m - m_1 - m_2) C^{j_1 j_2 j}_{m_1 m_2 m}\, |j_1 j_2 m_1 m_2> = 0$

Since the vectors $|j_1 j_2 m_1 m_2>$ are linearly-independent, this implies that,

Either, $C^{j_1 j_2 j}_{m_1 m_2 m} = 0, \quad m \neq (m_1 + m_2)$,

Or, $C^{j_1 j_2 j}_{m_1 m_2 m} = 0, \quad m = (m_1 + m_2)$,

Selection rule 2 may be derived as follows:

Maximum value of m

$$m = m_{max} \equiv (m_1 + m_2)_{max} = (j_1 + j_2)\,.$$

Thus, maximum value of j

$$j \equiv j_{max} \equiv m_{max} = (j_1 + j_2)\,.$$

The next lower value of

$$m = j_1 + j_2 - 1.$$

There are two states with this value of m, namely, $|m_1m_2> \equiv |j_1j_2 - 1>$ and $|j_1 - 1j_2>$. One of these belongs to $j = (j_1 + j_2)$ and the other to $j = j_1 + j_2 - 1$.

Similarly, there are three states with $m = j_1 + j_2 - 2$, the corresponding j- values being $(j_1 + j_2), (j_1 + j_2 - 1)$ and $j_1 + j_2 - 2$. In general, there are $(p + 1)$ states with $m = j_1 + j_2 - p$, the j- values being $(j_1 + j_2), (j_1 + j_2 - 1), \dots, (j_1 + j_2 - p)$.

The maximum value of p is given by

$$(j_2 - p_{max}) = (m_2)_{min} = -j_2, \quad if\ j_2 < j_1,$$

And by, $$(j_1 - p_{max}) = (m_1)_{min} = -j_1, \quad if\ j_1 < j_2.$$

That is, $p_{max} = 2_{j<}$, where $j_<$ is the lesser j_i.

Thus, $$j_{min} = j_1 + j_2 - p_{max}$$
$$= (j_1 - j_2), \quad \text{if} \quad j_1 > j_2$$
$$= (j_2 - j_1), \quad \text{if} \quad j_2 > j_1$$

Or, $$j_{min} = |j_1 - j_2|,$$

and, $$j = |j_1 - j_2|, |j_1 - j_2| + 1, \dots \dots \dots \dots \dots \dots \dots \dots \dots \dots \dots, (j_1 + j_2).$$

This method of obtaining the allowed values of j is illustrated in the following table for the case $j_1 = \frac{3}{2},\ j_2 = \frac{5}{2}$.

<u>Table</u>

j – values corresponding $j_1 = \frac{3}{2},\ j_2 = \frac{5}{2}. [p_{max} = 2_{j_1} = 3]$

m	m_1	m_2	j
$j_1 + j_2 = 4$	$j_1 = 3/2$	$j_2 = 5/2$	$m = 4$
$j_1 + j_2 - 1 = 3$	$j_1 = 3/2$ $j_1 - 1 = 1/2$	$j_2 - 1 = 3/2$ $j_2 = 5/2$	$j_1 + j_2 = 4;$ $j_1 + j_2 - 1 = 3;$
$j_1 + j_2 - 2 = 2$	$j_1 = 3/2$ $j_1 - 1 = 1/2$ $j_1 - 2 = -1/2$	$j_2 - 2 = 1/2$ $j_2 - 1 = 3/2$ $j_2 = 5/2$	$j_1 + j_2 = 4;$ $j_1 + j_2 - 1 = 3;$ $j_1 + j_2 - 2 = 2;$
$j_1 + j_2 - 3 = 1$	$j_1 = 3/2$ $j_1 - 1 = 1/2$ $j_1 - 2 = -1/2$ $j_1 - 3 = -3/2$	$j_2 - 3 = -1/2$ $j_2 - 2 = 1/2$ $j_2 - 1 = 3/2$ $j_2 = 5/2$	$j_1 + j_2 = 4;$ $j_1 + j_2 - 1 = 3;$ $j_1 + j_2 - 2 = 2;$ $j_1 + j_2 - 3 = 1;$

Exercise

1. Define generalized angular momentum operator.
2. Show that the components of the generalized angular momentum operators satisfy the following commutation relations:

$$a.\left[J_x, J_y\right] = i\hbar J_z \quad b.\left[J_y, J_z\right] = i\hbar J_x \quad c.\left[J_z, J_x\right] = i\hbar J_y$$

$$d.\ \vec{J} \times \vec{J} = i\hbar\, \vec{J} \qquad e.\left[J^2, J_x\right] = \left[J^2, J_x\right] = \left[J^2, J_z\right] = 0$$

3. Show that the components of the spin angular momentum operators satisfy the following commutation relations:

$$\left[S_x, S_y\right] = i\hbar S_z \quad \left[S_y, S_z\right] = i\hbar S_x \quad \& \quad \left[S_z, S_x\right] = i\hbar S_y$$

4. Define the Raising and Lowering operators of generalized angular momentum.
5. Write down Some properties of ladder operators J_+ [Raising] and J_-[Lowering].
6. Establish the commutation relations of Ladder operators [raising and lowering operators] and the product and add of raising and lowering operators:

$$a.\left[J_+, J_-\right] = 2\hbar J_z \quad b.\left[J_-, J_+\right] = \hbar J_- \quad c.\ \left[J_+, J_z\right] = -\hbar J_+ \quad d.\ \left[J_z, J_\pm\right] = \hbar J_\pm$$

$$e.\ \left[J^2, J_-\right] = 0 \qquad f.\ \left[J^2, J_+\right] = 0 \qquad g.\ J_+ J_- = J^2 - J_z^2 + \hbar J_Z$$

$$h.\ J_- J_+ = J^2 - J_z^2 - \hbar J_Z \quad i.\ J_+ J_- + J_- J_+ = 2\left(J_x^{\,2} + J_y^2\right)$$

7. Derive Ladder method.
8. Write the introduction to addition of generalized angular momentum and also calculate the addition of two generalized angular momenta.
9. To determine the common Eigen functions of J_1^2, J_2^2, J^2 and J_Z.
10. Define Clebsch–Gordan (CG) coefficients? What is the significance of Clebsch–Gordan (CG) coefficients.
11. Calculate the Calculation of Clebsch-Gordan coefficients for $J_1 = 1$ and $J_2 = 1/2$.
12. Calculate the Calculation of Clebsch-Gordan coefficients for $J_1 = 1/2$ and $J_2 = 1/2$.
13. Discuss the Selection Rules.
14. Explain the selection rules for simple harmonic oscillator.
15. For a state for which J_Z and J^2 are well- defined, show that

$$\langle J_x \rangle = \langle J_y \rangle = 0$$

$$\langle J_x^2 \rangle = \langle J_y^2 \rangle = \frac{1}{2}h^2(j(j+1) - m^2)$$

Chapter 11 Identical Particles and Spin

11.1 Identical Particles

If a physical system contains two identical particles [all electrons are identical, all protons are identical, all hydrogen atoms are identical] or more identical particles, there is no change in any of its observable quantities or properties or its evolution if we interchange any two identical particles of the system or more precisely, if roles of any two identical particles of the system are interchanged.

Two particles are said to be identical if all their intrinsic properties like mass, charge, spin angular momentum quantum number are exactly the same.

11.2 Spins and Statistics

We have seen that the symmetry character of a system of identical particles is a constant of motion. It is also ground that a given type of particle is associated with only one type of symmetry. Thus a system of electrons is always described by anti-symmetric wave function while a system of Pions is invariably described by a symmetric wave function. That is, *the symmetry character of the wave function is an intrinsic property of the particles.*

Now, the statistical properties of a system of a large number of particles depend on the available degrees of freedom for each value of the total energy of the system. For a system of distinguishable particles, every permutation of the particle gives rise to a different state. Such a system obeys classical of Boltzmann statistics. The statistics *obeyed by a system of identical particle with symmetric wave function is known as Bose-Einstein statistics* while *identical particle with anti-symmetric wave functions obey Fermi-Dirac statistics.* Since the symmetry character of the wave function is an intrinsic property of the particles, it follow that the statistics associated with a particle is also of an intrinsic nature. This fact justifies the classification of particles on the statistics obeyed by them. *Thus, particles obeying Bose-Einstein statistics are called Bosons while those obeying Fermi-Dirac Statistics are referred to as fermions.*

It is further found, and shown plausible in quantum field theory, that the statistics is intimately connected with the spin of the spin of the particles. *Bosons have integral (including zero) spin whereas particles with half-integral spin are fermions.* This correlation between spin and statistics applies not only to elementary particles but to composite particles (such as atoms and nuclei) as well. This understandable since a composite particle composed of fermions will have integral or half –integral spin according as the number of fermions is even or odd. In the former case, an interchange between two such particles is equivalent to an even number of interchanges of Fermions so that the wave function should be symmetric. Similarly, in the case the wave function should be anti-symmetric.

11.3 Symmetric and Anti-symmetric wave functions

Let us consider a system like hydrogen atom containing two identical particles. The complete wave function Ψ of the electron in a hydrogen atom can be expressed as the product of three separate wave functions, each describing that part of Ψ which is a function of one of the three coordinates r, θ, ϕ *i.e.*

$$\psi_{nlm}(r,\theta,\phi) = R(r)\Theta(\theta)\Phi(\phi) \quad [1]$$

$$\psi_{nlm}(r,\theta,\phi) = R(r)Y_{lm}(\theta,\Phi) \quad [2]$$

- The function $R(r)$[radial wave function] describes how the eigen function $\psi_{nlm}(r,\theta,\phi)$ varies along the line joining the electron and the proton, for given values of θ *and* ϕ
- The function $\Theta(\theta)$[zenith part] describes how the eigen function $\psi_{nlm}(r,\theta,\phi)$ varies with the zenith angle θ along a meridian on a sphere of radius r centered at the proton, for given values of r *and* ϕ
- The function $\Phi(\phi)$[azimuthal wave function] describes how the eigen function $\psi_{nlm}(r,\theta,\phi)$ varies with the azimuthal angle ϕ on a sphere of radius r centered at the proton, for given values of r *and* θ

The complete wave function $\psi(1,2,3,4,\dots\dots\dots\dots\dots\dots\dots\dots\dots\dots..,n)$ of a system of n particles in the same force field can be expressed as the product of the wave functions $\psi(1),\psi(2),\psi(3),\psi(4),\dots\dots\dots\dots\dots,\psi(n)$ of the individual particles. I.e.,

$$\psi(1,2,3,4,\dots\dots\dots,n) = \psi(1),\psi(2),\psi(3),\psi(4),\dots\dots\dots,\psi(n) \quad [3]$$

Let us suppose that one of the particles is in quantum state say M and other in state N. Because the particles are identical, it should make no difference in the probability density $\rho = \Psi\Psi^{\times} = |\Psi|^2$ of the system if the particles are exchanged, with the one in state M replacing the one in state N and vice-versa. Symbolically, we require that

$$|\Psi|^2(1,2) = |\Psi|^2(2,1) \quad [4]$$

Hence the wave function $\Psi(2,1)$, representing the exchanged particles, can be given by

$$\Psi(2,1) = \Psi(1,2) \quad \dots\dots\dots\dots\dots\dots\dots\dots\dots\dots\dots\dots\dots\dots\dots\dots \quad [5]$$ Symmetric

or, $$\Psi(2,1) = -\Psi(1,2) \quad \dots\dots\dots\dots\dots\dots\dots\dots\dots\dots\dots\dots\dots\dots\dots\dots \quad [6]$$ Anti-symmetric

The wave function of the system is not itself a measurable quantity, and so it can be altered in sign by the exchange of the particles. So, Wave functions unaffected by an exchange of particles are said to be symmetric, while those reversing sign upon such an exchange are said to be anti-symmetric. If particle 1 is in state M and particle 2 in state N, the wave function of the system can be written according to equation [3]

$$\Psi_I = \Psi_M(1)\Psi_N(2) \qquad [7]$$

While if particle 2 is in state M and particle 1 in state N, the wave function is

$$\Psi_{II} = \Psi_M(2)\Psi_N(1) \qquad [8]$$

We can say that the system spends the half time in the configuration whose wave function is Ψ_I and the other half in the configuration whose wave function is Ψ_{II}. Therefore, a linear combination of Ψ_I and Ψ_{II} is the proper description of the system. There are two such combinations possible, the symmetric one

$$\Psi_{Symmetric} = \frac{1}{\sqrt{2}}[\Psi_M(1)\Psi_N(2) + \Psi_M(2)\Psi_N(1)] \qquad [9]$$

And the anti-symmetric one

$$\Psi_{Anti-symmetric} = \frac{1}{\sqrt{2}}[\Psi_M(1)\Psi_N(2) - \Psi_M(2)\Psi_N(1)] \qquad [10]$$

The factor $\sqrt{2}$ is required to normalize $\Psi_{Symmetric}$ and $\Psi_{Anti\text{-}symmetric}$. Both Ψ_{Sy} and Ψ_{Asy} obey equation [4]

11.4 Construction of symmetric and anti-symmetric wave function

$\Psi(1,2)$ & $\Psi(2,1)$ are called *un-symmetrized* wave functions of an observable A of a system containing two identical particles. We now wish to construct symmetric and anti-symmetric wave function. Let A_{op} be the corresponding operator. Both $A(1,2)$ & $A_{op}(1,2)$ do not change at all if we interchange roles of the two identical particles:

$$A(1,2) = A(2,1), \quad A_{op}(1,2) = A_{op}(2,1)$$

Any observable must be symmetric function of the variables of the two identical particles. Example: classical Hamiltonian function of Helium atom is

$$H(1,2) = \frac{p_1^2}{2m} + \frac{p_2^2}{2m} + \frac{-2e^2}{r_1} + \frac{-2e^2}{r_2} + \frac{e^2}{|\vec{r_1} - \vec{r_2}|} \qquad [1]$$

And

$$H(2,1) = \frac{p_2^2}{2m} + \frac{p_1^2}{2m} + \frac{-2e^2}{r_2} + \frac{-2e^2}{r_1} + \frac{e^2}{|\vec{r_2} - \vec{r_1}|} \qquad [2]$$

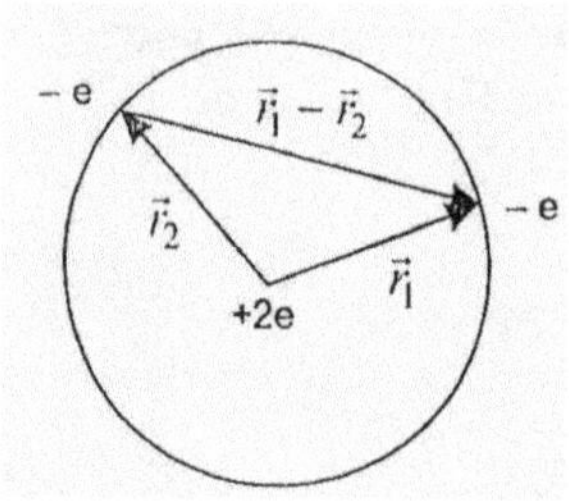

Comparing equations [1] and [2], we get $H(1,2) = H(2,1)$. And Hamiltonian operator is

$$H_{op}(1,2) = -\frac{\hbar^2}{2m}\nabla_1^2 - \frac{\hbar^2}{2m}\nabla_2^2 + \frac{-2e^2}{r_1} + \frac{-2e^2}{r_2} + \frac{e^2}{|\vec{r_1}-\vec{r_2}|} \quad [3]$$

And $$H_{op}(2,1) = -\frac{\hbar^2}{2m}\nabla_2^2 - \frac{\hbar^2}{2m}\nabla_1^2 + \frac{-2e^2}{r_2} + \frac{-2e^2}{r_1} + \frac{e^2}{|\vec{r_2}-\vec{r_1}|} \quad [4]$$

Comparing equations [3] and [4], we get

$$H_{op}(1,2) = H_{op}(2,1)$$

Let us write Eigen value equation as

$$A_{op}(1,2)\Psi(1,2) = a\,\Psi(1,2) \quad [5]$$

We can also write $$A_{op}(2,1)\Psi(2,1) = a\Psi(2,1)$$

Because, Eigen value remains unchanged in particle interchange.

Or, $$A_{op}(1,2)\Psi(2,1) = a\Psi(2,1) \text{ [Because } A_{op}(1,2) = A_{op}(2,1)] \quad [6]$$

Equations [5] and [6] show that $\Psi(1,2)$ and $\Psi(2,1)$ are degenerate Eigen functions of operator $A_{op}(1,2)$ corresponding to (same) Eigen value *a. This degeneracy is called exchange degeneracy*. Hence any linear combination of $\Psi(1,2)$ and $\Psi(2,1)$ is also an Eigen function of $A_{op}(1,2)$ belonging to the same Eigen value *a*.

$$\Psi_{symmetric} = \Psi(1,2) + \Psi(2,1) \quad [7]$$

And $$\Psi_{anti-symmetric} = \Psi(1,2) - \Psi(2,1) \quad [8]$$

Equation [7] & [8] are two linear combinations of particular interest. These are symmetric and anti-symmetric wave functions respectively with respect to interchange of roles of two identical particles. Equation [7] is symmetric, because interchange of roles of two identical particles changes $\Psi(1,2) + \Psi(2,1)$ to $\Psi(2,1) + \Psi(1,2)$ which are the same. Equation [8] is anti-symmetric, because interchange of roles of two identical particles changes $\Psi(1,2) - \Psi(2,1)$ to $\Psi(2,1) - \Psi(1,2) = -[\Psi(1,2) - \Psi(2,1)]$ which are opposite in sign. Thus using equation [7] and

[8] we can construct symmetric and anti-symmetric wave function if we have un-symmetrized wave functions $\Psi(1,2)$ and $\Psi(2,1)$. If un-symmetrized wave functions $\Psi(1,2)$ and $\Psi(2,1)$ are same, equation [8] shows that we cannot construct non-zero anti-symmetric wave function. $A_{op}\Psi_{sy}$ is symmetric and $A_{op}\Psi_{ansy}$ is anti-symmetric because A_{op} is symmetric with respect to particle interchange.

11.5 Construction of symmetric and anti-symmetric wave function for a system containing 3 identical particles

Let $\Psi(1,2,3)$ be the wave function of an observable A for a system containing 3 identical particles. By permuting identical particles, the following 6 un-symmetrized wave functions can be obtained I.e., $\Psi_1(1,2,3)$, $\Psi_2(1,3,2)$, $\Psi_3(2,1,3)$, $\Psi_4(2,3,1)$, $\Psi_5(3,1,2)$, $\Psi_6(3,2,1)$. Of the 6 wave functions, Ψ_2, Ψ_3, Ψ_6 result from original wave function $\Psi(=\Psi_1)$ by odd number (1) of permutation or particle interchange. The other three wave funtions Ψ_1, Ψ_4, Ψ_5 result from original wave function $\Psi(=\Psi_1)$ by even number (2) of permutations or particle interchanges.

Symmetric wave function Ψ_{sy} can be constructed as sum of all the 6 un-symmetrized wave functions I.e., $\Psi_{sy} = \Psi_1(1,2,3), \Psi_2(1,3,2), \Psi_3(2,1,3), \Psi_4(2,3,1), \Psi_5(3,1,2), \Psi_6(3,2,1)$. This is because interchange of any two identical particles of the system changes each one of the 6 Ψ's to another of them and the latter to the former so that the sum remains unchanged. Anti-symmetric wave function Ψ_{ansy} can be constructed as $\Psi_{asy} = (\Psi_2 + \Psi_3 + \Psi_6) - (\Psi_1 + \Psi_4 + \Psi_5)$ because interchange of any two identical particles changes $(\Psi_2 + \Psi_3 + \Psi_6)$ to $(\Psi_1 + \Psi_4 + \Psi_5)$ and $(\Psi_1 + \Psi_4 + \Psi_5)$ to $(\Psi_2 + \Psi_3 + \Psi_6)$. So that $(\Psi_2 + \Psi_3 + \Psi_6) - (\Psi_1 + \Psi_4 + \Psi_5)$ becomes $(\Psi_1 + \Psi_4 + \Psi_5) - (\Psi_2 + \Psi_3 + \Psi_6) = -[(\Psi_2 + \Psi_3 + \Psi_6) - (\Psi_1 + \Psi_4 + \Psi_5)]$, *I.e.,* $\Psi_{asy} = (\Psi_2 + \Psi_3 + \Psi_6) - (\Psi_1 + \Psi_4 + \Psi_5)$ changes sign.

11.6 Construction of symmetric and anti-symmetric wave function for a system containing n identical particles

Let $\Psi(1, 2, 3, \ldots\ldots\ldots, N)$ be the wave function of an observable A of a system containing N identical particles. By permuting identical particles, we can construct $N!$ un-symmetrized wave functions in total. Symmetric wave function Ψ_{sy} can be constructed as sum of all the $N!$ wave functions, because interchage of any two identical particles of the system changes one of the $N!$ Ψ's to another of them and the latter to the former. This leaves the sum unchanged.

Anti-symmetric wave function Ψ_{asy} can be constructed by adding together all permuted Ψ's among the $N!$. Ψ's that arise from original $\Psi(1,2,3,...,N)$ by means of an even number of interchanges of particles and subtracting from the sum the Ψ's among N!. Ψ's that arise from original $\Psi(1,2,3,...,N)$ by means of an odd number of interchange(s) of identical particles. Non-zero anti-symmetric wave function cannot be constructed if $\Psi(1,2,3,...,N)$ is unaltered by interchange of roles of any two identical particles.

11.7 The quantum mechanical property of an elementary particle: A brief description of spin

In an effort to account for both fine structure in spectral lines and the anomalous Zeeman Effect, S.A. Goudsmit and G.E. Uhlenbeck proposed in 1925 that the electron possesses an intrinsic angular momentum independent of any orbital angular momentum it might have and, associated with this angular momentum, a certain magnetic moment[22,23]. They had in mind a classical picture of an electron as a charged sphere spinning on its axis. The rotation involves angular momentum, and because the electron is negatively charged. It has a magnetic moment $\bar{\mu}_S$ opposite in direction to its angular momentum vector $\bar{L}_S$. Of course, the idea that electrons are spinning charged spheres is hardly in accord with quantum mechanics. In 1928, Dirac was able to show on the basis of a relativistic quantum-theoretical treatment that particles having the charge and mass of the electron must have just the intrinsic angular momentum and magnetic moment. Therefore, Spin can be uniquely defined as *"Spin is a quantum mechanical property of an elementary particle by virtue of which it shows magnetic dipole moment"*. For two bodies system like hydrogen atom the magnetic dipole moment can be is defined as[30]

$$\bar{\mu} = \gamma \bar{L} \qquad and \quad \gamma = \frac{e}{2m} \qquad [1]$$

Where γ the gyro- magnetic ratio, e is the charge of the electron and m is the mass of the electron.

11.8 Illustrate/Interpret Intrinsic Spin

- Samuel Goudsmit and George Uhlenbeck in Holland proposed that *the electron must have an intrinsic angular momentum* and therefore a magnetic moment
- Paul Ehrenfest showed that the surface of the spinning electron should be moving faster than the speed of light
- In order to explain experimental data, Goudsmit and Uhlenbeck proposed that the electron must have an intrinsic spin quantum number $s = \frac{1}{2}$

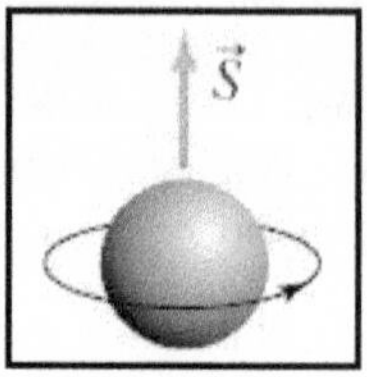

- The spinning electron reacts similarly to the orbiting electron in a magnetic field
- The magnetic spin quantum number m_s has only two values, $m_s = \pm\frac{1}{2}$
- The electron's spin will be either "up" or "down" and can never be spinning with its magnetic moment $\bar{\mu}_s$ exactly along the z axis
- The intrinsic spin angular momentum vector

$$|\bar{S}| = \sqrt{s(s+1)}\hbar = \sqrt{3/4\hbar}$$

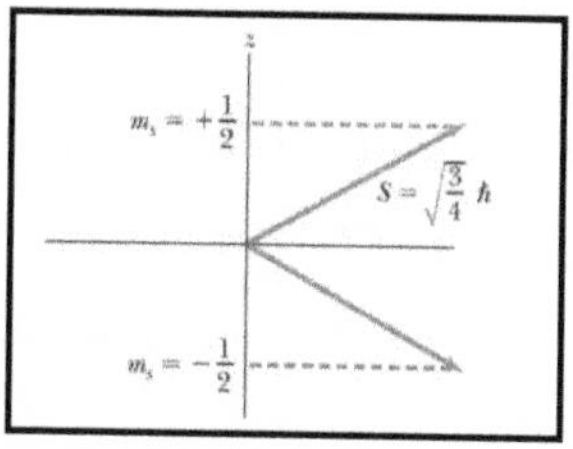

- The magnetic moment is $\bar{\mu}_s = -(e/m)\bar{S},\ or, -2\mu_B\,\bar{S}/\hbar$
- The coefficient of $\bar{S}/\hbar$ is $2\mu_B$ as with $\bar{L}$ is a consequence of theory of relativity
- The z component of $\bar{S}\ is\ S_z = m_s\hbar = \pm\frac{\hbar}{2}$
- The potential energy is quantized due to the magnetic quantum number m_l

11.9 Spin $\frac{1}{2}$ Particles and Spin Zero Particles

The spin part of the wave function could be a triplet or a singlet state depending on whether the spins of two particles are parallel or anti-parallel. The singlet state should be associated with the symmetric space function so that the corresponding cross-section is given by

$$\sigma_{\frac{1}{2}}^{1}(\theta) = \sigma_S(\theta).$$

Similarly, the triplet state is associated with the anti-symmetric space function. Therefore,

$$\sigma_{\frac{1}{2}}^{3}(\theta) = \sigma_A(\theta)\ .$$

The observed cross-section would be the weighted average of these two cases. That is,

$$\sigma_{\frac{1}{2}}^{1}(\theta) = \frac{1}{4}\sigma_S(\theta) + \frac{3}{4}\sigma_A(\theta)$$

$$= |f_k(\theta)|^2 + |f_k(\pi-\theta)|^2 - Re\,\{f_k^{\times}(\theta).f_k(\pi-\theta)\}$$

For $\theta = \frac{\pi}{2}$, this gives,

$$\sigma_{\frac{1}{2}}\left(\frac{\pi}{2}\right) = \left|f_k\left(\frac{\pi}{2}\right)\right|^2.$$

In this case the total wave function is symmetric so that,

$$\sigma_0(\theta) = \sigma_S(\theta),$$

And, $$\sigma_0\left(\frac{\pi}{2}\right) = \sigma_S\left(\frac{\pi}{2}\right) = 4\left|f_k\left(\frac{\pi}{2}\right)\right|^2$$

Also from Eq. ()

$$\sigma_{cl}\left(\frac{\pi}{2}\right) = 2\left|f_k\left(\frac{\pi}{2}\right)\right|^2$$

Thus, we have the result,

$$\sigma_0\left(\frac{\pi}{2}\right) : \sigma_{cl}\left(\frac{\pi}{2}\right) : \sigma_{\frac{1}{2}}\left(\frac{\pi}{2}\right) = 4:2:1$$

The result can be utilized to determine experimentally the spin of a particle like the Proton. In this case, the classical cross section is given by the Rutherford formula. The fact that the observed cross section at $\theta = \frac{\pi}{2}$ is approximately half of the value given by the Rutherford formula, indicates that the proton is a spin $\frac{1}{2}$ Particles.

11.10 Arbitrary spin state: Concept of spin up and spin down

Let us consider a particle having $j = 1/2$ and we do not know if it is in spin up state Ψ_1 or in spin down state Ψ_2. Thus we can write its state Ψ as

$$\Psi = c_1\Psi_1 + c_2\Psi_2 \quad [1]$$

$$= c_1\begin{pmatrix}1\\0\end{pmatrix} + c_2\begin{pmatrix}0\\1\end{pmatrix}$$

Where $\begin{pmatrix}1\\0\end{pmatrix}$ and $\begin{pmatrix}0\\1\end{pmatrix}$ are Eigen vectors of J_z. The Eigen values are $+\frac{1}{2}\hbar$ and $-\frac{1}{2}\hbar$ respectively.

Thus $$\Psi = \begin{pmatrix}c_1\\0\end{pmatrix} + \begin{pmatrix}0\\c_2\end{pmatrix} = \begin{pmatrix}c_1\\c_2\end{pmatrix}$$

From the normalization condition requires

$$(c_1^* \;\; c_2^*)\begin{pmatrix}c_1\\c_2\end{pmatrix} = 1,$$

Or, $$|c_1|^2 + |c_2|^2 = 1 \text{ as expected.}$$

$|c_1|^2$; is the probability that J_z is$+\frac{1}{2}\hbar$, I.e., the particle is in spin up state and $|c_2|^2$; is the probability that J_z is$-\frac{1}{2}\hbar$, I.e., the particle is in spin down state.

$$\langle J_z\rangle = (\Psi, J_z\Psi)$$

$$= (c_1^\times \;\; c_2^\times)\frac{1}{2}\hbar\begin{pmatrix}1 & 0\\0 & -1\end{pmatrix}\begin{pmatrix}c_1\\c_2\end{pmatrix} = \frac{1}{2}\hbar(c_1^\times \;\; c_2^\times)\begin{pmatrix}1 & 0\\0 & -1\end{pmatrix}\begin{pmatrix}c_1\\c_2\end{pmatrix}$$

$$= \frac{1}{2}\hbar(c_1^\times \;\; c_2^\times)\begin{pmatrix}c_1\\-c_2\end{pmatrix} = \frac{1}{2}\hbar(c_1^\times c_1 - c_2^\times c_2)$$

$$= \frac{1}{2}\hbar|c_1|^2 + \left(-\frac{1}{2}\hbar\right)|c_2|^2$$

$$\langle J_z\rangle = \left(\frac{1}{2}\hbar\right)(probability\ of\ spin\ up) + \left(-\frac{1}{2}\hbar\right)(probability\ of\ spin\ down).$$

This is a weighted average. Again,

$$\langle J^2\rangle = (\Psi, J^2\Psi)$$

$$= (c_1^\times \quad c_2^\times)\frac{3}{4}\hbar^2\begin{pmatrix}1 & 0\\0 & 1\end{pmatrix}\begin{pmatrix}c_1\\c_2\end{pmatrix} = \frac{3}{4}\hbar^2(c_1^\times \quad c_2^\times)\begin{pmatrix}1 & 0\\0 & 1\end{pmatrix}\begin{pmatrix}c_1\\c_2\end{pmatrix}$$

$$= \frac{3}{4}\hbar^2(c_1^\times \quad c_2^\times)\begin{pmatrix}c_1\\c_2\end{pmatrix} = \frac{3}{4}\hbar^2(c_1^\times c_1 - c_2^\times c_2)$$

$$= \frac{3}{4}\hbar^2(|c_1|^2 + |c_2|^2) = \frac{3}{4}\hbar^2(1) = \frac{3}{4}\hbar^2$$

This is because, $\frac{3}{4}\hbar^2$ is the only possible value of J^2 for $j = \frac{1}{2}$.

11.11 The Pauli Exclusion Principle: Basic concept

In 1925 Wolfgang Pauli discovered the fundamental principle that governs the electronic configurations of atoms having more than one electron. Each electron must have a different set of the quantum numbers like principal quantum number (n), orbital quantum number (l), magnetic quantum number (m_l), and spin quantum number (m_s). Pauli was led to this conclusion from a study of atomic spectra. It is impossible to determine the various states of an atom from its spectrum, and the quantum numbers of these states can be inferred. *Pauli Exclusion Principle states that "no two electrons in an atom can exist in the same quantum state."*[22]

In the absent state the quantum numbers of both electrons would be $n = 1,\ l = 0,\ m_l = 0,$ $m_s = \frac{1}{2}$; while in the state known to exist one of the electrons has $m_s = \frac{1}{2}$ and the other $m_s = -\frac{1}{2}$. Pauli showed that every unobserved atomic state involves two or more electrons with identical quantum numbers, and the exclusion principle is a statement of this empirical finding.

11.12 The Pauli Exclusion Principle: Mathematical ground

We have seen that the wave function of a system of spin-half particle is anti-symmetric. Consider a system of two spin-half particles. If $\Phi_{a_1}$and Φ_{a_2} denote the two quantum states available to the particles, then, $\Psi(1,2) = \Phi_{a_1}(1)\Phi_{a_2}(2)$, so that the wave function of the system is given by

$$\Psi_A(1,2) = \frac{1}{\sqrt{2!}}(\hat{1} - \hat{P}_{12})\Phi_{a_1}(1)\Phi_{a_1}(2)$$

$$= \frac{1}{\sqrt{2!}}\begin{vmatrix}\Phi_{a_1}(1) & \Phi_{a_1}(2)\\ \Phi_{a_1}(1) & \Phi_{a_1}(2)\end{vmatrix} \qquad [1]$$

Obviously, $\Psi_A(1,2)$ vanishes when $\Phi_{a_1} = \Phi_{a_2}$. That is, it is not possible for the two particles to occupy the same state. This is known as the Pauli Exclusion Principle.

Equation [1] is easily generalized to the case of n particles. In this care, the wave function will be a linear superposition of the n! functions that correspond to the n! permutation of the particles:

$$\Psi_A(1,2,\dots,n) = \frac{1}{\sqrt{n!}}\sum_p(-1)^{\pi(P)}\,\hat{P}\Phi_{a_1}(1)\Phi_{a_1}(2)\dots\Phi_{a_n}(n)$$

$$= \frac{1}{\sqrt{n!}}\begin{vmatrix}\Phi_{a_1}(1) & \Phi_{a_1}(2) & \dots & \Phi_{a_1}(n)\\ \Phi_{a_2}(n) & \Phi_{a_2}(2) & \dots & \Phi_{a_2}(n)\\ \Phi_{a_n}(1) & \Phi_{a_n}(1) & \dots & \Phi_{a_n}(n)\end{vmatrix} \qquad [2]$$

Here, $\hat{P}$ represents one of the n! permutations and $\pi(P)$ = the number of two particle exchange operators contained in $\hat{P}$. The determinant in [2] is known as the Slater determinant. Since the determinant vanishes when any two rows are identical, the exclusion principle follows.
For a system of bosons, on the other hand, the wave function is given by,

$$\Psi_S(1,2,\dots,n) = \frac{1}{\sqrt{n!\delta}}\sum_p\hat{P}\,\Phi_{a_1}(1)\Phi_{a_1}(2)\dots\Phi_{a_n}(n) \qquad [3]$$

Where δ is a factor which depends on the number of particles occupying the same state. Equation [3] does not vanish when any two α_i are equal. Thus, there is no exclusion principle for Bosons.

11.13 Spin up and spin down states

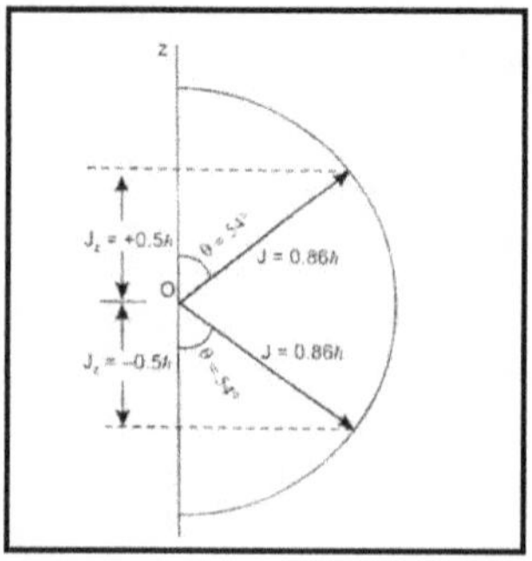

For $j = \frac{1}{2}$, we have $m = +\frac{1}{2}$ and $-\frac{1}{2}$. Thus $J_z = m\hbar = \pm\frac{1}{2}\hbar = \pm 0.5\hbar$.

$$J = \sqrt{j(j+1)}\hbar = \sqrt{\frac{1}{2}\left(\frac{1}{2}+1\right)}\,\hbar = \frac{\sqrt{3}}{2}\hbar \approx 0.86\hbar.$$

Thus there can be only two allowed orientations of j at an angle

$\cos^{-1}\left(\frac{\frac{1}{2}\hbar}{\frac{\sqrt{3}}{2}\hbar}\right) = \cos^{-1}\left(\frac{1}{\sqrt{3}}\right) = 54^\circ$ with the positive z-axis and with the negative z-axis. Although $\vec{J}$ cannot actually point along the positive or the negative z-axis, the two allowed orientations are still referred to as *spin-up* and *spin-down* states.

11.14 Pauli spin matrices

For $J = 1/2$, components of angular momentum $\vec{J}$ have the following matrix operators

$$J_x = \frac{\hbar}{2}\begin{pmatrix} 0 & 1 \\ 1 & 0 \end{pmatrix} = \frac{\hbar}{2}\sigma_x,\ J_y = \frac{\hbar}{2}\begin{pmatrix} 0 & -i \\ i & 0 \end{pmatrix} = \frac{\hbar}{2}\sigma_y,\ J_z = \frac{\hbar}{2}\begin{pmatrix} 1 & 0 \\ 0 & -1 \end{pmatrix} = \frac{\hbar}{2}\sigma_z$$

On the basis of common Eigen functions of J^2 and J_z. The *Pauli spin matrices* are

$$\sigma_x = \begin{pmatrix} 0 & 1 \\ 1 & 0 \end{pmatrix},\ \sigma_y = \begin{pmatrix} 0 & -i \\ i & 0 \end{pmatrix},\ \sigma_z = \begin{pmatrix} 1 & 0 \\ 0 & -1 \end{pmatrix}$$

11.15 Pauli matrices and their some of commutation relation

$$\sigma_x = \begin{pmatrix} 0 & 1 \\ 1 & 0 \end{pmatrix} \quad ; \quad \sigma_y = \begin{pmatrix} 0 & -i \\ i & 0 \end{pmatrix} \quad ; \quad \sigma_z = \begin{pmatrix} 1 & 0 \\ 0 & -1 \end{pmatrix}$$

The following commutation relations can be proved:

1. $[\sigma_x, \sigma_y] = 2i\sigma_z$ 2. $[\sigma_y, \sigma_z] = 2i\sigma_x$ 3. $[\sigma_z, \sigma_x] = 2i\sigma_y$ 4. $\sigma^2 = \sigma_x^2 + \sigma_y^2 + \sigma_z^2 = 3$
5. $[\sigma^2, \sigma_x] = 0$ 6. $[\sigma^2, \sigma_y] = 0$ 7. $[\sigma^2, \sigma_z] = 0$

1. $[\sigma_x, \sigma_y] = 2i\sigma_z$

$$\text{L.H.S} = [\sigma_x, \sigma_y] = \sigma_x\sigma_y - \sigma_y\sigma_x = \begin{pmatrix} 0 & 1 \\ 1 & 0 \end{pmatrix}\begin{pmatrix} 0 & -i \\ i & 0 \end{pmatrix} - \begin{pmatrix} 0 & -i \\ i & 0 \end{pmatrix}\begin{pmatrix} 0 & 1 \\ 1 & 0 \end{pmatrix}$$

$$= \begin{pmatrix} 0+i & 0+0 \\ 0+0 & -i+0 \end{pmatrix} - \begin{pmatrix} 0-i & 0+0 \\ 0+0 & -i+0 \end{pmatrix} = \begin{pmatrix} i & 0 \\ 0 & -i \end{pmatrix} - \begin{pmatrix} -i & 0 \\ 0 & i \end{pmatrix}$$

$$= \begin{pmatrix} 2i & 0 \\ 0 & -2i \end{pmatrix} = 2i\begin{pmatrix} 1 & 0 \\ 0 & -1 \end{pmatrix} = 2i\sigma_z$$

2. $[\sigma_y, \sigma_z] = 2i\sigma_x$

$$\text{L.H.S} = [\sigma_y, \sigma_z] = \sigma_y\sigma_z - \sigma_z\sigma_y = \begin{pmatrix} 0 & -i \\ i & 0 \end{pmatrix}\begin{pmatrix} 1 & 0 \\ 0 & -1 \end{pmatrix} - \begin{pmatrix} 1 & 0 \\ 0 & -1 \end{pmatrix}\begin{pmatrix} 0 & -i \\ i & 0 \end{pmatrix}$$

$$= \begin{pmatrix} 0+0 & 0+i \\ i+0 & 0+0 \end{pmatrix} - \begin{pmatrix} 0+0 & -i+0 \\ 0-i & 0+0 \end{pmatrix}$$

$$= \begin{pmatrix} 0 & i \\ i & 0 \end{pmatrix} - \begin{pmatrix} 0 & -i \\ -i & 0 \end{pmatrix} = \begin{pmatrix} 0 & 2i \\ 2i & 0 \end{pmatrix} = 2i\begin{pmatrix} 0 & 1 \\ 1 & 0 \end{pmatrix} = 2i\sigma_x$$

$$3.\ \text{L.H.S} = [\sigma_z, \sigma_x] = \sigma_z\sigma_x - \sigma_x\sigma_z = \begin{pmatrix} 1 & 0 \\ 0 & -1 \end{pmatrix}\begin{pmatrix} 0 & 1 \\ 1 & 0 \end{pmatrix} - \begin{pmatrix} 0 & 1 \\ 1 & 0 \end{pmatrix}\begin{pmatrix} 1 & 0 \\ 0 & -1 \end{pmatrix}$$

$$\Rightarrow \begin{pmatrix} 0+0 & 1+0 \\ 0-1 & 0+0 \end{pmatrix} - \begin{pmatrix} 0+0 & 0-1 \\ 1+0 & 0+0 \end{pmatrix} = \begin{pmatrix} 0 & 1 \\ -1 & 0 \end{pmatrix} - \begin{pmatrix} 0 & -1 \\ 1 & 0 \end{pmatrix}$$

$$\Rightarrow \begin{pmatrix} 0 & 2 \\ -2 & 0 \end{pmatrix} = 2\begin{pmatrix} 0 & 1 \\ -1 & 0 \end{pmatrix} = -2\begin{pmatrix} 0 & -1 \\ 1 & 0 \end{pmatrix} = -2i^2\begin{pmatrix} 0 & -1 \\ 1 & 0 \end{pmatrix} = 2i\begin{pmatrix} 0 & -i \\ i & 0 \end{pmatrix} = 2i\sigma_y$$

4. L.H.S $= \sigma_x^2 + \sigma_y^2 + \sigma_z^2 = \sigma_x\sigma_x + \sigma_y\sigma_y + \sigma_z\sigma_z$

$$\Rightarrow \begin{pmatrix} 0 & 1 \\ 1 & 0 \end{pmatrix}\begin{pmatrix} 0 & 1 \\ 1 & 0 \end{pmatrix} + \begin{pmatrix} 0 & -i \\ i & 0 \end{pmatrix}\begin{pmatrix} 0 & -i \\ i & 0 \end{pmatrix} + \begin{pmatrix} 1 & 0 \\ 0 & -1 \end{pmatrix}\begin{pmatrix} 1 & 0 \\ 0 & -1 \end{pmatrix}$$

$$\Rightarrow \begin{pmatrix} 1 & 0 \\ 0 & 1 \end{pmatrix} + \begin{pmatrix} 1 & 0 \\ 0 & 1 \end{pmatrix} + \begin{pmatrix} 1 & 0 \\ 0 & 1 \end{pmatrix} = 1+1+1=3$$

5. L.H.S $= [\sigma^2, \sigma_x]$

$= [\sigma_x^2 + \sigma_y^2 + \sigma_z^2, \sigma_x]$

$= [\sigma_x^2, \sigma_x] + [\sigma_y^2, \sigma_x] + [\sigma_z^2, \sigma_x]$

$= [\sigma_x\sigma_x, \sigma_x] + [\sigma_y\sigma_y, \sigma_x] + [\sigma_z\sigma_z, \sigma_x]$

$= (\sigma_x\sigma_x\sigma_x - \sigma_x\sigma_x\sigma_x) + (\sigma_y\sigma_y\sigma_x - \sigma_x\sigma_y\sigma_y) + (\sigma_z\sigma_z\sigma_x - \sigma_x\sigma_z\sigma_z)$

$= (\sigma_y\sigma_y\sigma_x - \sigma_x\sigma_y\sigma_y + \sigma_z\sigma_z\sigma_x - \sigma_x\sigma_z\sigma_z) + (\sigma_y\sigma_x\sigma_y - \sigma_y\sigma_x\sigma_y + \sigma_z\sigma_x\sigma_z - \sigma_z\sigma_x\sigma_z)$

$= \sigma_y(\sigma_y\sigma_x - \sigma_x\sigma_y) + (\sigma_y\sigma_x - \sigma_x\sigma_y)\sigma_y + \sigma_z(\sigma_z\sigma_x - \sigma_x\sigma_z) + (\sigma_z\sigma_x - \sigma_x\sigma_z)\sigma_z$

$= \sigma_y[\sigma_y, \sigma_x] + [\sigma_y, \sigma_x]\sigma_y + \sigma_z[\sigma_z, \sigma_x] + [\sigma_z, \sigma_x]\sigma_z$

$= \sigma_y(-2i\sigma_z) + (-2i\sigma_z)\sigma_y + \sigma_z(2i\sigma_y) + (2i\sigma_y)\sigma_y$

$= -2i\sigma_y\sigma_z - 2i\sigma_z\sigma_y + 2i\sigma_z\sigma_y + 2i\sigma_y\sigma_z$

$= 0$

Similarly, we can prove that $[\sigma^2, \sigma_y] = 0$ and $[\sigma^2, \sigma_z] = 0$.

11.16 Illustrative examples of the symmetry of the wave function in the dynamics of both bound and unbound systems of identical particles: The Helium atom

From the viewpoint of atomic properties, the He atom is a system of two electrons. Since electrons are fermions, the total wave function of the system must be anti-symmetric. However, the total wave function is the product of a space or orbital part and a spin part. Therefore, the space part alone could be symmetric or anti-symmetric depending on whether the spin part is anti-symmetric or symmetric. Since electron has spin $\frac{1}{2}$, the spin of a two electron system could be either 1 or 0. There are three states with spin 1 corresponding to the three values (1, 0 and -1) of the z-component of the spin. The spin-1 states are, hence, called the triplet states. We denote there by χ_1^3, χ_0^3 and χ_{-1}^3. Similarly, χ_0^1 represents the singlet state corresponding to spin 0. From equation () and () it follows that the triplet states are symmetric while the singlet state is anti-symmetric. Thus, out of the eight combinations possible with the four spin functions and the two (the symmetric and anti-symmetric) space functions, only four are permitted by the principle of indistinguishability of identical particles. These are:

$$\Psi_A\chi_0^1, \Psi_A\chi_1^3, \Psi_A\chi_0^3, \Psi_A\chi_{-1}^3 \quad [1]$$

Where,

$$\Psi_S(1,2) = \left[\frac{1}{2}\left(1+\delta_{\alpha\beta}\right)\right]^{\frac{1}{2}}\left(\hat{1}+\hat{P}_{12}\right)\Phi_\alpha(1)\Phi_\beta(2) \quad [2]$$

$$\Psi_A(1,2) = \left[\frac{1}{\sqrt{2}}\right]\left(\hat{1}-\hat{P}_{12}\right)\Phi_\alpha(1)\Phi_\beta(2) \quad [3]$$

Now, the Hamiltonian of the He atom can be written as

$$\hat{H} = \hat{H}_0 + \hat{H}_1, \quad [4]$$

Where,

$$\hat{H}_0 = \frac{-\hbar^2}{2\mu}\left(\nabla_1^2+\nabla_2^2\right) - 2e^2\left(\frac{1}{r_1}+\frac{1}{r_2}\right) \quad [5]$$

$$\hat{H}_1 = \frac{e^2}{r_{12}} \quad [6]$$

Here, r_1 and r_2 are the radial distances of the electrons from the center of the atom and r_{12} is the separation between the electrons. $\hat{H}_1$ Represents the mutual repulsion of the electrons and is small in comparison with $\hat{H}_0$. The contribution of $\hat{H}_1$ to the energy of the system can, therefore, be calculated using first order perturbation theory where the zero order wave functions (the eigenvectors of $\hat{H}_0$) are given by (). Since $\hat{H}_1$ does not contain the spin variables, the three triplet states would be degenerated. And the separation between the singlet and the triplet states arises entirely due to the different symmetry of the orbital part.

According to Eq. () the contribution to the energy due to $\hat{H}_1$, in the first order is equal to the expectation value of $\hat{H}_1$. Thus, for the singlet state,

$$\begin{aligned} E_S^{(1)} &= \langle\Psi_A\chi_0^1|\hat{H}_1|\Psi_S\chi_0^1\rangle \\ &= \langle\Psi_S|\hat{H}_1|\Psi_S\rangle \\ &= \left\{\frac{1}{(1+\delta_{\alpha\beta})}\right\}\left(J_{\alpha\beta}+K_{\alpha\beta}\right) \end{aligned} \quad [7]$$

With

$$J_{\alpha\beta} = \langle\Phi_\alpha(1)\Phi_\beta(2)|\hat{H}_1|\Phi_\alpha(1)\Phi_\beta(2)\rangle \quad [8]$$

$$K_{\alpha\beta} = \langle\Phi_\alpha(1)\Phi_\beta(2)|\hat{H}_1|\Phi_{\alpha\beta}(1)\Phi_\alpha(2)\rangle \quad [9]$$

Where expression () for Ψ_S and property of $\hat{P}_{12}$ are used.
Similarly, for the triplet states,

$$E_t^{(1)} = \langle \Psi_A | \widehat{H}_1 | \Psi_A \rangle = \left(J_{\alpha\beta} - K_{\alpha\beta} \right) \qquad [10]$$

$J_{\alpha\beta}$ and $K_{\alpha\beta}$ are positive since $\widehat{H}_1$ is positive definite. Therefore,

$$E_t^{(1)} < E_S^{(1)}$$

That is, the singlet state is always higher in energy that the triplet state.
For the ground state of He atom, $\alpha = \beta \equiv 1S$ (that is, n=1, l=0), so that $\Psi_A \equiv 0$.
Hence the only allowed state is $\Psi_S \chi_0^1$. Then,

$$E^{(1)} \equiv E_S^{(1)} = J_{(1S)^2} = \frac{5\, e^2}{4\, a_0} = 33.99\ eV;$$

Where a_0 is given by Eq. (). Also, $\frac{e^2}{a_0} = 27.19\ eV$. From Eq. (), we have,

$$E_{(1S)^2}^{(0)} = -4 \left(\frac{e^2}{a_0} \right) = -108.76 eV$$

So that,

$$\mathrm{E}_{(1S)^2} = E_{(1S)^2}^{(0)} + E_{(1S)^2}^{(1)} = -74.77\ eV$$

Similarly, we obtain the energies of some of the excited states, using the values,

$$J_{1S,2S} = 11.42\ eV;$$
$$K_{1S,2S} = 1.20\ eV$$
$$J_{1S,2P} = 13.22\ eV$$
$$K_{1S,2P} = 0.94\ eV$$

The energies so obtained are compared with the experimental values in table (). The spectroscopic notation $L_J^{(2S+1)}$ is used to specify a state, where S represents the spin and J the total angular momentum while L stands for the symbol that denotes the value of the orbital angular momentum according to the following scheme:

Symbol for $L \to S\ P\ D\ F \cdots\cdots\cdots\cdots\cdots\cdots\cdots\cdots\cdots\cdots\cdots\cdots\cdots\cdots$

Value of $L \to 0\ 1\ 2\ 3 \cdots\cdots\cdots\cdots\cdots\cdots\cdots\cdots\cdots\cdots\cdots\cdots\cdots\cdots\cdots$

Thus, the symmetry character of the wave function, arising from the identity of the electrons, provides an understanding of the qualitative features (such as the nature of the ground state and the difference in energy between the triplet and the singlet states belonging to a given configuration) of the low lying levels of the He atom.

11.17 Scattering of Identical Particles

In the scattering of identical particles, there are two situations indistinguishable from each other as shown in Fig. [11.16] . The incident particle and the target have equal and opposite velocities in the Centre-of-Mass system. In (a) the particle observed at the detector in the incident particle (particle number 1) while in (b) it is the recoiled target (particle number 2). The indistinguishable of the two situations is taken into account by using properly symmetrised wave functions for the calculation of the scattering cross-section.

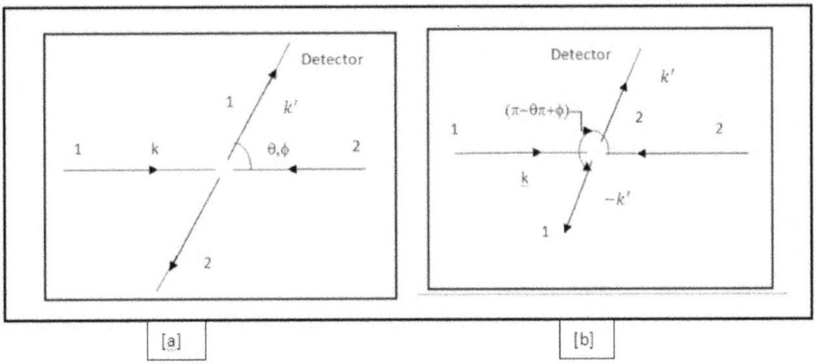

Fig. [11.17]: Shows scattering of identical particles in the centre of mass system

Now, for process (a), the asymptotic scattered wave function is

$$\Psi_a^{(S)}(k,k',r) \sim f_a(k,k')\frac{e^{ikr}}{r} \qquad [1]$$

Process (b) differs from process (a) in that particles 1 and 2 get interchanged at the time of scattering. Thus,

$$\Psi_b^{(S)}(k,k',r) \sim \Psi_a^{(S)}(k,-k',r) \qquad [2]$$

But, interchanging the two particles is equivalent to reversing the relative co- ordinate $r = r_1 - r_2$ and since, by definition, k' has the same direction as r, it is equivalent to reversing the direction of k'. Thus,

$$\hat{P}_{12}\Psi_b^{(S)}(k,k',r) = \Psi_a^{(S)}(k,-k',r) \qquad [3]$$

A properly symmetrised wave function is a symmetric or an anti-symmetric combination of $\Psi_a^{(S)}$ and $\Psi_b^{(S)}$:

$$\Psi_S^{(S)}(k,k',r) = \Psi_a^{(S)}(k,k',r) + \Psi_b^{(S)}(k,-k',r) \qquad [4]$$

$$\Psi_A^{(S)}(k,k',r) = \Psi_a^{(S)}(k,k',r) - \Psi_b^{(S)}(k,-k',r) \qquad [5]$$

Substituting from Equation [2] , [3] and [1] and dropping the labels a, we have,

$$\Psi_S^{(S)} \sim f_S(k,k')\left(\frac{e^{ikr}}{r}\right) \qquad [6]$$

$$\Psi_A^{(S)} \sim f_A(k,k')\left(\frac{e^{ikr}}{r}\right) \qquad [7]$$

Where,

$$f_S(k,k') = f(k,k') + f_S(k,-k') \qquad [8]$$

$$f_A(k,k') = f(k,k') - f_S(k,-k') \qquad [9]$$

The differential cross-section for the scattering of identical particles are then given by Eq.

$$\sigma_S(k,k') = |f_S(k,k')|^2$$

$$= |f(k,k')|^2 + |f(k,-k')|^2 + 2\, Re\,\{f^{\times}(k,k')\, f\,(k,-k')\} \qquad [10]$$

And $$\sigma_A(k,k') = |f_A(k,k')|^2$$

$$= |f(k,k')|^2 + |f(k,-k')|^2 - 2\, Re\,\{f^{\times}(k,k').f\,(k,-k')\} \qquad [11]$$

There are to be compared with the corresponding classical expression,

$$\sigma_{cl}(k,k') = \sigma(k,k') + \sigma(k,-k')$$

$$= |f(k,k')|^2 + |f(k,-k')|^2 \qquad [12]$$

In [10] and [11] Re { } represents the real part of the quantity within the bracket. We see that an important feature that distinguishes the quantum mechanical cross sections from the classical one, is the *Interference* between the scattering process (a) and (b).

In terms of the angles θ and ϕ that specify the direction of scattering with respect to the incident direction,

$$f(k,k') \equiv f_k(\theta,\phi),$$

And $$f(k,-k') \equiv f_k(\pi-\theta,\pi+\phi) \qquad [13]$$

Further, in the case of central forces, the scattering amplitude is independent of the angle ϕ. Then,

$$f(k,k') = f_k(\theta,\phi);\ f(k,-k') = f_k(\pi-\theta),$$

And we have,

$$\sigma_S(\theta) = |f_k(\theta) + f_k(\pi-\theta)|^2 \qquad [14]$$

$$\sigma_A(\theta) = |f_k(\theta) - f_k(\pi-\theta)|^2 \qquad [15]$$

$$\sigma_{cl}(\theta) = |f_k(\theta)|^2 + |f_k(\pi-\theta)|^2 \qquad [16]$$

Exercise

1. What is Identical Particles? Write an idea about Spins and Statistics.
2. Define Symmetric and Anti-symmetric wave functions.
3. Construct the symmetric and anti-symmetric wave function.
4. Construct the symmetric and anti-symmetric wave function for a system containing 3 identical particles.
5. Construct the symmetric and anti-symmetric wave function for a system containing n identical particles.
6. What is the quantum mechanical property of an elementary particle? Explain a brief description of spin.
7. Illustrate/Interpret Intrinsic Spin. Give an idea of Spin $\frac{1}{2}$ Particles and Spin Zero Particles.
8. What is an arbitrary spin state? Discuss the concept of spin up and spin down.
9. Write the Pauli Exclusion Principle. Explain the basic concept of the Pauli Exclusion Principle.
10. Obtain the mathematical ground of the Pauli Exclusion Principle.
11. Explain Spin up and spin down states.
12. Write Pauli spin matrices. Establish some of the following commutation relation of Pauli matrices:

    a. $[\sigma_x, \sigma_y] = 2i\sigma_z$ b. $[\sigma_y, \sigma_z] = 2i\sigma_x$ c. $[\sigma_z, \sigma_x] = 2i\sigma_y$ d. $\sigma^2 = \sigma_x^2 + \sigma_y^2 + \sigma_z^2 = 3$

    e. $[\sigma^2, \sigma_x] = 0$ f. $[\sigma^2, \sigma_y] = 0$ g. $[\sigma^2, \sigma_z] = 0$ h. $[\sigma^2, \sigma_y] = 0$ i. $[\sigma^2, \sigma_z] = 0$.

13. Interpret the concept of identical particles. Formulate the Pauli's principle on the basis of above concept.

Chapter 12 Quantum Dynamics: Schrödinger picture, Heisenberg picture and Dirac or Interaction picture

12.1 State vector of a system and state vector transformation

In classical mechanics, the set of 3N position Coordinates (x, y, z) & 3N momentum coordinates (P_x , P_y, P_z) describe the motion of a system of N particles. The combined position & momentum space is called phase space. [This mathematical space of 6N dimensions are called the phase space] This specification of a system of particles in terms of the position & momentum of each component particle at any instant of time is called a state of the system or its state vector[5]. *In quantum mechanics all the information about the physical state of a system is contained in a state vector.*

Let $\phi(t_0)$ be the state vector of the system at time t_0. The state vector $\phi(t)$ at a later time t then arises from $\phi(t_0)$ by the action of an operator T(t, t_0).

$$T \Longrightarrow \phi(t_0)\ \ \text{A} \xrightarrow[T(t,\,t_0)]{\hspace{8em}} \text{B}\ \ \phi(t)$$

Where $\phi(t_0)$ be the state vector at position A and $\phi(t)$ be the state vector at position B I.e.,

$$\phi(t) = T(t, t_0)\phi(t_0)$$

From the above equation the state vector has transformed from one state to another state by the action of an operator $T(t, t_0)$.

12.2Time development of Schrödinger, Heisenberg and Dirac or Interaction picture

There are three possible ways to describe the time development:

- Schrödinger picture of time development
- Heisenberg picture of time development
- Interaction or Dirac picture of time development

- **Schrödinger picture:** The form of the operator remains unchanged & the state vector changes with time. *I.e.,*

$$\Omega_S = \text{Constant}$$

&

$$\Psi_S = \Psi_S(t)$$

This is called Schrödinger picture of time development.

- **Heisenberg picture:** The form of the operator changes with time & the state vector is constant. *I.e.,*

$$\Omega_H = \Omega_H(t)$$

& $$\Psi_H = \text{Constant}$$

This is called Heisenberg picture of time development.

- **Interaction or Dirac picture:** Both the operator & state vector change with time. *I.e.,*

$$\Omega_I = \Omega_I\,(t)$$

& $$\Psi_I\,(t) = \Psi_I\,(t)$$

This is called Interaction picture or Dirac picture of time development. The interaction picture is a half way between the Schrödinger and Heisenberg pictures, and is particularly suited to develop the perturbation theory.

12.3 Dynamical equation of Schrödinger picture

Let $\phi(t_0)$ be the state vector of the system at time t_0. The state vector $\phi(t)$ at a later time t then arises from $\phi(t_0)$ by the action of an operator $T\,(t,\,t_0)$.

$$\phi(t) = T\,(t,t_0)\,\phi(t_0) \tag{1}$$

From the above equation the state vector has transformed from one state to another state by the action of an operator $T(t,t_0)$

$$T \Longrightarrow \phi(t_0)\ A \xrightarrow{\quad T\,(t,\,t_0)\quad} B\ \phi(t)$$

Where $\phi(t_0)$ be the state vector at position A and $\phi(t)$ be the state vector at position B. Now differentiating equation [1] with respect to t

$$\frac{\partial\varphi(t)}{\partial t} = \frac{\partial T}{\partial t}\varphi(t_0) + T\frac{\partial\varphi(t_0)}{\partial t} \tag{2}$$

In Schrödinger picture, the operator is constant. I.e., T = Constant. So equation [2] becomes

$$\frac{\partial\varphi(t)}{\partial t} = T\frac{\partial\varphi(t_0)}{\partial t} \tag{3}$$

From, time development Schrödinger equation

$$H\varphi(t) = i\hbar\frac{\partial}{\partial t}\varphi(t)$$

$$\frac{\partial\varphi(t)}{\partial t} = -\frac{i}{\hbar}H\varphi(t)$$

Therefore [3] becomes,

$$-\frac{i}{\hbar}H\varphi(t) = T\,\frac{\partial\varphi(t_0)}{\partial t}$$

$$-\frac{i}{\hbar}H\varphi(t) = \frac{\partial\varphi(t)}{\partial t} \quad \text{or,} \quad \frac{\partial\varphi(t)}{\partial t} + \frac{i}{\hbar}H\varphi(t) = 0$$

Or, $$\frac{\hbar}{i}\frac{\partial\varphi(t)}{\partial t} + H\varphi(t) = 0 \qquad [4]$$

The above equation is the dynamical equation of Schrödinger picture.

12.4 Solution of Schrödinger picture

Schrödinger equation can be written in terms of Ket notation

$$i\hbar\frac{\partial|\Psi_s(t)\rangle}{\partial t} = H|\Psi_s(t)\rangle$$

$$\frac{\partial|\Psi_s(t)\rangle}{|\Psi_s(t)\rangle} = -\frac{iH}{\hbar}\partial t$$

Now integrating the above equation on both sides

$$\ln|\Psi_s(t)\rangle = -\frac{i}{\hbar}Ht + C_1 \qquad [1]$$

When $t = 0$ and $\Psi_s(t) = \Psi_s(0)$; Then $\ln|\Psi_s(0)\rangle = C_1$

Putting the value of C_1 in equation [1]

$$\ln|\Psi_s(t)\rangle = -\frac{i}{\hbar}Ht + \ln|\Psi_s(0)\rangle$$

$$\ln|\Psi_s(t)\rangle - \ln|\Psi_s(0)\rangle = -\frac{i}{\hbar}Ht$$

$$\ln\frac{\Psi_s(t)}{\Psi_s(0)} = -\frac{i}{\hbar}Ht \; or, \; \frac{\Psi_s(t)}{\Psi_s(0)} = \exp\left(-\frac{i}{\hbar}Ht\right)$$

$$\Psi_s(t) = e^{-\frac{i}{\hbar}Ht}|\Psi_s(0)\rangle \qquad [2]$$

Equation [2] represents the solution of Schrödinger picture.

12.5 Necessary and sufficient condition of Schrödinger picture

The expectation value of the operator Ω_s at time t is,

$$<\Omega_s>_t = (\varphi(t), \Omega_s\,\varphi(t)) \qquad [1]$$

Now differentiating equation [1] with respect to t

Therefore, $$\frac{\partial <\Omega_s>_t}{\partial t} = \frac{\partial\varphi(t)}{\partial t}\Omega_s\varphi(t) + \varphi(t)\Omega_s\frac{\partial\varphi(t)}{\partial t} \qquad [2]$$

From the Schrödinger equation,

$$H\varphi(t) = E\varphi(t)$$

$$\Rightarrow H\varphi(t) = i\hbar\frac{\partial\varphi(t)}{\partial t} \quad \Rightarrow \frac{\partial\varphi(t)}{\partial t} = -\frac{i}{\hbar}H\varphi(t)$$

Therefore [2] becomes

$$\frac{\partial<\Omega_s>_t}{\partial t} = \left(-\frac{i}{\hbar}H\varphi(t), \Omega_s\ \varphi(t)\right) + \left(\varphi(t), \Omega_s\ (-\frac{i}{\hbar})H\ \varphi(t)\right)$$

$$[\text{But, } (\vec{a},\lambda\vec{b}) = \lambda(\vec{a},\vec{b}) \ \ and \ \ (\lambda\vec{a},\vec{b}) = \lambda^{\times}(\vec{a},\vec{b}) \quad [\text{ Here, } \lambda = -\frac{i}{\hbar}\ ]$$

$$\Rightarrow \frac{\partial<\Omega_s>_t}{\partial t} = \frac{i}{\hbar}(H\varphi(t),\ \Omega_s\ \varphi(t)) - \frac{i}{\hbar}(\varphi(t),\ \Omega_s\ H\varphi(t))$$

$$= \frac{i}{\hbar}\left(\varphi(t), H^*\Omega_s\ \varphi(t)\right) - \frac{i}{\hbar}\left(\varphi(t), \Omega_s H\ \varphi(t)\right) \ [\because (\Omega\ \underset{\rightarrow}{a}, \underset{\rightarrow}{b}) = (\underset{\rightarrow}{a}, \Omega^*\underset{\rightarrow}{b})]$$

Since, H is Hermitian operator, hence $H^* = H$

$$\therefore \frac{\partial<\Omega_s>_t}{\partial t} = \frac{i}{\hbar}(\varphi(t), H\Omega_s\ \varphi(t)) - \frac{i}{\hbar}(\varphi(t), \Omega_s\ H\phi(t))$$

$$= \frac{i}{\hbar}\left(\varphi(t), (H\Omega_s - \Omega_s H)\varphi(t)\right)$$

$$= \frac{i}{\hbar}\left(\varphi(t), [H,\Omega_s]\ \varphi(t)\right)$$

$$= \frac{i}{\hbar}[H,\Omega_s](\varphi(t),\varphi(t)) \quad \because (\vec{a},\lambda\vec{b}) = \lambda(\vec{a},\vec{b})$$

$$= \frac{i}{\hbar}[H,\Omega_s] \quad [\text{with } (\varphi(t),\varphi(t))] = 1]$$

$$\therefore \frac{\partial<\Omega_s>_t}{\partial t} = <\frac{i}{\hbar}[H,\Omega_s]> \qquad [2]$$

If the dynamical observables are constant of motion and not only they have any explicit time dependence but also their expectation value does not change in the course of time[1] I.e.,

$$\frac{\partial<\Omega_s>_t}{\partial t} = 0$$

Or, Ω_s = Constant

Thus *the necessary & sufficient condition for Schrödinger Picture is that its operator Ω_S commutes with Hamiltonian and is a constant of motion.*

12.6 Advantage and Disadvantage of Schrödinger picture

- If the dynamical observables of Schrödinger picture have no explicit time dependence, their expectation value does not change in the course of time. I.e.,

$$\frac{\partial < \Omega_s >_t}{\partial t} = 0 \quad Or, \quad \Omega_s = Constant$$

- At time t = 0; both Schrödinger and Heisenberg pictures are on the same platform I.e., the Schrödinger picture is coincide with the Heisenberg picture. So, $\psi_H = \psi_S$ or, $|\psi_H\rangle = |\psi_S\rangle$

12.7 Hamiltonian of a system can be regarded as the generator of infinitesimal canonical transformation

Let $\phi(t_0)$ be the state vector of the system at time t_0. The state vector $\phi(t_0)$ at a later time t then arises from $\phi(t_0)$ by the action of an operator $T(t, t_0)$[1] I.e.,

$$\phi(t_0) = T(t,t_0)\phi(t_0) \qquad [1]$$

Suppose at small time dt , the state vector becomes,

$$\varphi(t_0 + dt) = \varphi(t_0) + \frac{\partial \varphi(t_0)}{\partial t} dt + ... \qquad [Using\ Taylor\ series]$$

$$\Rightarrow \varphi(t_0 + dt) = \varphi(t_0) + \frac{\partial \varphi(t_0)}{\partial t} dt \qquad [2]$$

[Neglecting the higher order terms]

From the Schrödinger equation,

$$H\varphi(t_0) = E\varphi(t_0) \Rightarrow H\varphi(t_0) = i\hbar \frac{\partial}{\partial t}\varphi(t_0)$$

$$\Rightarrow \frac{\partial \varphi(t_0)}{\partial t} = -\frac{i}{\hbar} H\varphi(t_0) \qquad [3]$$

Hence [2] becomes,

$$\Rightarrow \varphi(t_0 + dt) = \varphi(t_0) + \frac{\partial \varphi(t_0)}{\partial t} dt$$

$$= \varphi(t_0) - \frac{i}{\hbar} Hdt\varphi(t_0) = [1 - \frac{i}{\hbar} Hdt]\varphi(t_0)$$

$$\Rightarrow \varphi(t_0 + dt) = [1 - \frac{i}{\hbar}(t - t_0)H]\varphi(t_0) \qquad [4]$$

Now, comparing [1] & [4]

$$T(t,t_0) = [1 - \frac{i}{\hbar}(t - t_0)H]$$

Therefore, *the Hamiltonian of a system can be regarded as the generator of the infinitesimal canonical transformation*[1].

12.8 Dynamical equation of Heisenberg picture

The forms of the operators are time dependent whereas the wave functions are time independent.

I.e.,

$$\Omega_H = \Omega_H(t) \quad \& \quad \Psi_H = \text{Constant}$$

The time dependence of the operators in the Heisenberg picture is obtained from the equality of the expectation value.

$$\left\langle \hat{\Omega}_H(t) \right\rangle = \left(\psi_H, \hat{\Omega}_H(t) \psi_H \right) \tag{7}$$

In bra and ket notation, the above equation will be

$$\left\langle \hat{\Omega}_H(t) \right\rangle = \left\langle \psi_H \middle| \hat{\Omega}_H(t) \middle| \psi_H \right\rangle \tag{8}$$

Again, the time dependence of the operators in Schrödinger picture is obtained from the expectation value

$$\begin{aligned} \left\langle \Omega_s \right\rangle_t &= \left(\psi_t(t), \hat{\Omega}_s \psi_s(t) \right) \\ &= \left\langle \psi_s(t) \middle| \hat{\Omega}_s \middle| \psi_s(t) \right\rangle \end{aligned} \tag{9}$$

Since the expectation value of the operators in the two pictures should be identical.

$$\left\langle \Psi_H \middle| \hat{\Omega}_H(t) \middle| \Psi_H \right\rangle = \left\langle \Psi_S(t) \middle| \hat{\Omega}_S \middle| \Psi_S(t) \right\rangle \tag{10}$$

From equation [2] in the section [10.4]

$$\left| \Psi_S(t) \right\rangle = e^{-\frac{iHt}{\hbar}} \left| \Psi_S(0) \right\rangle \tag{11}$$

$$\left| \Psi_S(t) \right\rangle = e^{-\frac{iHt}{\hbar}} \left| \Psi_H \right\rangle \tag{12}$$

Because, $\left| \Psi_S(0) \right\rangle$ actually defines the state vector in Heisenberg picture which does not change with time & retains its initial value as the time progresses. Therefore [10] becomes

$$\left\langle \Psi_H \middle| \hat{\Omega}_H(t) \middle| \Psi_H \right\rangle = \left\langle \Psi_H \middle| e^{\frac{iHt}{\hbar}} \hat{\Omega}_s e^{-\frac{iHt}{\hbar}} \middle| \Psi_H \right\rangle$$

$$\Rightarrow \left\langle \hat{\Omega}_H(t) \right\rangle = e^{\frac{iHt}{\hbar}} \hat{\Omega}_s e^{-\frac{iHt}{\hbar}} \tag{13}$$

$$\Rightarrow \frac{d\left\langle \hat{\Omega}_H(t) \right\rangle}{dt} = \frac{iH}{\hbar} e^{\frac{iHt}{\hbar}} \hat{\Omega}_s e^{-\frac{iHt}{\hbar}} + e^{\frac{iHt}{\hbar}} \hat{\Omega}_s \left(-\frac{iH}{\hbar} \right) e^{-\frac{iHt}{\hbar}}$$

$$\Rightarrow \frac{d\left\langle \hat{\Omega}_H(t) \right\rangle}{dt} = \frac{iH}{\hbar} \hat{\Omega}_H(t) + e^{\frac{iHt}{\hbar}} \hat{\Omega}_s \left(-\frac{iH}{\hbar} \right) e^{-\frac{iHt}{\hbar}}$$

$$\Rightarrow \frac{d\left\langle \hat{\Omega}_H(t) \right\rangle}{dt} = \frac{iH}{\hbar} \hat{\Omega}_H(t) - \hat{\Omega}_H(t) \frac{iH}{\hbar}$$

$$\Rightarrow \frac{d\left\langle \hat{\Omega}_H(t)\right\rangle}{dt} = \frac{i}{\hbar}\left(H\,\hat{\Omega}_H(t) - \hat{\Omega}_H(t)H \right)$$

$$\Rightarrow \frac{d\left\langle \hat{\Omega}_H(t)\right\rangle}{dt} = -\frac{1}{i\hbar}\left(H\,\hat{\Omega}_H(t) - \hat{\Omega}_H(t)H \right)$$

$$\Rightarrow i\hbar\frac{d\left\langle \hat{\Omega}_H(t)\right\rangle}{dt} = \left\langle \left(\hat{\Omega}_H(t)H - H\,\hat{\Omega}_H(t) \right)\right\rangle$$

$$\Rightarrow i\hbar\frac{d\left\langle \hat{\Omega}_H(t)\right\rangle}{dt} = \left\langle \left[\hat{\Omega}_H(t), H \right]\right\rangle \qquad [14]$$

Equation [14] is the Dynamical equation of Heisenberg picture.

12.9 Heisenberg equations of motion have the same form as Classical equations of motion: Beauty of Heisenberg picture[1]

The Hamiltonian operator for a Harmonic Oscillator,

$$H = \frac{P^2}{2m} + \frac{1}{2}kx^2 \Rightarrow \hat{H} = \frac{\hat{P}^2}{2m} + \frac{1}{2}\omega^2 m x^2 \Rightarrow \hat{H} = \frac{\hat{P}^2}{2m} + \frac{1}{2}m\omega^2 q^2 \quad [\text{Here } x = q]$$

For unit mass I.e., m = 1

$$\hat{H} = \frac{\hat{P}^2}{2} + \frac{1}{2}\omega^2 \hat{q}^2$$

$$\left.\begin{aligned} &\Rightarrow \frac{d\hat{H}}{dp} = \hat{p} \\ &\& \ \frac{d\hat{H}}{dq} = \omega^2 q \end{aligned}\right\} \qquad [2]$$

From restoring force equation

$$F = -kx = -kq$$

$$\Rightarrow m\frac{d^2q}{dt^2} = -kq$$

$$\Rightarrow \frac{d^2q}{dt^2} = -\frac{k}{m}q$$

$$\Rightarrow \frac{d^2q}{dt^2} = -\omega^2 q \quad [\because \omega = \sqrt{\frac{k}{m}}]$$

$$\Rightarrow m\frac{d^2q}{dt^2} = -m\omega^2 q \quad [\text{multipliying both sides by m}]$$

$$\Rightarrow \dot{\hat{p}} = -\omega^2 q \quad \text{[for unit mass, m = 1]}$$

Where, ω is the characteristic frequency.

Therefore [2] becomes,

$$\frac{d\hat{H}}{dq} = -\dot{\hat{p}} \quad \Rightarrow \dot{\hat{p}} = -\frac{dH}{dq}$$

So that *in Heisenberg picture, the equation of the basic observables is the same as those in classical equations of the motion*[1]. ***This is the beauty of Heisenberg picture.***

12.10 More difficult to solve the Heisenberg equations of motion

By the following way, Heisenberg picture is connected to the Schrödinger picture.

$$\hat{\varphi}(t) = T^{-1}(t,t_0)\varphi(t) \qquad [1\text{-a}]$$

$$\hat{\Omega}(t) = T^{-1}(t,t_0)\,\Omega T(t,t_0) \qquad [1\text{-b}]$$

Where Ω is the operator and φ is the state vector of Schrödinger picture and $\hat{\Omega}$ refer to the operator and $\hat{\varphi}$ state vector of Heisenberg picture. From equation [1-b], we have

$$\hat{\Omega}(t) = T^{-1}(t,t_0)\,\Omega T(t,t_0) \Rightarrow \frac{\partial \hat{\Omega}(t)}{\partial t} = \frac{\partial T^{-1}}{\partial t}\Omega T + T^{-1}\Omega\frac{\partial T}{\partial t} \qquad [1\text{-c}]$$

Again from equation [1-a], we have

$$\hat{\varphi}(t) = T^{-1}(t,t_0)\varphi(t) \Rightarrow \varphi(t) = \frac{\hat{\varphi}(t)}{T^{-1}(t,t_0)} \Rightarrow \varphi(t) = T(t,t_0)\hat{\varphi}(t)$$

$$\Rightarrow \frac{\partial \varphi(t)}{\partial t} = \frac{\partial T}{\partial t}\hat{\varphi}(t) + T\frac{\partial \hat{\varphi}(t)}{\partial t}$$

Since in Heisenberg picture, the state vector is time independent [constant].

I.e.,
$$\frac{\partial \hat{\varphi}(\mathrm{t})}{\partial \mathrm{t}} = 0$$

Therefore,
$$\frac{\partial \varphi(t)}{\partial t} = \frac{\partial T}{\partial t}\hat{\varphi}(t) \qquad [2]$$

From Schrödinger equation,

$$H\varphi(t) = E\varphi(t) \Rightarrow H\varphi(t) = i\hbar\frac{\partial}{\partial t}\varphi(t) \Rightarrow \frac{\partial \varphi(t)}{\partial t} = -\frac{i}{\hbar}H\varphi(t)$$

Therefore [2] becomes,

$$\frac{\partial T}{\partial t}\hat{\varphi}(t) + \frac{i}{\hbar}H\varphi(t) = 0 \Rightarrow \frac{\partial T}{\partial t}\hat{\varphi}(t) + \frac{i}{\hbar}HT\hat{\varphi}(t) = 0$$

$$\left[\text{Since, } \hat{\varphi}(t) = T^{-1}(t,t_0)\varphi(t) \Rightarrow \hat{\varphi}(t) = \frac{\varphi(t)}{T(t,t_0)} \Rightarrow \varphi(t) = T(t,t_0)\hat{\varphi}(t)\right]$$

$$\Rightarrow\left(\frac{\partial T}{\partial t}+\frac{i}{\hbar}HT\right)\phi(t)=0 \quad \Rightarrow\left(\frac{\partial T}{\partial t}+\frac{i}{\hbar}HT\right)=0\,;\ Since\ \phi(t)\neq 0$$

$$\Rightarrow\frac{\partial T}{\partial t}=-\frac{i}{\hbar}HT\Rightarrow\frac{\hbar}{i}\frac{\partial T}{\partial t}+HT=0 \qquad [3]$$

Now taking the complex conjugate of equation [3]

$$-\frac{\hbar}{\mathrm{i}}\frac{\partial \mathrm{T}^{\times}}{\partial \mathrm{t}}+\mathrm{H}^{\times}\mathrm{T}^{\times}=0$$

If H is Hermitian then $H^{\times}=H\ \&\quad T^{\times}=T^{-1}$

$$\Rightarrow-\frac{\hbar}{i}\frac{\partial T^{-1}}{\partial t}+HT^{-1}=0 \quad \Rightarrow\frac{\partial T^{-1}}{\partial t}=\frac{\hbar}{i}HT^{-1} \qquad [4]$$

Using [2] & (4) in equation [1-c]

$$\begin{aligned}\frac{\partial\hat{\Omega}(t)}{\partial t}&=\frac{\partial T^{-1}}{\partial t}\Omega T+T^{-1}\Omega\frac{\partial T}{\partial t}\\ &=\frac{i}{\hbar}T^{-1}H\Omega T+T^{-1}\Omega\left(-\frac{i}{\hbar}\right)HT\\ &=\frac{i}{\hbar}\left(T^{-1}H\Omega T-T^{-1}\Omega\, HT\right)\\ &=\frac{i}{\hbar}\left(T^{-1}HTT^{-1}\Omega T-T^{-1}\Omega\, TT^{-1}HT\right)\\ &=\frac{i}{\hbar}\left(\hat{H}'\,\hat{\Omega}'-\hat{\Omega}'\hat{H}'\right)=\frac{i}{\hbar}\left[\hat{\mathrm{H}}',\hat{\Omega}'\right]\end{aligned}$$

Since, a unitary transformation leaves the system completely unchanged.

$$\frac{\partial\hat{\Omega}(t)}{\partial t}=\frac{i}{\hbar}\left[\hat{\mathrm{H}},\hat{\Omega}\right] \qquad [5]$$

Heisenberg equations of motion are more difficult to solve because of their non-commutation of quantum mechanical operators.

12.11 Advantage and disadvantage of Heisenberg picture

❖ Advantage of Heisenberg picture

In Heisenberg picture, the equation of the basic observables is the same as those in classical equations of motion. *This is the beauty of Heisenberg picture.*

- The particular advantage of Heisenberg picture is that the equations of motion have a neat canonical form and allow for relativistic covariant descriptions

❖ **Disadvantage of Heisenberg picture**

- The disadvantage of Heisenberg picture is that we must solve a different and often complicated equation of motion for every basic observable
- Heisenberg equations of motion are more difficult to solve because of their non-commutation of quantum mechanical operator

12.12Dynamical equation of Dirac or Interaction picture

In this picture, both the operators & the wave functions are time dependent.

$$\Omega_I = \Omega_I(t) \quad \& \quad \Psi_I = \Psi_I(t)$$

Now, we split the Hamiltonian into two parts

$$H = H_0 + H'(t) \qquad [1]$$

Where, $H_0 =$ Unperturbed Hamiltonian

$H'(t) =$ Perturbation Hamiltonian

The time dependence of the operators in the interaction picture is obtained from the equality of the expectation value.

$$\langle \Psi_I(t)|\hat{\Omega}_I(t)|\Psi_I(t)\rangle = \langle \Psi_S(t)|\hat{\Omega}_S(t)|\Psi_S(t)\rangle \qquad [2]$$

The time dependence of the wave function is taken to be,

$$|\Psi_I(t)\rangle = e^{\frac{iH_0t}{\hbar}}|\Psi_S(t)\rangle \qquad [3]$$

Therefore [2] becomes,

$$\langle \Psi_s(t)|e^{-\frac{iH_0t}{\hbar}}\hat{\Omega}_I(t)e^{\frac{iH_0t}{\hbar}}|\Psi_s(t)\rangle = \langle \Psi_S(t)|\hat{\Omega}_S|\Psi_S(t)\rangle$$

$$\Rightarrow e^{-\frac{iH_0t}{\hbar}}\hat{\Omega}_I(t)e^{\frac{iH_0t}{\hbar}} = \hat{\Omega}_S \qquad [4]$$

$$\Rightarrow \left\langle \hat{\Omega}_I(t) \right\rangle = e^{\frac{iH_0t}{\hbar}}\hat{\Omega}_S e^{-\frac{iH_0t}{\hbar}}$$

$$\Rightarrow \frac{d\left\langle \hat{\Omega}_I(t)\right\rangle}{dt} = \frac{iH_0}{\hbar}e^{\frac{iH_0t}{\hbar}}\hat{\Omega}_S e^{-\frac{iH_0t}{\hbar}} + e^{\frac{iH_0t}{\hbar}}\hat{\Omega}_s\left(-\frac{iH_0}{\hbar}\right)e^{-\frac{iH_0t}{\hbar}}$$

$$\Rightarrow \frac{d\left\langle \hat{\Omega}_I(t)\right\rangle}{dt} = \frac{iH_0}{\hbar}\hat{\Omega}_I(t) - \frac{i}{\hbar}\hat{\Omega}_I(t)H_0$$

$$\Rightarrow \frac{d\left\langle \hat{\Omega}_I(t)\right\rangle}{dt} = \frac{i}{\hbar}\left(H_0\hat{\Omega}_I(t) - \hat{\Omega}_I(t)H_0\right)$$

$$\Rightarrow \frac{d\left\langle \Omega_I(t)\right\rangle}{dt} = -\frac{1}{i\hbar}\left(H_0\hat{\Omega}_I(t) - \hat{\Omega}_I(t)H_0\right)$$

$$\Rightarrow i\hbar\frac{d\left\langle \Omega_I(t)\right\rangle}{dt} = \left(\hat{\Omega}_I(t)H_0 - H_0\hat{\Omega}_I(t)\right)$$

$$\Rightarrow i\hbar\frac{d\left\langle \Omega_I(t)\right\rangle}{dt} = \left\langle\left[\Omega_I(t), H_0\right]\right\rangle \qquad [5]$$

Equation [5] is the Dynamical equation of Dirac or Interaction picture.

12.13 Highlight the features of three pictures in the description of the time evaluation of a microscopic system

- **Schrödinger Picture:** The form of the operators is time independent & the wave functions are time dependent. *I.e.,*

$$\Omega_s = \text{Constant} \ \& \ \Psi_s = \Psi_s(t)$$

The time evaluation of the wave functions is given by the Schrödinger equation

$$i\hbar\frac{\partial\left|\Psi_s(t)\right\rangle}{\partial t} = H\left|\Psi_s(\mathrm{t})\right\rangle \qquad [1]$$

The formal solution of [1] is,

$$\left|\Psi_s(t)\right\rangle = e^{-\frac{iHt}{\hbar}}\left|\Psi_s(0)\right\rangle \qquad [2]$$

The expectation value of any physical observable depends on time through the time dependence of the wave function.

$$\left\langle\Omega_s\right\rangle_t = \left(\psi_s(t), \hat{\Omega}_s\psi_s(t)\right) \qquad [3]$$

In bra and ket notation, the above equation will be

$$\left\langle\hat{\Omega}_s\right\rangle_t = \left\langle\psi_s(t)\right|\hat{\Omega}_s\left|\psi_s(t)\right\rangle \qquad [4]$$

Therefore, equation [4] becomes,

$$<\hat{\Omega}_s>_t = \left\langle\Psi_s(0)\right| e^{\frac{iHt}{\hbar}}\hat{\Omega}_s e^{-\frac{iHt}{\hbar}}\left|\Psi_s(0)\right\rangle$$

$$<\Omega_s>_t = e^{\frac{iHt}{\hbar}}\hat{\Omega}_s e^{-\frac{iHt}{t}} \qquad [5]$$

$$\Rightarrow \frac{\partial<\hat{\Omega}_s>_t}{\partial t} = \frac{iH}{\hbar}e^{\frac{iHt}{\hbar}}\hat{\Omega}_s e^{-\frac{iHt}{\hbar}} + e^{\frac{iHt}{\hbar}}\hat{\Omega}_s\left(-\frac{iH}{\hbar}\right)e^{-\frac{iHt}{\hbar}}$$

$$\Rightarrow \frac{\partial <\hat{\Omega}_s>_t}{\partial t} = \frac{iH}{\hbar} <\hat{\Omega}_s>_t - e^{\frac{iHt}{\hbar}} \hat{\Omega}_s e^{-\frac{iHt}{\hbar}} \frac{iH}{\hbar}$$

$$\Rightarrow \frac{\partial <\hat{\Omega}_s>_t}{\partial t} = \frac{i}{\hbar}\left(H <\hat{\Omega}_s>_t - <\hat{\Omega}_s>_t H\right)$$

$$\Rightarrow \frac{\partial <\hat{\Omega}_s>_t}{\partial t} = -\frac{1}{i\hbar}\left(H <\hat{\Omega}_s>_t - <\hat{\Omega}_s>_t H\right)$$

$$\Rightarrow i\hbar\frac{\partial <\hat{\Omega}_s>_t}{\partial t} = -\left(H <\hat{\Omega}_s>_t - <\hat{\Omega}_s>_t H\right)$$

$$\Rightarrow i\hbar\frac{\partial <\hat{\Omega}_s>_t}{\partial t} = \left(<\hat{\Omega}_s>_t H - H <\hat{\Omega}_s>_t\right)$$

$$\Rightarrow i\hbar\frac{\partial <\hat{\Omega}_s>_t}{\partial t} = \left\langle\left[\hat{\Omega}_s, H\right]\right\rangle \quad [6]$$

- **Heisenberg Picture:** The forms of the operators are time dependent whereas the wave functions are time independent. I.*e.,*

$$\Omega_H = \Omega_H(t) \quad \& \quad \Psi_H = \text{Constant}$$

The time dependence of the operators in the Heisenberg picture is obtained from the equality of the expectation value.

$$\left\langle\hat{\Omega}_H(t)\right\rangle = \left(\psi_H, \hat{\Omega}_H(t)\psi_H\right) \quad [7]$$

In bra and ket notation, the above equation will be

$$\left\langle\hat{\Omega}_H(t)\right\rangle = \left\langle\psi_H\middle|\hat{\Omega}_H(t)\middle|\psi_H\right\rangle \quad [8]$$

Again, the time dependence of the operators in Schrödinger picture is obtained from the expectation value

$$\begin{aligned}\langle\Omega_s\rangle_t &= \left(\psi_t(t), \hat{\Omega}_s\psi_s(t)\right)\\ &= \left\langle\psi_s(t)\middle|\hat{\Omega}_s\middle|\psi_s(t)\right\rangle\end{aligned} \quad [9]$$

Since the expectation value of the operator in the two pictures should be identical.

$$\left\langle\Psi_H\middle|\hat{\Omega}_H(t)\middle|\Psi_H\right\rangle = \left\langle\Psi_S(t)\middle|\hat{\Omega}_S\middle|\Psi_S(t)\right\rangle \quad [10]$$

From equation [2] $\left|\Psi_S(t)\right\rangle = e^{-\frac{iHt}{\hbar}}\left|\Psi_S(0)\right\rangle \Rightarrow \left|\Psi_S(t)\right\rangle = e^{-\frac{iHt}{\hbar}}\left|\Psi_H\right\rangle$ [11]

Because, $\left|\Psi_S(0)\right\rangle$ actually defines the state vector in Heisenberg picture which does not change with time & retains its initial value as time progresses.

Therefore [10] becomes

$$\langle \Psi_H | \hat{\Omega}_H(t) | \Psi_H \rangle = \langle \Psi_H | e^{\frac{iHt}{\hbar}} \hat{\Omega}_S e^{-\frac{iHt}{\hbar}} | \Psi_H \rangle$$

$$\Rightarrow \left\langle \hat{\Omega}_H(t) \right\rangle = e^{\frac{iHt}{\hbar}} \hat{\Omega}_S e^{-\frac{iHt}{\hbar}} \qquad [12]$$

$$\Rightarrow \frac{d\left\langle \hat{\Omega}_H(t) \right\rangle}{dt} = \frac{iH}{\hbar} e^{\frac{iHt}{\hbar}} \hat{\Omega}_S e^{-\frac{iHt}{\hbar}} + e^{\frac{iHt}{\hbar}} \hat{\Omega}_S \left(-\frac{iH}{\hbar}\right) e^{-\frac{iHt}{\hbar}}$$

$$\Rightarrow \frac{d\left\langle \hat{\Omega}_H(t) \right\rangle}{dt} = \frac{iH}{\hbar} \hat{\Omega}_H(t) + e^{\frac{iHt}{\hbar}} \hat{\Omega}_S e^{-\frac{iHt}{\hbar}} \left(-\frac{iH}{\hbar}\right)$$

$$\Rightarrow \frac{d\left\langle \hat{\Omega}_H(t) \right\rangle}{dt} = \frac{iH}{\hbar} \hat{\Omega}_H(t) - \hat{\Omega}_H(t) \frac{iH}{\hbar}$$

$$\Rightarrow \frac{d\left\langle \hat{\Omega}_H(t) \right\rangle}{dt} = \frac{i}{\hbar} \left(H \hat{\Omega}_H(t) - \hat{\Omega}_H(t) H \right)$$

$$\Rightarrow \frac{d\left\langle \hat{\Omega}_H(t) \right\rangle}{dt} = -\frac{1}{i\hbar} \left(H \hat{\Omega}_H(t) - \hat{\Omega}_H(t) H \right)$$

$$\Rightarrow i\hbar \frac{d\left\langle \hat{\Omega}_H(t) \right\rangle}{dt} = \left\langle \left(\hat{\Omega}_H(t) H - H \hat{\Omega}_H(t) \right) \right\rangle \Rightarrow i\hbar \frac{d\left\langle \hat{\Omega}_H(t) \right\rangle}{dt} = \left\langle \left[\hat{\Omega}_H(t), H \right] \right\rangle \qquad [13]$$

❖ **Interaction Picture:** In this picture, both the operators & the wave functions are time dependent.

$$\Omega_I = \Omega_I(t) \quad \& \quad \Psi_I = \Psi_I(t)$$

Now, we split the Hamiltonian into two parts

$$H = H_0 + H'(t) \qquad [14]$$

Where, $H_0 =$ Unperturbed Hamiltonian

$H'(t) =$ Perturbation Hamiltonian

The time dependence of the operators in the interaction picture is obtained from the equality of the expectation value.

$$\langle \Psi_I(t) | \hat{\Omega}_I(t) | \Psi_I(t) \rangle = \langle \Psi_S(t) | \hat{\Omega}_S(t) | \Psi_S(t) \rangle \qquad [15]$$

The time dependence of the wave function is taken to be,

$$|\Psi_I(t)\rangle = e^{\frac{iH_0 t}{\hbar}} |\Psi_S(t)\rangle \qquad [16]$$

Therefore [15] becomes,

$$\left\langle \Psi_s(t)\right| e^{-\frac{iH_0t}{\hbar}} \hat{\Omega}_I(t)\, e^{\frac{iH_0t}{\hbar}} \left|\Psi_s(t)\right\rangle = \left\langle \Psi_S(t)\right| \hat{\Omega}_S \left|\Psi_S(t)\right\rangle$$

$$\Rightarrow e^{-\frac{iH_0t}{\hbar}} \hat{\Omega}_I(t)\, e^{\frac{iH_0t}{\hbar}} = \hat{\Omega}_S \qquad [17]$$

$$\Rightarrow \left\langle \hat{\Omega}_I(t) \right\rangle = e^{\frac{iH_0t}{\hbar}} \hat{\Omega}_S\, e^{-\frac{iH_0t}{\hbar}} \qquad [18]$$

$$\Rightarrow \frac{d\left\langle \hat{\Omega}_I(t)\right\rangle}{dt} = \frac{iH_0}{\hbar} e^{\frac{iH_0t}{\hbar}} \hat{\Omega}_S\, e^{-\frac{iH_0t}{\hbar}} + e^{\frac{iH_0t}{\hbar}} \hat{\Omega}_s \left(-\frac{iH_0}{\hbar}\right) e^{-\frac{iH_0t}{\hbar}}$$

$$\Rightarrow \frac{d\left\langle \hat{\Omega}_I(t)\right\rangle}{dt} = \frac{iH_0}{\hbar} \hat{\Omega}_I(t) - \frac{i}{\hbar} \hat{\Omega}_I(t) H_0$$

$$\Rightarrow \frac{d\left\langle \hat{\Omega}_I(t)\right\rangle}{dt} = \frac{i}{\hbar}\left(H_0 \hat{\Omega}_I(t) - \hat{\Omega}_I(t) H_0 \right)$$

$$\Rightarrow \frac{d\left\langle \Omega_I(t)\right\rangle}{dt} = -\frac{1}{i\hbar}\left(H_0 \hat{\Omega}_I(t) - \hat{\Omega}_I(t) H_0 \right)$$

$$\Rightarrow i\hbar \frac{d\left\langle \Omega_I(t)\right\rangle}{dt} = \left(\hat{\Omega}_I(t) H_0 - H_0 \hat{\Omega}_I(t) \right)$$

$$\Rightarrow i\hbar \frac{d\left\langle \Omega_I(t)\right\rangle}{dt} = \left\langle \left[\Omega_I(t), H_0 \right] \right\rangle \ldots\ldots\ldots [19]$$

The expectation values remain the same in the different pictures although the operators and the state vectors are differ[11]. ***These are the Highlight features of the three pictures.***

12.14 Time development is governed solely by the interacting Hamiltonian

Time development of Schrödinger equation in ket form

$$i\hbar \frac{\partial \left|\Psi_S(t)\right\rangle}{\partial t} = H \left|\Psi_S(t)\right\rangle$$

$$\Rightarrow i\hbar \frac{\partial \left|\Psi_S(t)\right\rangle}{\partial t} = \left(H_0 + H'(t)\right)\left|\Psi_S(t)\right\rangle \qquad [1]$$

The time dependence of the wave function is taken to be

$$\left|\Psi_I(t)\right\rangle = e^{\frac{iH_0t}{\hbar}} \left|\Psi_S(t)\right\rangle \qquad [2]$$

$$\Rightarrow \frac{\partial \left|\Psi_I(t)\right\rangle}{\partial t} = \frac{\partial}{\partial t}\left[e^{\frac{iH_0t}{\hbar}} \left|\Psi_S(t)\right\rangle \right]$$

$$\Rightarrow \frac{\partial|\Psi_I(t)\rangle}{\partial t}=\frac{iH_0}{\hbar}e^{\frac{iH_0t}{\hbar}}|\Psi_S(t)\rangle+e^{\frac{iH_0t}{\hbar}}\frac{\partial|\Psi_S(t)\rangle}{\partial t}$$

$$\Rightarrow i\hbar\frac{\partial|\Psi_I(t)\rangle}{\partial t}=i\hbar\left\{\frac{iH_0}{\hbar}e^{\frac{iH_0t}{\hbar}}|\Psi_S(t)\rangle+\frac{1}{i\hbar}(H_0+H'(t))e^{\frac{iH_0t}{\hbar}}|\Psi_S(t)\rangle\right\}$$

$$\Rightarrow i\hbar\frac{\partial|\Psi_I(t)\rangle}{\partial t}=i\hbar\left\{\frac{iH_0}{\hbar}e^{\frac{iH_0t}{\hbar}}|\Psi_S(t)\rangle-\frac{iH_0}{\hbar}e^{\frac{iH_0t}{\hbar}}|\Psi_S(t)\rangle-\frac{iH'(t)}{\hbar}e^{\frac{iH_0t}{\hbar}}|\Psi_S(t)\rangle\right\}$$

$$\Rightarrow i\hbar\frac{\partial|\Psi_I(t)\rangle}{\partial t}=H'(t)e^{\frac{iH_0t}{\hbar}}|\Psi_S(t)\rangle=H'(t)|\Psi_I(t)\rangle$$

$$Finally,\ i\hbar\frac{\partial|\Psi_I(t)\rangle}{\partial t}=H'(t)|\Psi_I(t)\rangle \qquad [3]$$

The above equation shows that, the time development in governed solely by the interacting Hamiltonian[5]. This is the great advantage of Dirac Picture. From equation [19] in the section [10.12], the equation of motion for $\Omega_I(t)$ is determined by H_0 only. They are the equations of motion of the system without interactions. Hence in the Dirac Picture, the operators obey free equations of motion which is another advantage. Thus the evolution in time of the dynamical variables in the Dirac picture corresponds to free motion [Not influenced by $H'(t)$]. Whereas the time development of the state vector is given entirely by the interaction Hamiltonian $H'(t)$ I.e., by the dynamics of the system only.

12.15 Advantage and disadvantage of Dirac or Interaction picture

- The time development is governed solely by the interacting Hamiltonian. I.e.,

$$i\hbar\frac{\partial|\Psi_I(t)\rangle}{\partial t}=H'(t)|\Psi_I(t)\rangle$$

  This is a great advantage of Dirac Picture.
- In the Dirac or Interaction picture, the operators obey free equations of motion. I.e.,

$$i\hbar\frac{d\langle\Omega_I(t)\rangle}{dt}=\langle[\Omega_I(t),H_0]\rangle$$; Which is the another advantage.

- The Dirac Picture is widely used in Quantum Field Theory[**QFT**][5]
- The equation of motion for $\hat{\Omega}_I(t)$ is determined by H_0 only I.e., they are the equations of motion of the system without interactions
- The interaction picture is a half way between the Schrödinger and Heisenberg pictures, and is particularly suited to develop the perturbation theory.

❖ Disadvantage of Dirac or Interaction picture

- The disadvantage of Dirac or Interaction picture is that we must split the Hamiltonian into two parts I.e.,

$$H = H_0 + H'(t)$$

Where, $H_0 =$ Unperturbed Hamiltonian

$H'(t) =$ Perturbation Hamiltonian

12.16 Hamiltonian Operator in Heisenberg picture is independent of time

The Hamiltonian operator for Harmonic Oscillator,

$$\hat{H} = \frac{p^2}{2m} + \frac{1}{2}\omega^2 q^2$$

$$\Rightarrow \hat{H} = \frac{p^2}{2} + \frac{1}{2}\omega^2 q^2 \qquad [for\, unit\, mass\; m = 1] \qquad [1]$$

We have again,

$$\frac{d\hat{q}}{dt} = \frac{i}{\hbar}\left[\hat{H}, \hat{q}\right] = \frac{i}{\hbar}\left(\hat{H}\hat{q} - \hat{q}\hat{H}\right)$$

$$= \frac{i}{\hbar}\left\{\left(\frac{\hat{p}^2}{2} + \frac{1}{2}\omega^2\hat{q}^2\right)\hat{q} - \hat{q}\left(\frac{\hat{p}^2}{2} + \frac{1}{2}\omega^2\hat{q}^2\right)\right\}$$

$$= \frac{i}{\hbar}\left\{\frac{1}{2}\hat{p}^2\hat{q} + \frac{1}{2}\omega^2\hat{q}^2\hat{q} - \frac{1}{2}\hat{q}\hat{p}^2 - \frac{1}{2}\omega^2\hat{q}\hat{q}^2\right\}$$

$$= \frac{i}{2\hbar}\left\{\left(\hat{p}^2\hat{q} - \hat{q}\hat{p}^2\right) + \omega^2\left(\hat{q}^2\hat{q} - \hat{q}\hat{q}^2\right)\right\}$$

$$= \frac{i}{2\hbar}\left\{\left[\hat{p}^2, \hat{q}\right] + \omega^2\left[\hat{q}^2, \hat{q}\right]\right\}$$

$$= \frac{i}{2\hbar}\{[\hat{p}\hat{p}, \hat{q}] + \omega^2[\hat{q}\hat{q}, \hat{q}]\}$$

$$= \frac{i}{2\hbar}\{(\hat{p}\hat{p}\hat{q} - \hat{q}\hat{p}\hat{p}) + \omega^2(\hat{q}\hat{q}\hat{q} - \hat{q}\hat{q}\hat{q})\}$$

$$= \frac{i}{2\hbar}\left\{\hat{p}\left(\frac{\hbar}{i} + \hat{q}\hat{p}\right) - \left(\hat{p}\hat{q} - \frac{\hbar}{i}\right)\hat{p}\right\} \quad \left\{\because [\hat{p}, \hat{q}] = (\hat{p}\hat{q} - \hat{q}\hat{p}) = \frac{\hbar}{i}\right\}$$

$$= \frac{i}{2\hbar}\left\{\frac{\hbar}{i}\hat{p} + \hat{p}\hat{q}\hat{p} - \hat{p}\hat{q}\hat{p} + \frac{\hbar}{i}\hat{p}\right\}$$

$$= \frac{i}{2\hbar}\left(\frac{2\hbar}{i}\hat{p}\right)$$

$$\frac{d\hat{q}}{dt} = \hat{p} \qquad [2]$$

And $$\frac{d^2\hat{q}}{dt^2} = \frac{d\hat{p}}{dt} \qquad [3]$$

We have again,

$$\frac{d\hat{p}}{dt} = \frac{i}{\hbar}\left[\hat{H},\hat{p}\right]$$

$$= \frac{i}{\hbar}\left(\hat{H}\hat{p} - \hat{p}\hat{H}\right)$$

$$= \frac{i}{\hbar}\left\{\left(\frac{\hat{p}^2}{2} + \frac{1}{2}\omega^2\hat{q}^2\right)\hat{p} - \hat{p}\left(\frac{\hat{p}^2}{2} + \frac{1}{2}\omega^2\hat{q}^2\right)\right\}$$

$$= \frac{i}{\hbar}\left\{\frac{1}{2}\hat{p}^2\,\hat{p} + \frac{1}{2}\omega^2\hat{q}^2\hat{p} - \frac{1}{2}\hat{p}\,\hat{p}^2 - \frac{1}{2}\omega^2\,\hat{p}\hat{q}^2\right\}$$

$$= \frac{i}{\hbar}\left\{\frac{1}{2}\left(\hat{p}^2\,\hat{p} - \hat{p}\,\hat{p}^2\right) + \frac{1}{2}\omega^2\left(\hat{q}^2\,\hat{p} - \hat{p}\hat{q}^2\right)\right\}$$

$$= \frac{i}{2\hbar}\{(\hat{p}\hat{p}\hat{p} - \hat{p}\hat{p}\hat{p}) + \omega^2(\hat{q}\hat{q}\,\hat{p} - \hat{p}\hat{q}\hat{q})\}$$

$$= \frac{i}{2\hbar}\omega^2\left\{\hat{q}\left(\hat{p}\hat{q} - \frac{\hbar}{i}\right) - \left(\frac{\hbar}{i} + \hat{q}\,\hat{p}\right)\hat{q}\right\} \quad \left\{\because [\hat{p},\hat{q}] = (\hat{p}\,\hat{q} - \hat{q}\hat{p}) = \frac{\hbar}{i}\right\}$$

$$= \frac{i}{2\hbar}\omega^2\left\{\hat{q}\,\hat{p}\hat{q} - \frac{\hbar}{i}\hat{q} - \frac{\hbar}{i}\hat{q} - \hat{q}\,\hat{p}\hat{q}\right\}$$

$$= \frac{i}{2\hbar}\omega^2\left(-\frac{2\hbar}{i}\hat{q}\right)$$

$$\Rightarrow \frac{d\hat{p}}{dt} = -\omega^2\hat{q} \qquad [4]$$

Therefore, equation [3] becomes

$$\frac{d^2\hat{q}}{dt^2} = -\omega^2\hat{q} \quad \Rightarrow \ddot{\hat{q}} = -\omega^2\hat{q} \quad \Rightarrow \ddot{\hat{q}} + \omega^2\hat{q} = 0 \qquad [5]$$

This is exactly the same form as the classical equation of motion. The general solution of equation [5] is

$$\hat{q}(t) = \hat{q}_0\cos\omega t + \frac{\hat{p}_0}{\omega}\sin\omega t \qquad [6]$$

Where $\hat{q}_0$ & $\hat{p}_0$ are constant operators. At $t = 0$, equation [6] becomes,

$$\hat{q}(0) = \hat{q}_0 \Rightarrow \hat{q} = \hat{q}_0 \qquad [7]$$

Now differentiating equation [6] with respect to time

$$\frac{d\hat{q}(t)}{dt} = -\hat{q}_0\omega\sin\omega t + \hat{p}_0\cos\omega t$$

$$\Rightarrow \hat{p}(t) = -\hat{q}_0\omega\sin\omega t + \hat{p}_0\cos\omega t \qquad [8]$$

At $t = 0$, equation [8] becomes,

$$\hat{p}(0) = \hat{p}_0 \Rightarrow \hat{p} = \hat{p}_0 \qquad [9]$$

$$\therefore \hat{q}(t) = \hat{q}\cos\omega t + \frac{\hat{p}}{\omega}\sin\omega t \qquad [10]$$

$$\& \ \hat{p}(t) = -\hat{q}\omega \sin \omega t + \hat{p} \cos \omega t \qquad [11]$$

It will therefore be advantageous to set $t_0 = 0$, [because $\hat{\Omega}(t) = T^{-1}(t,t_0)\ \Omega\ T(t,t_0)$ & the initial condition $\mathrm{T}(\mathrm{t},\mathrm{t}_0) = 1$] the Heisenberg & Schrödinger operators coincide at $t = t_0 = 0$. Now denoting the fixed Schrödinger operators corresponding to $\hat{q}\ and\ \hat{p}$ by q and p hence, equations [10] and [11] become,

$$\hat{q}(t) = q\cos \omega t + \frac{p}{\omega}\sin \omega t \ \& \ \hat{p}(t) = -q\omega \sin \omega t + p\cos \omega t$$

Again from [1]

$$\hat{H} = \frac{p^2}{2} + \frac{1}{2}\omega^2 q^2 = \frac{1}{2}\left\{(-q\omega \sin \omega t + p\cos \omega t)^2 + \omega^2\left(\mathrm{q}\cos \omega t + \frac{p}{\omega}\sin \omega t\right)^2\right\}$$

$$\hat{H} = \frac{1}{2}\left\{\omega^2\left(q^2\cos^2 \omega t + \frac{p^2}{\omega^2}\sin^2 \omega t + 2\frac{pq}{\omega}\cos \omega t \sin \omega t\right)\right\}$$

$$+\frac{1}{2}\{q^2\omega^2 \sin^2 \omega t - 2pq\omega \sin \omega t \cos \omega t + p^2\cos^2 \omega t\}$$

$$\hat{H} = \frac{1}{2}\{\omega^2 q^2 \cos^2 \omega t + p^2 \sin^2 \omega t + 2pq\omega \sin \omega t \cos \omega t\}$$

$$+\frac{1}{2}\{q^2\omega^2 \sin^2 \omega t + p^2 \cos^2 \omega t - 2pq\omega \sin \omega t \cos \omega t\}$$

$$\hat{H} = \frac{1}{2}\{p^2(\sin^2 \omega + \cos^2 \omega) + \omega^2 q^2(\sin^2 \omega + \cos^2 \omega)\} = \frac{1}{2}(p^2 + \omega^2 q^2)$$

$$\hat{H} = \frac{1}{2}(p^2 + \omega^2 q^2)$$

Therefore, we can see from the above equation, $\hat{\mathrm{H}}$ is independent of time. If we have a conservation system, so that $\frac{d\hat{H}}{dt} = 0$

12.17 Connection between three pictures by unitary transformation

The pictures should be connected by Unitary Operator because the expectation value of any Operator remains unchanged in all pictures. As a result, the Operators transform between the pictures is a similarity of transformation induced unitary transformation.

12.18 State vector of Heisenberg picture is time independent

The state vector of the Heisenberg picture is independent of time, so the state vector is the same as that in the Schrodinger picture[1]. This can be proved in the followings way:

$$\hat{\varphi}(\mathrm{t}) = \mathrm{T}^{-1}(\mathrm{t},\mathrm{t}_0)\varphi(\mathrm{t})$$

$$= \mathrm{T}^{-1}(\mathrm{t},\mathrm{t}_0)\mathrm{T}(\mathrm{t},\mathrm{t}_0)\varphi(\mathrm{t}_0) \quad \text{where } \varphi(\mathrm{t}) = \mathrm{T}(\mathrm{t},\mathrm{t}_0)\varphi(\mathrm{t}_0)$$

Here, T^{-1} is the unitary operator which makes a transformation from t to t_0.

$$\therefore \quad \hat{\varphi}(t) = T(t_0, t) T(t, t_0)\, \varphi(t_0)$$

$$\Rightarrow \hat{\varphi}(t) = T(t_0, t_0)\varphi(t_0) \Rightarrow \hat{\varphi}(t) = \varphi(t_0);\ with\ \ T(t_0, t_0) = 1$$

$$Finally\ \ \hat{\varphi}(t) = \varphi(t_0)$$

12.19 Importance of unitary transformation

We can make a unitary transformation on both Ω & Ψ with the help of unitary operator.

$$\Omega \to \Omega' = U\,\Omega\,U^{-1} \quad \text{---------------------------------} [1]$$

$$\Psi \to \Psi' = U\,\Psi \quad \text{---------------------------------} [2]$$

Such a unitary transformation leaves the system completely unchanged. This can be proved by showing that the expectation value remains the same.

$$(\Psi', \Omega'\Psi') = (U\Psi, \Omega' U\Psi')$$

$$\Rightarrow (\Psi', \Omega'\Psi') = U^{\times}(\Psi, \Omega' U\Psi) \qquad \left[\because (\vec{a}, \lambda\vec{b}) = \lambda^{\times}(\bar{a}, \bar{b})\right]$$

$$\Rightarrow (\Psi', \Omega'\Psi') = (\Psi, U^{\times}\Omega' U\Psi) \qquad \left[(\bar{a}, \lambda)\bar{b} = \lambda(\bar{a}, \bar{b})\ \right]$$

$$\Rightarrow (\Psi', \Omega'\Psi') = (\Psi, U^{\times} U \Omega U^{-1} U\Psi)$$

$$\Rightarrow (\Psi', \Omega'\Psi') = (\Psi, \Omega\Psi) \quad \text{---------------------------------} [3]$$

From the above equation [3] it is clear that *unitary transformation leaves the system completely unchanged. It is the greatest importance of unitary transformation.*

Exercise

1. Explain State vector of a system and state vector transformation.
2. Write time development of Schrödinger, Heisenberg and Dirac or Interaction picture.
3. Determine the dynamical equation of Schrödinger picture.
4. Obtained the solution of Schrödinger picture.
5. What is the necessary and sufficient condition of Schrödinger picture?
6. Write the advantage and disadvantage of Schrödinger picture.
7. Show that Hamiltonian of a system can be regarded as the generator of infinitesimal canonical transformation.
8. Determine the dynamical equation of Heisenberg picture.
9. Show that Heisenberg equations of motion have the same form as Classical equations of motion.
10. What is the Beauty of Heisenberg picture?
11. What is the reason of the Heisenberg equations of motion are more difficult to solve?
12. Write the advantage and disadvantage of Heisenberg picture.
13. Determine the dynamical equation of Dirac or Interaction picture.
14. Highlight the features of three pictures in the description of the time evaluation of a microscopic system.
15. Prove that the expectation values remain the same in the different pictures although, the operators and the state vectors are differ.
16. Show that the time development is governed solely by the interacting Hamiltonian in Dirac or Interaction picture.
17. Write the advantage and disadvantage of Dirac or Interaction picture.
18. Show that the Hamiltonian Operator in Heisenberg picture is independent of time.
19. How three pictures are connected to each other by unitary transformation?
20. Show that the state vector of Heisenberg picture is time independent.
21. What is the importance of unitary transformation?

Chapter 13 Quantum Theory of Scattering

13.1 Differential Scattering cross- section and total scattering cross- section

A Schematic picture of a scattering experiment is shown below. We have a source of particles followed by a collimator (not shown in picture) which makes the beam of particles.

Consider a beam of particles moving along z-direction with a flux of N particles I.e., the number of incident particles crossing per unit time a unit area. The incident particles fall on the target system and the particles are scattering from the target in all directions and detected by the detector which is placed at angle (θ, ϕ) to the z-axis[5].

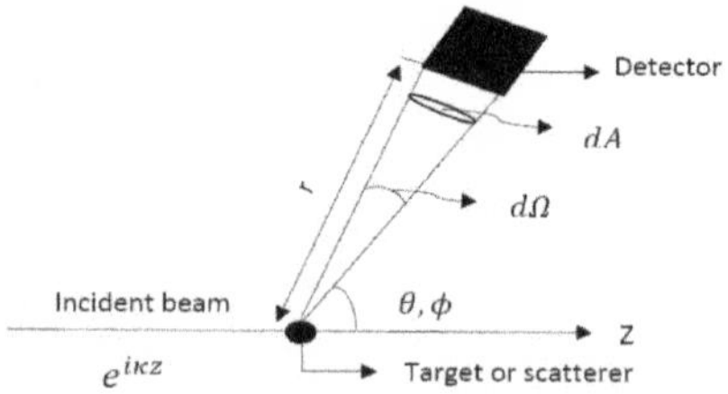

Fig.[13.1]: Shows, the scattering of an incident beam. The particles scattered into the solid angle $d\Omega$ are received by the detector.

If dN be the number of particles scattered per unit time into the solid angle $d\Omega$ located in the direction θ & φ with respect to the bombarding direction as polar axis; the $d\Omega$ will be proportional to [i] The incident flux and [ii] The solid angle $d\Omega$

$$dN \propto N d\Omega$$

$$\text{Or,} \quad dN = \sigma(\theta, \phi)\, N\, d\Omega \qquad [1]$$

Where, $\delta(\theta, \varphi)$ is the proportionality constant and has the dimension of an area. It can be regarded as the differential scattering cross-section of the incident beam through which all the particles pass that are scattered into the solid angle $d\Omega$ about θ & ϕ. From equation [1] which is defined

$$\sigma(\theta, \phi) d\Omega = \frac{dN}{N} \qquad [2]$$

$$= \frac{Numbers\ of\ particles\ scattered\ into\ the\ soild\ angle\ per\ unit\ time}{Insident\ intensity}$$

Therefore, total scattering cross-section

$$N_{scatt} = \int dN = \int \ \sigma(\theta, \phi)\, N\, d\Omega$$

Or, $$N_{scatt} = N \int \ \sigma(\theta, \phi)\, d\Omega = N\sigma_{tot}$$

Or, $$N_{scatt} = N\ \sigma_{tot} \qquad [3]$$

Where, $$\sigma_{tot} = \int \sigma(\theta, \phi)\, d\Omega \qquad [4]$$

$\therefore$ $$\sigma_{tot} = \frac{N_{scatt}}{N} \qquad [5]$$

The cross-section is usually measured in barns or milibarns. [1 barn $=10^{-24}$ cm^2 and 1 milibarns $=10^{-27}$ cm^2]

13.2 Dimension of differential scattering cross- section

From equation [1] in the section [13.1]

$$dN = \ \sigma N\, d\Omega$$

Where, dimension of $N = \frac{1}{L^2 T}$ & dimension of $dN = \frac{1}{T}$

& dimension of $d\Omega \ = 1$

$\therefore$ $$\sigma = \frac{dN}{N\, d\Omega} = \frac{T^{-1}}{L^{-2}T^{-1}} = L^2$$

Or, $$\sigma = \ L^2 \qquad [1]$$

The dimension of differential scattering cross- section is an area.

13.3 Typical elements of scattering and basic assumptions of scattering

Typical elements of scattering:

- Projectile
- Collimator
- Target
- Detector

Basic assumptions or condition for scattering:

- The width of the beam is kept narrow
- The intensity of the incident beam is assumed to be low enough
- The scattering is elastic
- Each scattering involves only one collision

13.4 Laboratory system and Centre of mass system

Let us consider two masses m_1 *and* m_2 and locate these masses by radii vectors $\boldsymbol{r_1}$ *and* $\boldsymbol{r_2}$ with respect to the origin. The radii $\boldsymbol{r_1}$ *and* $\boldsymbol{r_2}$ are called laboratory coordinates. Therefore, "A coordinate system in which the bombarded particles or target is initially at rest is known as laboratory system".

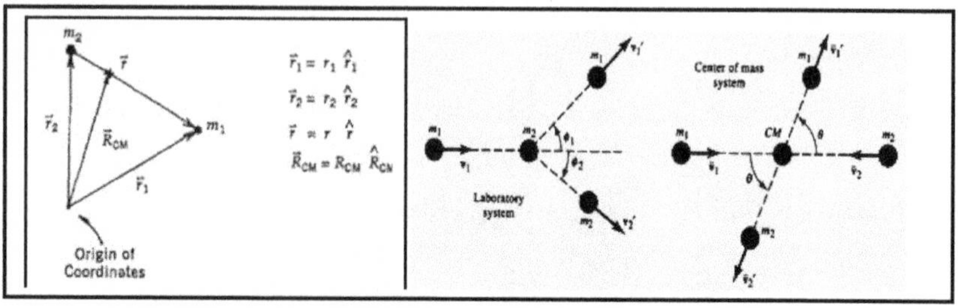

Fig.[13.4]: Shows, Laboratory system and Centre of mass system

Further, the two masses m_1 *and* m_2 can also be located with respect to the centre of mass by radii vectors $\boldsymbol{r_1'}$ *and* $\boldsymbol{r_2'}$ respectively. These are called centre of mass coordinates."A coordinate system in which the centre of mass of the two colliding particles is initially and always at rest is known as the centre of mass system."

13.5 Study of scattering experiment in Quantum mechanics and types of Scattering

- ### Study of Scattering experiment in Quantum mechanics

In modern Physics, scattering experiments and other analysis with the help of Quantum mechanics are the principal tools by means of which knowledge about Nuclear and elementary particles are obtained. The theory of scattering I.e., deflection of a particle by a force, centre is therefore very important. Thus in Nuclear Physics, we study scattering of Protons, Neutrons, Deuterons etc. from various target systems and gain information about detailed properties of Nuclear forces which are even now not very well-understood. Then in high energy Nuclear Physics or elementary particle Physics, Electrons, Protons, Mesons etc. are scattered from suitable target and analysis of the experimental data leads to the new knowledge about elementary particle systems.

- ### Types of Scattering

Elastic scattering: The collision in which the energy of the incident particle does not change after its interaction is called elastic collision and scattering as elastic scattering[31].

Inelastic scattering: When the energy of the incident particle is changed after its interaction, the scattering is known as inelastic scattering.

13.6 Relation between Differential Scattering cross- section and Scattering amplitude

The Schrödinger equation for scattering problems in terms of the relative coordinate between the target and the projectile is given by

$$H\,\psi(\vec{r}) = E\,\psi(\vec{r})$$

Or,
$$-\frac{\kappa^2}{2\mu}\,\nabla^2\psi(\vec{r}) + V\,\psi(\vec{r}) = E\,\psi(\vec{r}) \qquad [1]$$

In order to formulation the scattering phenomena, we note that we have a system of two particles, the target and the projectile. Denoting the mass of one of these say target be m_1 and of the other say projectile by m_2. Therefore the reduced mass

$$\mu = \frac{m_1 m_2}{m_1 + m_2} \qquad [2]$$

Or,
$$-\frac{\hbar^2}{2\mu}\left[\,\nabla^2\psi(\vec{r}) - \frac{2\mu}{\hbar^2}V\,\psi(\vec{r})\right] = E\,\psi(\vec{r})$$

Or,
$$\nabla^2\psi(\vec{r}) - \frac{2\mu}{\hbar^2}V\,\psi(\vec{r}) = -\frac{2\mu}{\hbar^2}E\,\psi(\vec{r})$$

Or,
$$\nabla^2\psi(\vec{r}) - \frac{2\mu}{\hbar^2}V\,\psi(\vec{r}) + \frac{2\mu}{\hbar^2}E\,\psi(\vec{r}) = 0 \qquad [3]$$

Or,
$$\nabla^2\psi(\vec{r}) + \kappa^2\,\psi(\vec{r}) - U(\vec{r})\,\psi(\vec{r}) = 0 \qquad [4]$$

Where,
$$k^2 = \frac{2\,\mu\,E}{\hbar^2} \qquad [5]$$

&
$$U(\vec{r}) = \frac{2\,\mu\,V(\vec{r})}{\hbar^2} \qquad [6]$$

In the absence of the target $V = 0$ and for the incident beam moving along z-axis, the equation [4] becomes

$$\nabla^2\psi(z) + \kappa^2\,\psi(z) = 0 \qquad [7]$$

Where the solution is

$$\psi_{free}(z) = e^{i\kappa z} \qquad [8]$$

Equation [8] is representing the free particles moving along z-axis.

Now in the presence of the target, we must solve the equation [4], such that the wave function $\psi(\vec{r})$ is the sum of an incident plane wave $e^{i\kappa z}$ and outgoing spherical wave $\frac{e^{i\kappa r}}{r}$ at a large distance from the target. Thus we impose the boundary condition that when the scattered particle is far away from the target I.e., r is very large; the solution of equation [4] must be of the form

$$\psi(\vec{r}) = e^{i\kappa z} + f(\theta)\frac{e^{i\kappa r}}{r} \qquad [9]$$

We note from the equation [9], the first term represents a particle moving in positive z- direction [along the polar axis $\theta = 0$, since $z = r\cos\theta$ and it is an infinite plane wave]. The second term represents a particle that is moving radially outward, its amplitude depends upon θ & ϕ and inversely proportional to r.

Now we calculate the incident flux or the current density carried by the incident wave. For this purpose, we use the general expression for the current density.

$$\vec{J} = \frac{\hbar}{2i\mu}[\psi^{\times}\vec{\nabla}\psi - \psi\,\vec{\nabla}\psi^{\times}] \qquad [10]$$

The magnitude of the incident flux density along z-direction is

$$J_{inci} = \frac{\hbar}{2i\mu}[\,e^{-ikz}(i\kappa)e^{ikz} - e^{ikz}(-i\kappa)\,e^{-ikz}] = \frac{\hbar}{2i\mu}\,2i\kappa = \frac{\hbar\,\kappa}{\mu} \qquad [11]$$

But $$\kappa^2 = \frac{2\,\mu\,E}{\hbar^2} \quad Or,\ E = \frac{1}{2\mu}\hbar^2\,\kappa^2 = \frac{1}{2}\mu v^2 = \frac{1}{2\mu}\hbar^2\kappa^2 = v = \frac{\hbar\,\kappa}{\mu} \qquad [12]$$

Therefore equation [11] becomes

$$J_{incident} = v \qquad [13]$$

And the magnitude of the scattered flux J_{sactt} for large r,

$$J_{sactt} = \frac{\hbar}{2i\mu}[e^{-i\kappa r}\frac{f^{\times}(\theta)}{r}\frac{f(\theta)}{r}(i\kappa)e^{i\kappa r} - e^{i\kappa r}\frac{f(\theta)}{r}\frac{f^{\times}(\theta)}{r}(-i\kappa)e^{-i\kappa r}]$$

$$= \frac{\hbar}{2i\mu}\left[\frac{|f(\theta)|^2}{r^2}(i\kappa) + \frac{|f(\theta)|^2}{r^2}\,i\kappa\right] = \frac{\hbar}{2i\mu}2\,|f(\theta)|^2\,\frac{i\kappa}{r^2}$$

$$J_{sactt} = \frac{\kappa\,\hbar}{\mu}\frac{|f(\theta)|^2}{r^2} = v\,\frac{|f(\theta)|^2}{r^2} \qquad [14]$$

Suppose the scattered particle are intercepted by a detector of an area dA at a distance r. The solid angle ($d\Omega$) subtended by the detector at the scattering center is

$$d\Omega = \frac{dA}{r^2} \quad Or, dA = r^2\,d\Omega \qquad [15]$$

The number of particles crossing the area dA per unit time is

$$J_{sactt} = v\,\frac{|f(\theta)|^2}{r^2}\,dA = v\frac{|f(\theta)|^2}{r^2}r^2\,d\Omega = v|f(\theta)|^2 d\Omega \qquad [16]$$

Therefore, the differential scattering cross-section $\sigma(\theta)$ is defined by

$$\sigma(\theta)d\Omega = \frac{number\ of\ particle\ scattered\ into\ d\Omega\ per\ unit\ time}{number\ of\ particle\ incident\ per\ unit\ area\ per\ unit\ time} = \frac{v\ |f(\theta)|^2}{v}\ d\Omega$$

$$\sigma(\theta) = |f(\theta)|^2 \qquad [17]$$

The differential scattering cross-section is equal to the absolute value of the scattering amplitude.

13.7 Bauer's formula

Schrödinger equation for free particle

$$\nabla^2\,\Psi + \frac{2mE}{\hbar^2}\,\Psi = 0 \text{ Or, } (\nabla^2 + k^2)\Psi = 0\text{ ; where } k^2 = \frac{2mE}{\hbar^2}$$

Solutions of Schrödinger equation free particle are

$$\Psi = e^{i\vec{k}.\vec{r}}$$

as well as

$$\Psi = \frac{u_l(\rho)}{\rho} Y_{l,m}(\theta,\varphi)$$

Here $\rho = kr$ and u_l satisfies the differential equation

$$\frac{d^2u_l}{d\rho^2} + \left[1 - \frac{l(l+1)}{\rho^2}\right]u_l = 0 \text{ with } u_l \sim \sin\theta\left(\rho - l\,\frac{\pi}{2}\right)$$

Let us write $e^{i\vec{k}.\vec{r}}$ as linear combination of the set of functions $\frac{u_l(\rho)}{\rho} Y_{lm}(\theta,\varphi)$ to get

$$e^{i\vec{k}.\vec{r}} = \sum_l \sum_m C_{l,m} \frac{u_l(\rho)}{\rho} Y_{l,m}(\theta,\varphi) \qquad [1]$$

Which is meaningful if we can evaluate the expansion coefficients $C_{l,m}$'s. For simplicity, let $\vec{k}$ be along the z-axis. As such $e^{i\vec{k}.\vec{r}} = e^{ikz} = e^{ikr\cos\theta} = e^{i\rho\cos\theta}$; where θ is a coordinate of (r,θ,φ) system i.e., spherical polar coordinate system. L.H.S. of equation [1] contains r and θ, not φ. Hence we should eliminate φ dependence of R.H.S of equation [1]. This can be done simply by setting m =0. Then $Y_{l,m}(\theta,\varphi) = P_l(\cos\theta)$ which is Legendre polynomial[20,21,24]. Then equation [1] becomes

$$e^{i\rho\cos\theta} = \sum_l \alpha_l \frac{u_l(\rho)}{\rho} P_l(\cos\theta)$$

Or,

$$e^{i\rho\mu} = \sum_l \alpha_l \frac{u_l(\rho)}{\rho} P_l(\mu) \qquad [2]$$

Where, $\mu = \cos\theta$. To evaluate α_l and put it in equation [2] to get what is called Bauer's formula. Equation [2] gives

$$\int_{-1}^{+1} e^{i\rho\mu} P_{l'}(\mu)d\mu = \sum_l \alpha_l \frac{u_l(\rho)}{\rho}\int_{-1}^{+1} P_l(\mu)P_{l'}(\mu)\,d\mu = \sum_l a_l \frac{u_l(\rho)}{\rho}\,\frac{2}{2l+1}\delta_{ll'} = a_{l'}\frac{u_{l'}(\rho)}{\rho}\,\frac{2}{2l'+1}$$

$$a_l \frac{u_l(\rho)}{\rho}\,\frac{2}{2l+1} = \int_{-1}^{+1} e^{i\rho\mu}\, P_l(\mu)d\mu$$

$$a_l \frac{2}{2l+1}\frac{u_l(\rho)}{\rho} = \int_{-1}^{+1} (P_l(\mu))(e^{i\rho\mu})d\mu$$

$$= P_\ell(\mu)\frac{e^{ipu}}{i\rho}\Big|_{-1}^{+1} - \int_{-1}^{+1} \boldsymbol{P}_l'(\mu)\frac{e^{i\rho\mu}}{i\rho}d\mu$$

$$= P_l(+1)\frac{e^{ip}}{i\rho} - P_l(-1)\frac{e^{-i\rho}}{i\rho}\left[P_\ell'(\mu)\frac{e^{i\rho\mu}}{(i\rho)^2}\Big|_{-1}^{+1} - \int_{-1}^{+1} P_\ell''(\mu)\frac{e^{i\rho\mu}}{(i\rho)^2}d\mu\right]$$

Using $\int_a^b uvdx = \left[u\int vdx - \int\left(\frac{du}{dx}\int vdx\right)dx\right]\Big|_a^b$

Or, $a_\ell \frac{2}{2\ell+1}\frac{u_\ell(\rho)}{\rho} \sim P_l(+1)\frac{e^{i\rho}}{i\rho} - P_\ell(-1)\frac{e^{-i\rho}}{i\rho}$; neglecting terms $\sim\frac{1}{\rho^n}$ with $n \geq 2$

$$\sim\frac{1}{i\rho}\left[(1)e^{i\rho} - (-1)^l e^{-i\rho}\right] \sim\frac{i}{i\rho}\left[e^{i\rho} - (e^{i\pi})^l\, e^{-i\rho}\right]$$

$$\sim\frac{1}{i\rho}e^{il\frac{\pi}{2}}\left[e^{i(\rho-l\frac{\pi}{2})} - e^{-i(\rho-l\frac{\pi}{2})}\right] \sim\frac{i}{i\rho}e^{il\frac{\pi}{2}}\,2i\sin\left(\rho - l\frac{\pi}{2}\right)$$

$\therefore \quad a_l \frac{2}{2\ell+1}\frac{u_\ell(\rho)}{\rho} \sim \frac{2}{\rho}e^{i\ell\frac{\pi}{2}}\sin\left(\rho - \ell\frac{\pi}{2}\right)$

Or, $a_\ell \frac{2}{2\ell+1}\frac{\sin\left(\rho-\ell\frac{\pi}{2}\right)}{\rho} = \frac{2}{\rho}e^{i\ell\frac{\pi}{2}}\sin\left(\rho - \ell\frac{\pi}{2}\right) \;\therefore u_\ell(\rho) \sim \sin\left(\rho - \ell\frac{\pi}{2}\right)$

Or $\frac{a_l}{2\ell+1} = \left(e^{i\,\pi/2}\right)^l = i^l$ $\qquad$ *Thus* $\boldsymbol{a_\ell = i^\ell(2\ell+1)}$ [3]

Which we can use in equation [2] and get

$$e^{ikz} = \sum_l i^\ell\,(2\ell+1)\frac{u_l(\rho)}{\rho}P_\ell(\cos\theta) \qquad [4]$$

$$\sim\sum_l i^\ell\,(2\ell+1)\frac{\sin\left(\rho-\ell\frac{\pi}{2}\right)}{\rho}P_\ell(\cos\theta) \qquad [5]$$

Equation [4] is Bauer's formula and equation [5] is its asymptotic form.

13.8 Partial wave method is suitable for low energy

Let us consider a particle with momentum $\hbar k$ [if no scattering force acts on it], would pass a distance S from the scattering center [This distance is called impact parameter], the angular momentum of the particle is kS . On the other hand, the angular momentum component in the lth partial wave is l. Hence $\rightarrow S = \frac{l}{k}$. But substantial scattering can be expected only if the impact parameter is smaller than the range of the scattering force. I.e., $S \leq a$. Where "a" is the range of potential energy.

So, $$S \le a \;\; Or, \frac{l}{k} \le a \;\; Or, l \le ka \;\therefore l \le \sqrt{\frac{2mE}{\hbar^2}}\, a \;\; Or, E \le ka \qquad [1]$$

When k is small I.e., low energy phenomena; V is large. The method of scattering, the solution in partial wave needs a number of l values. This number of l values increases with energy as indicated in the above equation. Therefore, the partial wave method is suitable for low energy[8,9].

13.9 Partial wave expansion of scattering amplitude and differential scattering cross-section

The partial wave method is applicable to spherically symmetric potentials. The spherically symmetric potentials are Coulomb potential, Gravitational potential, Yukawa potential and Gaussian potential etc. For such potentials, the angular momentum of the scattered particle is a constant of motion. Now we consider the Schrödinger equation in the presence of the potential

$$H\,\psi(\vec{r}) = E\,\psi(\vec{r})$$

$$[-\frac{\hbar^2}{2m}\,\nabla^2 + V(\vec{r})]\,\psi(\vec{r}) = E\,\psi(\vec{r})$$

$$\nabla^2\psi(\vec{r}) - \frac{2m}{\hbar^2}V(\vec{r})\,\psi(\vec{r}) = -\frac{2m}{\hbar^2}E\,\psi(\vec{r})$$

$$\nabla^2\psi(\vec{r}) - \frac{2m}{\hbar^2}V(\vec{r})\,\psi(\vec{r}) + \frac{2m}{\hbar^2}E\,\psi(\vec{r}) = 0$$

$$\nabla^2\psi(\vec{r}) + \kappa^2\,\psi(\vec{r}) - U(\vec{r})\,\psi(\vec{r}) = 0 \qquad [1]$$

Where, $$\kappa^2 = \frac{2\,\mu\,E}{\hbar^2} \qquad [2]$$

And $$U(\vec{r}) = \frac{2\,\mu\,V(\vec{r})}{\hbar^2} \qquad [3]$$

Therefore,

$$[\nabla^2 + \kappa^2 - U(\vec{r})]\,\psi(\vec{r}) = 0 \qquad [4]$$

Since V is a function of r only. We shall see the asymptotic form of the solution of the equation [4]. In the case of spherically symmetric potential, the wave function is independent of ϕ. Therefore the general solution of the equation [4] is

$$\psi(r,\theta,\phi) = \sum_{l=0}^{\infty} A_l\, R_l(r)\, P_l(\cos\theta) \qquad [5]$$

Where l = Orbital angular momentum quantum number.
A_l = Arbitrary constant.
$R_l(r)$ = The solution of the radial part of Schrödinger equation.
$P_l(\cos\theta)$ = The Legendre polynomial of order l.

The radial function $R_l(r)$ can be shown to satisfy the differential equation [1]

$$\frac{d^2R_l(r)}{dr^2} + \frac{2}{r}\frac{dR_l(r)}{dr} + \left[\kappa^2 - U(r) - \frac{l(l+1)}{r^2}\right] R_l(r) = 0 \qquad [5]$$

Equation [5] is the form of the spherical Bessels function and with the solution

$$R_l(r) = J_l(\kappa r) \tag{6}$$

In order to find the general nature of the asymptotic behavior, we consider r to be so large that U & l terms in equation [5] can be neglected. Hence the asymptotic solution of equation [5] is

$$R_l(r)(= J_l(\kappa r)) \underset{r\to\infty}{\longrightarrow} \frac{1}{\kappa r}\sin(\kappa r - \frac{l\pi}{2}) \tag{7}$$

In the presence of potential for large r, the asymptotic form of $R_l(r)$ is

$$R_l(r) \underset{r\to\infty}{\longrightarrow} \frac{1}{\kappa r}\sin(\kappa r - \frac{l\pi}{2} + \delta_l) \tag{8}$$

Where δ_l is the phase-shift of lth partial waves which is constant and depends on k.

Finally equation [5] becomes,

$$\psi(r,\theta,\phi) = \sum_{l=0}^{\infty} A_l \ P_l(\cos\theta)\ \frac{1}{\kappa r}\sin(\kappa r - \frac{l\pi}{2} + \delta_l)$$

Or, $$\psi(r,\theta,\phi) = \sum_{l=0}^{\infty} \frac{A_l \ P_l(\cos\theta)}{2i\kappa r}\left[e^{i\left(\kappa r - \frac{l\pi}{2} + \delta_l\right)} - e^{-i\left(\kappa r - \frac{l\pi}{2} + \delta_l\right)}\right] \tag{9}$$

Where $\left[\sin x = \frac{e^{ix} - e^{-ix}}{2i}\right]$; $e^{-i\theta} = \cos\theta - i\sin\theta$; $e^{i\theta} = \cos\theta + i\sin\theta$

Again in the scattering experiment, the wave function is the sum of the incident plane wave and an outgoing scattered wave I.e.,

$$\psi(r,\theta,\phi) = e^{i\kappa z} + \frac{f(\theta)}{r}e^{i\kappa r} \tag{10}$$

According to Bauer's formula

$$e^{i\kappa z} = e^{i\kappa r\cos\theta} = \sum_{l=0}^{\infty} A_l \ J_l(\kappa r)\ P_l(\cos\theta)$$

$$= \sum_{l=0}^{\infty}(2l+1)\ i^l\ P_l(\cos\theta)\ \frac{1}{\kappa r}\sin(\kappa r - \frac{l\pi}{2}) \quad ;\text{where } A_l = (2l+1)i^l$$

$$= \sum_{l=0}^{\infty}\frac{(2l+1)i^l}{\kappa r}\ P_l(\cos\theta)\sin(\kappa r - \frac{l\pi}{2})$$

$$= \sum_{l=0}^{\infty}\frac{(2l+1)i^l}{2i\kappa r}\ P_l(\cos\theta)\left[e^{i\left(\kappa r - \frac{l\pi}{2}\right)} - e^{-i\left(\kappa r - \frac{l\pi}{2}\right)}\right]$$

Therefore equation [10] becomes

$$\psi(r,\theta,\phi) = \sum_{l=0}^{\infty}\frac{(2l+1)i^l}{2i\kappa r}\ P_l(\cos\theta)\left[e^{i\left(\kappa r - \frac{l\pi}{2}\right)} - e^{-i\left(\kappa r - \frac{l\pi}{2}\right)}\right] + \frac{f(\theta)}{r}e^{i\kappa r} \tag{11}$$

From equation [9] and [11], equating the coefficient of $e^{-i\kappa r}$

$$-\sum_{l=0}^{\infty}\frac{A_l}{2\,i\,\kappa\,r}\ P_l(\cos\theta)\ e^{\frac{i\,l\pi}{2}}\,e^{-i\delta_l} = -\sum_{l=0}^{\infty}\frac{(2l+1)i^l}{2i\,\kappa\,r}\ P_l(\cos\theta)\ e^{\frac{i\,l\pi}{2}} \qquad [12]$$

Or, $\quad A_l e^{-i\delta_l} = (2l+1)i^l$

Or, $\quad A_l = (2l+1)i^l\, e^{i\delta_l} \qquad [13]$

Again equating the coefficient of $e^{i\kappa r}$

$$\sum_{l=0}^{\infty}\frac{(2l+1)i^l}{2i\,\kappa\,r}\ P_l(\cos\theta)\ e^{-\frac{i\,l\pi}{2}} + \frac{f(\theta)}{r} = \sum_{l=0}^{\infty}\frac{A_l}{2\,i\,\kappa\,r}\ P_l(\cos\theta)\ e^{-\frac{i\,l\pi}{2}}\ e^{i\delta_l} \qquad [14]$$

But we know that, $\quad i = \cos\frac{\pi}{2} + \mathrm{i}\sin\frac{\pi}{2} \quad$ Or, $\quad i = e^{\frac{i\pi}{2}}$ Or, $\ i^l = e^{\frac{i\pi l}{2}}$

Putting the value of A_l & i^l in the equation [14]

$$\sum_{l=0}^{\infty}\frac{(2l+1)}{2i\,\kappa\,r}\ P_l(\cos\theta)\ + \frac{f(\theta)}{r}\ = \ \sum_{l=0}^{\infty}\frac{(2l+1)\,e^{\frac{i\,l\pi}{2}}\ e^{i\delta_l}}{2\,i\,\kappa\,r}\ P_l(\cos\theta)\ e^{-\frac{i\,l\pi}{2}}\ e^{i\delta_l}$$

Or, $\quad \frac{f(\theta)}{r} = \sum_{l=0}^{\infty}\frac{(2l+1)}{2\,i\,\kappa\,r}\ P_l(\cos\theta)\,e^{2i\delta_l} - \sum_{l=0}^{\infty}\frac{(2l+1)}{2i\,\kappa\,r}\ P_l(\cos\theta)$

Or, $\quad \frac{f(\theta)}{r} = \frac{1}{2\,i\,\kappa\,r}\sum_{l=0}^{\infty}\ (2l+1)\ P_l\big(\cos\theta\big)\ \big[\,e^{2i\delta_l}-1\big]$

$$f(\theta)\ = \frac{1}{2\,i\,\kappa}\sum_{l=0}^{\infty}\ (2l+1)\ P_l\big(\cos\theta\big)\ \big[\,e^{2i\delta_l}-1\big] \qquad [15]$$

Where, $\quad e^{2i\delta_l} - 1 = \cos 2\delta_l + i\sin 2\delta_l - 1 = 1 - 2\sin^2\delta_l\ + i\sin 2\delta_l - 1$

$$= -2\sin^2\delta_l\ + i\sin 2\delta_l = i^2 2\sin^2\delta_l\ + i\,2\sin\delta_l\cos\delta_l$$

$$= 2\mathrm{i}\sin\delta_l\ (i\sin\delta_l + \cos\delta_l) = 2\mathrm{i}\sin\delta_l\, e^{i\delta_l}$$

Therefore, equation [15] becomes

$$f(\theta)\ = \frac{1}{2\,i\,\kappa}\sum_{l=0}^{\infty}\ (2l+1)\,P_l(\cos\theta) 2\mathrm{i}\sin\delta_l\, e^{i\delta_l}$$

$$= \frac{1}{\kappa}\sum_{l=0}^{l}\ (2l+1)\,P_l(\cos\theta)\sin\delta_l\ e^{i\delta_l} \qquad [16]$$

Equations [15] and [16] known as the partial wave expansion of the scattering amplitude. The quantity δ_l is called phase-shift for l^{th} partial wave.

Now we want to calculate the differential scattering cross-section:

We know the differential scattering cross-section

$$\sigma(\theta) = \ |f(\theta)|^2 = f(\theta) f^{\times}(\theta) \qquad [17]$$

Where, $f(\theta) = \frac{1}{\kappa}\sum_{l=0}^{\infty} (2l+1)\, P_l(\cos\theta) \sin\delta_l\; e^{i\delta_l}$

And $f^{\times}(\theta) = \frac{1}{\kappa}\sum_{l'=0}^{\infty} (2l'+1)\, P_{l'}(\cos\theta) \sin\delta_{l'}\; e^{-i\delta_{l'}}$

Hence equation [17] becomes

$$\sigma(\theta) = \frac{1}{\kappa^2}\sum_{l=0}^{\infty}\sum_{l'=0}^{\infty}(2l+1)\,(2l'+1)\, P_l(\cos\theta)\, P_{l'}(\cos\theta) \sin\delta_l\, \sin\delta_{l'}\; e^{i(\delta_l-\delta_{l'})}$$

Therefore, total scattering cross-section

$$\sigma_{tot} = \int \sigma(\theta) d\Omega = \int_0^{\pi}\int_0^{2\pi} \sigma(\theta)\, \sin\theta\, d\theta\, d\phi = 2\pi\int_0^{\pi} \sigma(\theta) \sin\theta\, d\theta$$

$$= 2\pi \int_0^{\pi} \frac{1}{\kappa^2}\sum_{l=0}^{\infty}\sum_{l'=0}^{\infty}(2l+1)\,(2l'+1) P_l(\cos\theta)\, P_{l'}(\cos\theta) \sin\delta_l\, \sin\delta_{l'}\; e^{i(\delta_l-\delta_{l'})} \sin\theta\, d\theta$$

From the orthogonality condition for Legendre polynomial

$$\int_0^{\pi} P_l(\cos\theta)\, P_{l'}(\cos\theta)\, \sin\theta\, d\theta = \frac{2}{2l+1}\, \delta_{ll'} \qquad [18]$$

$$\therefore \quad \sigma_{tot} = \frac{2\pi}{\kappa^2}\sum_{l=0}^{\infty}\sum_{l'=0}^{\infty}(2l+1)\,(2l'+1) \sin\delta_l\, \sin\delta_{l'}\; e^{i(\delta_l-\delta_{l'})}\, \frac{2}{2l+1}\, \delta_{ll'}$$

For $l = l' \;;\; \delta_{ll'} = 1$

$$\therefore \quad \sigma_{tot} = \frac{2\pi}{\kappa^2}\sum_{l=0}^{\infty}(2l+1)\, \sin^2\delta_l\, 2 = \frac{4\pi}{\kappa^2}\sum_{l=0}^{\infty}(2l+1)\, \sin^2\delta_l \qquad [19]$$

$$\sigma_{tot} = \frac{4\pi}{\kappa^2}\, \sin^2\delta_0 \quad ; \text{ for } \; l = 0 \qquad [20]$$

Equation [19] is the total differential scattering cross-section of the partial wave expansion.

13.10 Optical theorem

Optical theorem states that - *the imaginary part of the forward scattering amplitude is equal to $\frac{\kappa}{4\pi}$ times the total scattering cross-section.* Let us prove the optical theorem by using the partial wave expansion.

We know the partial wave expansion of the scattering amplitude

$$f(\theta) = \frac{1}{\kappa}\sum_{l=0}^{\infty} (2l+1)\, P_l(\cos\theta) \sin\delta_l\; e^{i\delta_l} \qquad [1]$$

$$\& \quad f^{\times}(\theta) = \frac{1}{\kappa}\sum_{l'=0}^{\infty} (2l'+1)\, P_{l'}(\cos\theta) \sin\delta_{l'}\; e^{-i\delta_{l'}} \qquad [2]$$

The total scattering cross-section

$$\sigma_{tot} = 2\pi \int_0^{\pi} \sigma(\theta) \sin\theta \, d\theta = 2\pi \int_0^{\pi} f(\theta) f^{\times}(\theta) \sin\theta \, d\theta \quad [3]$$

$$2\pi \int_0^{\pi} \frac{1}{\kappa^2} \sum_{l=0}^{\infty} \sum_{l'=0}^{\infty} (2l+1)(2l'+1) P_l(\cos\theta) P_{l'}(\cos\theta) \sin\delta_l \sin\delta_{l'} \, e^{i(\delta_l - \delta_{l'})} \sin\theta \, d\theta$$

From the orthogonality condition for Legendre polynomial

$$\int_0^{\pi} P_l(\cos\theta) P_{l'}(\cos\theta) \sin\theta \, d\theta = \frac{2}{2l+1} \delta_{ll'} \quad [4]$$

$$\therefore \quad \sigma_{tot} = \frac{2\pi}{\kappa^2} \sum_{l=0}^{\infty} \sum_{l'=0}^{\infty} (2l+1)(2l'+1) \sin\delta_l \sin\delta_{l'} \, e^{i(\delta_l - \delta_{l'})} \frac{2}{2l+1} \delta_{ll'}$$

For $\quad l = l' \; ; \; \delta_{ll'} = 1$

$$\therefore \quad \sigma_{tot} = \frac{2\pi}{\kappa^2} \sum_{l=0}^{\infty} (2l+1) \sin^2\delta_l \, 2 = \frac{4\pi}{\kappa^2} \sum_{l=0}^{\infty} (2l+1) \sin^2\delta_l \quad [5]$$

Now putting $\theta = 0$; in equation [1]

$$f(0) = \frac{1}{\kappa} \sum_{l=0}^{\infty} (2l+1) P_l(\cos 0) \sin\delta_l \, e^{i\delta_l}$$

But $P_l(\cos 0) = P_l(1) = 1$; for all l.

$$\text{Or, } f(0) = \frac{1}{\kappa} \sum_{l=0}^{\infty} (2l+1)(\cos\delta_l + i\sin\delta_l) \sin\delta_l \quad [6]$$

On the other hand, the imaginary part of the scattering amplitude is given by

$$Im\, f(0) = \frac{1}{\kappa} \sum_{l=0}^{\infty} (2l+1) \sin^2\delta_l \quad [7]$$

Finally, from equation [5] gives

$$\sigma_{tot} = \frac{4\pi}{\kappa^2} \sum_{l=0}^{\infty} (2l+1) \sin^2\delta_l = \frac{4\pi}{\kappa} \frac{1}{\kappa} \sum_{l=0}^{\infty} (2l+1) \sin^2\delta_l = \frac{4\pi}{\kappa} Im\, f(0)$$

$$\sigma_{tot} = \frac{4\pi}{\kappa} Im\, f(0$$

$$\text{Or, } \quad Im\, f(0) = \frac{\kappa}{4\pi} \sigma_{tot} \quad [8]$$

The imaginary part of the forward scattering amplitude is equal to $\frac{\kappa}{4\pi}$ *times the total scattering cross-section. This is known as the optical theorem.*

13.11 Phase- shift is called the meeting ground of theory and experiment and Phase Shift in terms of potential

The physical interest of the scattering experiment is to know the interactions between two particles such as a projectile and a target. We want to know what kind of potential does the target bear. Experimentalist find differential scattering cross-section [σ] from the detector and we can evaluate phase-shift [δ] from scattering cross-section [σ]. However, the theoretician assumes different values of potentials $V(r)$ and then calculated phase-shift [δ] from $V(r)$. If two results coincide or nearly equal; the assumption of potentials will be correct and Phase-shift is called the meeting ground of the theory and experiment[5,4].

- **Phase Shift in terms of potential**

We know the radial equation of the scattering problem for a spherically symmetric partial $V = V(r)$ is

$$\frac{d^2R_l(r)}{dr^2} + \frac{2}{r}\frac{dR_l(r)}{dr} + \left[\kappa^2 - U(r) - \frac{l(l+1)}{r^2}\right] R_l(r) = 0$$

Or, $$\frac{1}{r^2}\frac{d}{dr}\left(r^2\frac{dR_l(r)}{dr}\right) + \left[\kappa^2 - U(r) - \frac{l(l+1)}{r^2}\right] R_l(r) = 0 \qquad [1]$$

The above equation is the form of spherical Bessel function and with the solution

$$R_l^{(0)}(r) = J_l(\kappa r) \qquad [2]$$

Where, $\kappa^2 = \frac{2mE}{\hbar^2}$ & $U(r) = \frac{2m}{\hbar^2}V(r)$

In the absence of the potential $V(r) = 0$

$$\therefore \quad \frac{1}{r^2}\frac{d}{dr}\left(r^2\frac{R_l^{(0)}(r)}{dr}\right) + \left[\kappa^2 - \frac{l(l+1)}{r^2}\right] R_l^{(0)}(r) = 0 \qquad [3]$$

Multiplying [1] by $R_l^{(0)}$ and [3] by R_l & subtracting

$$\frac{R_l^{(0)}}{r^2}\frac{d}{dr}\left(r^2\frac{dR_l}{dr}\right) + R_l^{(0)}\left[\kappa^2 - U(r) - \frac{l(l+1)}{r^2}\right] R_l - \frac{R_l}{r^2}\frac{d}{dr}\left(r^2\frac{dR_l^{(0)}}{dr}\right) - R_l\left[\kappa^2 - \frac{l(l+1)}{r^2}\right] R_l^{(0)} = 0$$

Or, $$\frac{R_l^{(0)}}{r^2}\frac{d}{dr}\left(r^2\frac{dR_l}{dr}\right) + R_l^{(0)}\,U(r)\,R_l - \frac{R_l}{r^2}\frac{d}{dr}\left(r^2\frac{dR_l^{(0)}}{dr}\right) = 0$$

Now multiplying both sides by r^2 and integrating over r from 0 to ∞

$$\int_0^\infty\left[R_l^{(0)}\frac{d}{dr}\left(r^2\frac{dR_l}{dr}\right) - R_l\frac{d}{dr}\left(r^2\frac{dR_l^{(0)}}{dr}\right)\right]dr = \int_0^\infty r^2 R_l^{(0)}U(r)\,R_l\,dr$$

Or, $$\left[R_l^{(0)}\,r^2\frac{dR_l}{dr} - R_l r^2\frac{dR_l^{(0)}}{dr}\right]_{r=0}^{\infty} = \int_0^\infty r^2 R_l^{(0)}\,U(r)\,R_l\,dr \qquad [4]$$

The left hand side is zero for the lower limit. For the upper limit $r = \infty$, we put the asymptotic relations for $R_l^{(0)}$ & R_l in the left hand side of equation [4]

$$R_l^{(0)} \xrightarrow{r\to\infty} \frac{1}{\kappa r}\sin\left(\kappa r - \frac{l\pi}{2}\right) \;\;\&\;\; R_l \xrightarrow{r\to\infty} \frac{1}{\kappa r}\sin\left(\kappa r - \frac{l\pi}{2} + \delta_l\right)$$

Therefore equation [4] becomes

$$\frac{1}{\kappa r}\sin(\kappa r-\frac{l\pi}{2})r^2\frac{d}{dr}[\frac{1}{\kappa r}\sin(\kappa r-\frac{l\pi}{2}+\delta_l)]-\frac{1}{\kappa r}\sin(\kappa r-\frac{l\pi}{2}+\delta_l)r^2\frac{d}{dr}[\frac{1}{\kappa r}\sin(\kappa r-\frac{l\pi}{2})]$$

$$=\int_{r=0}^{\infty} r^2\, R_l^{(0)}\; U(r)\; R_l\, dr$$

Or, $$\frac{\sin(\kappa r-\frac{l\pi}{2})}{\kappa r}\, r^2\left[\frac{\cos\left(\kappa r-\frac{l\pi}{2}+\delta_l\right)}{\kappa r}\kappa-\frac{\sin\left(\kappa r-\frac{l\pi}{2}+\delta_l\right)}{\kappa r^2}\right]-\frac{\sin(\kappa r-\frac{l\pi}{2}+\delta_l)}{\kappa r}\, r^2\left[\frac{\cos\left(\kappa r-\frac{l\pi}{2}\right)}{\kappa r}\kappa-\frac{\sin\left(\kappa r-\frac{l\pi}{2}\right)}{\kappa r^2}\right]$$

$$=\int_0^{\infty} r^2\, R_l^{(0)}\; U(r)\; R_l\, dr$$

Or, $$\frac{1}{\kappa}[\sin(\kappa r-\frac{l\pi}{2})\cos\left(\kappa r-\frac{l\pi}{2}+\delta_l\right)-\sin\left(\kappa r-\frac{l\pi}{2}+\delta_l\right)\cos\left(\kappa r-\frac{l\pi}{2}\right)]$$

$$=\int_0^{\infty} r^2 R_l^{(0)} U(r)\; R_l\, dr$$

Or, $\frac{1}{\kappa}[\sin(\kappa r-\frac{l\pi}{2}-\kappa r+\frac{l\pi}{2}-\delta_l)=\int_0^{\infty} r^2\, R_l^{(0)}\; U(r)\; R_l\, dr$

Or, $-\frac{1}{\kappa}[\sin\;\delta_l]=\int_0^{\infty} r^2\, R_l^{(0)}\; U(r)\; R_l\, dr$

Or, $\sin\;\delta_l=-\kappa\int_0^{\infty} J_l(\kappa r)U(r)\, R_l(r)r^2 dr$; where $R_l^{(0)}(r)=J_l(\kappa r)$

If we start with a potential $\lambda\, U(r)$; the lth phase-shift is given by

$$sin\delta_l=-\kappa\lambda\int_0^{\infty} J_l(\kappa r)U(r)\; R_l(r)\, r^2 dr \qquad [5]$$

Let us suppose that λ is a small parameter and we can expand δ_l in powers of λ

$$\delta_l=\lambda\delta_l^{(1)}+\lambda\delta_l^{(2)}+\cdots\cdots\cdots \qquad [6]$$

Then from partial wave analysis the scattering amplitude is

$$f(\theta)=\frac{1}{2i\,\kappa}\sum_{l=0}^{\infty}(2l+1)\,P_l(\cos\theta)\sin\delta_l\;\;[e^{2i\delta_l}-1]$$

$$=\frac{1}{2i\,\kappa}\sum_{l=0}^{\infty}(2l+1)\,P_l(\cos\theta)\sin\delta_l\;\;[1+2i\delta_l+\frac{(2i\delta_l)^2}{2!}\cdots\cdots\cdots-1]$$

$$=\frac{1}{2i\,\kappa}\sum_{l=0}^{\infty}(2l+1)\,P_l(\cos\theta)\,2i[\,\delta_l+i{\delta_l}^2+\cdots\cdots\cdots]$$

$$=\frac{1}{\kappa}\sum_0^{\infty}(2l+1)P_l(cos\theta)\left[\left(\lambda\delta_l^{(1)}+\lambda^2\,\delta_l^{(2)}+\cdots\right)+i\left(\lambda\delta_l^{(1)}+\lambda^2\,\delta_l^{(2)}+\cdots\cdots\right)^2+\cdots\cdots\right]$$

Keeping only the term only involving λ,

$\therefore$ $$f(\theta)=\frac{1}{\kappa}\sum_{l=0}^{\infty}(2l+1)\,P_l(\cos\theta)\,\lambda\;\delta_l^{(1)}$$

$$f(\theta) = \frac{\lambda}{\kappa}\sum_{l=0}^{\infty} (2l+1)\, P_l(\cos\theta)\ \delta_l^{(1)} \qquad [7]$$

Also we can write

$$\sin\delta_l = \sin[\ \lambda\delta_l^{(1)} + \lambda^2\,\delta_l^{(2)} + \cdots\cdots\cdots\cdots\cdots\cdots\cdots\cdots]$$
$$\approx \sin\ \lambda\delta_l^{(1)} \approx \lambda\delta_l^{(1)} \qquad [8]$$

Now comparing equation (5) and (8) we have

$$\lambda\delta_l^{(1)} = -\kappa\lambda\int_0^{\infty} J_l(\kappa r)U(r)R_l(r)\, r^2 dr$$

Or,
$$\delta_l^{(1)} = -\kappa\int_0^{\infty} J_l(\kappa r)\,U(r)\ R_l(r)\, r^2 dr$$

Again we replace $R_l(r)$ by $R_l^{(0)}(r)$ Then

$$\delta_l^{(1)} = -\kappa\int_0^{\infty} [J_l(\kappa r)]^2\, U(r)\, r^2 dr \qquad [\because R_l^{(0)}(r) = J_l(\kappa r)] \qquad [9]$$

Equation [9] indicates the integral representation of phase-shift. We also note that the phase shift δ_l is positive for attractive potential [$V(r)$ negative] and it is negative for repulsive potential [$V(r)$ positive].

Now from equation [7], to the 1st order in λ:

$$f'(\theta) = \frac{\lambda}{\kappa}\sum_{l=0}^{\infty} (2l+1)\, P_l(\cos\theta)\, [-\kappa\int_0^{\infty}[J_l(\kappa r)]^2 U(r)\, r^2 dr\ ]$$

$$f'(\theta) = -\lambda\sum_{l=0}^{\infty} (2l+1)\, P_l(\cos\theta)\int_0^{\infty}[J_l(\kappa r)]^2 U(r)\, r^2 dr \qquad [10]$$

Now using this expression

$$\frac{\sin qr}{qr} = \sum_{l=0}^{\infty} (2l+1)\, P_l(\cos\theta)\ [J_l(\kappa r)]^2 P_l(\cos\theta)\ ;\text{Where, } q = 2\kappa\sin\frac{\theta}{2} \qquad [11]$$

Therefore [10] can be written

$$f'(\theta) = -\lambda\int_{r=0}^{\infty}\frac{\sin qr}{qr}\, U(r)\, r^2 dr \qquad [12]$$

Equation [12] is known as the Born-approximation and $\mathbf{f}'(\theta)$ is scattering amplitude in Born-approximation.

13.12 Scattering amplitude is independent of azimuthal angle

The Born approximation for the scattering amplitude

$$f(\theta,\phi) = -\frac{1}{4\pi}\int e^{i\,\bar{\kappa}.\bar{r}'}\, U(\bar{r}')\, d^3\mathbf{r}' \qquad [1]$$

The above equation greatly applicable for a central force I.e., the potential depends on the distance $\mathbf{r}'$ but not on the direction $\bar{\mathbf{r}}'$. So $U(\bar{\mathbf{r}}') = U(\mathbf{r}')$. Also considering that the direction of $\bar{\kappa}$ along the polar axis (angles). Then

$$e^{i\bar{\kappa}.\bar{r}'} = e^{i\kappa r'\cos\theta} \qquad [2]$$

$\therefore$
$$f(\theta,\phi) = -\frac{1}{4\pi}\int_{r=0}^{\infty}\int_{\theta=0}^{\pi}\int_{\phi=0}^{2\pi} e^{i\kappa r'\cos\theta}\; U(r')\, r'^2 \sin\theta\, dr' d\theta\, d\phi$$

Put $z = \cos\theta$ $or, dz = -\sin\theta\, d\theta$; When $\theta = 0,\ z = 1,\ \&$ When $\theta = \pi,\ z = -1$

$$f(\theta,\phi) = -\frac{1}{4\pi}\, 2\pi \int_{+1}^{-1}\int_{r=0}^{\infty} e^{i\kappa r' z}\; U(r')\,(-dz) r'^2\, dr'$$

$$= -\frac{1}{2}\int_{+1}^{-1}\int_{r=0}^{\infty} e^{i\kappa r' z}\, dz\, U(r')\, r'^2\, dr'$$

$$= -\frac{1}{2}\int_{r=0}^{\infty}\left[\frac{e^{i\kappa r' z}}{i\kappa r'}\right]_{-1}^{1} U(r')\, r'^2\, dr'$$

$$= -\frac{1}{2i\kappa r'}\int_{r=0}^{\infty}\left[e^{i\kappa r'} - e^{-i\kappa r'}\right] U(r')\, r'^2\, dr'$$

$$= -\frac{1}{\kappa}\int_{r=0}^{\infty}\left[\frac{e^{i\kappa r'} - e^{-i\kappa r'}}{2i}\right] U(r')\, r' dr'$$

$$f(\theta,\phi) = -\frac{1}{\kappa}\int_{r=0}^{\infty} \sin\kappa r'\, U(r')\, r' dr' \qquad [3]$$

The L.H.S. of equation [3] is independent of ϕ. So rearranging equation [3] is

$$f(\theta) = -\frac{1}{\kappa}\int_{r=0}^{\infty} \sin\kappa r'\, U(r')\, r' dr' \qquad [4]$$

Consider the equation from the section [13.11]

$$f'(\theta) = -\lambda\int_{r=0}^{\infty} \frac{\sin\kappa r}{\kappa r}\; U(r)\, r^2\, dr$$

$$= -\frac{\lambda}{\kappa}\int_{r=0}^{\infty} \sin\kappa\, r\; U(r)\; r\; dr \qquad [5]$$

Now comparing equation [4] with the equation [5] and noted that the scattering amplitude of both equations are independent of ϕ.

13.13 S-matrix and T-matrix

In physics, the **S-matrix** or **scattering matrix** relates the initial state and the final state of a physical system undergoing a scattering process. It is used in quantum mechanics, scattering theory and quantum field theory (QFT). More formally, in the context of QFT, the S-matrix is defined as the unitary matrix connecting sets of asymptotically free particle states (the *in-states* and the *out-states*) in the Hilbert space of physical states. *Asymptotically free* then means that the state has this appearance in either the distant past or the distant future. **S-matrix theory** was a proposal for replacing local quantum field theory as the basic principle of elementary particle physics. S-matrix theory was largely abandoned by physicists in the 1970s, as quantum chromo dynamics was recognized to solve the problems of strong interactions within the framework of field theory. But in the guise of string theory, S-matrix theory is still a popular approach to the problem of quantum gravity.

While the S-matrix may be defined for any background (space time) that is asymptotically solvable and has no event horizons, it has a simple form in the case of the Minkowski space. In this special case, the Hilbert space is a space of irreducible unitary representations of the inhomogeneous Lorentz group (the Poincare group); the S-matrix is the evolution operator between time equal to minus infinity (the distant past), and time equal to plus infinity (the distant future). It is defined only in the limit of zero energy density (or infinite particle separation distance). It can be shown that if a quantum field theory in Minkowski space has a mass gap, the state in the asymptotic past and in the asymptotic future are both described by Fock spaces.

T-Matrix

The incident and scattered electric field are expanded into spherical vector wave functions (SVWF), which are also encountered in Mie scattering. They are the fundamental solutions of the vector Helmholtz equation and can be generated from the scalar fundamental solutions in spherical coordinates, the spherical Bessel functions of the first kind and the spherical Hankel Functions. Accordingly, there are two linearly independent sets of solutions denoted as M^1, N^1 and M^3, N^3, respectively. They are also called regular and propagating spherical vector wave functions [SVWFs], respectively. With this, we can write the incident field as

$$E_{inc} = \sum_{n=1}^{\infty} \sum_{m=-n}^{n} a_{mn} M_{mn}^1 + b_{mn} N_{mn}^1$$

The scattered field is expanded into radiating spherical vector wave functions [SVWFs]:

$$E_{scatt} = \sum_{n=1}^{\infty} \sum_{m=-n}^{n} c_{mn} M_{mn}^3 + d_{mn} N_{mn}^3$$

The T-Matrix relates the expansion coefficients of the incident field to those of the scattered field.

$$\begin{pmatrix} a_{mn} \\ b_{mn} \end{pmatrix} = T \begin{pmatrix} c_{mn} \\ d_{mn} \end{pmatrix}$$

The T-Matrix is determined by the scatterer shape and material and for given incident field allows to calculate the scattered field. The **T-matrix method** is a computational technique of light scattering by no spherical particles originally formulated by P. C. Waterman (1928-2012) in 1965. The technique is also known as null field method and extended boundary technique method (EBCM). In the method, matrix elements are obtained by matching boundary conditions for solutions of Maxwell equations.

13.14 Relation between K_l , S_l and T_l matrices and the scattering amplitude in terms of T_l matrix

We know that in the absence of potential the regular solution of the radial equation is

$$R_l^{(0)}(r) = A_l J_l(\kappa r) \tag{1}$$

And its asymptotic form is

$$R_l^{(0)}(r) \to J_l(\kappa r) \xrightarrow{r\to\infty} \frac{1}{\kappa r}\sin(\kappa r - \frac{l\pi}{2}) \tag{2}$$

In presence of the potential, the general solution of the radial equation is

$$R_l(r) = C_l J_l(\kappa r) + D_l n_l(\kappa r) \tag{3}$$

Which has the asymptotic form

$$R_l^{(0)}(r) \xrightarrow{r\to\infty} \frac{1}{\kappa r}\sin(\kappa r - \frac{l\pi}{2} + \delta_l) \tag{4}$$

In equation [3] $n_l(\kappa r)$ is the Newmann function and the asymptotic form of the Newmann function

$$n_l(\kappa r) \xrightarrow{r\to\infty} \frac{1}{\kappa r}\cos\,(\kappa r - \frac{l\pi}{2}) \tag{5}$$

Using all the asymptotic relations in equation [3]

$$\frac{1}{\kappa r}\sin(\kappa r - \frac{l\pi}{2} + \delta_l) = C_l \frac{1}{\kappa r}\sin(\kappa r - \frac{l\pi}{2}) + D_l \frac{1}{\kappa r}\cos(\kappa r - \frac{l\pi}{2})$$

Or,
$$\sin(\kappa r - \frac{l\pi}{2})\cos\delta_l + \cos\left(\kappa r - \frac{l\pi}{2}\right)\sin\delta_l = C_l\sin(\kappa r - \frac{l\pi}{2}) + D_l\cos(\kappa r - \frac{l\pi}{2})$$

Equating the coefficients of $\sin(\kappa r - \frac{l\pi}{2})$ and $\cos\left(\kappa r - \frac{l\pi}{2}\right)$, we get

$$\cos\delta_l = C_l \; and \; \sin\delta_l = D_l$$

Or,
$$\tan\delta_l = \frac{D_l}{C_l} \tag{6}$$

The quantity $\tan\delta_l$ is called the κ_l matrix and therefore

$$\kappa_l = \tan\delta_l = \frac{D_l}{C_l} \qquad [7]$$

Again we consider the second independent solution as

$$R_l(r) = C_l' J_l(\kappa r) + D_l' h_l^{(1)}(\kappa r) \qquad [8]$$

Here $h_l^{(1)}$ is called Hankel function of order 1, and its asymptotic form is

$$h_l^{(1)}(\kappa r) = \frac{1}{\kappa r} e^{i(\kappa r - \frac{l\pi}{2})} \qquad [9]$$

Using all the asymptotic relations in [8]

$$\frac{1}{\kappa r}\sin(\kappa r - \frac{l\pi}{2} + \delta_l) = C_l' \frac{1}{\kappa r}\sin(\kappa r - \frac{l\pi}{2}) + D_l' \frac{1}{\kappa r} e^{i(\kappa r - \frac{l\pi}{2})}$$

Or, $$\sin(\kappa r - \frac{l\pi}{2})\cos\delta_l + \cos\left(\kappa r - \frac{l\pi}{2}\right)\sin\delta_l$$

$$= C_l' \sin(\kappa r - \frac{l\pi}{2}) + D_l'[\cos(\kappa r - \frac{l\pi}{2}) + i\sin\left(\kappa r - \frac{l\pi}{2}\right)]$$

Equating the coefficient of $\sin\left(\kappa r - \frac{l\pi}{2}\right)$ and $\cos(\kappa r - \frac{l\pi}{2})$

$$\cos\delta_l = C_l' + \mathrm{i}\, D_l' \quad \& \quad \sin\delta_l = D_l'$$

$$\tan\delta_l = \frac{D_l'}{C_l' + \mathrm{i}\, D_l'} \qquad [10]$$

But the T matrix is defined as

$$T_l = \frac{D_l'}{C_l'} \qquad [11]$$

Now from [10]

$$\frac{\sin\delta_l}{\cos\delta_l} = \frac{D_l'}{C_l' + \mathrm{i}\, D_l'} \quad \text{or,} \quad \sin\delta_l\, C_l' + \mathrm{i}\sin\delta_l\, D_l' = D_l'\cos\delta_l$$

Or, $$D_l'(\cos\delta_l - i\sin\delta_l) = C_l'\sin\delta_l \quad or, \quad D_l'\, e^{-i\delta_l} = C_l' \sin\delta_l$$

Or, $$\frac{D_l'}{C_l'} = \sin\delta_l\, e^{i\delta_l} \qquad [12]$$

Or, $$T_l = \frac{D_l'}{C_l'} = \sin\delta_l\, e^{i\delta_l} \qquad [13]$$

Finally, we can write the radial function $R_l(r)$ in the form

$$R_l(r) \to C_l'' h_l^{(1)}(\kappa r) + D_l'' h_l^{(2)}(\kappa r) \qquad [14]$$

Here $h_l^{(2)}(\kappa r)$ is the Hankel function of order 2 and its asymptotic form is

$$h_l^{(2)}(\kappa r) \to -\frac{1}{\kappa r} e^{-i(\kappa r - \frac{l\pi}{2})}$$

Putting all the asymptotic relations in [14]

$$\frac{1}{\kappa r}\sin(\kappa r - \frac{l\pi}{2} + \delta_l) = C_l''\frac{1}{\kappa r}e^{i(\kappa r - \frac{l\pi}{2})} - D_l''\frac{1}{\kappa r}e^{-i(\kappa r - \frac{l\pi}{2})}$$

Or, $\sin(\kappa r - \frac{l\pi}{2})\cos\delta_l + \cos\left(\kappa r - \frac{l\pi}{2}\right)\sin\delta_l$

$$= C_l''[\cos(\kappa r - \frac{l\pi}{2}) + i\sin\left(\kappa r - \frac{l\pi}{2}\right)] - D_l''[\cos(\kappa r - \frac{l\pi}{2}) - i\sin\left(\kappa r - \frac{l\pi}{2}\right)]$$

Equating the coefficients of $\sin\left(\kappa r - \frac{l\pi}{2}\right)$ and $\cos(\kappa r - \frac{l\pi}{2})$

$\therefore$ $\cos\delta_l = iC_l'' + i\,D_l''$ *and* $\sin\delta_l = C_l'' - D_l''$

Therefore, $$\tan\delta_l = \frac{C_l''-D_l''}{iC_l''+\mathrm{i}\,D_l''} \qquad [15]$$

The S_l matrix is defined as $$S_l = \frac{C_l''}{D_l''} \qquad [16]$$

Now from [15]

$$iC_l''\sin\delta_l + \mathrm{i}\,D_l''\sin\delta_l = C_l''\cos\delta_l - D_l''\cos\delta_l$$

Or, $$C_l''(\cos\delta_l - i\sin\delta_l) = D_l''(\cos\delta_l + i\sin\delta_l)$$

Or, $$C_l''e^{-i\delta_l} = D_l''e^{i\delta_l} \text{ or, } \frac{C_l''}{D_l''} = e^{2i\delta_l}$$

Therefore, the S_l matrix is

$$S_l = \frac{C_l''}{D_l''} = e^{2i\delta_l} \qquad [17]$$

Rewriting the S_l, T_l and κ_l matrices

$$\kappa_l = \tan\delta_l,\ T_l = \sin\delta_l\,e^{i\delta_l} \text{ And } S_l = e^{2i\delta_l}$$

Now $$S_l = \frac{e^{i\delta_l}}{e^{-i\delta_l}} \text{ Or, } S_l - 1 = \frac{e^{i\delta_l}}{e^{-i\delta_l}} - 1$$

Or, $$S_l - 1 = \frac{e^{i\delta_l}-e^{-i\delta_l}}{e^{-i\delta_l}} = \frac{\cos\delta_l+i\sin\delta_l-\cos\delta_l+i\sin\delta_l}{e^{-i\delta_l}}$$

Or, $$S_l - 1 = 2i\,\sin\delta_l\,e^{i\delta_l} = 2i\left(\sin\delta_l\,e^{i\delta_l}\right) = 2i\ T_l \qquad [18]$$

Equation [18] represents the relation between S_l and T_l matrices. Again

$$S_l = \frac{e^{i\delta_l}}{e^{-i\delta_l}} = \frac{\cos\delta_l+i\sin\delta_l}{\cos\delta_l-i\sin\delta_l} = \frac{1+i\ \tan\delta_l}{1-i\ \tan\delta_l}$$

Or, $$S_l = \frac{1+i\kappa_l}{1-i\kappa_l} \qquad [19]$$

Equation [19] is the relation between S_l and κ_l matrices.

Now from the scattering amplitude

$$f(\theta) = \frac{1}{2i\,\kappa}\sum_{l=0}^{\infty} (2l+1)\, P_l(\cos\theta)\, [e^{2i\delta_l} - 1]$$

$$= \frac{1}{2i\,\kappa}\sum_{l=0}^{\infty} (2l+1)\, P_l(\cos\theta)\, [\, S_l \,- 1]$$

$$= \frac{1}{2i\,\kappa}\sum_{l=0}^{\infty} (2l+1)\, P_l(\cos\theta)\, [2i\, T_l]$$

Or, $$f(\theta) = \frac{1}{\kappa}\sum_{l=0}^{\infty} (2l+1)\, T_l\, P_l(\cos\theta) \qquad [20]$$

Equation [20] represents the scattering amplitude in terms of T_l matrix.

13.15 Ramsauer–Townsend effect and Townsend discharge: A unique description

❖ Description

Ramsauer-Townsend Effect,[also sometimes called the Ramsauer Effect or the Townsend Effect] is a physical phenomenon involving the scattering of low energy electrons by atoms of a noble gas. [Because noble gas atoms have relatively high first ionization energy and the electrons do not carry enough energy to cause excited electronic states, ionization and excitation of the atom are unlikely, and the probability of elastic scattering over all angles is approximately equal to the probability of collision.] Since its explanation requires the wave theory of Quantum Mechanics, it demonstrates the need for physical theories more sophisticated than those of Newtonian Mechanics. When an electron moves through a gas, its interaction with the gas atoms cause scattering to occur. These interactions are classified as inelastic if they cause excitation or ionization of the atom to occur elastic if they do not. The effect is named for Carl Ramsauer (1879-1955) and John Sealy Townsend (1868-1957), who is independently, studied the collision between atoms and low energy electrons in the early 1920s.

❖ Definitions

Ramsauer-Townsend observed that for slow moving electrons in Argon, Krypton or Xenon, the probability of collision between the electrons and gas atoms obtains a minimum value for electrons with a certain amount of kinetic energy (about 1eV for Xenon gas). This is the Ramsauer -Townsend Effect. Physically, the Ramsauer -Townsend Effect may be thought of as a diffraction of the electron around the rare- gas atom, in which the wave function inside the atom is distorted in just such a way that it fits on smoothly to an undistorted wave function outside.

❖ Ramsauer–Townsend effect: From partial wave expansion

From partial wave expansion, the scattering amplitude can be written

$$f(\theta) = \frac{1}{2 i \kappa} \sum_{l=0}^{\infty} (2l+1)\, P_l(\cos\theta)\ [e^{2i\delta_l} - 1]$$

An attractive potential might be strong enough $[l = 0]$; so that all other phase-shifts are negligibly small when $\delta_0 = 180^0$. The scattering amplitude $f(\theta)$ then vanishes for all (θ) and there is no scattering. This is the explanation of the Ramsauer -Townsend Effect, the extremely low minimum observed in the scattering cross-section of electrons by a rare gas at about 0.7 eV bombarding energy.

❖ **Townsend discharge**:

Sir John Sealy Edward Townsend, was an Irish mathematical physicist who conducted various studies concerning the electrical conduction of gases (concerning the kinetics of electrons and ions) and directly measured the electrical charge. He was a Wykeham Professor of physics at Oxford University. The phenomenon of the electron avalanche was discovered by him, and is known as the Townsend discharge.

13.16 Ramsaur-Townsend Effect: For Square well potential

The square well potential is

$$V(r) = \begin{cases} -V_0 & for\ r < a \\ 0 & for\ r > a \end{cases}$$

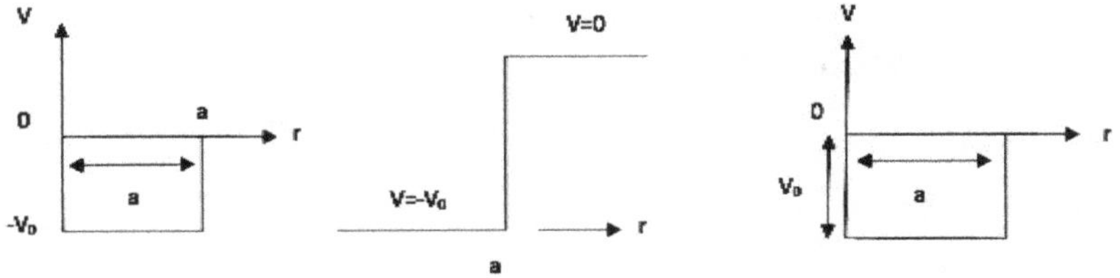

The radial equation has the form

$$\frac{d^2R_l}{dr^2} + \frac{2}{r}\frac{dR_l}{dr} + \left[\kappa^2 + U_0 - \frac{l(l+1)}{r^2}\right] R_l = 0 \quad ; \quad \text{For } r < a \qquad [1]$$

$$\& \quad \frac{d^2R_l}{dr^2} + \frac{2}{r}\frac{dR_l}{dr} + \left[\kappa^2 - \frac{l(l+1)}{r^2}\right] R_l = 0 \quad ; \quad \text{For } r > a \qquad [2]$$

For $l = 0$; Equation [1] and [2] become

$$\frac{d^2R_0}{dr^2} + \frac{2}{r}\frac{dR_0}{dr} + [\kappa_0{}^2]\, R_0 = 0 \quad ; \quad \text{For } r < a \qquad [3]$$

$$\& \frac{d^2R_0}{dr^2} + \frac{2}{r}\frac{dR_0}{dr} + [\kappa^2\,]\,R_0 = 0 \qquad ; \quad \text{For } r > a \qquad [4]$$

Where, $\kappa_0{}^2 = \kappa^2 + U_0$ [5]

Let $U_0 = r\,R_0$

$$\frac{dU_0}{dr} = R_0 + r\frac{dR_0}{dr} \text{ and } \frac{d^2U_0}{dr^2} = \frac{dR_0}{dr} + \frac{dR_0}{dr} + r\frac{d^2R_0}{dr^2} = 2\frac{dR_0}{dr} + r\frac{d^2R_0}{dr^2}$$

Or, $2\frac{dR_0}{dr} + r\frac{d^2R_0}{dr^2} = \frac{d^2U_0}{dr^2}$ $Or,$ $\frac{d^2R_0}{dr^2} = \frac{1}{r}\frac{d^2U_0}{dr^2} - \frac{2}{r}\frac{dR_0}{dr}$

Using this in equation [3] and [4]

$$\frac{1}{r}\frac{d^2U_0}{dr^2} - \frac{2}{r}\frac{dR_0}{dr} + \frac{2}{r}\frac{dR_0}{dr} + \kappa_0^2\frac{U_0}{r_0} = 0 \quad \text{Or,} \quad \frac{1}{r}\frac{d^2U_0}{dr^2} + \kappa_0^2\frac{U_0}{r} = 0$$

Or, $\frac{d^2U_0}{dr^2} + \kappa_0^2 U_0 = 0$; for r < a [6]

Similarly, we get $\frac{d^2U_0}{dr^2} + \kappa^2 U_0 = 0$; for r > a [7]

The corresponding solutions are

$U_{01} = A\sin\kappa_0 r$ for x < a [8]

& $U_{02} = A\sin(\kappa r + \delta_0)$ for x > a [9]

When $r = a$; using the boundary condition

$$U_{01}\,|_{r=a} = U_{02}\,|_{r=a} \text{ and } \left.\frac{dU_{01}}{dr}\right|_{r=a} = \left.\frac{dU_{02}}{dr}\right|_{r=a}$$

$\therefore A\sin\kappa_0 a = A\sin(\kappa a + \delta_0)$ [10]

And $A\kappa_0\cos\kappa_0 a = A\,k\cos(\kappa a + \delta_0)$ [11]

$$\therefore \quad \frac{\tan\kappa_0 a}{\kappa_0} = \frac{\tan(\kappa a + \delta_0)}{\kappa} \; Or, \; \kappa\tan\kappa_0 a = \kappa_0\left[\frac{\tan\kappa a + \tan\delta_0}{1 - \tan\kappa a\,\tan\delta_0}\right]$$

Or, $\kappa\tan\kappa_0 a - \kappa\tan\kappa_0 a\,\tan\kappa a\tan\delta_0 = \kappa_0\tan\kappa a + \kappa_0\tan\delta_0$

Or, $\tan\delta_0\,[\kappa_0 + \kappa\tan\kappa_0 a\,\tan\kappa a] = \kappa\tan\kappa_0 a - \kappa_0\tan\kappa a$

$$\text{Or, } \tan\delta_0 = \frac{\kappa\tan\kappa_0 a - \kappa_0\tan\kappa a}{[\kappa_0 + \kappa\tan\kappa_0 a\,\tan\kappa a]} = \frac{\frac{\kappa}{\kappa_0}\tan\kappa_0 a - \tan\kappa a}{\left[1 + \frac{\kappa}{\kappa_0}\tan\kappa_0 a\,\tan\kappa a\right]} \qquad [12]$$

For very low energy we can replace $\tan ka \approx \kappa a$ [$\because \kappa a \ll 1$]. The denominator $1 + \frac{\kappa}{\kappa_0}\kappa_0 a\;\kappa a = 1 + (\kappa a)^2 \approx 1$

Therefore equation [12] becomes

$$\tan\delta_0 \simeq \frac{\kappa}{\kappa_0}\tan\kappa_0 a - \kappa a = \kappa a\left[\frac{\tan\kappa_0 a}{\kappa_0 a} - 1\right]$$

But $\sin^2\delta_0 = \frac{\tan^2\delta_0}{1+\tan^2\delta_0}$ $or,$ $\sin\delta_0 = \frac{\tan\delta_0}{\sqrt{1+\tan^2\delta_0}}$

Here
$$\sin\delta_0 = \frac{\kappa a\left[\frac{\tan\kappa_0 a}{\kappa_0 a} - 1\right]}{\sqrt{1+\kappa^2 a^2\left[\frac{\tan\kappa_0 a}{\kappa_0 a} - 1\right]^2}} \quad [13]$$

For very low energy, the second term of the denominator in equation [13] can be neglected. Hence

$$\sin\delta_0 \simeq \kappa a\left[\frac{\tan\kappa_0 a}{\kappa_0 a} - 1\right] \quad [14]$$

We know, the total scattering cross-section for $l = 0$ is

$$\sigma_{tot} = \frac{4\pi}{\kappa^2}\sin^2\delta_0 = \frac{4\pi}{\kappa^2}\kappa^2 a^2\left[\frac{\tan\kappa_0 a}{\kappa_0 a} - 1\right]^2 \quad [15]$$

$$\sigma_{tot} = 4\pi a^2\left[\frac{\tan\kappa_0 a}{\kappa_0 a} - 1\right]^2 \quad [16]$$

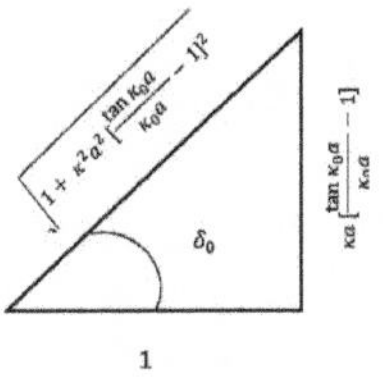

The scattering vanishes when $\tan\kappa_0 a = \kappa_0 a$ $[\delta_0 = 0\ or\ 180^\circ]$; *this is called Ramsour-Townsend effect.* While if $\kappa_0 a$ is an odd mulitiple of $\frac{\pi}{2}$ the cross-section becomes infinite. Thus if $\kappa_0 a$ is an odd multiple of $\frac{\pi}{2}$ the cross section is very large and we have a resonance scattering[17].

13.17 Neglect the higher order terms of Born approximation and validity of Born approximation

In the first Born approximation, there is only one scattering and in the higher order Born approximation there are multiples scattering. If the potential energy is weaker compare to the high energy incident beam then we can neglect the higher order terms of Born approximation, because higher order terms consist of the product of multiple potentials. So in the integrand, the potential falls off faster than $\frac{1}{r}$. That's why we can neglect the higher order terms of Born approximation.

- **Validity of Born approximation:** We know the 1st Born- approximation

$$\psi_{\kappa}^{+(1)}(r) = e^{\iota\kappa z} - \frac{1}{4\pi}\int \frac{e^{\iota\kappa|\bar{r}-\bar{r}'|}}{|\bar{r}-\bar{r}'|} U(\bar{r}')\, e^{\iota\kappa z'} d^3r' \qquad [1]$$

Where the first term is the incident plane wave and the second term is assumed in this approximation to be the total scattered wave. This implies the assumption that the scattered wave is only a small addition to the incident wave. Thus a sufficient condition for the validity of Born approximation

$$|\psi_{sactt}(\bar{r})| \ll \left|e^{\iota\kappa z}\right| = 1 \qquad [2]$$

The scattered wave $\psi_{sactt}(\bar{r})$ is largest at the center of the potential hence we must have

$$|\psi_{sactt}(0)| \ll 1 \qquad [3]$$

From equation [1] we get the scattered wave

$$\psi_{sactt}(\bar{r}) = -\frac{1}{4\pi}\int \frac{e^{\iota\kappa|\bar{r}-\bar{r}'|}}{|\bar{r}-\bar{r}'|} U(\bar{r}')\, e^{\iota\kappa z'} d^3r'$$

Thus,

$$|\psi_{sactt}(0)| = \left|\frac{1}{4\pi}\int \frac{e^{\iota\kappa r'}}{r'} U(\bar{r}')\, e^{\iota\kappa z'} d^3r'\right|$$

$$= \left|\frac{1}{4\pi}\int_0^{\infty}\int_0^{\pi}\int_0^{2\pi} \frac{e^{\iota\kappa r'}}{r'}\; U(\bar{r}')\, e^{\iota\kappa r'\cos\theta}\, r'^2 \sin\theta\, dr'\, d\theta\, d\phi\right|$$

Now put $x = \cos\theta \quad \therefore\ dx = -\sin\theta\, d\theta$; When $\theta = 0,\ x = 1;\quad \theta = \pi,\ \ x = -1$

$$\therefore \psi_{sactt}(0) = \left|\frac{1}{4\pi}\, 2\pi \int_0^{\infty}\int_{+1}^{-1} r' e^{\iota\kappa r'}\; U(\bar{r}')\, e^{\iota\kappa r' x}\, dx\;\; dr'\right|$$

$$= \left|\frac{1}{2}\int_0^{\infty}\int_{+1}^{-1} r' e^{\iota\kappa r'}\; U(\bar{r}')\, e^{\iota\kappa r' x}\, dx\;\; dr'\right|$$

$$= \left|\frac{1}{2}\int_0^{\infty} r' e^{\iota\kappa r'}\; U(\bar{r}') \left[\frac{e^{\iota\kappa r' x}}{\iota\kappa r'}\right]_{-1}^{1} dr'\right|$$

$$= \left|\frac{1}{2}\int_0^{\infty} r' e^{\iota\kappa r'}\; U(\bar{r}')\; \frac{1}{\iota\kappa r'}\left[e^{\iota\kappa r'} - e^{-\iota\kappa r'}\right] dr'\right|$$

$$= \left|\frac{1}{2}\int_0^{\infty} r' e^{\iota\kappa r'}\; U(\bar{r}')\; \frac{2i}{\iota\kappa r'}\left[\frac{e^{\iota\kappa r'} - e^{-\iota\kappa r'}}{2i}\right] dr'\right|$$

$$= \left| \frac{1}{2} \int_0^\infty r' e^{i\kappa r'} U(\bar{r}') \frac{2i}{i\kappa r'} \sin \kappa r' \, dr' \right|$$

$$= \left| \frac{1}{\kappa} \int_0^\infty e^{i\kappa r'} U(\bar{r}') \sin \kappa r' \, dr' \right|$$

$$= \left| \frac{2\mu}{\hbar^2 \kappa} \int_0^\infty e^{i\kappa r'} V(\bar{r}') \sin \kappa r' \, dr' \right| \quad [\because U(r) = \frac{2\mu}{\hbar^2} V(\bar{r}')]$$

Hence the validity condition

$$|\psi_{sactt}(0)| \ll 1 \quad or, \left| \frac{2\mu}{\hbar^2 \kappa} \int_0^\infty e^{i\kappa r'} V(\bar{r}') \sin \kappa r' dr' \right| \ll 1$$

The above condition can be satisfied when κ is large or when V is small. *Thus Born approximation is valid either at large energy or with a weak potential.*

13.18 Born approximation is applicable for high energy and validity criterion of Born approximation for real potential

Consider a particle with momentum $\hbar k$ [if no scattering force acts on it], would pass a distance s from the scattering center [This distance is called impact parameter], the angular momentum of the particle is ks . If the potential energy has a range "a" and an average strength V_0; then the first Born approximation

$$\therefore \psi_{Scatt}(0) = \frac{\mu V_0 a}{\hbar^2 k} = \frac{V_0 a}{\hbar v} \left[\frac{\mu V_0 a}{\hbar^2 k} = \frac{\mu V_0 a}{\hbar k.\hbar} = \frac{V_0 a}{\hbar v} \; ; \; where\; p = \mu v \;\; Or, \frac{\mu}{p} (= \frac{\mu}{\hbar k}) = \frac{1}{v}\right] \quad [1]$$

Where v is the velocity of the particle.

On the other hand the angular momentum component in the lth partial wave is l. Hence $\rightarrow s = \frac{l}{k}$

But substantial scattering can be expected only if the impact parameter is greater than the range of the scattering force. I.e., $s \geq a$ Where "a" is the range of potential energy. So,

$$s \geq a \;\; Or, \frac{l}{k} \geq a \;\; Or, l \geq ka \; \therefore l \geq \sqrt{\frac{2mE}{\hbar^2}}\, a \;\; I.e.,, ka \geq E \quad [2]$$

When k is large I.e., for high energy limit $ka \geq 1$ [high energy phenomena]; the potential energy V is small. Thus Born approximation can be used with confidence [In comparison with interaction energy] with high incident energies. *Thus Born approximation is valid either at large energy or with a weak potential.* Note that the potential energy is strong enough i.e. $\frac{\mu V_0 a}{\hbar^2 k} = \frac{V_0 a}{\hbar v} \gg 1$; *The Born approximation cannot be used in low energy limit* [for low energy limit $ka \leq 1\, means\;\; s \leq a \;\; Or, \frac{l}{k} \leq a \;\; Or, l \leq ka \; \therefore l \leq \sqrt{\frac{2mE}{\hbar^2}}\, a \;\; I.e., ka \leq E$]

- **Validity criterion of Born approximation for real potential**

In the 1st Born approximation, the scattering amplitude for the real potential

$V(r) = (-V_0 e^{-\frac{r}{a}})$ with spherically symmetric can be written as

$$f(\theta) = -\frac{2\mu}{\hbar^2\kappa}\int_{r=0}^{\infty} r\,\sin\kappa r\,V(r)\,dr = -\frac{2\mu}{\hbar^2\kappa}\int_{r=0}^{\infty} r\,\sin\kappa r\,(-V_0 e^{-\frac{r}{a}})\,dr$$

$$= \frac{2\mu V_0}{\hbar^2\kappa}\int_{r=0}^{\infty} r\,\sin\kappa r\;e^{-\frac{r}{a}}\,dr$$

$$= \frac{2\mu V_0}{\hbar^2\kappa}\int_{r=0}^{\infty} re^{-\frac{r}{a}}\left[\frac{e^{i\kappa r}-e^{-i\kappa r}}{2i}\right]dr$$

$$= \frac{2\mu V_0}{2i\,\hbar^2\kappa}\int_{r=0}^{\infty}\left[re^{-r\left(-i\kappa+\frac{1}{a}\right)} - re^{-r\left(i\kappa+\frac{1}{a}\right)}\right]dr$$

We know the standard integral $\int_0^{\infty} x^n\,e^{-ax}dx\;\frac{n!}{a^{n+1}}$

$$= \frac{2\mu V_0}{2i\,\hbar^2\kappa}\left[\frac{1}{\left(-i\kappa+\frac{1}{a}\right)^2} - \frac{1}{\left(i\kappa+\frac{1}{a}\right)^2}\right] = \frac{\mu V_0}{i\,\hbar^2\kappa}\left[\frac{a^2}{(1-i\kappa a)^2} - \frac{a^2}{(1+i\kappa a)^2}\right]$$

$$= \frac{\mu V_0 a^2}{i\,\hbar^2\kappa}\left[\frac{1+2i\kappa a-\kappa^2a^2-1+2i\kappa a+\kappa^2a^2}{(1-i\kappa a)^2(1+i\kappa a)^2}\right] = \frac{\mu V_0 a^2}{i\,\hbar^2\kappa}\;\frac{4i\kappa a}{(1+\kappa^2a^2)^2}$$

$$\therefore\; f(\theta) = \frac{4\mu V_0}{\hbar^2}\;\frac{a^3}{(1+\kappa^2a^2)^2}$$

Hence the differential scattering cross – section

$$\sigma(\theta) = |f(\theta)|^2 = \left|\frac{4\mu V_0}{\hbar^2}\;\frac{a^3}{(1+\kappa^2a^2)^2}\right|^2 = \frac{16\mu^2V_0^2}{\hbar^4}\;\frac{a^6}{(1+\kappa^2a^2)^4}$$

$$= \frac{16\mu^2V_0^2a^6}{\hbar^4}\left(1+4\kappa_0^2\sin^2\frac{\theta}{2}a^2\right)^{-4}\quad\left[\because\;\kappa = 2\kappa_0\sin\frac{\theta}{2}\right]$$

The total scattering cross – section

$$\sigma_{tot} = 2\pi\int_0^{\pi}\sigma(\theta)\sin\theta\,d\theta = 2\pi\int_0^{\pi}\frac{16\mu^2V_0^2a^6}{\hbar^4}\left(1+4\kappa_0^2a^2\sin^2\frac{\theta}{2}\right)^{-4}\sin\theta\,d\theta$$

$$= \frac{32\pi\mu^2V_0^2a^6}{\hbar^4}\int_0^{\pi}\left(1+4\kappa_0^2a^2\sin^2\frac{\theta}{2}\right)^{-4}2\sin\frac{\theta}{2}\cos\frac{\theta}{2}d\theta$$

$$= \frac{64\pi\mu^2V_0^2a^6}{\hbar^4}\;\frac{1}{4\kappa_0^2a^2}\int_0^{\pi}\left(1+4\kappa_0^2a^2\sin^2\frac{\theta}{2}\right)^{-4}4\kappa_0^2a^2\sin\frac{\theta}{2}\cos\frac{\theta}{2}d\theta$$

Let, $z = 1+4\kappa_0^2a^2\sin^2\frac{\theta}{2}$; $dz = 4\kappa_0^2a^2\,2\sin\frac{\theta}{2}\cos\frac{\theta}{2}\,\frac{d\theta}{2}$

$dz = 4\kappa_0^2a^2\sin\frac{\theta}{2}\cos\frac{\theta}{2}d\theta$; When $\theta = 0$, $z = 1$; When $\theta = \pi$, $z = (1+4\kappa_0^2a^2)$

Therefore,

$$\sigma_{tot} = \frac{16\pi\mu^2{V_0}^2a^4}{\hbar^4{\kappa_0}^2} \int_0^{1+4{\kappa_0}^2a^2} z^{-4}\, dz$$

$$= \frac{16\pi\mu^2{V_0}^2a^4}{\hbar^4{\kappa_0}^2}\left[\frac{z^{-3}}{-3}\right]_0^{1+4{\kappa_0}^2a^2} = \frac{16\pi\mu^2{V_0}^2a^4}{3\hbar^4{\kappa_0}^2}[1-(1+4{\kappa_0}^2a^2)^{-3}]$$

$$= \frac{16\pi\mu^2{V_0}^2a^4}{3\hbar^4{\kappa_0}^2}\left[1-\frac{1}{(1+4{\kappa_0}^2a^2)^3}\right]$$

Now from the condition for the validity of Born approximation

$$\therefore \psi_{sactt}(0) = \frac{\mu}{\hbar^2\,\kappa}\left|\int_0^\infty V(r)\,(e^{2i\kappa r}-1)\,dr\right|$$

$$= \frac{\mu\,V_0}{\hbar^2\,\kappa}\left|\int_0^\infty e^{-\frac{r}{a}}\,(e^{2i\kappa r}-1)\,dr\right|$$

$$= \frac{\mu\,V_0}{\hbar^2\,\kappa}\left|\int_0^\infty \left(e^{-r(\frac{1}{a}-2i\kappa)} - e^{-\frac{r}{a}}\right)dr\right|$$

$$= \frac{\mu\,V_0}{\hbar^2\,\kappa}\left|\left[\frac{e^{-r(\frac{1}{a}-2i\kappa)}}{-(\frac{1}{a}-2i\kappa)}\right]_0^\infty - \left[\frac{e^{-\frac{r}{a}}}{-\frac{1}{a}}\right]_0^\infty\right| = \frac{\mu\,V_0}{\hbar^2\,\kappa}\left|\left[\frac{-1}{-(\frac{1}{a}-2i\kappa)} - \frac{-1}{-\frac{1}{a}}\right]\right|$$

$$= \frac{\mu\,V_0}{\hbar^2\,\kappa}\left|\left[\frac{a}{(1-2i\kappa a)} - a\right]\right| = \frac{\mu\,V_0 a}{\hbar^2\,\kappa}\left|\left[\frac{1}{(1-2i\kappa a)} - 1\right]\right|$$

$$= \frac{\mu\,V_0 a}{\hbar^2\,\kappa}\left|\left[\frac{1}{-(2i\kappa a-1)} - 1\right]\right| = \frac{\mu\,V_0 a}{\hbar^2\,\kappa}\left|\left[\frac{2i\kappa a}{(2i\kappa a-1)}\right]\right| = \frac{2\mu\,V_0 a^2}{\hbar^2}\left|\left[\frac{i}{(2i\kappa a-1)}\right]\right|$$

$$\therefore \psi_{sactt}(0) = \frac{2\mu\,V_0 a^2}{\hbar^2}\frac{1}{\sqrt{(1+4\kappa^2a^2)}} \qquad [1]$$

• **Case – 1**: For low energy limit. I.e., $E \ll \kappa a \ll 1$

Therefore [1] becomes

$$|\psi_{sactt}(0)| = \frac{2\mu\,V_0 a^2}{\hbar^2}$$

Thus for the validity of Born's approximation in low energy

$$|\psi_{sactt}(0)| = \frac{2\mu\,V_0 a^2}{\hbar^2} \qquad [2]$$

• **Case – 2**: For high energy limit I.e., $E \gg \kappa a \gg 1$

$$|\psi_{sactt}(0)| = \frac{2\mu\,V_0 a^2}{\hbar^2}\,\frac{1}{2\kappa a} = \frac{\mu\,V_0\,a}{\hbar^2\,\kappa}$$

$$= \frac{\mu V_0 a}{\hbar^2 k} = \frac{\mu V_0 a}{\hbar k.\hbar} = \frac{V_0 a}{\hbar v}\ ;\ where\ p = \mu v\ \ Or, \frac{\mu}{p}(= \frac{\mu}{\hbar k}) = \frac{1}{v}$$

Thus for the validity of Born's approximation in high energy

$$|\psi_{sactt}(0)| = \frac{\mu\,V_0\,a}{\hbar^2\,\kappa} = \frac{V_0 a}{\hbar v} \ll 1 \qquad [3]$$

Thus for the validity of Born's approximation in low energy

$$|\psi_{sactt}(0)| = \frac{2\mu\, V_0 a^2}{\hbar^2} \ll 1 \qquad [4]$$

From equation [2] it shows that if the potential is strong enough to bind a particle I.e., $\frac{2\mu\, V_0 a^2}{\hbar^2} \gg 1$, then the Born approximation can not be used in low energy limit. And from equation [3], if the potential energy has a range "a" And an average strength V_0 ; then the first Born approximation is likely to give reliable values if

$$\frac{\mu\, V_0\, a}{\hbar^2\, \kappa} \ll 1 = \frac{\mu\, V_0\, a}{\hbar v} \ll 1 \qquad [5]$$

Equation [5] is the condition that represents generally satisfied in high energy limit. From this we note that Born approximation can be used with confidence with high incident energies.

13.19 Yukawa potential and differential Scattering cross-section for Yukawa potential

The force between a Neutron and a Proton can be described by a potential energy function of the form $V(r) = -V_0 \frac{e^{-r/r_o}}{r/r_o}$; this form arises in the Meson theory of Nuclear forces and is called Yukawa potential[5].

- **Differential Scattering cross- section for Yukawa potential**

In the first Born approximation the scattering amplitude for the potential with spherical symmetric can be written as

$$f(\theta) = -\frac{1}{\kappa}\int_{r=0}^{\infty} \sin\kappa r'\; U(r')\, r'\, dr' = -\int_{r=0}^{\infty} \frac{\sin qr}{qr}\; U(r)\, r^2\, dr$$

$$= -\frac{1}{\kappa}\int_{r=0}^{\infty} \sin\kappa r\, U(r)\, r\, dr = -\frac{2\mu}{\hbar^2\kappa}\int_{r=0}^{\infty} r\, \sin\kappa r\; V(r)\, dr$$

$$= -\frac{2\mu}{\hbar^2\kappa}\int_{r=0}^{\infty} r\, \sin\kappa r(-\, V_0\, \frac{e^{-\alpha r}}{\alpha r})dr = \frac{2\mu\, V_0}{\hbar^2\kappa\,\alpha}\int_{r=0}^{\infty} \sin\kappa r\; e^{-\alpha r}\, dr$$

$$= \frac{2\mu\, V_0}{\hbar^2\kappa\,\alpha}\int_{r=0}^{\infty}\left[\frac{e^{i\kappa r}-e^{-i\kappa r}}{2i}\right] e^{-\alpha r}\, dr$$

$$= \frac{2\mu\, V_0}{\hbar^2\kappa\,\alpha}\frac{1}{2i}\int_{r=0}^{\infty} [\, e^{-(\alpha-i\kappa)r} - e^{-(\alpha+i\kappa)r}\,]\, dr = \frac{\mu\, V_0}{\hbar^2\kappa\,\alpha\, i}\left[\frac{e^{-(\alpha-i\kappa)r}}{-(\alpha-i\kappa)} - \frac{e^{-(\alpha+i\kappa)r}}{-(\alpha+i\kappa)}\right]_0^{\infty}$$

$$= \frac{\mu\, V_0}{\hbar^2\kappa\,\alpha\, i}\left[\frac{1}{(\alpha-i\kappa)} - \frac{1}{(\alpha+i\kappa)}\right] = \frac{\mu\, V_0}{\hbar^2\kappa\,\alpha\, i}\left[\frac{2i\kappa}{(\alpha-i\kappa)\,(\alpha+i\kappa)}\right]$$

$$f(\theta) = \frac{2\,\mu\, V_0}{\hbar^2\,\alpha}\frac{1}{\alpha^2+\kappa^2}$$

Hence the differential scattering cross – section

$$\sigma(\theta) = |f(\theta)|^2 = \left|\frac{2\,\mu\,V_0}{\hbar^2\,\alpha}\;\frac{1}{\alpha^2+\kappa^2}\right|^2 = \frac{4\,\mu_0^2\,V_0{}^2}{\hbar^4\,\alpha^2}\frac{1}{\left(\alpha^2+4\kappa_0{}^2\sin^2\frac{\theta}{2}\right)^2}\;[\because\;\kappa = 2\kappa_0\,\sin\frac{\theta}{2}]$$

$$= \frac{4\,\mu_0^2\,V_0{}^2}{\hbar^4\,\alpha^2}\left(\alpha^2 + 4\kappa_0{}^2\sin^2\frac{\theta}{2}\right)^{-2}$$

Hence the total scattering cross – section

$$\sigma_{tot} = \int \sigma(\theta)d\Omega = \int_0^{\pi}\int_0^{2\pi}\sigma(\theta)\,\sin\theta\;d\theta\;d\phi = 2\pi\int_0^{\pi}\;\sigma(\theta)\sin\theta\;d\theta$$

$$= 2\pi\;\frac{4\,\mu_0^2\,V_0{}^2}{\hbar^4\,\alpha^2}\int_0^{\pi}\;\left(\alpha^2 + 4\kappa_0{}^2\sin^2\frac{\theta}{2}\right)^{-2}\;\sin\theta\;d\theta$$

$$= \frac{8\pi\mu_0^2\,V_0{}^2}{\hbar^4\,\alpha^2}\int_0^{\pi}\;\left(\alpha^2 + 4\kappa_0{}^2\sin^2\frac{\theta}{2}\right)^{-2}\;2\sin\frac{\theta}{2}\;\cos\frac{\theta}{2}\,d\theta$$

$$= \frac{16\pi\mu_0^2\,V_0{}^2}{4\kappa_0{}^2\,\hbar^4\,\alpha^2}\int_0^{\pi}\;\left(\alpha^2 + 4\kappa_0{}^2\sin^2\frac{\theta}{2}\right)^{-2}\;4\kappa_0{}^2\sin\frac{\theta}{2}\;\cos\frac{\theta}{2}\,d\theta$$

Let, $z = \alpha^2 + 4\kappa_0{}^2\sin^2\frac{\theta}{2}\;\;;dz = \;4\kappa_0{}^2\,2\sin\frac{\theta}{2}\;\cos\frac{\theta}{2}\,\frac{d\theta}{2} = 4\kappa_0{}^2\;\sin\frac{\theta}{2}\;\cos\frac{\theta}{2}\,d\theta$

$$\therefore\;\sigma_{tot} \;=\; \frac{16\pi\mu_0^2\,V_0{}^2}{4\kappa_0{}^2\,\hbar^4\,\alpha^2}\;\int_0^{\pi}z^{-2}\,dz = \frac{16\pi\mu_0^2\,V_0{}^2}{4\kappa_0{}^2\,\hbar^4\,\alpha^2}\left[\frac{\left(\alpha^2+4\kappa_0{}^2\sin^2\frac{\theta}{2}\right)^{-1}}{-1}\right]_0^{\pi}$$

$$= \frac{16\pi\mu_0^2\,V_0{}^2}{4\kappa_0{}^2\,\hbar^4\,\alpha^2}\left[-\left(\alpha^2+4\kappa_0{}^2\right)^{-1} + (\alpha^2)^{-1}\right] = \frac{16\pi\mu_0^2\,V_0{}^2}{4\kappa_0{}^2\,\hbar^4\,\alpha^2}\left[\frac{1}{\alpha^2} - \frac{1}{\alpha^2+4\kappa_0{}^2}\right]$$

$$= \frac{16\pi\mu_0^2\,V_0{}^2}{4\kappa_0{}^2\,\hbar^4\,\alpha^2}\;\frac{4\kappa_0{}^2}{\alpha^2(\alpha^2+4\kappa_0{}^2)} = \frac{16\pi\mu_0^2\,V_0{}^2}{\hbar^4\,\alpha^4}\;\frac{1}{(\alpha^2+4\kappa_0{}^2)}$$

13.20 Green's functions: A brief description

Green's functions are named after the British mathematician George Green, who first developed the concept in the 1830s. In the modern study of linear partial differential equations, Green's functions are studied largely from the point of view of fundamental solutions instead [In mathematics, a Green's function is the impulse response of an inhomogeneous linear differential equation defined on a domain, with specified initial conditions or boundary conditions].

❖ Advanced and retarded Green's functions

Sometimes the Green's function can be split into a sum of two functions. One with the variable positive (+) and the other with the variable negative (-).These are the advanced and retarded Green's functions, and when the equation under study depends on time, one of the parts is causal and the other anti-causal. In these problems usually the causal part is the important one. Through the superposition principle for linear operator problems, the convolution of a Green's function with an arbitrary function $f(x)$ on that domain is the solution to the inhomogeneous differential equation for $f(x)$. In other words, given a linear ordinary differential equation (ODE), L(solution)

= source, one can first solve $L(\text{green}) = \delta_s$, for each s, and realizing that, since the source is a sum of delta functions, the solution is a sum of Green's functions as well, by linearity of L.

❖ Definition and uses

A Green's function, $G(x,s)$, of a linear differential operator $L = L(x)$ acting on distributions over a subset of the Euclidean space $\mathbb{R}^n$, at a point s, is any solution of

$$LG(x,s) = \delta(s-x) \qquad [1]$$

where δ is the Dirac delta function. This property of a Green's function can be exploited to solve differential equations of the form

$$Lu(x) = f(x) \qquad [2]$$

If the kernel of L is non-trivial, then the Green's function is not unique. However, in practice, some combination of symmetry, boundary conditions and/or other externally imposed criteria will give a unique Green's function. Green's functions may be categorized, by the type of boundary conditions satisfied, by a Green's function number. Also, Green's functions in general are distributions, not necessarily functions of a real variable.

As a side note, the Green's function as used in physics is usually defined with the opposite sign, instead, that is,

$$LG(x,s) = \delta(x-s) \qquad [3]$$

This definition does not significantly change any of the properties of the Green's function. If the operator is translation invariant, that is, when L has constant coefficients with respect to x, then the Green's function can be taken to be a convolution operator, that is,

$$G(x,s) = G(x-s) \qquad [4]$$

In this case, the Green's function is the same as the impulse response of linear time-invariant system theory.

Loosely speaking, if such a function G can be found from the operator L, then, if we multiply the equations [1] for the Green's by f(s), and then integrate with respect to s, we obtain

$$\int LG(x,s)f(s)ds = \int \delta(s-x)\, f(s)ds = f(x) \qquad [5]$$

The R.H.S of equation [5] is given by the equation [2] to be equal to $Lu(x)$ thus

$$Lu(x) = \int LG(x,s)f(s)ds \qquad [6]$$

Because the operator $L = L(x)$ is linear and acts on the variable x alone [not on the variable of integration s], one may take the operator L outside of the integration on the R.H.S, yielding

$$Lu(x) = L(\int G(x,s)f(s)ds) \qquad [7]$$

which suggests,

$$u(x) = \int G(x,s)f(s)ds \qquad [8]$$

Thus, one may obtain the function $u(x)$ through knowledge of the Green's function in equation [1] and the source term on the R.H.S in equation [2]. This process relies upon the linearity of the operator L.

The primary use of Green's functions in mathematics is to solve non-homogeneous boundary value problems. In modern theoretical physics, Green's functions are also usually used as propagators in Feynman diagrams; the term Green's function is often further used for any correlation function.

Under many-body theory, the term is also used in physics, specifically in quantum field theory, aerodynamics, aero acoustics, electrodynamics, seismology and statistical field theory, to refer to various types of correlation functions, even those that do not fit the mathematical definition. *In quantum field theory, Green's functions take the roles of propagators.*

Green's functions are also useful tools in solving wave equations and diffusion equations. In quantum mechanics, the Green's function of the Hamiltonian is a key concept with important links to the concept of density of states.

13.21 Born approximation and its applications

The *Born approximation* is named after Max *Born* who proposed this *approximation* in early days of quantum theory development. Generally, in scattering theory and in particular in quantum mechanics, the Born approximation consists of taking the incident field in place of the total field as the driving field at each point in the scatterer. It is the perturbation method applied to scattering by an extended body. It is accurate if the scattered field is small compared to the incident field on the scatterer.

- **Applications**

The Born approximation is used in several different physical contexts. In neutron scattering, the first-order Born approximation is almost always adequate, except for neutron optical phenomena like internal total reflection in a neutron guide, or grazing-incidence small-angle scattering. The Born approximation has also been used to calculate conductivity in bilayer graphene and to approximate the propagation of long-wavelength waves in elastic media. The same ideas have also been applied to studying the movements of seismic waves through the earth.

For example, the scattering of radio waves by a light Styrofoam column can be approximated by assuming that each part of the plastic is polarized by the same electric field that would be present at that point without the column, and then calculating the scattering as a radiation integral over that polarization distribution.

13.22 Green function and first Born approximation

Consider the Schrödinger equation in the presence of the potential

$$H\,\psi(\vec{r}) = E\,\psi(\vec{r})$$

$$\text{Or, } [-\frac{\kappa^2}{2m}\,\nabla^2 + V(\vec{r})]\,\psi(\vec{r}) = E\,\psi(\vec{r})$$

$$\text{Or, } \nabla^2\psi(\vec{r}) - \frac{2m}{\hbar^2}V(\vec{r})\,\psi(\vec{r}) = -\frac{2m}{\hbar^2}E\,\psi(\vec{r})$$

$$\text{Or, } \nabla^2\psi(\vec{r}) - \frac{2m}{\hbar^2}V(\vec{r})\,\psi(\vec{r}) + \frac{2m}{\hbar^2}E\,\psi(\vec{r}) = 0$$

$$\text{Or, } [\nabla^2 + \kappa^2 - U(\vec{r})\,]\psi(\vec{r}) = 0 \qquad [1]$$

Where, $\kappa^2 = \frac{2\,\mu\,E}{\hbar^2}$ *and* $U(\vec{r}) = \frac{2\,\mu\,V(\vec{r})}{\hbar^2}$ [2]

Therefore equation [1] can be written as

$$[\nabla^2 + \kappa^2\,]\,\psi(\vec{r}) = U(\vec{r})\psi(\vec{r}) \qquad [4]$$

We have to solve the auxiliary problem of the scattering

$$[\nabla^2 + \kappa^2\,]\;G(\vec{r},\vec{r}\,') = \delta(\vec{r} - \vec{r}\,') \qquad [5]$$

Here $G(\vec{r},\vec{r}\,')$ & $\delta(\vec{r} - \vec{r}\,')$are the Green's function & the Dirac-delta function respectively. In terms of $(\vec{r},\vec{r}\,')$, the solution of [4] can be written as

$$\psi(\vec{r}) = \phi(\vec{r}) + \int G(\vec{r},\vec{r}\,')\;U(\vec{r}\,')\psi(\vec{r}\,')d\vec{r}\,' \qquad [6]$$

Where $\phi(\vec{r})$ is the free particle solution of the Schrödinger equation I.e., the solution of the Schrödinger equation

$$[\nabla^2 + \kappa^2\,]\phi(\vec{r}) = 0 \qquad [7]$$

The solution of this equation is the travelling wave type solution & is

$$\phi(\vec{r}) = e^{i\,\bar{\kappa}.\vec{r}} \qquad [8]$$

Taking Fourier analysis of Green's function $G(\vec{r},\vec{r}\,')$ & Dirac-delta function $\delta(\vec{r} - \vec{r}\,')$:

$$G(\vec{r},\vec{r}\,') = \frac{1}{(2\pi)^3}\int g(\vec{\kappa}\,')\,e^{i\,\bar{\kappa}'.(\vec{r}-\vec{r}\,')}\;d\vec{\kappa}\,' \qquad [9]$$

$$\delta(\vec{r} - \vec{r}\,') = \frac{1}{(2\pi)^3}\int e^{i\,\bar{\kappa}'.(\vec{r}-\vec{r}\,')}\;d\vec{\kappa}\,' \qquad [10]$$

Putting these in [5]

$$[\nabla^2 + \kappa^2\,]\;\frac{1}{(2\pi)^3}\int g(\vec{\kappa}\,')\,e^{i\,\bar{\kappa}'.(\vec{r}-\vec{r}\,')}\;d\vec{\kappa}\,' = \frac{1}{(2\pi)^3}\int e^{i\,\bar{\kappa}'.(\vec{r}-\vec{r}\,')}\;d\vec{\kappa}\,'$$

Or, $\frac{1}{(2\pi)^3}\int [\nabla^2 + \kappa^2]\, g(\vec{\kappa}')\, e^{i\vec{\kappa}'.(\vec{r}-\vec{r}')}\, d\vec{\kappa}' = \frac{1}{(2\pi)^3}\int e^{i\vec{\kappa}'.(\vec{r}-\vec{r}')}\, d\vec{\kappa}'$

Or, $\frac{1}{(2\pi)^3}\int [-\kappa'^2 + \kappa^2]\, g(\vec{\kappa}')\, e^{i\vec{\kappa}'.(\vec{r}-\vec{r}')}\, d\vec{\kappa}' = \frac{1}{(2\pi)^3}\int e^{i\vec{\kappa}'.(\vec{r}-\vec{r}')}\, d\vec{\kappa}'$

Or, $\left[-\kappa'^2 + \kappa^2\right] g(\vec{\kappa}') = 1$ $Or,\ g(\vec{\kappa}') = \frac{1}{-\left[\kappa'^2 - \kappa^2\right]}$

Now from [9] $G(\vec{r},\vec{r}') = -\frac{1}{(2\pi)^3}\int \frac{e^{i\vec{\kappa}'.(\vec{r}-\vec{r}')}}{\left[\kappa'^2 - \kappa^2\right]}\, d\vec{\kappa}'$

Putting $\vec{R} = \vec{r} - \vec{r}'$ [11]

Therefore, $G(\vec{r},\vec{r}') = -\frac{1}{(2\pi)^3}\int \frac{e^{i\vec{\kappa}'.\vec{R}}}{\left[\kappa'^2 - \kappa^2\right]}\, d\vec{\kappa}'$ Choosing $\vec{R}$ as the polar axis, then

$$G(\vec{r},\vec{r}') = -\frac{1}{(2\pi)^3}\int_{\theta=0}^{\pi}\int_{\phi=0}^{2\pi}\int_{\kappa'=0}^{\infty} \frac{e^{i\kappa'.R\cos\theta}}{\left[\kappa'^2 - \kappa^2\right]}\, \kappa'^2\, d\kappa' \sin\theta\, d\theta\, d\phi$$

$$= -\frac{1}{(2\pi)^3}\, 2\pi \int_{\kappa'=0}^{\infty}\left[\int_{\theta=0}^{\pi} \frac{e^{i\kappa'.R\cos\theta}}{\left[\kappa'^2 - \kappa^2\right]}\sin\theta\, d\theta\right] \kappa'^2\, d\kappa'$$

Let $i\kappa' R\cos\theta = z$ $or, dz = -i\kappa' R\sin\theta\, d\theta$

If $\theta = 0,\ z = i\kappa' R$ and if $\theta = \pi,\ z = -i\kappa' R$

$$G(\vec{r},\vec{r}') = -\frac{1}{(2\pi)^2}\int_{\kappa'=0}^{\infty}\left[\int_{i\kappa' R}^{-i\kappa' R} \frac{e^z}{\left[\kappa'^2 - \kappa^2\right]}\frac{dz}{(-i\kappa' R)}\right] \kappa'^2\, d\kappa'$$

Or, $G(\vec{r},\vec{r}') = \frac{1}{(2\pi)^2}\int_{\kappa'=0}^{\infty} \frac{1}{-iR\kappa'(\kappa'^2 - \kappa^2)}\left[\int_{i\kappa' R}^{-i\kappa' R} e^z\, dz\right] \kappa'^2\, d\kappa'$

$$= \frac{1}{(2\pi)^2}\int_{\kappa'=0}^{\infty} \frac{1}{\kappa'(\kappa'^2 - \kappa^2)}\left[e^z\right]_{i\kappa' R}^{-i\kappa' R} \kappa'^2\, d\kappa'$$

$$= \frac{1}{4\pi^2 iR}\int_{\kappa'=0}^{\infty} \frac{1}{(\kappa'^2 - \kappa^2)}\left[e^{-i\kappa' R} - e^{i\kappa' R}\right] \kappa'\, d\kappa'$$

$$= \frac{1}{4\pi^2 iR}\,\frac{1}{2}\int_{\kappa'=-\infty}^{\infty} \frac{e^{-i\kappa' R} - e^{i\kappa' R}}{(\kappa'^2 - \kappa^2)}\, \kappa'\, d\kappa'$$

$$= \frac{1}{8\pi^2 iR}\left[\int_{\kappa'=-\infty}^{\infty} \frac{e^{-i\kappa' R}}{(\kappa'^2 - \kappa^2)}\, \kappa'\, d\kappa' - \int_{\kappa'=-\infty}^{\infty} \frac{e^{i\kappa' R}}{(\kappa'^2 - \kappa^2)}\, \kappa'\, d\kappa'\right.$$

Using $\kappa' \to -\kappa'$ in the first integral then $d\kappa' \to -d\kappa'$ & the limit $-\infty \to +\infty$ & $+\infty \to -\infty$

$$G(\vec{r},\vec{r}') = \frac{1}{8\pi^2 iR}\left[+\int_{\kappa'=+\infty}^{-\infty}\frac{e^{i\kappa' R}}{(\kappa'^2-\kappa^2)}\kappa'\,d\kappa' - \int_{\kappa'=-\infty}^{+\infty}\frac{e^{i\kappa' R}}{(\kappa'^2-\kappa^2)}\kappa'\,d\kappa'\right.$$

Or, $G(\vec{r},\vec{r}') = \frac{1}{8\pi^2 iR}\left[-\int_{\kappa'=-\infty}^{+\infty}\frac{e^{i\kappa' R}}{(\kappa'^2-\kappa^2)}\kappa'\,d\kappa' - \int_{\kappa'=-\infty}^{+\infty}\frac{e^{i\kappa' R}}{(\kappa'^2-\kappa^2)}\kappa'\,d\kappa'\right.$

Or, $G(\vec{r},\vec{r}') = -\frac{1}{4\pi^2 iR}\int_{\kappa'=-\infty}^{+\infty}\frac{e^{i\kappa' R}}{(\kappa'^2-\kappa^2)}\kappa'\,d\kappa'$ [12]

The integrand has poles (i goes to infinity at $\kappa' = \pm k$. Now applying Couchy-Reimann Residue theorem performing the integration for $\kappa' = +k$.

$$G^+(\vec{r},\vec{r}') = -\frac{1}{4\pi^2 iR}2\pi i\sum Residue = -\frac{1}{4\pi^2 iR}2\pi i\left[\frac{\kappa' e^{i\kappa' R}}{(\kappa'+\kappa)}\right]_{\kappa'=\kappa}$$

$$= -\frac{1}{4\pi^2 iR}2\pi i\frac{\kappa e^{i\kappa R}}{2\kappa} = -\frac{1}{4\pi R}e^{i\kappa R} \qquad [13]$$

Again applying Couchy-Reimann Residue theorem performing the integration for $\kappa' = -k$.

$$G^-(\vec{r},\vec{r}') = -\frac{1}{4\pi^2 iR}2\pi i\sum Residue = -\frac{1}{4\pi^2 iR}2\pi i\left[\frac{\kappa' e^{i\kappa' R}}{(\kappa'-\kappa)}\right]_{\kappa'=\kappa}$$

$$= -\frac{1}{4\pi^2 iR}2\pi i\frac{-\kappa' e^{i\kappa R}}{(-\kappa-\kappa)} = -\frac{1}{4\pi R}e^{-i\kappa R} \qquad [14]$$

Here $G^+(\vec{r},\vec{r}')$ & $G^-(\vec{r},\vec{r}')$ are called the advanced & retarded Green's function respectively. Now the corresponding solutions for $\psi(\vec{r})$, from [6]

$$\psi^\pm(\vec{r}) = e^{i\vec{\kappa}.\vec{r}} + \int G^\pm(\vec{r},\vec{r}')\,U(\vec{r}')\psi^\pm(\vec{r}')d\vec{r}' = e^{i\vec{\kappa}.\vec{r}} - \frac{1}{4\pi}\int\frac{e^{\pm i\kappa R}}{R}U(\vec{r}')\psi^\pm(\vec{r}')d\vec{r}'$$

$$\psi^\pm(\vec{r}) = e^{i\vec{\kappa}.\vec{r}} - \frac{1}{4\pi}\int\frac{e^{\pm i\kappa|\vec{r}-\vec{r}'|}}{|\vec{r}-\vec{r}'|}U(\vec{r}')\psi^\pm(\vec{r}')d\vec{r}' \qquad [15]$$

We can write,

$|\vec{r}-\vec{r}'|^2 = r^2 - 2\,\vec{r}.\vec{r}' + r'^2 = r^2(1 - \frac{2\,\vec{r}.\vec{r}'}{r^2} + \frac{r'^2}{r^2})$ If $r \gg r'$, then

$$|\vec{r}-\vec{r}'|^2 = r^2(1 - \frac{2\vec{r}}{r}\frac{\vec{r}'}{r}) = r^2(1 - \frac{2\hat{r}.\vec{r}'}{r})$$

$$|\vec{r}-\vec{r}'| = r(1 - \frac{2\hat{r}.\vec{r}'}{r})^{\frac{1}{2}}$$

$$= r\left(1 - \frac{1}{2}\frac{2\hat{r}.\vec{r}'}{r}\right) \; [expanding\ Binomially]$$

$$|\vec{r}-\vec{r}'| = r - \hat{r}.\vec{r}'$$

Now using this in [15] & dropping the 2nd term for the denominator

$$\psi^{\pm}(\vec{r}) = e^{\,i\,\vec{K}.\vec{r}} - \frac{1}{4\pi}\int \frac{e^{\pm\, i\,\kappa(r-\hat{r}.\vec{r}')}}{r}\, U(\vec{r}')\psi^{\pm}(\vec{r}')d\vec{r}'$$

Or, $$\psi^{\pm}(\vec{r}) = e^{\,i\,\vec{K}.\vec{r}} - \frac{e^{\pm\, i\,\kappa r}}{4\pi r}\int e^{\mp\, i\,\kappa\,\hat{r}.\vec{r}'}\, U(\vec{r}')\psi^{\pm}(\vec{r}')d\vec{r}' \qquad [16]$$

But this can be written as

$$\psi^{\pm}(\vec{r}) = e^{\,i\,\vec{K}.\vec{r}} + f_{\kappa}(\theta,\phi)\frac{e^{\pm\, i\,\kappa r}}{r} \qquad [17]$$

Comparing this [17] with [16], the scattering amplitude is

$$f_{\kappa}(\theta,\phi) = -\frac{1}{4\pi}\int e^{\mp\, i\,\kappa\,\hat{r}.\vec{r}'}\, U(\vec{r}')\psi^{\pm}(\vec{r}')d\vec{r}'$$

$$f_{\kappa}(\theta,\phi) = -\frac{1}{4\pi} < \psi_{\kappa}^{-}|U|\psi_{\kappa}^{+} > \qquad [18]$$

From equation [18], *the scattering amplitude is seemed to be the matrix element of the scattering potential between the actual scattering state* ψ_{κ}^{+} *of the whole system and the particular scattered final free state* ψ_{κ}^{-}.

13.23 Lippmann–Schwinger equation

The Lippmann–Schwinger equation (named after Bernard Lippmann and Julian Schwinger) is one of the most used equations to describe particle collisions or, more precisely, scattering in quantum mechanics. It may be used in scattering of molecules, atoms, neutrons, photons or any other particles and is important mainly in atomic, molecular, and optical physics, nuclear physics and particle physics, but also for seismic scattering problems in geophysics. It relates the scattered wave function with the interaction that produces the scattering (the scattering potential) and therefore allows calculation of the relevant experimental parameters (scattering amplitude and cross sections).

The most fundamental equation to describe any quantum phenomenon, including scattering, is the Schrödinger equation. In physical problems, this differential equation must be solved with the input of an additional set of initial and/or boundary conditions for the specific physical system studied. The Lippmann–Schwinger equation is equivalent to the Schrödinger equation plus the typical boundary conditions for scattering problems. In order to embed the boundary conditions, the Lippmann–Schwinger equation must be written as an integral equation. For scattering problems, the Lippmann–Schwinger equation is often more convenient than the original Schrödinger equation.

The Lippmann–Schwinger equation's general form is (in reality, two equations are shown below, one for the plus[+]sign and other for the minus[-]sign):

$$|\,\psi_{\kappa}^{\pm}\rangle = |\phi_{\kappa}\rangle + \frac{1}{E-H_0 \pm i\epsilon}\, V|\,\psi_{\kappa}^{\pm}\rangle$$

The potential energy V describes the interaction between the two colliding systems. The Hamiltonian H_0 describes the situation in which the two systems are infinitely far apart and do not interact. Its eigen functions are $|\phi_\kappa\rangle$ and its eigen values are the energies E. Finally, $i\epsilon$ is a mathematical technicality necessary for the calculation of the integrals needed to solve the equation and has no physical meaning.

13.24 Usage the Lippmann–Schwinger equation

The Lippmann–Schwinger equation is useful in a very large number of situations involving two-body scattering. For three or more colliding bodies it does not work well because of mathematical limitations; Faddeev equations may be used instead. However, there are approximations that can reduce a many-body problem to a set of two-body problems in a variety of cases. For example, in a collision between electrons and molecules, there may be tens or hundreds of particles involved. But the phenomenum may be reduced to a two-body problem by describing all the molecule constituent particle potentials together with a pseudo potential. In these cases, the Lippmann–Schwinger equations may be used.

13.25 Derivation of Lippmann-Schwinger equation

We apply the Green's function technique to solve the Schrödinger equation for scattering problem. The Schrödinger equation

$$[\nabla^2 + \frac{2m}{\hbar^2}(E - V)]\psi = 0$$

Or, $$[\nabla^2 + \frac{2mE}{\hbar^2} - \frac{2mV(\vec{r})}{\hbar^2}]\psi(\vec{r}) = 0$$

Where, $$\kappa^2 = \frac{2\,\mu\,E}{\hbar^2} \quad \& \quad U(\vec{r}) = \frac{2\,\mu\,V(\vec{r})}{\hbar^2}$$

Or, $$[\nabla^2 + \kappa^2 - U(\vec{r})\,]\,\psi(\vec{r}) = 0$$

$$[\nabla^2 + \kappa^2\,]\,\psi(\vec{r}) = U(\vec{r})\psi(\vec{r}) \qquad [1]$$

For large $\vec{r}$, the wave function is given by

$$\psi(\vec{r}) \xrightarrow{r\to\infty} e^{i\kappa z} + f(\theta)\frac{e^{i\kappa r}}{r} \qquad [2]$$

Where the 1st term is the incident plane wave and the 2nd term is an outgoing spherically scattered wave. Using Green's function technique, we can write,

$$[\nabla^2 + \kappa^2\,]\;G(\vec{r},\vec{r}') = \delta(\vec{r} - \vec{r}') \qquad [3]$$

And the Green's function for the outgoing wave

$$G^{+}(\vec{r},\vec{r}') = \frac{1}{(2\pi)^3} \int_{-\infty}^{+\infty} \frac{e^{i\vec{\kappa}'.\bar{R}}}{\kappa^2 - \kappa'^2 + i\epsilon}\, d\vec{\kappa}' \qquad [4]$$

Where, $\bar{R} = (\vec{r} - \vec{r}')$ [5]

Using the Green's function, the solution of [1] can be written as

$$\psi_{\kappa}^{+}(\vec{r}) = \phi_{\kappa}(\vec{r}) + \int G^{+}(\vec{r},\vec{r}')\, U(\vec{r}')\psi_{\kappa}^{+}(\vec{r}')d\vec{r}' \qquad [6]$$

Where $\phi_{\kappa}(\vec{r})$ is the solution of the homogeneous equation (I.e., the free particle solution)

$$[\nabla^2 + \kappa^2]\, \phi_{\kappa}(\vec{r}) = 0 \qquad [7]$$

We can easily prove this

$$[\nabla^2 + \kappa^2]\, \psi_{\kappa}^{+}(\vec{r}) = [\nabla^2 + \kappa^2]\, [\phi_{\kappa}(\vec{r}) + \int G^{+}(\vec{r},\vec{r}')\, U(\vec{r}')\psi_{\kappa}^{+}(\vec{r}')d\vec{r}']$$

$$= [\nabla^2 + \kappa^2]\, \phi_{\kappa}(\vec{r}) + \int\, [\nabla^2 + \kappa^2] G^{+}(\vec{r},\vec{r}')\, U(\vec{r}')\psi_{\kappa}^{+}(\vec{r}')d\vec{r}']$$

Or, $[\nabla^2 + \kappa^2]\, \psi_{\kappa}^{+}(\vec{r}) = 0 + \int \delta(\vec{r} - \vec{r}')\, U(\vec{r}')\psi_{\kappa}^{+}(\vec{r}')d\vec{r}'$

But from the definition of Dirac-delta function

$$\int \delta(x-a)\, f(x)dx = f(a) \quad \& \quad \delta(-x) = \delta(x)$$

$\therefore [\nabla^2 + \kappa^2]\, \psi_{\kappa}^{+}(\vec{r}) = U(\vec{r}')\psi_{\kappa}^{+}(\vec{r}')$; Which is just the equation [1]. Hence the solution [6] is justified. Thus the solution of [1] with outgoing & incoming waves are given by

$$\psi^{\pm}(\vec{r}) = \phi_{\kappa}(\vec{r}) + \frac{1}{(2\pi)^3} \int \frac{e^{i\vec{\kappa}'.\bar{R}}}{\kappa^2 - \kappa'^2 \pm i\epsilon}\, U(\vec{r}')\psi_{\kappa}^{\pm}(\vec{r}')d\vec{\kappa}'d\vec{r}'$$

$$\psi^{\pm}(\vec{r}) = \phi_{\kappa}(\vec{r}) + \frac{1}{(2\pi)^3} \int \frac{e^{i\vec{\kappa}'.\bar{R}}}{\kappa^2 - \kappa'^2 \pm i\epsilon}\, \frac{2m\, V(\vec{r}')}{\hbar^2}\psi_{\kappa}^{\pm}(\vec{r}')d\vec{\kappa}'d\vec{r}'$$

$$= \phi_{\kappa}(\vec{r}) + \frac{1}{(2\pi)^3} \int \frac{e^{i\vec{\kappa}'.\bar{R}}}{\frac{\hbar^2\kappa^2}{2m} - \frac{\hbar^2\kappa'^2}{2m} \pm i\epsilon\frac{\hbar^2}{2m}}\, V(\vec{r}')\psi_{\kappa}^{\pm}(\vec{r}')d\vec{\kappa}'d\vec{r}'$$

$$= \phi_{\kappa}(\vec{r}) + \frac{1}{(2\pi)^3} \int \frac{e^{i\vec{\kappa}'.\bar{R}}}{E(\kappa) - E(\kappa') \pm i\epsilon}\, V(\vec{r}')\psi_{\kappa}^{\pm}(\vec{r}')d\vec{\kappa}'d\vec{r}' \qquad [8]$$

Where, $E(\kappa) = \frac{\hbar^2\kappa^2}{2m}$ [9]

Equation [8] is the integral equation which determines the complete wave function of the scattering problem. Now using Dirac Ket and Bra notation, $\phi_{\kappa}(\vec{r})$ is the general representative of the general state $|\phi_\kappa>$ specified by in the coordinate basis $|\vec{r}>$ I.e.,

$$\phi_\kappa(\vec{r}) = <\vec{r}|\phi_\kappa> = <\vec{r}|\kappa> = \frac{1}{(2\pi)^{\frac{3}{2}}} e^{i\,\vec{\kappa}\,.\vec{r}}$$

Where $\phi_\kappa(\vec{r})$ is the normalized incident plane wave & in this notation the Green's function becomes

$$G^{\pm}(\vec{r},\vec{r}') = \frac{1}{(2\pi)^3}\int \frac{e^{i\vec{\kappa}'.\vec{R}}}{E(\kappa)-E(\kappa')\pm i\epsilon}\,d\vec{\kappa}' = \frac{1}{(2\pi)^3}\int \frac{e^{i\vec{\kappa}'.(\vec{r}-\vec{r}')}}{E(\kappa)-E(\kappa')\pm i\epsilon}\,d\vec{\kappa}'$$

$$= \int \frac{\frac{1}{(2\pi)^{\frac{3}{2}}}e^{i\vec{\kappa}'.\vec{r}}\ \frac{1}{(2\pi)^{\frac{3}{2}}}e^{-i\vec{\kappa}'\vec{r}'}}{E(\kappa)-E(\kappa')\pm i\epsilon}\,d\vec{\kappa}' = \int \frac{<\vec{r}|\vec{\kappa}'>\ <\vec{\kappa}'|\vec{r}'>}{E(\kappa)-E(\kappa')\pm i\epsilon}\,d\vec{\kappa}'$$

$$G^{\pm}(\vec{r},\vec{r}') = <\vec{r}|\int \frac{d\vec{\kappa}'\,|\vec{\kappa}'><\vec{\kappa}'|}{E(\kappa)-E(\kappa')\pm i\epsilon}|\vec{r}'> \qquad [11]$$

But the operator G_κ is defined

$$G_{\vec{\kappa}}^{\pm} = \int d\vec{\kappa}'\,\frac{|\vec{\kappa}'><\vec{\kappa}'|}{E(\kappa)-E(\kappa')\pm i\epsilon} \qquad [12]$$

Hence [11] becomes

$$G^{\pm}(\vec{r},\vec{r}') = <\vec{r}|G_{\vec{\kappa}}^{\pm}|\vec{r}'> \qquad [13]$$

Using [10] & [13] we can write [8] as

$$\psi^{\pm}(\vec{r}) = <\vec{r}|\,\psi_{\vec{\kappa}}^{\pm}> = <\vec{r}|\phi_\kappa> + \int <\vec{r}|G_{\vec{\kappa}}^{\pm}|\vec{r}'>V(\vec{r})\ <\vec{r}|\,\psi_{\vec{\kappa}}^{\pm}>d\vec{r}'$$

$$<\vec{r}|\,\psi_{\vec{\kappa}}^{\pm}> = <\vec{r}|\phi_\kappa> + \int <\vec{r}|G_{\vec{\kappa}}^{\pm}|\vec{r}'><\vec{r}|\,V\psi_{\vec{\kappa}}^{\pm}>d\vec{r}'$$

This can be written as

$$\psi_{\vec{\kappa}}^{\pm} = \phi_\kappa + G_{\vec{\kappa}}^{\pm}V\psi_{\vec{\kappa}}^{\pm} \qquad [14]$$

Equation [14] is known as Lippmann –Schwinger's equation.

We know the free particle Hamiltonian $H_0 = -\frac{\hbar^2}{2m}\nabla^2$ & $E = E(\kappa) = \frac{\hbar^2\kappa^2}{2m}$. Now we can write

$$\frac{E+\frac{\hbar^2}{2m}\nabla^2 \pm i\epsilon}{E-\frac{\hbar^2\kappa'^2}{2m}\pm i\epsilon}\,e^{i\vec{\kappa}'.\vec{r}} = \frac{E-\frac{\hbar^2\kappa'^2}{2m}\pm i\epsilon}{E-\frac{\hbar^2\kappa'^2}{2m}\pm i\epsilon}\,e^{i\vec{\kappa}'.\vec{r}} = e^{i\vec{\kappa}'.\vec{r}}$$

$\therefore$

$$\frac{1}{E+\frac{\hbar^2}{2m}\nabla^2 \pm i\epsilon}\,e^{i\vec{\kappa}'.\vec{r}} = \frac{1}{E-\frac{\hbar^2\kappa'^2}{2m}\pm i\epsilon}\,e^{i\vec{\kappa}'.\vec{r}}$$

Or,

$$\frac{1}{E-\frac{\hbar^2\kappa'^2}{2m}\pm i\epsilon}\,e^{i\vec{\kappa}'.\vec{r}} = \frac{1}{E-H_0\pm i\epsilon}\,e^{i\vec{\kappa}'.\vec{r}}$$

Or, $\frac{1}{E-\frac{\hbar^2\kappa'^2}{2m}\pm i\epsilon} = \frac{1}{E-H_0 \pm i\epsilon}$

Now from [12] $G_\kappa^\pm = \frac{1}{E-\frac{\hbar^2\kappa'^2}{2m}\pm i\epsilon} = \frac{1}{E-H_0 \pm i\epsilon}$

Hence [14] becomes

$$\psi_\kappa^\pm = \phi_\kappa + \frac{1}{E-H_0 \pm i\epsilon} V\psi_\kappa^\pm \qquad [15]$$

13.26 Formal solution of Lippmann-Schwinger equation

We know the Lippmann-Schwinger equation

$$\psi_\kappa^+ = \phi_\kappa + \frac{1}{E-H_0+i\epsilon} V\psi_\kappa^+ \qquad [1]$$

We know the operator relation is

$$\frac{1}{A} - \frac{1}{B} = \frac{1}{A}(B-A)\frac{1}{B} = \frac{1}{B}(B-A)\frac{1}{A} \qquad [2]$$

Let $A = E - H + i\epsilon \quad \& \quad B = E - H_0 + i\epsilon$

$$\frac{1}{E-H+i\epsilon} - \frac{1}{E-H_0+i\epsilon} = \frac{1}{E-H+i\epsilon}(E - H_0 + i\epsilon - E + H - i\epsilon)\frac{1}{E-H_0+i\epsilon}$$

$$= \frac{1}{E-H+i\epsilon}(-H_0 + H)\frac{1}{E-H_0+i\epsilon}$$

$$= \frac{1}{E-H+i\epsilon}(-H_0 + H_0 + V)\frac{1}{E-H_0+i\epsilon}$$

$$= \frac{1}{E-H+i\epsilon}(-H_0 + H_0 + V)\frac{1}{E-H_0+i\epsilon}$$

$$= \frac{1}{E-H+i\epsilon} V \frac{1}{E-H_0+i\epsilon}$$

Or, $\frac{1}{E-H_0+i\epsilon} = \frac{1}{E-H+i\epsilon} - \frac{1}{E-H+i\epsilon} V \frac{1}{E-H_0+i\epsilon}$

Using this in equation [1]

$$\psi_\kappa^+ = \phi_\kappa + \left(\frac{1}{E-H+i\epsilon} - \frac{1}{E-H+i\epsilon} V \frac{1}{E-H_0+i\epsilon}\right) V\psi_\kappa^+$$

$$\psi_\kappa^+ = \phi_\kappa + \frac{1}{E-H+i\epsilon} V\psi_\kappa^+ - \frac{1}{E-H+i\epsilon} V \frac{1}{E-H_0+i\epsilon} V\psi_\kappa^+$$

$$\psi_\kappa^+ = \phi_\kappa + \frac{1}{E-H+i\epsilon} V\psi_\kappa^+ - \frac{1}{E-H+i\epsilon} V(\psi_\kappa^+ - \phi_\kappa) \text{ ; using [1]}$$

$$\psi_\kappa^+ = \phi_\kappa + \frac{1}{E-H+i\epsilon} V\phi_\kappa \qquad [3]$$

This is the formal solution of Lippmann-Schwinger equation. Again using the second identity of [2]

$$\frac{1}{E-H+i\epsilon}-\frac{1}{E-H_0+i\epsilon}=\frac{1}{E-H_0+i\epsilon}(E-H_0+i\epsilon-E+H-i\epsilon)\frac{1}{E-H+i\epsilon}=\frac{1}{E-H_0+i\epsilon}V\frac{1}{E-H+i\epsilon}$$

$$\frac{1}{E-H+i\epsilon}=\frac{1}{E-H_0+i\epsilon}\left(1+V\frac{1}{E-H+i\epsilon}\right) \qquad [3a]$$

Hence from [3]

$$\psi_\kappa^+=\phi_\kappa+\frac{1}{E-H_0+i\epsilon}\left(1+V\frac{1}{E-H+i\epsilon}\right)V\phi_\kappa \quad \text{; using equation [3a]}$$

We now introduced an operator $T=\left(1+V\frac{1}{E-H+i\epsilon}\right)V$; is called the transition operator. Then

$$\psi_\kappa^+=\phi_\kappa+\frac{1}{E-H_0+i\epsilon}\,T\,\phi_\kappa \qquad [4]$$

This is the solution of the Lippmann-Schwinger equation in terms of T operator. Here

$$T=\left(V+V\frac{1}{E-H+i\epsilon}V\right)=\left(V+V\frac{1}{E-H_0+i\epsilon}\left(1+V\frac{1}{E-H+i\epsilon}\right)V\right)=V+V\frac{1}{E-H_0+i\epsilon}T \text{ ; from [3a]}$$

This is also a Lippmann-Schwinger equation in terms of T operator. The Lippmann-Schwinger equation can be expanded in powers of V by the simple method of iteration. From [4]

$$\psi_\kappa^\pm=\phi_\kappa+\frac{1}{E-H_0+i\epsilon}\left(V+V\frac{1}{E-H+i\epsilon}V\right)\phi_\kappa=\phi_\kappa+\frac{1}{E-H_0+i\epsilon}V\phi_\kappa+V\phi_\kappa\frac{1}{E-H+i\epsilon}V$$

$$=\phi_\kappa+\frac{1}{E-H_0+i\epsilon}V(\phi_\kappa+V\phi_\kappa\frac{1}{E-H+i\epsilon})=\phi_\kappa+\frac{1}{E-H_0+i\epsilon}V(\phi_\kappa+V\psi_\kappa^\pm\frac{1}{E-H_0+i\epsilon})$$

$$=\phi_\kappa+\frac{1}{E-H_0\pm i\epsilon}V\phi_\kappa+\frac{1}{E-H_0\pm i\epsilon}V\frac{1}{E-H_0\pm i\epsilon}V\psi_\kappa^\pm)$$

$$\psi_\kappa^\pm=\phi_\kappa+\frac{1}{E-H_0\pm i\epsilon}V\phi_\kappa+\frac{1}{E-H_0\pm i\epsilon}V\frac{1}{E-H_0\pm i\epsilon}V\phi_\kappa+\cdots\cdots\cdots \qquad [6]$$

In the coordinate basis, this can be written as

$$\psi_\kappa^\pm(\vec{r})=\phi_\kappa(\vec{r})+\int G_\kappa^\pm(\vec{r},\vec{r}')\,U(\vec{r}')\,\phi_\kappa(\vec{r}')d\vec{r}' + \int G_\kappa^\pm(\vec{r},\vec{r}')\,U(\vec{r}')\,G_\kappa^\pm(\vec{r},\vec{r}'')\,U(\vec{r}'')\phi_\kappa(\vec{r}'')d\vec{r}'d\vec{r}''+\cdots\cdots\cdots \qquad [7]$$

This is known as the Von-Neumann series. Taking approximation, we can write [7] using Green's function

$$\psi_\kappa^+(\vec{r})=\phi_\kappa(\vec{r})+\int G_\kappa^+(\vec{r},\vec{r}')\,U(\vec{r}')\,\phi_\kappa(\vec{r}')d\vec{r}'=\phi_\kappa(\vec{r})-\frac{1}{4\pi}\int\frac{e^{i\kappa|\vec{r}-\vec{r}'|}}{|\vec{r}-\vec{r}'|}\,U(\vec{r}')\,\phi_\kappa(\vec{r}')d\vec{r}' \qquad [8]$$

Now, $|\vec{r}-\vec{r}'|=\sqrt{r^2-2\vec{r}.\vec{r}'+r'^2}=r\,(1-\frac{2\vec{r}.\vec{r}'}{r^2}+\frac{r'^2}{r^2})^{\frac{1}{2}}\simeq r\,(1-\frac{2\hat{r}.\vec{r}'}{r})^{\frac{1}{2}}$

$$\simeq r\left(1-\frac{1}{2}2\frac{\hat{r}.\vec{r}'}{r}\right)\simeq r-\hat{r}.\vec{r}'$$

Using this in [8] & neglecting the $\hat{r}.\vec{r}'$ in the denominator

$$\psi_\kappa^+(\vec{r})=\phi_\kappa(\vec{r})-\frac{1}{4\pi}\int\frac{e^{i\kappa(r-\hat{r}.\vec{r}')}}{r}\,U(\vec{r}')\,\phi_\kappa(\vec{r}')d\vec{r}'=\phi_\kappa(\vec{r})-\frac{e^{i\kappa r}}{4\pi r}\int e^{-i\kappa\,\hat{r}.\vec{r}'}\,U(\vec{r}')\,\phi_\kappa(\vec{r}')d\vec{r}'$$

Let $\vec{\kappa}'=\kappa\,\hat{r}$. Then, $\psi_\kappa^+(\vec{r})=\phi_\kappa(\vec{r})-\frac{e^{i\kappa r}}{4\pi r}\int e^{-i\vec{\kappa}'.\vec{r}'}\,U(\vec{r}')\,\phi_\kappa(\vec{r}')d\vec{r}'$

But $\phi_\kappa(\vec{r})=\frac{1}{(2\pi)^{\frac{3}{2}}}\,e^{i\vec{\kappa}.\vec{r}}$

$$\psi_\kappa^+(\vec{r})=\frac{1}{(2\pi)^{\frac{3}{2}}}\left[e^{i\vec{\kappa}.\vec{r}}-\frac{e^{i\kappa r}}{4\pi r}\right]\int e^{i(\vec{\kappa}-\vec{\kappa}').\vec{r}'}\,U(\vec{r}')d\vec{r}' \qquad [9]$$

Comparing this with $\psi_\kappa(\vec{r})=e^{i\kappa z}+f(\theta)\frac{e^{i\kappa r}}{r}$ [10]

Comparing [9] & [10], the scattering amplitude

$$f(\theta)=-\frac{1}{4\pi}\int e^{i(\vec{\kappa}-\vec{\kappa}').\vec{r}'}\,U(\vec{r}')d\vec{r}' \qquad [11]$$

Equation [11] is the Born-approximation scattering amplitude.

Now we introduce the vector $\vec{q}=\vec{\kappa}-\vec{\kappa}'$ & $q=2\kappa\sin\frac{\theta}{2}$

We can write the Born-approximation scattering amplitude as

$$f(\theta)=-\frac{1}{4\pi}\int e^{i\vec{q}.\vec{r}'}\,U(\vec{r}')d\vec{r}'=-\int_{r=0}^{\infty}\frac{\sin qr}{qr}\,U(\vec{r}')\,r^2dr\ [\because d\vec{r}'=r^2dr\sin\theta\,d\theta\,d\phi]$$

And the differential scattering cross-section is

$$\frac{d\sigma}{d\Omega}=|f(\theta)|^2=\left|\int_{r=0}^{\infty}\frac{\sin qr}{qr}\,U(\vec{r}')\,r^2dr\right|$$

13.27 Scattering amplitude is seemed to be the matrix element of the interacting potential

We know the first Born-approximation

$$\psi_\kappa^{+(1)}(\vec{r})=e^{i\kappa z}-\frac{1}{4\pi}\int\frac{e^{i\kappa|\vec{r}-\vec{r}'|}}{|\vec{r}-\vec{r}'|}\,U(\vec{r}')\,e^{i\kappa z'}d^3\vec{r}' \qquad [1]$$

Where the first term is the incident plane wave & the second term is assumed in this approximation to be the total scattered wave. The scattered wave is only a small addition to the incident wave. To put $e^{i\kappa z'}=\psi_\kappa^{+(0)}(\vec{r}')$ is the zeroth approximation on the R.H.S. of equation [1]

$$\therefore \quad \psi_\kappa^{+(1)}(\vec{r}) = e^{i\kappa z} - \frac{1}{4\pi}\int \frac{e^{i\kappa|\vec{r}-\vec{r}'|}}{|\vec{r}-\vec{r}'|} U(\vec{r}')\, \psi_\kappa^{+(0)}(\vec{r}')\, d^3\vec{r}' \qquad [2]$$

The next Born-approximation will be

$$\psi_\kappa^{+(2)}(\vec{r}) = e^{i\kappa z} - \frac{1}{4\pi}\int \frac{e^{i\kappa|\vec{r}-\vec{r}'|}}{|\vec{r}-\vec{r}'|} U(\vec{r}')\, \psi_\kappa^{+(1)}(\vec{r}) d^3\vec{r}'$$

$$\psi_\kappa^{+(2)}(\vec{r}) = e^{i\kappa z} - \frac{1}{4\pi}\left[\int \frac{e^{i\kappa|\vec{r}-\vec{r}'|}}{|\vec{r}-\vec{r}'|} U(\vec{r}') d^3\vec{r}' \left(e^{i\kappa z'} - \int \frac{e^{i\kappa|\vec{r}-\vec{r}''|}}{|\vec{r}-\vec{r}''|} U(\vec{r}'') e^{i\kappa z''} d^3\vec{r}''\right)\right]$$

Similarly, the nth Born-approximation

$$\psi_\kappa^{+(n)}(\vec{r}) = e^{i\kappa z} - \frac{1}{4\pi}\int \frac{e^{i\kappa|\vec{r}-\vec{r}'|}}{|\vec{r}-\vec{r}'|} U(\vec{r}')\, \psi_\kappa^{+(n-1)}(\vec{r}')\, d^3\vec{r}'$$

From equation [1], the scattering amplitude

$$f(\theta) = -\frac{1}{4\pi}\left[\int \frac{e^{i\kappa|\vec{r}-\vec{r}'|}}{|\vec{r}-\vec{r}'|} U(\vec{r}') e^{i\kappa z'} d^3\vec{r}' \qquad [3]$$

Now $|\vec{r}-\vec{r}'| = \sqrt{r^2 - 2\vec{r}.\vec{r}' + r'^2} = r\left(1 - \frac{2\vec{r}.\vec{r}'}{r^2} + \frac{r'^2}{r^2}\right)^{\frac{1}{2}} \simeq r\left(1 - \frac{2\vec{r}.\vec{r}'}{r^2}\right)^{\frac{1}{2}}$

$$\simeq r\left(1 - \frac{1}{2}\frac{2\hat{r}.\vec{r}'}{r^2}\right) \simeq r\left(1 - \frac{r\,\hat{r}.\vec{r}'}{r^2}\right) \simeq r - \hat{r}.\vec{r}'$$

$$f(\theta) = -\frac{e^{i\kappa r}}{4\pi r}\int e^{i(\bar{\kappa}-\vec{\kappa}').\vec{r}'}\, U(\vec{r}') d^3\vec{r}'$$

Now we introduced the notation for momentum transfer vector as $\bar{\kappa} = \overline{\kappa_0} - \hat{\bar{\kappa}}$

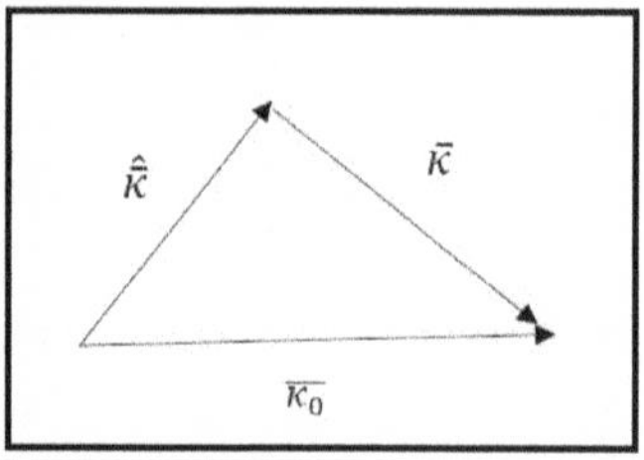

Note: $\hat{r} = \frac{\bar{r}}{|\bar{r}|}$; $e^{i(\bar{k}-\bar{k}').\bar{r}'} = e^{i\bar{q}.\bar{r}'} = e^{i\bar{k}.\bar{r}'} = e^{i\kappa r'\cos\theta} = e^{i\kappa z'}$

and $\hat{\bar{k}} + \bar{k} = \overline{k_0} \rightarrow \bar{k} = \overline{k_0} - \hat{\bar{k}}$ $\therefore$ $e^{i\kappa z'} = e^{i\kappa r'\cos\theta} = e^{i\bar{k}.\bar{r}'} = e^{i\bar{q}.\bar{r}'} = e^{i\bar{k}.\bar{r}'} = e^{i\left(\overline{k_0} - \hat{\bar{k}}\right).\bar{r}'}$

Therefore equation [1] becomes

$$\psi_\kappa^{+(1)}(\vec{r}) = e^{i\kappa z} - \frac{e^{i\kappa r}}{r}\frac{1}{4\pi}\int e^{i(\overline{\kappa_0}-\hat{\bar{\kappa}}).\vec{r}'}\, U(\vec{r}') d^3\vec{r}' \qquad [3a]$$

Comparing equation [2] with the following

$$\psi_\kappa(\vec{r}) = e^{i\kappa z} + f(\theta)\frac{e^{i\kappa r}}{r} \quad [3b]$$

Comparing equation [3a] & [3b], the scattering amplitude

$$f(\theta) = -\frac{1}{4\pi}\int e^{i(\overline{\kappa_0}-\widehat{\kappa}).\vec{r}'}\, U(\vec{r}')d^3\vec{r}' = -\frac{\mu}{2\pi\hbar^2}\int e^{-i\,\widehat{\kappa}.\vec{r}'}\, V(\vec{r}')\, e^{i\overline{\kappa_0}.\vec{r}'}\, d^3\vec{r}' \quad [4]$$

$$= -\frac{\mu}{2\pi\hbar^2} < e^{-i\widehat{\kappa}.\vec{r}'} | V(\vec{r}') | e^{i\overline{\kappa_0}.\vec{r}'} > = -\frac{\mu}{2\pi\hbar^2} < f | V | i > \quad [5]$$

Also equation [4] can be written as

$$f(\theta) = -\frac{1}{4\pi}\int e^{-i\widehat{\kappa}.\vec{r}'}\, U(\vec{r}')\, e^{i\overline{\kappa_0}.\vec{r}'}\, d^3\vec{r}' = -\frac{1}{4\pi} < e^{-i\widehat{\kappa}.\vec{r}'} | V(\vec{r}') | e^{i\overline{\kappa_0}.\vec{r}'} >$$

$$= -\frac{1}{4\pi} < \psi_\kappa^- | V(\vec{r}') | \psi_\kappa^+ > \quad [6]$$

Now we noted from equation [6] *the scattering amplitude is seemed to be the matrix element of the scattering or interacting potential between the incident plane wave and the outgoing or emerging plane wave*[12].

13.28 Scattering amplitude is seemed to be the Fourier transformation of the interacting potential

We know the first Born-approximation

$$\psi_\kappa^{+(1)}(\vec{r}) = e^{i\kappa z} - \frac{1}{4\pi}\int \frac{e^{i\kappa|\vec{r}-\vec{r}'|}}{|\vec{r}-\vec{r}'|}\, U(\vec{r}')\, e^{i\kappa z'} d^3\vec{r}' \quad [1]$$

Where the first term is the incident plane wave & the second term is assumed in this approximation to be the total scattered wave. The scattered wave is only a small addition to the incident wave.

$e^{i\kappa z'} = \psi_\kappa^{+(0)}(\vec{r}')$ is the zeroth approximation on the R.H.S. of equation [1]

$$\psi_\kappa^{+(1)}(\vec{r}) = e^{i\kappa z} - \frac{1}{4\pi}\int \frac{e^{i\kappa|\vec{r}-\vec{r}'|}}{|\vec{r}-\vec{r}'|}\, U(\vec{r}')\, \psi_\kappa^{+(0)}(\vec{r}')\, d^3\vec{r}' \quad [2]$$

The next Born-approximation will be

$$\psi_\kappa^{+(2)}(\vec{r}) = e^{i\kappa z} - \frac{1}{4\pi}\left[\int \frac{e^{i\kappa|\vec{r}-\vec{r}'|}}{|\vec{r}-\vec{r}'|}\, U(\vec{r}')d^3\vec{r}'\left(e^{i\kappa z'} - \int \frac{e^{i\kappa|\vec{r}-\vec{r}''|}}{|\vec{r}-\vec{r}''|}\, U(\vec{r}'')e^{i\kappa z''}d^3\vec{r}'\right)\right]$$

Similarly, the nth Born-approximation

$$\psi_\kappa^{+(n)}(\vec{r}) = e^{i\kappa z} - \frac{1}{4\pi}\int \frac{e^{i\kappa|\vec{r}-\vec{r}'|}}{|\vec{r}-\vec{r}'|}\, U(\vec{r}')\, \psi_\kappa^{+(n-1)}(\vec{r}')\, d^3\vec{r}'$$

From equation [1], the scattering amplitude

$$f(\theta) = -\frac{1}{4\pi}\left[\int \frac{e^{i\kappa|\vec{r}-\vec{r}'|}}{|\vec{r}-\vec{r}'|}\, U(\vec{r}')e^{i\kappa z'}d^3\vec{r}'\right.$$

Now, $|\vec{r}-\vec{r}'| = \sqrt{r^2 - 2\vec{r}.\vec{r}' + r'^2} = r\,(1 - \frac{2\vec{r}.\vec{r}'}{r^2} + \frac{r'^2}{r^2})^{\frac{1}{2}} \simeq r\,(1 - \frac{2\hat{r}.\vec{r}'}{r^2})^{\frac{1}{2}}$

$$\simeq r\,(1 - \frac{1}{2}.\frac{2\,\hat{r}.\vec{r}'}{r^2}) \simeq r\,(1 - \frac{r\,\hat{r}.\vec{r}'}{r^2}) \simeq r - \hat{r}.\vec{r}'$$

Therefore, $f(\theta) = -\frac{e^{i\kappa r}}{4\pi}\int e^{i(\vec{\kappa}-\vec{\kappa}').\vec{r}'}\,U(\vec{r}')d^3\vec{r}'$

Therefore equation [1] becomes

$$\psi_\kappa^{+(1)}(\vec{r}) = e^{i\kappa z} - \frac{e^{i\kappa r}}{r}\frac{1}{4\pi}\int e^{i(\overline{\kappa_0}-\hat{\vec{\kappa}}).\vec{r}'}\,U(\vec{r}')\,d^3\vec{r}' \qquad [3a]$$

Comparing equation [2] with the following

$$\psi_\kappa(\vec{r}) = e^{i\kappa z} + f(\theta)\frac{e^{i\kappa r}}{r} \qquad [3b]$$

Comparing equation [3a] & [3b], the scattering amplitude

$$f(\theta) = -\frac{1}{4\pi}\int e^{i(\overline{\kappa_0}-\hat{\vec{\kappa}}).\vec{r}'}\,U(\vec{r}')d^3\vec{r}' \qquad [4]$$

$$= -\frac{\mu}{2\pi\hbar^2}\int e^{-i\hat{\vec{\kappa}}.\vec{r}'}\,V(\vec{r}')\,e^{i\overline{\kappa_0}.\vec{r}'}\,d^3\vec{r}' = -\frac{\mu}{2\pi\hbar^2} < e^{-i\hat{\vec{\kappa}}.\vec{r}'}\big|\,V(\vec{r}')\big|e^{i\overline{\kappa_0}.\vec{r}'} > \qquad [5]$$

$$= -\frac{\mu}{2\pi\hbar^2} < f\,|\,V\,|\,i> = -\frac{1}{4\pi}\int e^{-i\hat{\vec{\kappa}}.\vec{r}'}\,U(\vec{r}')\,e^{i\overline{\kappa_0}.\vec{r}'}\,d^3\vec{r}'$$

$$= -\frac{1}{4\pi} < e^{-i\hat{\vec{\kappa}}.\vec{r}'}\big|\,V(\vec{r}')\big|e^{i\overline{\kappa_0}.\vec{r}'} > = -\frac{1}{4\pi} < \psi_{\bar{\kappa}}^{+}\big|\,V(\vec{r}')\big|\psi_{\bar{\kappa}}^{-} > \qquad [6]$$

Now we noted from equation [6] *the scattering amplitude is seemed to be the matrix element of the scattering or interacting potential between the incident plane wave and the outgoing or emerging plane wave.* From equation [4]

$$f(\theta) = -\frac{1}{4\pi}\int e^{i(\overline{\kappa_0}-\hat{\vec{\kappa}}).\vec{r}'}\,U(\vec{r}')d^3\vec{r}'$$

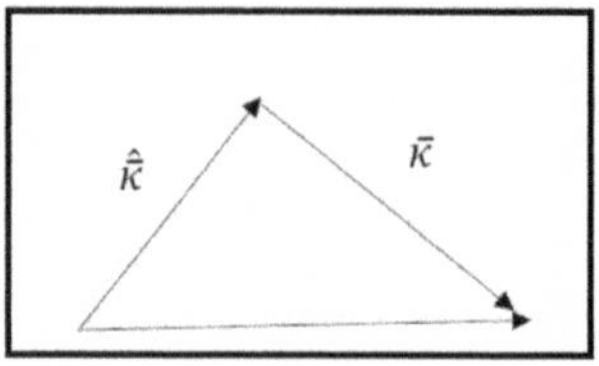

Now we introduced the notation for momentum transfer vector as $\bar{\kappa} = \overline{\kappa_0} - \hat{\vec{\kappa}}$

Hence $f(\theta) = -\frac{1}{4\pi}\int e^{i\bar{\kappa}.\vec{r}'}\,U(\vec{r}')d^3\vec{r}' = -\frac{\mu}{2\pi\hbar^2}\int e^{i\bar{\kappa}.\vec{r}'}\,U(\vec{r}')d^3\vec{r}'$ [7]

Equation [7] shows that in the first Born-approximation, *the scattering amplitude is seemed to be the Fourier transformation of the scattering or interacting potential*

Exercise

1. Define differential scattering cross- section and total scattering cross- section.
2. What are typical elements of scattering and basic assumptions of scattering? Write the dimension of differential scattering cross- section.
3. Explain Laboratory system and Centre of mass system.
4. Write the importance of study of scattering experiment in Quantum mechanics and the types of Scattering.
5. Obtain the relation between differential Scattering cross- section and scattering amplitude.
6. Derivation of Bauer's formula.
7. Explain Partial wave method is suitable for low energy.
8. Derive Partial wave expansion of scattering amplitude and differential scattering cross-section.
9. State and explain the Optical theorem.
10. Why Phase- shift is called the meeting ground of theory and experiment? Write down the Phase Shift in terms of potential.
11. Show that the scattering amplitude is independent of azimuthal angle.
12. Explain S-matrix and T-matrix. Find the relation between K_l , S_l and T_l matrices and the scattering amplitude in terms of T_l matrix.
13. Write a unique description of Ramsauer–Townsend effect and Townsend discharge.
14. Write a brief description of Ramsaur-Townsend Effect for Square well potential

$$V(r) = \begin{cases} -V_0 & for\ r < a \\ 0 & for\ r > a \end{cases}$$

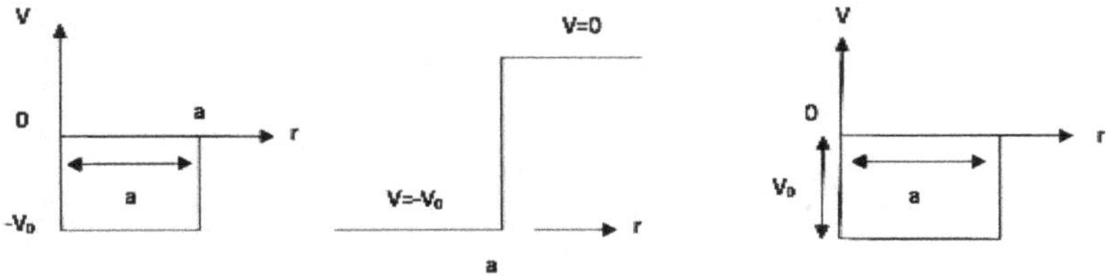

15. Why we neglect the higher order terms of Born approximation? Discuss the validity of Born approximation.
16. Why Born approximation is applicable for high energy? Write the validity criterion of Born approximation for real potential.
17. What is the spherically symmetric potential? Define Yukawa potential and find the differential scattering cross-section for Yukawa potential.

18. Write a brief description of Green's function. What do you mean by Advanced and retarded Green's functions?
19. What is Born approximation and write its applications?
20. What is Green function and find the first Born approximation?
21. Show that the scattering amplitude is seemed to be the matrix element of the scattering potential between the actual scattering state ψ_{κ}^{+} of the whole system and the particular scattered final free state ψ_{κ}^{-}.
22. What is the Lippmann–Schwinger equation? What is its importance/usage in Quantum Mechanics?
23. Derivation of Lippmann-Schwinger equation.
24. Write down the Formal solution of Lippmann-Schwinger equation.
25. Show that Scattering amplitude is seemed to be the matrix element of the interacting potential.
26. Show that Scattering amplitude is seemed to be the matrix element of the scattering or interacting potential between the incident plane wave and the outgoing or emerging plane wave.
27. Show that scattering amplitude is seemed to be the Fourier transformation of the interacting potential.
28. For the rigid sphere of radius a, show that the scattering cross- section is given by $\sigma = 4\pi a^2$
29. Describe the method of partial wave for scattering. By the method of partial waves, show that the total scattering cross- section $\sigma = \frac{4\pi}{k^2}\sum_{l=0}^{\infty}(2l+1)\, sin^2\delta_l$
30. Determine, in the Born approximation, the effective cross- section of scattering for a spherical potential well

    $$V = V_0 \text{ for } r < a \text{ and } V = 0 \text{ for } r > a$$

31. Find the scattering amplitude and total differential scattering cross- section by using the Gaussian potential $V(r) = -V_0 e^{-\frac{r^2}{a^2}}$
32. Find the scattering amplitude and total differential scattering cross- section by using the exponential potential $V(r) = V_0 e^{-r/a}$
33. Find the scattering amplitude and total differential scattering cross- section by using the screened coulomb potential $V(r) = -\frac{ZZ'e^2}{r}e^{-r/a}$; where $= \frac{1}{\alpha}$, a is the atomic radius. Where, the length $a = \frac{1}{\alpha}$ may be considered as the range of the potential.
34. Find the scattering amplitude and total differential scattering cross- section by using the Yukawa potential $V(r) = -V_0 \frac{e^{-\alpha r}}{\alpha r}$

Chapter 14 Relativistic Quantum Mechanics

14.1 The Klein-Gordon equation

We know, the time dependent Schrödinger equation

$$\left[-\frac{\hbar^2}{2m}\nabla^2 + V(r)\right]\psi = i\hbar\frac{\partial\psi}{\partial t} \qquad [1]$$

The above equation is valid for low energy system, because it is not Lorentz invariant & it cannot describe the basic law of nature. Also the above equation is first order in time & second order in space. So it is not symmetrical in space and time. To put Schrödinger equation on a relativisting footing, we start with relativistic energy momentum relation

$$E^2 = {m_0}^2c^4 + p^2c^2 \qquad [2]$$

Introducing the operator, $E = i\hbar\frac{\partial}{\partial t}$ $and\ \bar{p} = -i\hbar\bar{\nabla}$ [3]

Therefore equation [2] becomes

$$\left(i\hbar\frac{\partial}{\partial t}\right)^2 = m_0^2c^4 + \left(-i\hbar\vec{\nabla}\right)^2c^2 \Rightarrow \left(i\hbar\frac{\partial}{\partial t}\right)^2\Psi = \left[m_0^2c^4 + \left(-i\hbar\vec{\nabla}\right)^2c^2\right]\Psi$$

$$\Rightarrow -\hbar^2\frac{\partial^2\Psi}{\partial t^2} = (m_0^2c^4 - \hbar^2\nabla^2c^2)\Psi \qquad \Rightarrow \hbar^2\nabla^2c^2 - m_0^2c^4\,\Psi - \hbar^2\frac{\partial^2\Psi}{\partial t^2} = 0$$

$$\Rightarrow \nabla^2\Psi - \frac{1}{c^2}\frac{\partial^2\Psi}{\partial t^2} - \frac{m_0^2c^2}{\hbar^2}\Psi = 0 \quad \Rightarrow \left[\nabla^2 - \frac{1}{c^2}\frac{\partial^2}{\partial t^2} - \frac{m_0^2c^2}{\hbar^2}\right]\Psi = 0$$

$$\Rightarrow \left[\quad - \frac{m_0^2c^2}{\hbar^2}\right]\Psi = 0 \qquad [4]$$

Where D'Alembertian operator $\quad = \nabla^2 - \frac{1}{c^2}\frac{\partial^2}{\partial t^2}$ [5]

& *put* $k = \frac{m_0c}{\hbar}$; Therefore equation [4] finally becomes,

$$(\quad - k^2)\Psi = 0 \qquad [6]$$

Equation [6] is known as Klein-Gordon or Schrödinger- Gordon, or Klein-Fock equation for free particle.

14.2 Drawbacks of Schrodinger equation and merit of Klein- Gordon equation

We know, the time dependent Schrodinger equation

$$\left[-\frac{\hbar^2}{2m}\nabla^2 + V(r)\right]\psi = i\hbar\frac{\partial\psi}{\partial t} \qquad [1]$$

Drawbacks of Schrödinger equation

- The above equation is valid for low energy system
- Because it is not Lorentz invariant & it cannot describe the basic law of nature
- Also the above equation is first order in time & second order in space
- So it is not symmetrical in space and time
- To put Schrodinger equation on a relativisting footing for making second order in time & second order in space

Merit of Klein- Gordon equation

We know *Klein-Gordon [K-G]* equation for free particle

$$\left[\nabla^2 - \frac{1}{c^2}\frac{\partial^2}{\partial t^2} - \frac{m_0^2 c^2}{\hbar^2}\right]\Psi = 0$$

$$\Rightarrow\left[\quad - \frac{m_0^2 c^2}{\hbar^2}\right]\Psi = 0$$

- This equation is an important one for further development of relativistic field theories
- *Klein-Gordon[K-G] equation* is valid for high energy system. I.e., the equation is valid for certain classes of particles such as the π and K mesons, which are strongly interacting
- It is Lorentz invariant & it can describe the basic law of nature
- Also the above equation is second order in time & second order in space
- So it is symmetrical in space and time
- Klein-Gordon equation is the second-order differential equation
- The Klein-Gordon equation, it turns out is the correct one for spinless particles & not for electrons

14.3 Development of Dirac equation

We know, the time dependent Schrodinger equation

$$\left[-\frac{\hbar^2}{2m}\nabla^2 + V(r)\right]\Psi = i\hbar\frac{\partial\Psi}{\partial t} \qquad [1]$$

The above equation is valid in low energy system, because it is not Lorentz invariant & it cannot describe the basic law of nature. Also the above equation is first order in time & second order in space. So it is not symmetrical in space and time. Now put Schrödinger equation on a relativistic footing. We start with relativistic energy momentum relation

$$E^2 = m_0^2 c^4 + p^2 c^2 \qquad [2]$$

Introducing the operator, $E = i\hbar\frac{\partial}{\partial t}$ & $\vec{P} = -i\hbar\vec{\nabla}$ [3]

Therefore, equation [2] becomes

$$\left(i\hbar\frac{\partial}{\partial t}\right)^2 = m_0^2c^4 + \left(-i\hbar\vec{\nabla}\right)^2c^2 \quad \Rightarrow \left(i\hbar\frac{\partial}{\partial t}\right)^2\Psi = \left[m_0^2c^4 + \left(-i\hbar\vec{\nabla}\right)^2c^2\right]\Psi$$

$$\Rightarrow -\hbar^2\frac{\partial^2\Psi}{\partial t^2} = (m_0^2c^4 - \hbar^2\nabla^2c^2)\Psi \Rightarrow \hbar^2\nabla^2c^2 - m_0^2c^4\,\Psi - \hbar^2\frac{\partial^2\Psi}{\partial t^2} = 0$$

$$\Rightarrow \nabla^2\Psi - \frac{1}{c^2}\frac{\partial^2\Psi}{\partial t^2} - \frac{m_0^2c^2}{\hbar^2}\Psi = 0 \Rightarrow \left[\nabla^2 - \frac{1}{c^2}\frac{\partial^2}{\partial t^2} - \frac{m_0^2c^2}{\hbar^2}\right]\Psi = 0$$

$$\Rightarrow \left[\quad - \frac{m_0^2c^2}{\hbar^2}\right]\Psi = 0$$

Where D'Alembertian operator $\quad = \nabla^2 - \frac{1}{c^2}\frac{\partial^2}{\partial t^2}$ [5]

& put $k = \frac{m_0c}{\hbar}$; Therefore equation [4] finally becomes,

$$\left(\quad - k^2\right)\Psi = 0 \qquad [6]$$

Equation [6] is known as Klein-Gordon [K-G] equation for free particle. Let us look for a plane-wave solution of the K-G equation in the form

$$\Psi(x,t) = N\exp{}^{i}\!/\!_{\hbar}\,(\vec{p}.\vec{x} - Et) \qquad [7]$$

[a plane wave solution $\psi = e^{i(\overline{k}.\overline{r} - \omega t)}$; therefore equation [6] gives

$$(ik)^2 - \frac{1}{c^2}(-i\omega)^2 = \frac{m_0^2c^2}{\hbar^2} \text{ or, } k^2 + \frac{\omega^2}{c^2} = \frac{m_0^2c^2}{\hbar^2}$$

Or, $\quad -\hbar^2k^2c^2 + \hbar^2\omega^2 = m_0^2c^4$ or, $\hbar\omega = \pm\sqrt{\hbar^2k^2c^2 + m_0^2c^4}$

Allowed values of energy are $\hbar\omega = \pm\sqrt{\hbar^2k^2c^2 + m_0^2c^4}$ or, $E = \pm\sqrt{p^2c^2 + m_0^2c^4}$]

For a given momentum $\vec{p}$, there are both positive and negative energy is

$$E = \pm\sqrt{m_0^2c^4 + p^2c^2} \qquad [8]$$

To understand the meaning of the negative energy solution, we derive in the usual manner a conservation law for the probability current. Now from the Schrödinger relativistic equation

$$\left(\nabla^2 - \frac{1}{c^2}\frac{\partial^2}{\partial t^2}\right)\Psi = \frac{m_0^2c^2}{\hbar^2}\Psi \qquad [9]$$

Multiplying both sides by $\Psi^{\times}$ from the left,

$$\Psi^{\times}\left(\nabla^2 - \frac{1}{c^2}\frac{\partial^2}{\partial t^2}\right)\Psi = \frac{m_0^2c^2}{\hbar^2}\Psi^{\times}\Psi \qquad [10]$$

Taking complex conjugate of equation [9] & multiplying by Ψ from the right, we have,

$$\left(\nabla^2 - \frac{1}{c^2}\frac{\partial^2}{\partial t^2}\right)\Psi^\times \Psi = \frac{m_0^2 c^2}{\hbar^2}\Psi^\times \Psi \qquad [11]$$

Subtracting [11] from [10], we get

$$\Psi^\times\left(\nabla^2 - \frac{1}{c^2}\frac{\partial^2}{\partial t^2}\right)\Psi - \left(\nabla^2 - \frac{1}{c^2}\frac{\partial^2}{\partial t^2}\right)\Psi^\times \Psi = 0$$

$$\left[\Psi^\times\left(\nabla^2\Psi\right) - \left(\nabla^2\Psi^\times\right)\Psi\right] + \frac{1}{c^2}\left[\frac{\partial^2\Psi^\times}{\partial t^2}\Psi - \Psi^\times\frac{\partial^2\Psi}{\partial t^2}\right] = 0$$

$$\vec{\nabla}\left[\Psi^\times\left(\vec{\nabla}\Psi\right) - \left(\vec{\nabla}\Psi^\times\right)\Psi\right] + \frac{1}{c^2}\left[\frac{\partial^2\Psi^\times}{\partial t^2}\Psi - \Psi^\times\frac{\partial^2\Psi}{\partial t^2}\right] = 0$$

$$\vec{\nabla}\left[\frac{\hbar}{2im}\{\Psi^\times\left(\vec{\nabla}\Psi\right) - \left(\vec{\nabla}\Psi^\times\right)\Psi\}\right] + \frac{\partial}{\partial t}\left[\frac{\hbar^2}{2imc^2}\{\frac{\partial\Psi^\times}{\partial t}\Psi - \Psi^\times\frac{\partial\Psi}{\partial t}\}\right] = 0$$

$$\vec{\nabla}\,\vec{J} + \frac{\partial\rho}{\partial t} = 0 \qquad [12]$$

Where $\vec{J} = \frac{\hbar}{2im}\left[\Psi^\times\left(\vec{\nabla}\Psi\right) - \left(\vec{\nabla}\Psi^\times\right)\Psi\right]$ [13]

; is the current density.

And $\rho = \left[\frac{\hbar}{2imc^2}\left\{\frac{\partial\Psi^\times}{\partial t}\Psi - \Psi^\times\frac{\partial\Psi}{\partial t}\right\}\right]$

Or, $\rho = \left[-\frac{i\hbar}{2mc^2}\left\{\Psi^\times\frac{\partial\Psi}{\partial t} - \frac{\partial\Psi^\times}{\partial t}\Psi\right\}\right]$ [14]

; is the probability density.

Equation [12] is known as equation of continuity & is invariant with respect to Lorentz transformation. Now consider about the probability density

$$\rho = \frac{\hbar}{2imc^2}\left[\frac{\partial\Psi^\times}{\partial t}\Psi - \Psi^\times\frac{\partial\Psi}{\partial t}\right] \Rightarrow \frac{-i\hbar}{2mc^2}\left[\frac{\partial\Psi^\times}{\partial t}\Psi - \Psi^\times\frac{\partial\Psi}{\partial t}\right]$$

$$= \frac{1}{2mc^2}\left[\left(-i\hbar\frac{\partial\Psi^\times}{\partial t}\right)\Psi + \Psi^\times\left(i\hbar\frac{\partial\Psi}{\partial t}\right)\right] \Rightarrow \frac{1}{2mc^2}\left[\left(i\hbar\frac{\partial\Psi}{\partial t}\right)^\times\Psi + \Psi^\times\left(i\hbar\frac{\partial\Psi}{\partial t}\right)\right]$$

$$= \frac{1}{2mc^2}\left[\left(E^\times\Psi^\times\right)\Psi + \Psi^\times\left(E\Psi\right)\right] \Rightarrow \frac{1}{2mc^2}\left[E\left(\Psi^\times\Psi\right) + E\left(\Psi^\times\Psi\right)\right] \qquad [15]$$

$$= \frac{1}{2mc^2}2E(\Psi^\times\Psi) \qquad [16]$$

$$\rho = \frac{E}{mc^2}\left[\Psi^\times\Psi\right] \qquad [17]$$

E is of the order of mc^2 and the probability density $\rho = \Psi^\times\Psi$ as in non-relativistic Quantum

Mechanics. Since, E can be positive and negative I.e., the sign of ρ is the sign of the energy. As the energy of the particle may be positive or negative, hence the probability density will be positive or negative. But the probability density should always be positive. Schrödinger tried to solve the situation by multiplying [13] and [14] by the charge of the particle (e) & thus defining the probability density & the probability current density[17]

$$\rho = \frac{i\hbar e}{2mc^2}\left(\Psi^{\times}\frac{\partial \Psi}{\partial t} - \frac{\partial \Psi^{\times}}{\partial t}\Psi \right) \qquad [18]$$

$$\vec{J} = \frac{\hbar e}{2im}\left(\Psi^{\times}(\vec{\nabla}\Psi) - (\vec{\nabla}\Psi^{\times})\Psi \right)$$

Since charge can be positive or negative, and there is no difficulty in interpretation. To overcome all the difficulties & drawbacks those were faces by Schrödinger & Klein–Gordon, Dirac proposed a new type of equation which will be first order in space & time coordinates & hence are the same footing [symmetric] expressed by

$$(\gamma_{\mu}\partial^{\mu} + k)\Psi = 0 \qquad [19]$$

The above equation is Lorentz covariance & has a positive definite probability density. However, the above equation also encounter the problem of negative energy which is successfully overcome by the introduction of hole theory. *In Dirac theory, there is only one drawbacks, its hole interpretation cannot be applicable for Bosons which are not obeying the Pauli exclusion principle.*

14.4 Electron spins for Dirac equation

We know the Eigen value equation

$$H\Psi = E\Psi$$

$$\Rightarrow i\hbar\frac{\partial \Psi}{\partial t} = C\vec{\alpha}.(-i\hbar\vec{\nabla})\Psi + m_0c^2\beta\Psi$$

$$\Rightarrow i\hbar\frac{\partial \Psi}{\partial t} = (C\vec{\alpha}.\vec{P} + m_0c^2\beta)\Psi \qquad [1]$$

So that the Hamiltonian operator for the Dirac equation is

$$H_D = C\vec{\alpha}.\vec{P} + m_0c^2\beta \qquad [2]$$

If we have a potential $V(\vec{r})$; the Hamiltonian operator for the Dirac equation is

$$H_D = C\vec{\alpha}.\vec{P} + m_0c^2\beta + V(\vec{r}) \qquad [3]$$

Classically L_x, the x- component of the orbital angular momentum is a constant of motion in a central field. We shall now check weather this so in Dirac theory too or not. We know that if any operator commutes with Hamiltonian(H) it is said to be a constant of motion. Let us examine this, for x- component of orbital angular momentum[17,10]

$$i\hbar\frac{\partial \hat{L}_x}{\partial t}=\left[\hat{L}_x,\hat{H}\right]=\hat{L}_x\hat{H}-\hat{H}\hat{L}_x$$

$$=\hat{L}_x\left\{c\left(\alpha_x P_x+\alpha_y P_y+\alpha_z P_z\right)+m_0 c^2\beta+V(x)\right\}-\left\{c\left(\alpha_x P_x+\alpha_y P_y+\alpha_z P_z\right)+m_0 c^2\beta+V(x)\right\}\hat{L}_x$$

Since L_x commutes with every quantity in H except P_y and P_z . Hence

$$i\hbar\frac{\partial \hat{L}_x}{\partial t}=c\left\{\alpha_y\left(\hat{L}_x P_y-P_y\hat{L}_x\right)+\alpha_z\left(\hat{L}_x P_z-P_z\hat{L}_x\right)\right\} \quad [4]$$

Consider $\hat{L}_x P_y-P_y\hat{L}_x=\left(yP_z-zP_y\right)P_y-P_y\left(yP_z-zP_y\right)$

$$=\left(y\frac{\hbar}{i}\frac{\partial}{\partial z}-z\frac{\hbar}{i}\frac{\partial}{\partial y}\right)\frac{\hbar}{i}\frac{\partial}{\partial y}-\frac{\hbar}{i}\frac{\partial}{\partial y}\left(y\frac{\hbar}{i}\frac{\partial}{\partial z}-z\frac{\hbar}{i}\frac{\partial}{\partial y}\right)$$

$$=\left(\frac{\hbar}{i}\right)^2\left\{y\frac{\partial}{\partial z}-z\frac{\partial}{\partial y}\right\}\frac{\partial}{\partial y}-\left(\frac{\hbar}{i}\right)^2\left\{\frac{\partial}{\partial y}\left(y\frac{\partial}{\partial z}-z\frac{\partial}{\partial y}\right)\right\}$$

$$=\left(\frac{\hbar}{i}\right)^2\left(-\frac{\partial}{\partial z}\right)=i\hbar P_z$$

Similarly, $\hat{L}_x P_z-P_z\hat{L}_x=-i\hbar P_y$

Finally equation [4] becomes

$$i\hbar\frac{\partial \hat{L}_x}{\partial t}=c\left\{\alpha_y\left(i\hbar P_z\right)+\alpha_z\left(-i\hbar P_y\right)\right\}=-i\hbar c\left(\alpha_z P_y-\alpha_y P_z\right)$$

$$\frac{\partial \hat{L}_x}{\partial t}=-c\left(\alpha_z P_y-\alpha_y P_z\right) \quad [5]$$

Hence in the Dirac theory, in contradiction to Classical Mechanics, the x- component of the orbital angular momentum of an electron moving in a central electrostatic field is not a constant of motion. I.e., L does not commute with H. So we must find another operator such that the sum of this operator and L is then a constant of motion[17] and can be interpreted as the angular momentum. I.e.,

The quantity $(\vec{L}+\frac{\hbar}{2}\vec{\sigma}')$ commutes with H or $(\vec{L}+\frac{\hbar}{2}\vec{\sigma}')$ is constant of motion. This can be taken as that as total angular momentum (i.e. $\vec{J}=\vec{L}+\vec{S}$). We refer to the operator $\vec{S}=\frac{\hbar}{2}\vec{\sigma}'$ as the spin angular momentum of the electron.

Consider,

$$i\hbar\frac{\partial \sigma_x'}{\partial t}=\sigma_x'\hat{H}-\hat{H}\sigma_x'; \quad Where \quad \sigma_x'=\begin{pmatrix}\sigma_x & 0\\ 0 & \sigma_x\end{pmatrix}$$

$$=\sigma_x'\{c(\alpha_x p_x+\alpha_y p_y+\alpha_z p_z)+m_0c^2\beta+V(x)\}-\{c(\alpha_x p_x+\alpha_y p_y+\alpha_z p_z)+m_0c^2\beta+V(x)\}\sigma_x'$$

But σ_x' commutes with every component but not with α_y & α_z.

$$\therefore i\hbar\frac{d}{dt}\sigma_x'=c(\sigma_x'\alpha_y-\alpha_y\sigma_x')p_y+c(\sigma_x'\alpha_z-\alpha_z\sigma_x')p_z \quad [6]$$

$$\text{Now}\ \ \sigma'_x\beta-\beta\sigma'_x=\begin{pmatrix}\sigma_x & 0\\ 0 & \sigma_x\end{pmatrix}\begin{pmatrix}1 & 0\\ 0 & -1\end{pmatrix}-\begin{pmatrix}1 & 0\\ 0 & -1\end{pmatrix}\begin{pmatrix}\sigma_x & 0\\ 0 & \sigma_x\end{pmatrix}=\begin{pmatrix}\sigma_x & 0\\ 0 & -\sigma_x\end{pmatrix}-\begin{pmatrix}\sigma_x & 0\\ 0 & -\sigma_x\end{pmatrix}=0$$

$$\therefore \sigma'_x\alpha_x-\alpha_x\sigma'_x=\begin{pmatrix}\sigma_x & 0\\ 0 & \sigma_x\end{pmatrix}\begin{pmatrix}0 & \sigma_x\\ \sigma_x & 0\end{pmatrix}-\begin{pmatrix}0 & \sigma_x\\ \sigma_x & 0\end{pmatrix}\begin{pmatrix}\sigma_x & 0\\ 0 & \sigma_x\end{pmatrix}=\begin{pmatrix}0 & \sigma^2{}_x\\ \sigma^2{}_x & 0\end{pmatrix}-\begin{pmatrix}0 & \sigma^2{}_x\\ \sigma^2{}_x & 0\end{pmatrix}=0$$

$$\therefore \sigma'_x\alpha_y-\alpha_y\sigma'_x=\begin{pmatrix}\sigma_x & 0\\ 0 & \sigma_x\end{pmatrix}\begin{pmatrix}0 & \sigma_y\\ \sigma_y & 0\end{pmatrix}-\begin{pmatrix}0 & \sigma_y\\ \sigma_y & 0\end{pmatrix}\begin{pmatrix}\sigma_x & 0\\ 0 & \sigma_x\end{pmatrix}=\begin{pmatrix}0 & \sigma_x\sigma_y\\ \sigma_x\sigma_y & 0\end{pmatrix}-\begin{pmatrix}0 & \sigma_y\sigma_x\\ \sigma_y\sigma_x & 0\end{pmatrix}$$

$$=\begin{pmatrix}0 & \sigma_x\sigma_y-\sigma_y\sigma_x\\ \sigma_x\sigma_y-\sigma_y\sigma_x & 0\end{pmatrix}=\begin{pmatrix}0 & 2i\sigma_z\\ 2i\sigma_z & 0\end{pmatrix}=2i\begin{pmatrix}0 & \sigma_z\\ \sigma_z & 0\end{pmatrix}=2i\alpha_z$$

$$\&\quad \sigma'_x\alpha_z-\alpha_z\sigma'_x=\begin{pmatrix}\sigma_x & 0\\ 0 & \sigma_x\end{pmatrix}\begin{pmatrix}0 & \sigma_z\\ \sigma_z & 0\end{pmatrix}-\begin{pmatrix}0 & \sigma_z\\ \sigma_z & 0\end{pmatrix}\begin{pmatrix}\sigma_x & 0\\ 0 & \sigma_x\end{pmatrix}$$

$$=\begin{pmatrix}0 & \sigma_x\sigma_z\\ \sigma_x\sigma_z & 0\end{pmatrix}-\begin{pmatrix}0 & \sigma_z\sigma_x\\ \sigma_z\sigma_x & 0\end{pmatrix}$$

$$=\begin{pmatrix}0 & \sigma_x\sigma_z-\sigma_z\sigma_x\\ \sigma_x\sigma_z-\sigma_z\sigma_x & 0\end{pmatrix}=\begin{pmatrix}0 & -2i\sigma_y\\ -2i\sigma_y & 0\end{pmatrix}$$

$$=-2i\begin{pmatrix}0 & \sigma_y\\ \sigma_y & 0\end{pmatrix}=-2i\alpha_y$$

Therefore, [6] becomes

$$\therefore i\hbar\frac{d}{dt}\sigma'_x=c\,2i\alpha_z p_y+c(-2i\alpha_y)p_z=2ic(\alpha_z p_y-\alpha_y p_z)$$

$$\frac{d}{dt}(\frac{\hbar}{2}\sigma'_x)=c(\alpha_z p_y-\alpha_y p_z) \qquad [7]$$

Now adding [5] and [7], we get

$$\frac{d\hat{L}_x}{dt}+\frac{d}{dt}(\frac{\hbar}{2}\sigma'_x)=-c(\alpha_z p_y-\alpha_y p_z)+c(\alpha_z p_y-\alpha_y p_z)=0$$

$$\frac{d}{dt}(\hat{L}_x+\frac{\hbar}{2}\sigma'_x)=0 \quad \text{Thus } (\hat{L}_x+\frac{\hbar}{2}\sigma'_x)H-H(\hat{L}_x+\frac{\hbar}{2}\sigma'_x)=0$$

Hence the quantity $(\vec{L}+\frac{\hbar}{2}\vec{\sigma}')$ commutes with H or $(\vec{L}+\frac{\hbar}{2}\vec{\sigma}')$ is constant of motion. This can be taken as that as total angular momentum (i.e. $\vec{J}=\vec{L}+\vec{S}$) .We refer to the operator $\vec{S}=\frac{\hbar}{2}\vec{\sigma}'$ as the spin angular momentum of the electron.

14.5 Comparative study of Schrodinger, Klein-Gordon and Dirac equations

Schrodinger equation:

- The Schrodinger equation is valid in low energy system

- It is not Lorentz invariant & it cannot describe the basic law of nature
- Also the above equation is first order in time & second order in space
- So it is not symmetrical in space and time
- To put Schrodinger equation on a relativisting footing for making first order in time & second order in space

Klein-Gordon equation:

- This equation is an important one for further development of relativistic field theories
- *Klein-Gordon [K-G] equation* is valid in high energy system. I.e. the equation is valid for certain classes of particles such as the π and K mesons, which are strongly interacting
- It is Lorentz invariant & it can describe the basic law of nature
- Also *Klein-Gordon [K-G] equation* is second order in time & space
- So it is symmetrical in space and time
- Klein-Gordon equation is the second-order differential equation
- The Klein-Gordon equation it turns out is the correct one for spinless particles & not for electrons

Dirac equation:

- The above equation is Lorentz covariance & has a positive definite probability density
- However, the above equation also encounters the problem of negative energy which is successfully overcome by the introduction of hole theory
- *In Dirac theory, there is only one drawback its hole interpretation cannot be applicable for Bosons which are not obeying the Pauli Exclusion Principle*

14.6 Solution of Dirac equation: For free particle

We know, the Dirac equation for free particle

$$\left(\gamma_\mu \partial_\mu + k\right) = 0 \qquad [1]$$

For a general plane-wave solution of the Dirac equation, we write

$\psi(x,t) = uexp\left[\left(\frac{-i}{\hbar}\right)(Et - px)\right]$; For the sake of simplicity, we restricted to a one-dimensional motion in the Z- direction I.e., we consider $\psi(x_0, t) = \psi(x_3, t)$ and we try to ersatz

$$\psi(x,t) = u\; exp\left[\left(\frac{-i}{\hbar}\right)(Et - p_3 x_3)\right]\; Or,\;\; \psi(x,t) = u\; exp\left[\left(\frac{i}{\hbar}\right)(p_3 x_3 - Et)\right]$$

Where p & E are numbers not operators, and u is a (4×1) column vector,

$$u = \begin{pmatrix} u_1 \\ u_2 \\ u_3 \\ u_4 \end{pmatrix}$$

The components of the Dirac spinor (u) does not depend on x & t. [coordinate &time]

Therefore [1] becomes

$$(\gamma_\mu \partial_\mu + k)u \exp\left[\frac{i}{\hbar}(p_3x_3 - Et)\right] \to (\gamma_\mu \partial_\mu + k)u \exp\left[\frac{i}{\hbar}(p_3x_3)\right]\exp\left[(-\frac{i}{\hbar}Et)\right] = 0$$

$$\Rightarrow \left[\frac{i}{\hbar}\gamma_3 p_3 + \frac{1}{iC}\gamma_4(-\frac{i}{\hbar})E + \frac{m_0C}{\hbar}\right]u \exp\left[\frac{i}{\hbar}(p_3x_3)\right]\exp\left[(-\frac{i}{\hbar}Et)\right] = 0$$

$$\Rightarrow \left[\frac{i}{\hbar}\gamma_3 p_3 - \frac{\gamma_4 E}{C\hbar} + \frac{m_0C}{\hbar}\right]u = 0 \Rightarrow \left(\frac{i}{\hbar}\right)\left[\gamma_3 p_3 + \frac{i}{C}\gamma_4 E - i\,m_0C\right]u = 0$$

$$\Rightarrow \frac{i}{\hbar}\left(\gamma_3 p_3 + \frac{i}{C}\gamma_4 E - i\,m_0C\right)u = 0 \quad \Rightarrow \left(\gamma_3 p_3 + \frac{i}{C}\gamma_4 E - i\,m_0C\right)u = 0 \qquad [3]$$

As Dirac operators γ's are (4×4) matrices. Dirac operators must have four components I.e., the wave function $\Psi(r)$ itself is a matrix with four rows & one column. I.e.,

$$\Psi(r) = \begin{pmatrix} \Psi_1 \\ \Psi_2 \\ \Psi_3 \\ \Psi_4 \end{pmatrix};\ \text{Also we have } \gamma_3 = \begin{pmatrix} 0 & 0 & i & 0 \\ 0 & 0 & 0 & -i \\ -i & 0 & 0 & 0 \\ 0 & i & 0 & 0 \end{pmatrix} \quad \& \quad \gamma_4 = \begin{pmatrix} 0 & 0 & 1 & 0 \\ 0 & 0 & 0 & 1 \\ 1 & 0 & 0 & 0 \\ 0 & 1 & 0 & 0 \end{pmatrix}$$

$$\therefore [3] \text{ becomes} \left[P_3\begin{pmatrix} 0 & 0 & i & 0 \\ 0 & 0 & 0 & -i \\ -i & 0 & 0 & 0 \\ 0 & i & 0 & 0 \end{pmatrix} + \frac{i}{c}E\begin{pmatrix} 0 & 0 & 1 & 0 \\ 0 & 0 & 0 & 1 \\ 1 & 0 & 0 & 0 \\ 0 & 1 & 0 & 0 \end{pmatrix} - im_0C\begin{pmatrix} 1 & 0 & 0 & 0 \\ 0 & 1 & 0 & 0 \\ 0 & 0 & 1 & 0 \\ 0 & 0 & 0 & 1 \end{pmatrix}\right]\begin{pmatrix} u_1 \\ u_2 \\ u_3 \\ u_4 \end{pmatrix} = 0$$

$$\Rightarrow \left[\begin{pmatrix} 0 & 0 & iP_3 & 0 \\ 0 & 0 & 0 & -iP_3 \\ -iP_3 & 0 & 0 & 0 \\ 0 & iP_3 & 0 & 0 \end{pmatrix} + \begin{pmatrix} 0 & 0 & \frac{iE}{c} & 0 \\ 0 & 0 & 0 & \frac{iE}{c} \\ \frac{iE}{c} & 0 & 0 & 0 \\ 0 & \frac{iE}{c} & 0 & 0 \end{pmatrix} + \begin{pmatrix} -im_0c & 0 & 0 & 0 \\ 0 & -im_0c & 0 & 0 \\ 0 & 0 & -im_0c & 0 \\ 0 & 0 & 0 & -im_0c \end{pmatrix}\right]\begin{pmatrix} u_1 \\ u_2 \\ u_3 \\ u_4 \end{pmatrix} = 0$$

$$\Rightarrow \begin{pmatrix} iP_3 u_3 & +\dfrac{iE}{c}u_3 & -im_0cu_1 \\ -iP_3 u_4 & +\dfrac{iE}{c}u_4 & -im_0cu_2 \\ -iP_3 u_1 & +\dfrac{iE}{c}u_1 & -im_0cu_3 \\ iP_3 u_2 & +\dfrac{iE}{c}u_2 & -im_0cu_4 \end{pmatrix} = 0$$

From the above matrix we write four simultaneous equations,

$$\left.\begin{aligned} (P_3 + \frac{E}{c})u_3 - m_0cu_1 &= 0 \\ (-P_3 + \frac{E}{c})u_4 - m_0cu_2 &= 0 \\ (-P_3 + \frac{E}{c})u_1 - m_0cu_3 &= 0 \\ (P_3 + \frac{E}{c})u_2 - m_0cu_4 &= 0 \end{aligned}\right\} \qquad [4]$$

Equation [4] has non-zero solution only if the determinant of the coefficient vanishes. I.e.,

$$\begin{vmatrix} -m_0c & 0 & P_3 + \dfrac{E}{c} & 0 \\ 0 & -m_0c & 0 & -P_3 + \dfrac{E}{c} \\ -P_3 + \dfrac{E}{c} & 0 & -m_0c & 0 \\ 0 & P_3 + \dfrac{E}{c} & 0 & -m_0c \end{vmatrix} = 0$$

Now expanding the determinant

$$-m_0C \begin{vmatrix} -m_0C & 0 & -P_3 + \dfrac{E}{c} \\ 0 & -m_0C & 0 \\ P_3 + \dfrac{E}{c} & 0 & -m_0C \end{vmatrix} + (P_3 + \frac{E}{c}) \begin{vmatrix} 0 & -m_0C & -P_3 + \dfrac{E}{c} \\ -P_3 + \dfrac{E}{c} & 0 & 0 \\ 0 & P_3 + \dfrac{E}{c} & -m_0C \end{vmatrix} = 0$$

$$\Rightarrow -m_0C\left[-m_0\,c(m_0^2c^2 - 0) + (\frac{E}{c} - P_3)\left\{0 - (-m_0\,c)(P_3 + \frac{E}{c})\right\}\right]$$

$$+(P_3 + \frac{E}{c})\left[-m_0\,c\left\{0 - (-m_0\,c)(\frac{E}{c} - P_3)\right\} + (\frac{E}{c} - P_3)\left\{(\frac{E}{c} + P_3)(\frac{E}{c} - P_3) - 0\right\}\right] = 0$$

$$\Rightarrow m_0^{\ 4}c^4 - m_0^{\ 2}c^2(\frac{E^2}{c^2} - P_3^{\ 2}) - m_0^{\ 2}c^2(\frac{E^2}{c^2} - P_3^{\ 2}) + (\frac{E^2}{c^2} - P_3^{\ 2})^2 = 0$$

$$\Rightarrow m_0^{\ 4}c^4 - m_0^{\ 2}E^2 + m_0^{\ 2}c^2P_3^2 - m_0^{\ 2}E^2 + m_0^{\ 2}c^2P_3^2 + (\frac{E^2}{c^2} - P_3^{\ 2})^2 = 0$$

$$\Rightarrow m_0^{\ 4}c^4 - 2m_0^{\ 2}E^2 + 2m_0^{\ 2}c^2P_3^2 + (\frac{E^2}{c^2} - P_3^{\ 2})^2 = 0$$

$$\Rightarrow \left(m_0^{\ 2}c^2\right)^2 - 2m_0^{\ 2}c^2(\frac{E^2}{c^2} - P_3^{\ 2}) + (\frac{E^2}{c^2} - P_3^{\ 2})^2 = 0$$

$$\Rightarrow (m_0^{\ 2}c^2 - \frac{E^2}{c^2} + P_3^{\ 2})^2 = 0 \quad \Rightarrow (m_0^{\ 2}c^4 - E^2 + P_3^{\ 2}c^2)^2 = 0$$

$$\Rightarrow (-E^2 + P_3^2c^2 + m_0^{\ 2}c^4)^2 = 0 \quad \Rightarrow E^2 - P_3^2c^2 - m_0^{\ 2}c^4 = 0$$

$$\Rightarrow E^2 - P_3^2c^2 = m_0^{\ 2}c^4 \Rightarrow 1 = \frac{m_0^{\ 2}c^4}{E^2 - P_3^2c^2}$$

$$\Rightarrow E^2 = \frac{E^2m_0^{\ 2}c^4}{E^2 - P_3^2c^2} \Rightarrow E^2 = \frac{m_0^{\ 2}c^4}{1 - \frac{P_3^2c^2}{E^2}}$$

$$\Rightarrow E^2 = \frac{m_0^{\ 2}c^4}{1-\beta^2} \qquad Where \quad \beta = \frac{P_3c}{E} \quad \Rightarrow E = \pm\frac{m_0^{\ 2}c^4}{(1-\beta^2)^{\frac{1}{2}}} \qquad [5]$$

With $E = +\frac{m_0c^2}{(1-\beta^2)^{\frac{1}{2}}}$; There are two linear independent solutions, which are conveniently written as

$u_1 = 1 \quad \& \quad u_2 = 0 \quad [put\ in\ eqn(4)]$

$$u_3(P_3 + \frac{E}{c}) - m_0C = 0 \qquad \Rightarrow u_3 = \frac{m_0c}{P_3 + \frac{E}{c}}$$

$$= \frac{m_0c}{\frac{E}{c}(1 + \frac{P_3C}{E})} = \frac{m_0c^2}{\frac{E}{c}(1+\beta)} = \frac{m_0c^2}{\frac{m_0c^2}{(1-\beta^2)^{\frac{1}{2}}}(1+\beta)} = \frac{\sqrt{1-\beta^2}}{1+\beta} = \frac{\sqrt{(1+\beta)(1-\beta)}}{1+\beta} = \frac{\sqrt{(1-\beta)}}{\sqrt{(1+\beta)}}$$

$u_4 = 0$

Again when $u_1 = 0 \quad \& \quad u_2 = 1\,[put\ in\ eqn\ (4)]$

$$(-P_3 + \frac{E}{c})u_4 - m_0c = 0$$

$$\Rightarrow u_4 = \frac{m_0c}{\frac{E}{c} - P_3} = \frac{m_0c}{\frac{E}{c}(1 - \frac{P_3c}{E})} = \frac{m_0c^2}{E(1-\beta)}$$

$$= \frac{m_0c^2}{\frac{m_0c^2}{(1-\beta^2)^{\frac{1}{2}}}(1-\beta)} = \frac{(1-\beta^2)^{\frac{1}{2}}}{(1-\beta)} = \frac{\sqrt{(1+\beta)}}{\sqrt{(1-\beta)}}$$

And $u_3 = 0$

Therefore, the two linear independent solutions are conveniently written a

$$u = \begin{pmatrix} u_1 \\ u_2 \\ u_3 \\ u_4 \end{pmatrix} \Rightarrow u^I = \begin{pmatrix} 1 \\ 0 \\ \frac{\sqrt{(1-\beta)}}{\sqrt{(1+\beta)}} \\ 0 \end{pmatrix} \;\&\; u^{II} = \begin{pmatrix} 1 \\ 0 \\ 0 \\ \frac{\sqrt{(1+\beta)}}{\sqrt{(1-\beta)}} \end{pmatrix}$$

With $E = -\frac{m_0c^2}{\sqrt{(1-\beta^2)}}$; there are two linear independent solutions, which are conveniently written as When $u_3 = 1$ and $u_4 = 0$ [put in equation(4)]

$$(P_3 + \frac{E}{c}) - m_0cu_1 = 0$$

$$\Rightarrow u_1 = \frac{\frac{E}{c} + P_3}{m_0c} = \frac{\frac{E}{c}(1 + \frac{P_3C}{E})}{m_0c} = \frac{E(1+\beta)}{m_0c^2} = \frac{1+\beta}{m_0c^2}\left(\frac{-m_0c^2}{\sqrt{1-\beta^2)}}\right) = -\frac{\sqrt{(1+\beta)}}{\sqrt{(1-\beta)}} \;\&\; u_2 = 0$$

Again When $u_3 = 0$ and $u_4 = 1$ [put in equation(4)]

$$\therefore (P_3 + \frac{E}{c})u_2 - m_0C = 0 \Rightarrow u_2 = \frac{m_0c}{P_3 + \frac{E}{c}} = \frac{m_0c}{\frac{E}{c}(1 + \frac{P_3c}{E})} = \frac{m_0c^2}{E(1+\beta)}$$

$$\Rightarrow u_2 = \left(\frac{m_0c^2}{(1+\beta)}\right)\left(-\frac{\sqrt{(1-\beta^2)}}{m_0c^2}\right) = -\frac{\sqrt{(1-\beta)}}{\sqrt{(1+\beta)}} \;\&\; u_4 = 0$$

Therefore, the two independent solutions are

$$u^{III} = \begin{pmatrix} -\frac{\sqrt{(1+\beta)}}{\sqrt{(1-\beta)}} \\ 0 \\ 1 \\ 0 \end{pmatrix} \;\&\; u^{IV} = \begin{pmatrix} 0 \\ -\frac{\sqrt{(1-\beta)}}{\sqrt{(1+\beta)}} \\ 0 \\ 1 \end{pmatrix}$$

Finally, four linear independent solutions are

$$u=\begin{pmatrix}u_1\\u_2\\u_3\\u_4\end{pmatrix}\Rightarrow u^{I}=\begin{pmatrix}1\\0\\\frac{\sqrt{(1-\beta)}}{\sqrt{(1+\beta)}}\\0\end{pmatrix}\ \&\ u^{II}=\begin{pmatrix}1\\0\\0\\\frac{\sqrt{(1+\beta)}}{\sqrt{(1-\beta)}}\end{pmatrix}\ u^{III}=\begin{pmatrix}-\frac{\sqrt{(1+\beta)}}{\sqrt{(1-\beta)}}\\0\\1\\0\end{pmatrix}\ \&\ u^{IV}=\begin{pmatrix}0\\-\frac{\sqrt{(1-\beta)}}{\sqrt{(1+\beta)}}\\0\\1\end{pmatrix}$$

14.7 Dirac's relativistic wave equation: For free particle

From first order time dependent Schrodinger wave equation

$$i\hbar\frac{\partial\Psi}{\partial t}=\hat{H}\Psi \qquad [1]$$

For free particle Hamiltonian operator H_{op} should be linear in P_x, P_y and P_z . So that the components of the momentum can be represented by the equivalents

$$p_x=\frac{i}{\hbar}\frac{\partial}{\partial x}=-i\hbar\frac{\partial}{\partial x}\ ;\quad p_y=\frac{i}{\hbar}\frac{\partial}{\partial y}=-i\hbar\frac{\partial}{\partial y}\ ;\quad p_z=\frac{i}{\hbar}\frac{\partial}{\partial z}=-i\hbar\frac{\partial}{\partial z} \qquad [2]$$

So that the Hamiltonian operator for the Dirac equation is

$$H_D=C\vec{\alpha}.\vec{P}+m_0c^2\beta \qquad [3]$$

If we have a potential $V(\vec{r})$; the Hamiltonian operator for the Dirac equation is

$$H_D=C\vec{\alpha}.\vec{P}+m_0c^2\beta+V(\vec{r}) \qquad [4]$$

The Hamiltonian operator for the Dirac equation without potential $V(\vec{r})$ I.e., for free particle

$$\hat{H}=C\alpha_x\hat{p}_x+C\alpha_y\hat{p}_y+C\alpha_z\hat{p}_z+\beta m_0C^2 \qquad [5]$$

Therefore, equation [1] becomes

$$\left.\begin{aligned}i\hbar\frac{\partial\Psi}{\partial t}&=(C\alpha_x\hat{p}_x+C\alpha_y\hat{p}_y+C\alpha_z\hat{p}_z)\Psi+\beta m_0C^2\Psi\\ i\hbar\frac{\partial\Psi}{\partial t}&=(c\vec{\alpha}.\vec{p})\Psi+\beta m_0C^2\Psi\end{aligned}\right\} \qquad [6]$$

Where $\hat{\vec{p}}=-i\hbar\vec{\nabla}$ [7]

Equation [6] is Dirac's relativistic wave equation for free particle.

Form Dirac's relativistic wave equation some significant observations are:

- m_0c^2 and pc have the dimension of energy
- Relativistic Hamiltonian can include rest mass of energy m_0c^2
- Dimension of the term βm_0c^2 in Hamiltonian operator will not remain energy if α_x ,α_y ,α_z and β must not contain p_x , p_y , p_z
- α_x ,α_y ,α_z and β commute with p_x , p_y , p_z
- α_x ,α_y ,α_z and β must not contain x, y, z, t. Otherwise, space and time coordinates are not on the same footing in Dirac equation.
- Hamiltonian operator will contain space or time dependent potential. Otherwise, it will not be valid for free particle

14.8 Dirac matrices

We know that The Hamiltonian operator of any system can be regarded as the total energy of the system. I.e.,

$$\hat{H}_D = \hat{E} \quad \text{or,} \quad \hat{H}_D\,\hat{H}_D = \hat{E}\,\hat{E} \qquad [1]$$

So that the Hamiltonian operator for the Dirac equation is

$$H_D = c\vec{\alpha}.\vec{P} + m_0c^2\beta \qquad [2]$$

If we have a potential $V(\vec{r})$; the Hamiltonian operator for the Dirac equation is

$$H_D = c\vec{\alpha}.\vec{P} + m_0c^2\beta + V(\vec{r}) \qquad [3]$$

The Hamiltonian operator for the Dirac equation without potential $V(\vec{r})$ I.e., for free particle

$$\hat{H}_D = c\alpha_x\hat{p}_x + c\alpha_y\hat{p}_y + c\alpha_z\hat{p}_z + \beta\, m_0\, c^2 \qquad [4]$$

Then equation [1] becomes

$$[c(\alpha_x\hat{p}_x + \alpha_y\hat{p}_y + \alpha_z\hat{p}_z) + \beta m_0c^2][c(\alpha_x\hat{p}_x + \alpha_y\hat{p}_y + \alpha_z\hat{p}_z) + \beta m_0c^2] = p_{op}{}^2c^2 + m_0{}^2c^4$$

Or, $c\alpha_x\hat{p}_x c\alpha_x\hat{p}_x + c\alpha_x\hat{p}_x c\alpha_y\hat{p}_y + c\alpha_x\hat{p}_x c\alpha_z\hat{p}_z + c\alpha_x\hat{p}_x\beta m_0c^2 + c\alpha_y\hat{p}_y c\alpha_x\hat{p}_x + c\alpha_y\hat{p}_y c\alpha_y\hat{p}_y$

$+ c\alpha_y\hat{p}_y c\alpha_z\hat{p}_z + c\alpha_y\hat{p}_y\beta m_0c^2 + c\alpha_z\hat{p}_z c\alpha_x\hat{p}_x + c\alpha_z\hat{p}_z c\alpha_y\hat{p}_y + c\alpha_z\hat{p}_z c\alpha_z\hat{p}_z + c\alpha_z\hat{p}_z\beta m_0c^2$

$+ \beta m_0c^2 c\alpha_x\hat{p}_x + \beta m_0c^2 c\alpha_y\hat{p}_y + \beta m_0c^2 c\alpha_z\hat{p}_z + \beta m_0c^2\beta m_0c^2$

$= \hat{p}_x\hat{p}_x c^2 + \hat{p}_x\hat{p}_y c^2 + \hat{p}_z\hat{p}_z c^2 + (m_0c^2)^2$

Or, $c^2\alpha_x\alpha_x\hat{p}_x\hat{p}_x + c^2\alpha_x\alpha_y\hat{p}_x\hat{p}_y + c^2\alpha_x\alpha_z\hat{p}_x\hat{p}_z + c\alpha_x\beta\,\hat{p}_x m_0c^2$

$+ c^2\alpha_y\alpha_x\hat{p}_y\hat{p}_x + c^2\alpha_y\alpha_y\hat{p}_y\hat{p}_y + c^2\alpha_y\alpha_z\hat{p}_y\hat{p}_z + c\alpha_y\beta\,\hat{p}_y m_0c^2$

$+ c^2\alpha_z\alpha_x\hat{p}_z\hat{p}_x + c^2\alpha_z\alpha_y\hat{p}_z\hat{p}_y + c^2\alpha_z\alpha_z\hat{p}_z\hat{p}_z + c\alpha_z\beta\,\hat{p}_z m_0c^2$

$+ c\beta\,\alpha_x\hat{p}_x m_0c^2 + c\beta\alpha_y\hat{p}_y m_0c^2 + c\beta\alpha_z\hat{p}_z m_0c^2 + \beta\,\beta\left(m_0c^2\right)^2$

$$= \hat{p}_x\hat{p}_x c^2 + \hat{p}_y\hat{p}_y c^2 + \hat{p}_z\hat{p}_z c^2 + \left(m_0c^2\right)^2 \qquad [5]$$

Since $\alpha_x, \alpha_y, \alpha_z, \beta$ commute with $\hat{p}_x, \hat{p}_y, \hat{p}_z$ and m_0c^2 and c are the constants. Equating coefficients of $\hat{p}_x\hat{p}_x, \hat{p}_y\hat{p}_y, \hat{p}_z\hat{p}_z$ and $m_0^2\, c^4$ of two sides of equation [5],We get

$$\left.\begin{aligned} c^2 &= c^2\alpha_x\alpha_x \\ c^2 &= c^2\alpha_y\alpha_y \\ c^2 &= c^2\alpha_z\alpha_z \\ \beta^2 &= 1 \end{aligned}\right\}. \qquad [6]$$

Thus we can easily write from equation [6]

$$\alpha_x^2 = \alpha_y^2 = \alpha_z^2 = \beta^2 = 1 \qquad [7]$$

Again equating coefficients of $\hat{p}_x\hat{p}_x$, $\hat{p}_y\hat{p}_y$, $\hat{p}_z\hat{p}_z$ of two sides of equation [5],We get

$$\left.\begin{aligned} c^2\alpha_x\alpha_y + c^2\alpha_y\alpha_x &= 0 \\ c^2\alpha_y\alpha_z + c^2\alpha_z\alpha_y &= 0 \\ c^2\alpha_z\alpha_x + c^2\alpha_x\alpha_z &= 0 \end{aligned}\right\} \qquad [8]$$

Thus we can easily write from equation [7]

$$\left.\begin{aligned} \alpha_x\alpha_y + \alpha_y\alpha_x &= 0 \\ \alpha_y\alpha_z + \alpha_z\alpha_y &= 0 \\ \alpha_z\alpha_x + \alpha_x\alpha_z &= 0 \end{aligned}\right\} \qquad [9]$$

Again equating coefficients of $\hat{p}_x, \hat{p}_y \text{and } \hat{p}_z$ of two sides of equation [5], we get

$$\left.\begin{aligned} c\alpha_x\beta\, m_0c^2 + c\beta\,\alpha_x m_0c^2 &= 0 \\ c\alpha_y\beta\, m_0c^2 + c\beta\,\alpha_y m_0c^2 &= 0 \\ c\alpha_z\beta\, m_0c^2 + c\beta\,\alpha_z m_0c^2 &= 0 \end{aligned}\right\} \qquad [10]$$

Thus we can easily write from equation [9]

$$\left.\begin{aligned} \alpha_x\beta + \beta\alpha_x &= 0 \\ \alpha_y\beta + \beta\alpha_y &= 0 \\ \alpha_z\beta + \beta\alpha_z &= 0 \end{aligned}\right\} \qquad [11]$$

From equations [7], [9] and [11]; we find that

$$\left[\alpha_i, \alpha_j\right] \neq 0 \text{ and } \left[\alpha_i, \beta\right] \neq 0; \quad \text{For } i \neq j; \text{ where } i = j = x, y, z$$

We have understood that $\alpha_x, \alpha_y, \alpha_z, \beta$ are not numbers but matrices. So $\alpha_x, \alpha_y, \alpha_z, \beta$ matrices are defined by

$$\alpha_x = \begin{pmatrix} 0 & 0 & 0 & 1 \\ 0 & 0 & 1 & 0 \\ 0 & 1 & 0 & 0 \\ 1 & 0 & 0 & 0 \end{pmatrix}, \quad \alpha_y = \begin{pmatrix} 0 & 0 & 0 & -i \\ 0 & 0 & i & 0 \\ 0 & -i & 0 & 0 \\ i & 0 & 0 & 0 \end{pmatrix}, \quad \alpha_z = \begin{pmatrix} 0 & 0 & 1 & 0 \\ 0 & 0 & 0 & -1 \\ 1 & 0 & 0 & 0 \\ 0 & -1 & 0 & 0 \end{pmatrix}, \quad \beta = \begin{pmatrix} 1 & 0 & 0 & 0 \\ 0 & 1 & 0 & 0 \\ 1 & 0 & -1 & 0 \\ 0 & 0 & 0 & -1 \end{pmatrix},$$

The four matrices are Hermitian ($A_{mn} = A^{\times}_{nm}$) and are often summarized as

$$\alpha_x = \begin{pmatrix} 0 & \sigma_x \\ \sigma_x & 0 \end{pmatrix}, \quad \alpha_y = \begin{pmatrix} 0 & \sigma_y \\ \sigma_y & 0 \end{pmatrix}, \quad \alpha_z = \begin{pmatrix} 0 & \sigma_z \\ \sigma_z & 0 \end{pmatrix}, \quad \beta = \begin{pmatrix} 1 & 0 \\ 0 & -1 \end{pmatrix},$$

Where

- $\sigma_x, \sigma_y, \sigma_z$ are (2×2) array of elements of Pauli spin matrices (for j=1/2)
- 0 stands for (2×2) array $\begin{pmatrix} 0 & 0 \\ 0 & 0 \end{pmatrix}$

- 1stands for (2×2) array $\begin{pmatrix} 1 & 0 \\ 0 & 1 \end{pmatrix}$
- -1 stands for (2×2) array $\begin{pmatrix} -1 & 0 \\ 0 & -1 \end{pmatrix}$

14.9 Dirac's relativistic equation in matrix form

We know the Eigen value equation

$$H\Psi = E\Psi$$

$$\Rightarrow i\hbar\frac{\partial\Psi}{\partial t} = C\vec{\alpha}.(-i\hbar\vec{\nabla})\Psi + m_0c^2\beta\Psi$$

$$\Rightarrow i\hbar\frac{\partial\Psi}{\partial t} = (C\vec{\alpha}.\vec{P} + m_0c^2\beta)\Psi \qquad [1]$$

Hamiltonian operator for the Dirac equation is

$$H_D = C\vec{\alpha}.\vec{P} + m_0c^2\beta \qquad [2]$$

If we have a potential $V(\vec{r})$; the Hamiltonian operator for the Dirac equation is

$$H_D = C\vec{\alpha}.\vec{P} + m_0c^2\beta + V(\vec{r}) \qquad [3]$$

The Hamiltonian operator for the Dirac equation without potential $V(\vec{r})$ I.e., for free particle

$$\hat{H} = C\alpha_x\hat{p}_x + C\alpha_y\hat{p}_y + C\alpha_z\hat{p}_z + \beta m_0 C^2 \qquad [4]$$

Dirac relativistic equation

$$i\hbar\frac{\partial\Psi}{\partial t} = (C\vec{\alpha}.\hat{\vec{p}} + \beta m_0c^2)\Psi$$

Or, $$i\hbar\frac{\partial\Psi}{\partial t} = c\alpha_x\hat{p}_x\Psi + c\alpha_y\hat{p}_y\Psi + c\alpha_z\hat{p}_z\Psi + \beta m_0c^2\Psi$$

Or, $$i\hbar\frac{\partial\Psi}{\partial t} = c\hat{p}_x\alpha_x\Psi + c\hat{p}_y\alpha_y\Psi + c\hat{p}_z\alpha_z\Psi + m_0c^2\beta\Psi \qquad [5]$$

Since $\alpha_x, \alpha_y, \alpha_z$ commute with $\hat{p}_x, \hat{p}_y, \hat{p}_z$ and $\alpha_x, \alpha_y, \alpha_z, \beta$ are (4×4) matrices, and α_x, α_y, α_z, β operate on Ψ as shown in equation [5]. Thus Ψ must have 4 elements I.e.,

$$\psi = \begin{pmatrix} \psi_1 \\ \psi_2 \\ \psi_3 \\ \psi_4 \end{pmatrix} \text{ Or, } i\hbar\frac{\partial\Psi}{\partial t} = \begin{pmatrix} i\hbar\frac{\partial\Psi_1}{\partial t} \\ i\hbar\frac{\partial\Psi_2}{\partial t} \\ i\hbar\frac{\partial\Psi_3}{\partial t} \\ i\hbar\frac{\partial\Psi_4}{\partial t} \end{pmatrix}$$

From equation [4], we consider the first, second, third and fourth terms of R.H.S separately.

$$c\hat{p}_x\alpha_x\Psi = c\hat{p}_x\begin{pmatrix}0&0&0&1\\0&0&1&0\\0&1&0&0\\1&0&0&0\end{pmatrix}\begin{pmatrix}\psi_1\\\psi_2\\\psi_3\\\psi_4\end{pmatrix}=\begin{pmatrix}0&0&0&c\hat{p}_x\\0&0&c\hat{p}_x&0\\0&c\hat{p}_x&0&0\\c\hat{p}_x&0&0&0\end{pmatrix}\begin{pmatrix}\psi_1\\\psi_2\\\psi_3\\\psi_4\end{pmatrix}$$

$$c\hat{p}_y\alpha_y\Psi = c\hat{p}_y\begin{pmatrix}0&0&0&-i\\0&0&i&0\\0&-i&0&0\\i&0&0&0\end{pmatrix}\begin{pmatrix}\psi_1\\\psi_2\\\psi_3\\\psi_4\end{pmatrix}=\begin{pmatrix}0&0&0&-ic\hat{p}_y\\0&0&ic\hat{p}_y&0\\0&-ic\hat{p}_y&0&0\\ic\hat{p}_y&0&0&0\end{pmatrix}\begin{pmatrix}\psi_1\\\psi_2\\\psi_3\\\psi_4\end{pmatrix}$$

$$c\hat{p}_z\alpha_z\Psi = c\hat{p}_z\begin{pmatrix}0&0&1&0\\0&0&0&-1\\1&0&0&0\\0&-1&0&0\end{pmatrix}\begin{pmatrix}\psi_1\\\psi_2\\\psi_3\\\psi_4\end{pmatrix}=\begin{pmatrix}0&0&c\hat{p}_z&0\\0&0&0&-c\hat{p}_z\\c\hat{p}_z&0&0&0\\0&-c\hat{p}_z&0&0\end{pmatrix}\begin{pmatrix}\psi_1\\\psi_2\\\psi_3\\\psi_4\end{pmatrix}$$

$$\mathrm{m_0c^2}\beta\Psi = m_0c^2\begin{pmatrix}1&0&0&0\\0&1&0&0\\0&0&-1&0\\0&0&0&-1\end{pmatrix}\begin{pmatrix}\psi_1\\\psi_2\\\psi_3\\\psi_4\end{pmatrix}=\begin{pmatrix}m_0c^2&0&0&0\\0&m_0c^2&0&0\\0&0&-m_0c^2&0\\0&0&0&-m_0c^2\end{pmatrix}\begin{pmatrix}\psi_1\\\psi_2\\\psi_3\\\psi_4\end{pmatrix}$$

Using the above expressions equation [4] becomes,

$$\begin{pmatrix}i\hbar\frac{\partial\Psi_1}{\partial t}\\i\hbar\frac{\partial\Psi_2}{\partial t}\\i\hbar\frac{\partial\Psi_3}{\partial t}\\i\hbar\frac{\partial\Psi_4}{\partial t}\end{pmatrix}=\begin{pmatrix}0&0&0&c\hat{p}_x\\0&0&c\hat{p}_x&0\\0&c\hat{p}_x&0&0\\c\hat{p}_x&0&0&0\end{pmatrix}\begin{pmatrix}\psi_1\\\psi_2\\\psi_3\\\psi_4\end{pmatrix}+\begin{pmatrix}0&0&0&-\mathrm{ic}\hat{\mathrm{p}}_\mathrm{y}\\0&0&\mathrm{ic}\hat{\mathrm{p}}_\mathrm{y}&0\\0&-\mathrm{ic}\hat{\mathrm{p}}_\mathrm{y}&0&0\\\mathrm{ic}\hat{\mathrm{p}}_\mathrm{y}&0&0&0\end{pmatrix}\begin{pmatrix}\psi_1\\\psi_2\\\psi_3\\\psi_4\end{pmatrix}+\begin{pmatrix}m_0c^2&0&0&0\\0&m_0c^2&0&0\\0&0&-m_0c^2&0\\0&0&0&-m_0c^2\end{pmatrix}\begin{pmatrix}\psi_1\\\psi_2\\\psi_3\\\psi_4\end{pmatrix}$$

Or,
$$\begin{pmatrix}i\hbar\frac{\partial\Psi_1}{\partial t}\\i\hbar\frac{\partial\Psi_2}{\partial t}\\i\hbar\frac{\partial\Psi_3}{\partial t}\\i\hbar\frac{\partial\Psi_4}{\partial t}\end{pmatrix}=\begin{pmatrix}m_0c^2&0&c\hat{p}_z&c(\hat{p}_x-i\hat{p}_y)\\0&m_0c^2&c(\hat{p}_x+i\hat{p}_y)&-c\hat{p}_z\\c\hat{p}_z&c(\hat{p}_x-i\hat{p}_y)&-m_0c^2&0\\c(\hat{p}_x+i\hat{p}_y)&-c\hat{p}_z&0&-m_0c^2\end{pmatrix}\begin{pmatrix}\psi_1\\\psi_2\\\psi_3\\\psi_4\end{pmatrix}\qquad [6]$$

Equation [6] is the matrix form of Dirac's relativistic wave equation. From the above matrix we write four simultaneous equations,

$$i\hbar\frac{\partial\Psi_1}{\partial t}\ m_0c^2\psi_1 + 0\psi_2 + c\hat{p}_z\psi_3 + c(\hat{p}_X - i\hat{p}_y)\psi_4 \quad [7]$$

$$i\hbar\frac{\partial\Psi_2}{\partial t} = 0\psi_1 + m_0c^2\psi_2 + c(\hat{p}_X + i\hat{p}_y)\psi_3 - c\hat{p}_z\psi_4 \quad [8]$$

$$i\hbar\frac{\partial\Psi_3}{\partial t} = c\hat{p}_z\psi_1 + c(\hat{p}_X - i\hat{p}_y)\psi_2 - m_0c^2\psi_3 + 0\psi_4 \quad [9]$$

$$i\hbar\frac{\partial\Psi_4}{\partial t} = c(\hat{p}_X + i\hat{p}_y)\psi_1 - c\hat{p}_z\psi_2 + 0\psi_3 - m_0c^2\psi_4 \quad [10]$$

Equations [7], [8], [9] and [10] give Dirac's relativistic wave equation.

14.10 Plain wave solution of Dirac's relativistic wave equation

Let us try plane wave solution of Dirac's relativistic wave equation

$\psi_j = u_j \exp\left[i(\vec{k}.\vec{r} - \omega t\right]$; where j = 1, 2, 3, 4 to the system offollowing 4 equations

$$i\hbar\frac{\partial\Psi_1}{\partial t}\ m_0c^2\psi_1 + 0\psi_2 + c\hat{p}_z\psi_3 + c(\hat{p}_X - i\hat{p}_y)\psi_4 \quad [1]$$

$$i\hbar\frac{\partial\Psi_2}{\partial t} = 0\psi_1 + m_0c^2\psi_2 + c(\hat{p}_X + i\hat{p}_y)\psi_3 - c\hat{p}_z\psi_4 \quad [2]$$

$$i\hbar\frac{\partial\Psi_3}{\partial t} = c\hat{p}_z\psi_1 + c(\hat{p}_X - i\hat{p}_y)\psi_2 - m_0c^2\psi_3 + 0\psi_4 \quad [3]$$

$$i\hbar\frac{\partial\Psi_4}{\partial t} = c(\hat{p}_X + i\hat{p}_y)\psi_1 - c\hat{p}_z\psi_2 + 0\psi_3 - m_0c^2\psi_4 \quad [4]$$

Since ψ_j is plane wave, u_j does not depend on space coordinates and time. As such

$$\frac{\partial\Psi_j}{\partial x} = ik_x\Psi_j,\quad \frac{\partial\Psi_j}{\partial y} = ik_y\Psi_j,\quad \frac{\partial\Psi_j}{\partial z} = ik_z\Psi_j \quad \text{and} \quad \frac{\partial\Psi_j}{\partial t} = -i\omega\Psi_j,$$

From equation [1]

$$i\hbar(-i\omega)\Psi_1 = m_0c^2\psi_1 + 0\psi_2 + c\frac{\hbar}{i}(ik_z)\psi_3 + c\frac{\hbar}{i}(ik_x - iik_y)\psi_4$$

From equation [1] $\hbar\omega u_1 = m_0c^2u_1 + 0u_2 + c\hbar k_z u_3 + c(\hbar k_x - i\hbar k_y)u_4$

From equation [2] $\hbar\omega u_2 = 0u_1 + m_0c^2u_2 + c(\hbar k_x + i\hbar k_y)u_3 - c\hbar k_z u_4$

From equation [3] $\hbar\omega u_3 = c\hbar k_z u_1 + c(\hbar k_x - i\hbar k_y)u_2 - m_0c^2u_3 + 0u_4$

From equation [4] $\hbar\omega u_4 = c(\hbar k_x + i\hbar k_y)u_1 - c\hbar k_z u_2 + 0u_3 - m_0c^2u_4$

Thus we have rearranged the above equations in the following way

$$(\hbar\omega - m_0c^2)u_1 - 0u_2 - c\hbar k_z u_3 - c(\hbar k_x - i\hbar k_y)u_4 = 0 \qquad [5]$$

$$0u_1 + (\hbar\omega - m_0c^2)u_2 - c(\hbar k_x + i\hbar k_y)u_3 + c\hbar k_z u_4 = 0 \qquad [6]$$

$$-c\hbar k_z u_1 - c(\hbar k_x - i\hbar k_y)u_2 + (\hbar\omega + m_0c^2)u_3 - 0u_4 = 0 \qquad [7]$$

$$-c(\hbar k_x + i\hbar k_y)u_1 + c\hbar k_z u_2 - 0u_3 + (\hbar\omega + m_0c^2)u_4 = 0 \qquad [8]$$

Equation [5] to [8] has non-zero solution only if the determinant of the coefficient vanishes. I.e.,

$$\begin{vmatrix} \hbar\omega - m_0c^2 & 0 & -c\hbar k_z & -c(\hbar k_x - i\hbar k_y) \\ 0 & \hbar\omega - m_0c^2 & -c(\hbar k_x + i\hbar k_y) & c\hbar k_z \\ -c\hbar k_z & -c(\hbar k_x - i\hbar k_y) & (\hbar\omega + m_0c^2) & 0 \\ -c(\hbar k_x + i\hbar k_y) & c\hbar k_z & 0 & \hbar\omega + m_0c^2 \end{vmatrix} = 0$$

Simplification gives $E = \hbar\omega = \pm\sqrt{\hbar^2k^2c^2 + m_0{}^2C^4}$ same as that obtained from Klein-Gordon equation. E Can be $\geq +m_0{}^2C^4$ and $\leq -m_0{}^2C^4$ but E cannot be in the range $-m_0c^2$ to $+m_0c^2$. If we use $\hbar\omega = \pm\sqrt{\hbar^2k^2c^2 + m_0{}^2C^4}$ in equation [5], [6], [7], [8] we get a system of 4 equations having 4 unknowns u_1, u_2, u_3, u_4. If we solve the 4 equations, we get 2 sets of values of u_1, u_2, u_3, u_4. Thus we get 2 more column matrices for Ψ which are

$$\Psi = \begin{pmatrix} u_1\, e^{i(\vec{k}.\vec{r} - \omega t)} \\ u_2\, e^{i(\vec{k}.\vec{r} - \omega t)} \\ u_3\, e^{i(\vec{k}.\vec{r} - \omega t)} \\ u_4\, e^{i(\vec{k}.\vec{r} - \omega t)} \end{pmatrix}.$$

These are 2 Dirac spinors for +ive energy.

If we use $\hbar\omega = -\sqrt{\hbar^2k^2c^2 + m_0{}^2C^4}$ in equation [5], [6], [7] and [8], we get a system of 4 equations having 4 unknowns u_1, u_2, u_3, u_4. If we solve the 4 equations, we get 2 sets of values of u_1, u_2, u_3, u_4. Thus we get 2 more column matrices for Ψ which are

$$\Psi = \begin{pmatrix} u_1\, e^{i(\vec{k}.\vec{r}-\omega t)} \\ u_2\, e^{i(\vec{k}.\vec{r}-\omega t)} \\ u_3\, e^{i(\vec{k}.\vec{r}-\omega t)} \\ u_4\, e^{i(\vec{k}.\vec{r}-\omega t)} \end{pmatrix}.$$

These are 2 Dirac spinors for –ive energy.

14.11 Equation of continuity from Dirac's Hamiltonian

If we have a potential $V(\vec{r})$; the Hamiltonian operator for the Dirac equation is

$$H_D = C\,\vec{\alpha}.\vec{P} + m_0 c^2 \beta + V(\vec{r}) \qquad [1]$$

The Hamiltonian operator for the Dirac equation without potential $V(\vec{r})$, then Dirac equation is

$$i\hbar\frac{\partial\Psi}{\partial t} = C\,\vec{\alpha}.\hat{p}\Psi + \beta m_0 c^2\, \Psi = \frac{\hbar}{i} C\,\vec{\alpha}.\vec{\nabla}\Psi + \beta m_0 c^2\, \Psi \qquad [2]$$

Where, $\alpha_x, \alpha_y, \alpha_z, \beta$ are 4 by 4 matrices, Ψ is a column matrix.

$$\Psi = \begin{pmatrix} \Psi_1 \\ \Psi_2 \\ \Psi_3 \\ \Psi_4 \end{pmatrix}$$

The Hermitian conjugate [adjoint] matrix $\Psi^\dagger = (\Psi_1^\times, \Psi_2^\times, \Psi_3^\times, \Psi_4^\times)$; which is one row four coloumn matrix. Therefore,

$$\Psi^\dagger\Psi = (\Psi_1^\times, \Psi_2^\times, \Psi_3^\times, \Psi_4^\times)\begin{pmatrix} \Psi_1 \\ \Psi_2 \\ \Psi_3 \\ \Psi_4 \end{pmatrix} = \Psi_1^\times\Psi_1 + \Psi_2^\times\Psi_2 + \Psi_3^\times\Psi_3 + \Psi_4^\times\Psi_4.$$

From equation [2]

$$\Psi^\dagger(i\hbar\frac{\partial\Psi}{\partial t}) = \Psi^\dagger(\frac{\hbar}{i} C\,\vec{\alpha}.\vec{\nabla}\Psi + \beta m_0 c^2\, \Psi) = \frac{\hbar}{i} C\Psi^\dagger\, \vec{\alpha}.\vec{\nabla}\Psi + m_0 c^2 \Psi^\dagger \beta\Psi \qquad [3]$$

Again from equation [2]

$$\left(i\hbar\frac{\partial\Psi}{\partial t}\right)^\dagger = \left(\frac{\hbar}{i} C\,\vec{\alpha}.\vec{\nabla}\Psi + \beta m_0 c^2\, \Psi\right)^\dagger$$

$$\text{Or,} \quad -i\hbar\frac{\partial\Psi^\dagger}{\partial t} = -\frac{\hbar}{i} C\,(\vec{\nabla}\Psi^\dagger).\vec{\alpha} + m_0 c^2 \Psi^\dagger \beta$$

Here $(cA)^\dagger = c^\times A^\dagger$ and $(AB)^\dagger = B^\dagger A^\dagger$

$\alpha_x^\dagger = \alpha_x, \quad \alpha_y^\dagger = \alpha_y, \quad \alpha_z^\dagger = \alpha_z, \quad \beta^\dagger = \beta$ because $\alpha_x, \alpha_y, \alpha_z, \beta$ are Hermitian.

Or, $\left(-i\hbar\dfrac{\partial\Psi^\dagger}{\partial t}\right)\Psi = \left(-\dfrac{\hbar}{i}C(\vec{\nabla}\Psi^\dagger).\vec{\alpha} + m_0c^2\Psi^\dagger\beta\right)\Psi$

Or, $-i\hbar\dfrac{\partial\Psi^\dagger}{\partial t}\Psi = -\dfrac{\hbar}{i}C(\vec{\nabla}\Psi^\dagger).\vec{\alpha}\Psi + m_0c^2\Psi^\dagger\beta\Psi$ [4]

Subtracting [4] from [3]

$$i\hbar(\Psi^\dagger\frac{\partial\Psi}{\partial t} + \frac{\partial\Psi^\dagger}{\partial t}\Psi) = \frac{\hbar}{i}C\Psi^\dagger\,\vec{\alpha}.\vec{\nabla}\Psi + \frac{\hbar}{i}C(\vec{\nabla}\Psi^\dagger).\vec{\alpha}\Psi$$

Or,

$$i\hbar\frac{\partial}{\partial t}(\Psi^\dagger\Psi) = \frac{\hbar}{i}c[\Psi^\dagger\,\vec{\alpha}.(\vec{\nabla}\Psi) + (\vec{\nabla}\Psi^\dagger).\vec{\alpha}\Psi]$$

$$i\hbar\frac{\partial}{\partial t}(\Psi^\dagger\Psi) = \frac{\hbar}{i}c[\Psi^\dagger\vec{\nabla}.\vec{\alpha}(\Psi) + (\vec{\nabla}\Psi^\dagger).\vec{\alpha}\Psi]$$

Since $\vec{\alpha}$ commutes with $\vec{\nabla}$ and $\vec{\alpha}$ commutes with $\hat{\vec{p}}$

$$\frac{-\hbar}{i}\frac{\partial}{\partial t}(\Psi^\dagger\Psi) = \frac{\hbar}{i}c\vec{\nabla}.[(\Psi^\dagger)(\vec{\alpha}\Psi)]$$

$\vec{\nabla}.(c\Psi^\dagger\vec{\alpha}\Psi) = -\dfrac{\partial}{\partial t}(\Psi^\dagger\Psi)$ [5]

$\vec{\nabla}.\vec{J} + \dfrac{\partial\rho}{\partial t} = 0$ [6]

Equation [6] is known as equation of continuity . We put $\rho = \Psi^\dagger\Psi$ is the probability density and $\vec{S}\left[\vec{J}\right] = c\Psi^\dagger\vec{\alpha}\Psi$ is the probability current density.

Where $\vec{J} = \dfrac{\hbar}{2im}\left[\Psi^\times\left(\vec{\nabla}\Psi\right) - \left(\vec{\nabla}\Psi^\times\right)\Psi\right]$; is the probability current density. [7]

And $\rho = \left[\dfrac{\hbar}{2imc^2}\left\{\dfrac{\partial\Psi^\times}{\partial t}\Psi - \Psi^\times\dfrac{\partial\Psi}{\partial t}\right\}\right]$

Or, $\rho = \left[-\dfrac{i\hbar}{2mc^2}\left\{\Psi^\times\dfrac{\partial\Psi}{\partial t} - \dfrac{\partial\Psi^\times}{\partial t}\Psi\right\}\right]$ [8]

; is the probability density.

Equation [6] is known as equation of continuity & is invariant with respect to Lorentz transformation.

14.12 Covariance of Dirac equation

Consider two observers in Lorentz frames are Σ & Σ', the requirement is that the scalar quantities should remain the same in two systems while the four vectors must transform like the space-time coordinates. Although the wave functions for the two observers need not be the same, the recipe [direction for making something] for writing down the Dirac equation must be the same for both observers. The Dirac equations in Σ & Σ' frames are to be[10]

$$(\gamma_\mu \partial_\mu + k)\Psi = 0 \qquad [1A]$$

$$(\gamma_\mu \frac{\partial}{\partial x_\mu} + k)\Psi(x_\mu) = 0 \qquad [1B]$$

$$\& \quad (\gamma_\nu \partial'_\nu + k)\Psi' = 0 \qquad [2A]$$

$$\Rightarrow \quad (\gamma_\nu \frac{\partial}{\partial x'_\nu} + k)\Psi'(x'_\nu) = 0 \qquad [2B]$$

Notice that $\gamma[(4\times 4)\, matrices]$ are the same in both frames [they define the recipe & ought not to be changed]. The coordinates x & x' are related through

$$x'_\nu = a^\mu_\nu x_\mu = a_{\mu\nu} x_\mu \qquad [3]$$

a^μ_ν *or* $a_{\mu\nu} x_\mu$ = *Operator of Lorentz transformatiom*

and x'_ν *is the prime coordinate* x_μ *is the unprime coordinate.*

& the inverse transformation

$$x_\mu = a^\nu_\mu x'_\nu \qquad [4]$$

With restrictions

$$a^\mu_\lambda a^\nu_\mu = \delta^\nu_\lambda \qquad \left[a_{\mu\lambda} a_{\nu\mu} = \delta_{\nu\lambda}\right] \qquad [5]$$

Thus we have

$$\frac{\partial \Psi}{\partial x_\mu} = \frac{\partial \Psi}{\partial x'_\nu}\frac{\partial x'_\nu}{\partial x_\mu} = \frac{\partial \Psi}{\partial x'_\nu} a^\mu_\nu \qquad [6]$$

In analogy, the transformation of four vectors, we expect that $\Psi(x_\mu)$ and $\Psi'(x'_\nu)$ be related by a linear transformation

$$\Psi'(x'_\nu) = S\Psi(x_\mu) \qquad [7]$$

Where S is a (4 by 4) matrix which depends only the nature of Lorentz transformation an the inverse relation therefore

$$\Psi'(x'_\nu) = S\Psi(x_\mu)$$

$$\Psi(x_\mu) = S^{-1}\Psi'(x'_\nu) \qquad [8]$$

Hence equation [1] becomes

$$(\gamma_\mu \frac{\partial}{\partial x_\mu} + k)\Psi(x_\mu) = 0$$

$$(\gamma_\mu a^\mu_\nu \frac{\partial}{\partial x'_\nu} + k)S^{-1}\Psi'(x'_\nu) = 0$$

Multiply on the left by S, we have

$$(a_\nu^\mu S\gamma_\mu S^{-1}\frac{\partial}{\partial x'_\nu}+SkS^{-1})\Psi'(x'_\nu)=0 \quad [9]$$

Equation [9] is identical with [2b] provided

$$\gamma_\nu = a_\nu^\mu S\gamma_\mu S^{-1} \quad [10]$$

Multiplying both sides by a_ρ^ν

$$a_\rho^\nu\gamma_\nu = a_\rho^\nu a_\nu^\mu S\gamma_\mu S^{-1}$$

$$= \delta_\rho^\mu S\gamma_\mu S^{-1}[\text{when } \mu=\rho, \text{then } \delta_\rho^\mu\ \delta_{\mu\rho} \text{ or } = 1]$$

$$a_\rho^\nu\gamma_\nu = S\gamma_\mu S^{-1}$$

But $\gamma'_\nu = a_\rho^\nu\gamma_\nu$ [11]

$$\therefore \gamma'_\nu = S\gamma_\mu S^{-1} \quad [12]$$

There [9] becomes $(a_{\mu\nu}\gamma'_\nu\frac{\partial}{\partial x'_\nu}+k)\Psi'(x'_\nu)=0$

$$(a_{\mu\nu}a_{\nu\rho}\gamma_\nu\frac{\partial}{\partial x'_\nu}+k)\Psi'(x'_\nu)=0$$

$$(\delta_{\mu\rho}\gamma_\nu\frac{\partial}{\partial x'_\nu}+k)\Psi'(x'_\nu)=0$$

$$(\gamma_\nu\frac{\partial}{\partial x'_\nu}+k)\Psi'(x'_\nu)=0[\text{when } \mu=\rho, \text{then } \delta_{\mu\rho}=1]$$

The above equation is the same form in $\sum'$ frame. Hence the Dirac equation is covariant under Lorentz transformation.

14.13 Concept of negative energy state solution of Dirac equation

Can you neglect the negative energy state solution of Dirac Equation? No, in Relativistic Quantum Mechanics we can't neglect the negative energy state solution of Dirac equation. The negative energy solutions refer to the motion of a new kind of particles having mass of an electron and opposite charge. Such particles are observed experimentally and are called positrons. It has a great importance in Relativistic Quantum Mechanics; such as pair production & pair annihilations.

14.14 Relativistic Effect in quantum mechanics

When does Relativistic Effect come in Quantum mechanics? When the velocity of the particle is 10% of the velocity of light then the relativistic effect comes to describe the nature of particle. Since an electron moves with 1/3 velocity of light, so the properties of the electron cannot be describe by the Schrödinger equation. Because in Schrödinger equation, the space &time coordinates are not on the same footing. In 1928, Dirac constructs a relativistic equation which was suitable for a description of the properties of electrons & other spin 1/2 particles. Hence the Dirac equation contains first order time derivative & first order space derivative. Since the special theory of relativity requires that time & space coordinates are on the same footing.

14.15 Arguments of Dirac theory

- The non-relativistic Schrödinger equation should be a relativistic form of an equation, because non-relativistic Schrödinger equation is not a basic law of nature.
- The correct relativistic form of an equation should have a space & time symmetry.
- The correct relativistic form of an equation should have Lorentz transformation invariance.
- The correct relativistic form of an equation should have a first degree equation in time & also first order in space derivative.

14.16 Particles are obeying Klein-Gordon and Dirac equations

What types of particles do the Klein-Gordon equation obey ?

- The spin of full integral type particles obeys the Klein-Gordon equation. For example: Π mesons, K-mesons & η-mesons which are strongly interacting.

What types of particle does the Dirac equation obey?

- The spin of half integral type particles I.e., Fermions obey the Dirac equation. For example: electron, positron, antineutrino etc.

14.17 Failure of non-relativistic theory

- The non-relativistic theory fails to explain the spin of an elementary system, such as the electron
- This theory cannot give a satisfactory amount of high energy phenomena such as pair-production
- The non-relativistic quantum theory is not invariant under Lorentz transformation
- The non-relativistic Schrödinger equation is not symmetry in space & time & cannot describe the basic law of nature

14.18 Difficulties of Schrödinger, Klein-Gordon and Dirac equations

❖ **Schrödinger equation**

We know, the time dependent Schrödinger equation,

$$[-\frac{\hbar^2}{2m}\nabla^2+V(r)]\Psi=i\hbar\frac{\partial\Psi}{\partial t}$$

The above equation is valid in low energy system.

- Because it is not Lorentz invariant & it cannot describe the basic law of nature
- Also the above equation is first order in time & second order in space
- So it is not symmetrical in space and time
- To put Schrodinger equation on a relativisting footing for making first order in time & second order in space.

Again the probability density $\rho = \frac{E}{mc^2}[\Psi^{\times}\Psi]$; E is of the order of mc^2; so we find that the probability density $\rho = \Psi^{\times}\Psi$ as in non- relativistic Quantum Mechanics. Since, E can be positive and negative I.e., the sign of ρ is the sign of the energy. As the energy of the particle may be positive or negative, the probability density will be positive or negative. But the probability density should always be positive. Schrödinger tried to solve the situation by multiplying by the charge of the particle (e) & thus defining the probability density & the probability current density[9]

$$\rho = \frac{i\hbar e}{2mc^2}\left(\Psi^{\times}\Psi\frac{\partial\Psi}{\partial t} - \frac{\partial\Psi^{\times}}{\partial t}\Psi\right) \qquad [3]$$

$$\vec{J} = \frac{\hbar e}{2im}\left(\Psi^{\times}(\vec{\nabla}\Psi) - (\vec{\nabla}\Psi^{\times})\Psi\right)$$

Since charge can be positive or negative, there is no difficulty in interpretation. To overcome all the difficulties & drawbacks those were faced by Schrödinger & Klein–Gordon.

❖ **Klein-Gordon[K-G] equation:**

We know Klein-Gordon [K-G] equation for free particle.

$$[\nabla^2 - \frac{1}{c^2}\frac{\partial^2}{\partial t^2} - \frac{m_0^2c^2}{\hbar^2}]\Psi = 0 \Rightarrow [\quad - \frac{m_0^2c^2}{\hbar^2}]\Psi = 0 \qquad [1]$$

- The above *Klein-Gordon [K-G] equation* is valid in high energy system. i.e. the equation is valid for certain classes of particles such as the π and K mesons, which are strongly interacting
- Klein-Gordon equation is the second-order differential equation.
- The Klein-Gordon equation, it turns out, is the correct one for spinless particles & not for electrons

❖ **Dirac equation:**

Dirac proposed a new type of equation which will be first order in space & time coordinates & hence are the same footing [symmetric] expressed by

$$(\gamma_\mu \partial^\mu + k)\Psi = 0 \qquad [1]$$

- the above equation also encounters the problem of negative energy which is successfully overcome by the introduction of hole theory.
- In Dirac theory, there is only one drawback, its hole interpretation cannot be applicable for Bosons which are not obeying the Pauli exclusion principle.

14.19 Dirac's hole theory

In 1930, Dirac attempted to remedy this difficulty by proposing the so called hole theory. Observing that spin of 1/2 integral particles obey the Pauli-exclusion principle. He suggested that in normal ground state of nature all possible negative energy levels are already occupied. Hence whatever interactions they experience, positive energy electrons cannot fall down into the negative energy states. The Pauli-principle forbids two identical particles in the same state. Thus stability of positive energy electrons can be explained. From Dirac relativistic equation, there are two regions of the continuous energy spectrum of a free electron with energy $E = \pm\sqrt{m_0^2 c^4 + p^2 c^2}$. The minimum energy will be when p = 0. When these direct electrons are at rest, the positive & the negative energy state is separated by a gap $\approx 2m_0 c^2$.

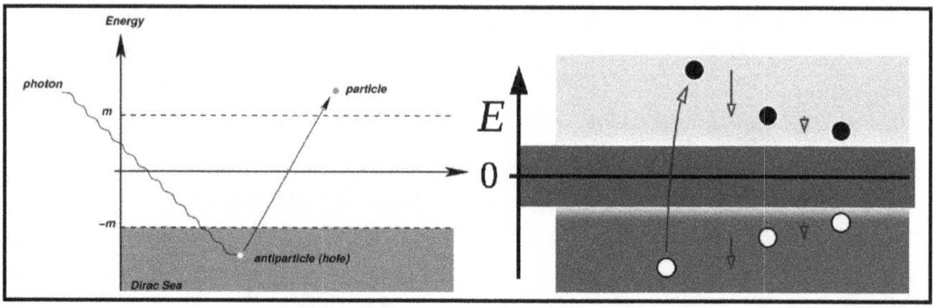

Fig.[14.19(a)]: Shows Dirac's hole theory

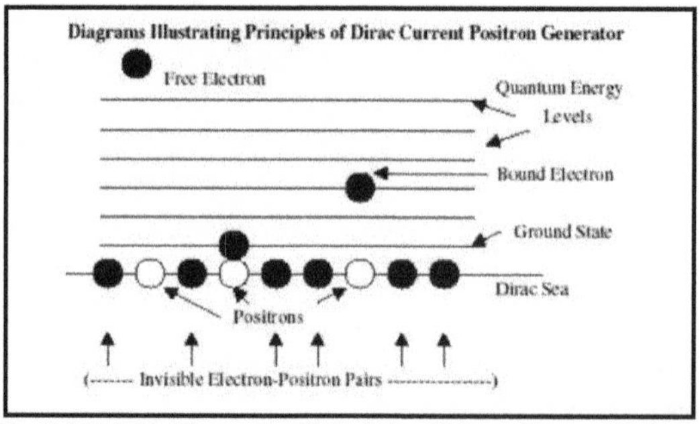

Fig.[14.19(b)]: Illustrating Principle of Dirac Current Positron Generator

It will become possible to 'raise' one negative energy electron from the vacuum into a positive energy state when we invest an amount of energy $2m_0c^2$ of the order of MeV. At the same time a "hole" will be produced in the sea of negative energy electrons[13,14].

It can be proven that such a 'hole' behaves, for all purposes exactly as a particle with the same mass as that of an electron but having an opposite sign of electric charge. Hence it is identified with "positron" discovered by Anderson in cosmic ray researches. Thus in this case "electron-positron" pair is created.

Positron created in this way will, however, disappear quickly. The fact is that in this formulation the existence of a positron is equivalent to the existence of a 'hole' in the negative energy sea. So an ordinary electron will very soon fall down into the empty negative energy state. Both the electron & the hole will disappear & the energy equivalent will be regained in the form of electromagnetic radiation. Thus processes of pair creation & pair annihilation have been indeed observed. The 'hole' theory has been reformulated as to describe electron & positron in a completely symmetric manner.

14.20 Comparative study between Fermions and Bosons

- Fermions are quantum mechanical particle. They are obeying the Pauli Exclusion Principle. The spin of Fermions are half integral particles [I.e., $\frac{1}{2}, \frac{3}{2}, \dots \dots \dots .. etc.$]. Fermions obey the Dirac equation or Fermi- Dirac statistics. For example: electron, positron, antineutrino etc. They have non- symmetrical wave functions.

- Bosons are quantum mechanical particle. They are not obeying the Pauli Exclusion Principle. The spin of Bosons are full integral particles [I.e., $0, 1, 3, \dots \dots \dots .. etc.$]. Bosons obey the Bose- Einstein statistics. For example: Photon, $\prod$ mesons, K-mesons & η-mesons etc. which are strongly interacting. They have symmetrical wave functions.

Exercise

1. Write down the Klein-Gordon [KG] or Klein-Fock equation for free particle.
2. Prove that ($\quad - k^2)\psi = 0$;Where D'Alembertian operator $\quad = \nabla^2 - \frac{1}{c^2}\frac{\partial^2}{\partial t^2}$ and put $k = \frac{m_0}{\hbar}$
3. What are the drawbacks of Schrodinger equation and merit of Klein- Gordon equation?
4. Discuss and explain the development of Dirac equation. What is the drawback in Dirac theory?
5. In Dirac theory, there is only one drawback, its hole interpretation cannot be applicable for Bosons which are not obeying the Pauli exclusion principle, explain it.
6. Show that the operator $\vec{S} = \frac{\hbar}{2}\vec{\sigma}'$ as the spin angular momentum of the electron for Dirac equation.
7. Establish the comparative study of Schrodinger, Klein-Gordon and Dirac equations.
8. Write the Solution of Dirac equation for free particle.
9. Find the Dirac's relativistic wave equation for a free particle.
10. What are the significances that you have observed in Dirac's relativistic wave equation?
11. Find out the Dirac matrix $\alpha_x, \alpha_y, \alpha_z$ *and* β
12. Obtain the Dirac's relativistic equation in matrix form.
13. Write the Plain wave solution of Dirac's relativistic wave equation.
14. Find the equation of continuity from Dirac's Hamiltonian?
15. Prove that $\overline{\nabla}.\bar{J} + \frac{\partial \rho}{\partial t} = 0$; ρ is the probability density and $\bar{J}$ is probability current density.
16. Show that the equation of continuity is invariant with respect to Lorentz transformation.
17. Show that the Dirac equation is covariant under Lorentz transformation.
18. Why cannot we neglect negative energy state solution of Dirac equation?
19. Explain the Relativistic effect in quantum mechanics.
20. Write some arguments of Dirac theory.
21. What type of Particles is obeying Klein-Gordon and Dirac equations?
22. What are the failures of non-relativistic theory?
23. Write down difficulties of Schrödinger, Klein-Gordon and Dirac equations.
24. Explain Dirac's hole theory.
25. Write a Comparative study between Fermions and Bosons.
26. Give a simple derivation of Klein- Gordon [KG] equation. Discuss the difficulties historically associated with the interaction of this equation and how they have been overcome.
27. Using relativistic Dirac equation to show that electron is endowed with spin $\frac{1}{2}$.

28. Obtain the relation between spin and magnetic moment of an electron on the basis of Dirac electron theory.

Chapter 15 The Variational method: An approximation method

15.1 Significance of an approximation method and Importance of variation method

When the system contains only a single electron, the exact solution of the Schrödinger equation is possible as in case of Hydrogen like atom. However, if several electrons are involved, the exact solution is not possible, approximation methods are used.

The harmonic oscillator is one of the few systems in quantum mechanics which allows us to find exact solutions both for the Eigen values and Eigen states. For essentially all other systems it is necessary to make use of approximation methods in order to calculate these quantities from a given Hamiltonian. There are basically two different classes of such methods. There are a number of such methods like variation method & perturbation method. The first [variation method] is especially suited for investigations of the ground state and the lowest excited states. In the other scheme one assumes that the real system is approximately equal to another system which allows exact solutions. Then using perturbation theory one can systematically calculate corrections to this lowest and often crude approximation. This method can be applied to any state and makes it possible to calculate any quantity as accurately as is needed.

❖ Importance of variation method

The variation method is one of the most powerful methods of approximation & it is invariably used when all other methods fail. In many cases, time independent perturbation theory can be applied as there is no unperturbed Hamiltonian Ho whose solutions can be used to calculate expectation values of the perturbation Hamiltonian In other words, "*When the division of the total Hamiltonian into an unperturbed part & a perturbation is impossible, it is Convenient to use the variation method of approximation.* We can use more the variation method for evaluating the energies of the ground state of hydrogen atom & Helium atom etc. & the first few discrete states of a bound system.

15.2 Principle of variational method

We know, the expectation value of the Hamiltonian in any state is ψ

$$\langle H\rangle = \frac{(\psi,H\psi)}{(\psi,\psi)} \quad [1]$$

Where the scalar product $(\psi, H\psi) = \int \psi^{\times} H\psi \, d\tau$ [2]

From [1] the quantity $\langle H\rangle$ is the upper limit to the energy E_0 of the ground state of the system. Instead of wave fraction ψ, we use the exact wave function ψ_0 of the ground state. The expectation value $\langle H\rangle$ then equals H_0. So the ground state energy

$$(\psi_0, H_0\psi_0) = (\psi_0, E_0\psi_0) = E_0(\psi_0, \psi_0)$$

$$E_0 = \frac{(\psi_0,H_0\psi_0)}{(\psi_0,\psi_0)} \quad [2]$$

When ψ is not equal to ψ_0 we may expand ψ in terms of the complete set of orthogonal Eigen functions $\psi_0, \psi_1, \psi_2 \dots\dots\dots\dots\dots\dots\dots\dots\dots\dots\dots \psi_n$ of H belonging to the Eigen values $E_0, E_1, E_2 \dots\dots\dots\dots\dots\dots\dots\dots\dots\dots\dots\dots\dots\dots\dots E_n$ respectively. Thus

$$\psi = \sum_n a_n \psi_n \qquad [3]$$

If we normalize ψ to unity, then $\sum_n a_n^\times a_n = 1 \quad \therefore \left[\int a_n^\times a_n d\tau = (a_n, a_n) = 1\right]$

From equation [1] and [3]

$$\langle H \rangle = \frac{(\psi, H\psi)}{(\psi,\psi)} = \frac{(\sum_n a_n\psi_n, H\sum_m a_m\psi_m)}{(\sum_n a_n\psi_n, \sum_m a_m\psi_m)}$$

$$= \frac{(\sum_n a_n^\times \sum_m a_m(\psi_n, H\psi_m))}{(\sum_n a_n^\times \sum_m a_m(\psi_n,\psi_m))} = \frac{(\sum_n a_n^\times \sum_m a_m(\psi_n, E_m\psi_m))}{(\sum_n a_n^\times \sum_m a_m(\psi_n,\psi_m))}$$

$$= \frac{(\sum_n a_n^\times \sum_m a_m E_m(\psi_n,\psi_m))}{(\sum_n a_n^\times \sum_m a_m(\psi_n,\psi_m))} = \frac{(\sum_n a_n^\times \sum_m a_m E_m \delta_{nm})}{(\sum_n a_n^\times \sum_m a_m \delta_{nm})}$$

$$= \frac{(\sum_n a_n^\times \sum_m a_m E_m)}{(\sum_n a_n^\times \sum_m a_m \delta_{nm})} = \frac{(\sum_n a_n^\times a_n E_n)}{(\sum_n a_n^\times a_n)} \quad [\text{When } m = n \therefore \delta_{nm} = 1]$$

$$= (\sum_n a_n^\times a_n E_n) \qquad [4]$$

Or, $\langle H \rangle - E_0 = (\sum_n a_n^\times a_n E_n) - E_0(\sum_n a_n^\times a_n) \qquad [\because \sum_n a_n^\times a_n = 1]$

Or, $\langle H \rangle - E_0 = \sum_n a_n^\times a_n (E_n - E_0)$

But $(E_n - E_0) \geq 0 \; and \sum_n a_n^\times a_n \geq 0 \,; Hence \; \langle H \rangle - E_0 \geq 0$

Finally $\langle H \rangle - E_0 \geq$ Or, $\langle H \rangle \geq E_0$ [5]

Therefore, *the expectation value of the Hamiltonian in any state ψ gives the upper bound to the ground state energy*. This result is the basis of the variation method and is known as Variation Principle[5].

15.3 Ground state energy of hydrogen atom by using variation method

The trial wave function $\psi = exp(-\alpha r/r_0)$ [1]

Let the normalized trial wave function $\psi = Aexp(-\alpha r/r_0)$ [2]

For the ground state of Hydrogen atom, the potential energy will be $V = -\frac{e^2}{r}$

This is spherically Symmetric. Hence the Hamiltonian operator $\hat{H}$ will be

$$\hat{H} = -\frac{\hbar^2}{2\mu}\nabla^2 - \frac{e^2}{r}$$

In spherical polar coordinate, the Laplacian operator ∇^2 can be expressed as

$$\nabla^2 = \frac{1}{r^2}\frac{\partial}{\partial r}(r^2\frac{\partial}{\partial r}) + \frac{1}{r^2 \sin\theta}\frac{\partial}{\partial\theta}(\sin\theta\frac{\partial}{\partial\theta}) + \frac{1}{r^2\sin^2\theta}\frac{\partial^2}{\partial\phi^2}$$

As the energy occurs in the radial part of the Schrödinger equation so we consider the radial part

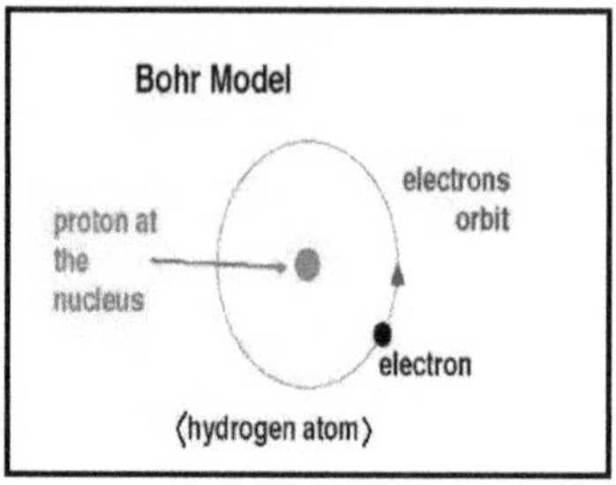

Fig.[15.3]: Two body system: Hydrogen atom

of the Laplacian operator ∇^2

$$\nabla^2 = \frac{1}{r^2}\frac{\partial}{\partial r}(\mathrm{r}^2\frac{\partial}{\partial r}) = \frac{1}{r^2}(2r\frac{\partial}{\partial r} + r^2\frac{\partial^2}{\partial r^2}) = \frac{2}{r}\frac{\partial}{\partial r} + \frac{\partial^2}{\partial r^2}$$

Therefore, $$\hat{H} = -\frac{\hbar^2}{2\mu}\nabla^2 - \frac{e^2}{r} = -\frac{\hbar^2}{2\mu}\left(\frac{2}{r}\frac{\partial}{\partial r} + \frac{\partial^2}{\partial r^2}\right) - \frac{e^2}{r}$$

$$\hat{H}\psi = \left[-\frac{\hbar^2}{2\mu}\nabla^2 - \frac{e^2}{r}\right]\psi = -\frac{\hbar^2}{2\mu}\left(\frac{2}{r}\frac{\partial}{\partial r} + \frac{\partial^2}{\partial r^2}\right)\psi - \frac{e^2}{r}\psi$$

$$\hat{H}\psi = \left[-\frac{\hbar^2}{2\mu}\left(\frac{2}{r}\frac{\partial}{\partial r} + \frac{\partial^2}{\partial r^2}\right) - \frac{e^2}{r}\right]A\exp(-\alpha r/r_0)$$

$$H\psi = -\frac{\hbar^2}{2\mu}\left(-\frac{2\alpha}{rr_0}\right)A\exp(-\alpha r/r_0) + \left(\frac{\alpha}{r_0}\right)^2 A\exp(-\alpha r/r_0) - \frac{e^2}{r}A\exp(-\alpha r/r_0)$$

Also, $$\int \Psi^\times \Psi d\tau = \int_0^\infty\int_0^\pi\int_0^{2\pi} A^\times \exp(-\alpha r/r_0) A\exp(-\alpha r/r_0) r^2 \sin\theta\, d\theta\, d\phi\, dr$$

$$= |A|^2 (4\pi)\int_0^\infty r^2 \exp(-2\alpha r/r_0)\, dr$$

But we know the standard integral $\int_0^\infty x^n e^{-\alpha x} dx = \frac{n!}{\alpha^{n+1}}$

$$\therefore \int \psi^\times \psi d\tau = |A|^2 (4\pi)\frac{2!}{\left(\frac{2\alpha}{r_0}\right)^{2+1}} = |A|^2 (4\pi)\frac{2}{\frac{8\alpha^3}{r_0^3}} = |A|^2 4\pi 2\frac{r_0^3}{8\alpha^3} = |A|^2 \frac{\pi r_0^3}{\alpha^3}$$

$$\int \psi^\times \psi d\tau = |A|^2 \frac{\pi r_0^3}{\alpha^3} \quad Or,\ 1 = |A|^2 \frac{\pi r_0^3}{\alpha^3} \quad Or,\ |A|^2 \frac{\pi r_0^3}{\alpha^3} = 1 \ \therefore |A| = \left(\frac{\alpha^3}{\pi r_0^3}\right)^{\frac{1}{2}}$$

Hence the normalized wave function

$$\Psi = A\exp(-\alpha r / r_0) == \left(\frac{\alpha^3}{\pi r_0^3}\right)^{\frac{1}{2}} \exp(-\alpha r / r_0)$$

According to variation method

$$E = \frac{\int \psi^\times \hat{H}\psi d\tau}{\int \psi^\times \psi d\tau} = \frac{1}{\int \psi^\times \psi d\tau}\int \psi^\times \hat{H}\psi d\tau$$

$$= \frac{1}{|A|^2 \frac{\pi r_0^3}{\alpha^3}}[\int_0^\infty\int_0^\pi\int_0^{2\pi} A^\times \exp(-\alpha r / r_0)\left(-\frac{\hbar^2}{2\mu}\right)\left(-\frac{2\alpha}{r r_0}\right) A\exp(-\alpha r / r_0)$$

$$+\left(-\frac{\hbar^2}{2\mu}\right)\left(\frac{\alpha}{r_0}\right)^2 A^\times \exp(-\alpha r / r_0) A\exp(-\alpha r / r_0)$$

$$-\frac{e^2}{r} A^\times \exp(-\alpha r / r_0) A\exp(-\alpha r / r_0)] \, r^2 \sin\theta \, d\theta \, d\phi \, dr$$

$$= \frac{|A|^2}{|A|^2 \frac{\pi r_0^3}{\alpha^3}} 4\pi \int_0^\infty [\left(-\frac{\hbar^2}{2\mu}\right)\left(-\frac{2\alpha}{r r_0}\right)\exp(-2\alpha r / r_0) r^2 dr$$

$$-\left(\frac{\hbar^2}{2\mu}\right)\left(\frac{\alpha}{r_0}\right)^2 \exp(-2\alpha r / r_0) r^2 dr - \frac{e^2}{r}\exp(-2\alpha r / r_0) r^2 dr]$$

$$= \frac{4\alpha^3}{r_0^3}\left[\frac{\hbar^2}{2\mu}\frac{2\alpha}{r_0}\int_0^\infty \mathrm{r}\exp(-2\alpha r / r_0) dr - \left(\frac{\hbar^2}{2\mu}\right)\left(\frac{\alpha}{r_0}\right)^2 \int_0^r \mathrm{r}^2 \exp(-2\alpha r / r_0) dr - e^2 \int_0^\infty r\exp(-2\alpha r / r_0)\right]$$

$$= \frac{4\alpha^3}{r_0^3}\left[\frac{\hbar^2}{2\mu}\frac{2\alpha}{r_0}\frac{1}{\left(\frac{2\alpha}{r_0}\right)^2} - \left(\frac{\alpha}{r_0}\right)^2 \frac{\hbar^2}{2\mu}\frac{2!}{\left(\frac{2\alpha}{r_0}\right)^3} - e^2 \frac{1}{\left(\frac{2\alpha}{r_0}\right)^2}\right]\left[\because \int_0^\infty x^n e^{-\alpha x} dx = \frac{n!}{\alpha^{n+1}}\right]$$

$$= \frac{4\alpha^3}{r_0^3}\left[\frac{\hbar^2}{2\mu}\frac{2\alpha}{r_0}\left(\frac{r_0}{2\alpha}\right)^2 - \left(\frac{\alpha}{r_0}\right)^2 \frac{2\hbar^2}{2\mu}\left(\frac{r_0}{2\alpha}\right)^3 - e^2\left(\frac{r_0}{2\alpha}\right)^2\right]$$

$$= \frac{4\alpha^3}{r_0^3}\left[\frac{\hbar^2}{2\mu}\left(\frac{r_0}{2\alpha} - \frac{r_0}{4\alpha}\right) - e^2 \frac{r_0^2}{4\alpha^2}\right] = \frac{4\alpha^3}{r_0^3}\left[\frac{\hbar^2}{2\mu}\frac{r_0}{4\alpha} - e^2 \frac{r_0^2}{4\alpha^2}\right] = \frac{\hbar^2}{2\mu}\frac{\alpha^2}{r_0^2} - e^2 \frac{\alpha}{r_0}$$

We shall choose α in such a manner that to give the minimum energy.

$$\frac{dE}{d\alpha}=0\ or,\ \frac{\hbar^2}{2\mu}\frac{2\alpha}{r_0^2}-\frac{e^2}{r_0}=0\quad or,\frac{\hbar^2\alpha}{\mu r_0^2}=\frac{e^2}{r_0}\qquad or,\alpha=\frac{\mu e^2 r_0}{\hbar^2}$$

Where, $r_0=\frac{\hbar^2}{\mu e^2}$ is the Bohr radius. Therefore, $\alpha=\frac{\mu e^2}{\hbar^2}\frac{\hbar^2}{\mu e^2}=1$

$$E_{minimum}=\frac{\hbar^2}{2\mu}\frac{\alpha^2}{r_0^2}-e^2\frac{\alpha}{r_0}=\frac{\hbar^2}{2\mu}\frac{1}{r_0^2}-\frac{e^2}{r_0}[\because \alpha=1]=\frac{\hbar^2}{2\mu}\frac{1}{r_0}\frac{\mu e^2}{\hbar^2}-\frac{e^2}{r_0}=\frac{e^2}{2r_0}-\frac{e^2}{r_0}=-\frac{e^2}{2r_0}$$

$$\boxed{E_{minimum}=-\frac{e^2}{2r_0}=-\frac{e^2}{2}\frac{me^2}{\hbar^2}=-\frac{me^4}{2\hbar^2}}$$

And the ground state wave function

$$\Psi=A\exp(-\alpha r/r_0)=\left(\frac{\alpha^3}{\pi r_0^3}\right)^{\frac{1}{2}}\exp(-\alpha r/r_0)\ Or,\Psi_0=\frac{1}{\sqrt{\pi r_0^3}}\exp(-r/r_0)[\because \alpha=1]$$

15.4 Many-electron atom

The observed spectrum of the hydrogen atom is not much different from what follows from the solution of the Schrödinger equation in a pure Coulomb potential. Relativistic corrections are small since they are given by powers of the fine structure constant. And effects due to the spin of the electron are also small for the same reason.

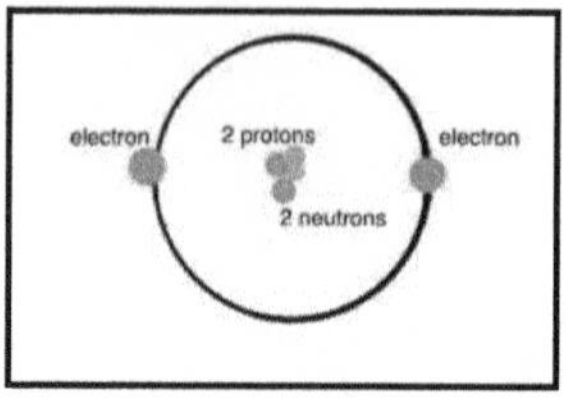

Fig[15.4]: Shows many electron atom (*In atoms with more than one electron)*

In atoms with more than one electron, a new perturbation enters. This is the Coulomb interaction between the atoms. For the lightest atoms this is not small since it is of the same order as the interaction with the central nucleus. Standard perturbation theory will therefore not be very useful for these atoms. Also for heavier atoms the mutual interaction between the atoms plays a central role. But when the number of electrons is suffiently large, we will see that it is possible to approximate all these non-central interactions with a new, effective potential which will have central symmetry. Even if the actual values of atomic of the energy levels must now be found by numerical methods, the symmetry of the system allows for a straight forward classification of the different energy levels which will occur.

Here the spins of the electrons play a decisive role since the Pauli principle contains the full wave function to be completely anti-symmetric in all the variables describing the electrons. The important role of the electron spins we already see in the simplest many-electron atom, i.e. the helium atom with two electrons. Historically, this was one of the first important challenges of the new quantum mechanics which replaced the original, semi-classical ideas proposed by Bohr.

15.5 Ground state energy of Helium atom

The Helium atom consists of a nucleus of charge $+Ze$ (where $Z = 2$). Therefore, the potential energy of the system

$$V = V_1 + V_2 + V_3 = -\frac{Ze^2}{r_1} - \frac{Ze^2}{r_2} + \frac{e^2}{r_{12}} \qquad [1]$$

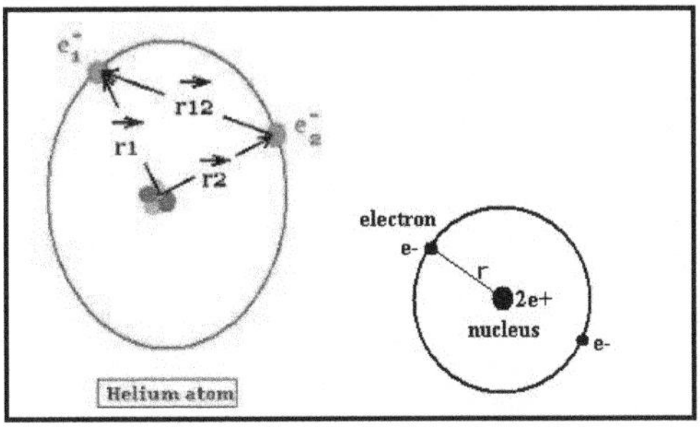

$$V_1 = -\frac{Z\,e^2}{r_1}$$

$$V_2 = -\frac{Z\,e^2}{r_2}$$

$$V_3 = -\frac{e^2}{r_{12}}$$

$Z' =$ *effective nuclear charge*
or = effective atomic number

Fig.[15.4a]: Shows many electrons atom: Helium atom

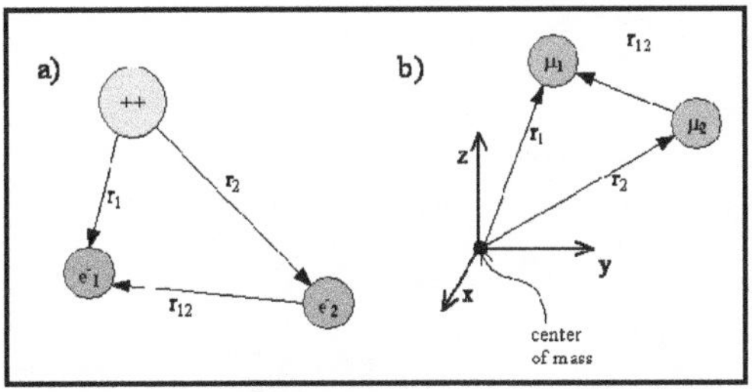

Fig.[15.4b]: Shows many electrons atom: Helium atom in coordinate system

$$Given\ \psi(r_1, r_2) = \left(\frac{Z^3}{\pi a_0}\right) e^{-\frac{zr_1}{a_0}} e^{-\frac{zr_2}{a_0}} \qquad [2]$$

Writing the constant z as the variable parameter z′; where z′ is the effective atomic number[17,5]. [The calculation of energy can be greatly increased in accuracy by considering the quantity z as a variable parameter z′]. Therefore, the ground state energy of Helium atom

$$E = \int \psi^{\times} H \psi d\tau \qquad [3]$$

Here the Hamiltonian operator of Helium atom

$$H = -\frac{\hbar^2}{2m}\left(\nabla_1^2 + \nabla_2^2\right) + V = -\frac{\hbar^2}{2m}\left(\nabla_1^2 + \nabla_2^2\right) - \frac{Z\,e^2}{r_1} - \frac{Z\,e^2}{r_2} + \frac{e^2}{r_{12}}$$

Therefore equation [3] becomes

$$E=\iint \Psi^{\times}\left[-\frac{\hbar^2}{2m}\left(\nabla_1^2+\nabla_2^2\right)-\frac{Ze^2}{r_1}-\frac{Ze^2}{r_2}+\frac{e^2}{r_{12}}\right]\Psi\, d\tau_1 d\tau_2 \qquad [4]$$

From Schrödinger equation

$$\nabla^2\Psi+\frac{2m}{\hbar^2}(E-V)\Psi=0$$

$$or,\quad -\frac{\hbar^2}{2m}\nabla_1^2\Psi+V\Psi=E\Psi$$

$$or,\quad -\frac{\hbar^2}{2m}\nabla_1^2\Psi-\frac{z'e^2}{r_1}=z'^2E_H\Psi$$

Writing the constant z as a variable parameter z′ where z′ is the effective atomic number.

$$\left[where\ E_H=-\frac{e^2}{2r_0}=-\frac{e^2}{2}.\frac{me^2}{\hbar^2}=-\frac{me^4}{2\hbar^2}\right]$$

$$\therefore\ -\frac{\hbar^2}{2m}\nabla_1^2\Psi-\frac{z'e^2}{r_1}\Psi=\frac{z'^2me^4}{2\hbar^2}\Psi$$

$$or,\quad -\frac{\hbar^2}{2m}\nabla_1^2\Psi-\frac{z'e^2}{r_1}\Psi=-z'^2E_H\Psi$$

$$or,\quad -\frac{\hbar^2}{2m}\nabla_1^2\Psi=-z'^2E_H\Psi+\frac{z'e^2}{r_1}\Psi$$

$$similarly,\quad -\frac{\hbar^2}{2m}\nabla_2^2\Psi=-z'^2E_H\Psi+\frac{z'e^2}{r_2}\Psi$$

Therefore $-\frac{\hbar^2}{2m}\left(\nabla_1^2+\nabla_2^2\right)=-z'^2E_H\Psi+\frac{z'e^2}{r_1}\Psi-z'^2E_H\Psi+\frac{z'e^2}{r_2}\Psi$ [5]

Now using equations [5] in equation [4], we have

$$E=\iint\Psi^{\times}\left[-z'^2E_H+\frac{z'e^2}{r_1}-z'^2E_H+\frac{z'e^2}{r_2}-\frac{ze^2}{r_1}-\frac{ze^2}{r_2}+\frac{e^2}{r_{12}}\right]\Psi\, d\tau_1 d\tau_2$$

$$=\iint\Psi^{\times}\left[-2z'^2E_H+\frac{e^2}{r_1}(z'-z)+\frac{e^2}{r_2}(z'-z)+\frac{e^2}{r_{12}}\right]\Psi\, d\tau_1 d\tau_2$$

$$=\iint\Psi^{\times}\left[-2z'^2E_H+(z'-z)e^2\left(\frac{1}{r_1}+\frac{1}{r_2}\right)+\frac{e^2}{r_{12}}\right]\Psi\, d\tau_1 d\tau_2 \qquad [6]$$

Taking the integration in the 1st term of R.H.S

$$I_1=\iint\Psi^{\times}\Psi\, d\tau_1 d\tau_2=1 \qquad [7]$$

Taking the integration in the 2nd term of R.H.S

$$I_2=\iint\Psi^{\times}\left(\frac{1}{r_1}+\frac{1}{r_2}\right)\Psi\, d\tau_1 d\tau_2=\iint\Psi^{\times}\frac{2\Psi}{r_1}d\tau_1\quad[\because r_1\ \ r_2]\quad=2\int\frac{\Psi^{\times}\Psi}{r_1}d\tau_1 \qquad [8]$$

$$Now\qquad \Psi(r_1)=\sqrt{\frac{Z'^3}{\pi a_0^3}}\exp(-z'r_1/a_0)$$

$$I_2 = 2\int_0^\infty \int_0^\pi \int_0^{2\pi} \left(\frac{z'^3}{\pi a_0^3}\right)^{\frac{1}{2}} \exp(-z'r_1/a_0)\left(\frac{z'^3}{\pi a_0^3}\right)^{\frac{1}{2}} \exp(-z'r_1/a_0)\frac{1}{r_1}.r_1^2\, dr_1 \sin\theta\, d\theta\, d\varphi$$

$$= 2(4\pi)\frac{Z'^3}{\pi a_0^3}\int_0^\infty \exp(-2z'r_1/a_0)\, r_1\, dr_1$$

$$Let \quad x = \frac{2z'r_1}{a_0} \quad \therefore dx = \frac{2z'}{a_0}dr_1 \quad or,\ r_1 = \frac{a_0}{2z'}x \quad \&\ dr_1 = \frac{a_0}{2z'}dx$$

$$I_2 = (8\pi)\frac{Z'^3}{\pi a_0^3}\int_0^\infty e^{-x}\frac{a_0}{2z'}\, x\, \frac{a_0}{2z'}dx = \frac{Z'^3}{\pi a_0^3}(8\pi)\frac{a_0^2}{4z'^3}\int_0^\infty x e^{-x}\, dx$$

$$= \frac{2z'}{a_0}\left[\because \int_0^\infty x^n e^{-\alpha x}\, dx = \frac{n!}{(\alpha)^{n+1}}\right]$$

$$\therefore I_2 = 2z'\frac{2E_H}{e^2}\left[\because E_H = \frac{e^2}{2a_0}\right] \quad I_2 = \frac{4z'}{e^2}E_H \qquad [9]$$

Again taking the integration in the 3rd term of R.H.S of equation [6]

$$I_3 = \int\int \Psi^\times \frac{1}{r_{12}}\Psi\, d\tau_1 d\tau_2$$

$$= \iint\left(\frac{z'^3}{\pi a_0^3}\right)^2 \exp(-z'r_1/a_0)\exp(-z'r_2/a_0)\frac{1}{r_{12}}\exp(-z'r_1/a_0)\exp(-z'r_2/a_0)d^3r_1\, d^3r_2$$

$$= \left(\frac{z'^3}{\pi a_0^3}\right)^2 \iint \frac{1}{r_{12}}\exp\left[-2z'(r_1 + r_2/a_0)\right]d^3r_1\, d^3r_2$$

$$let\ \frac{2z'r_1}{a_0} = R_1 \ \&\ \frac{2z'r_2}{a_0} = R_2\ \ d^3r_1 = \left(\frac{a_0}{2z'}\right)^3 d^3R_1\ and\ d^3r_2 = \left(\frac{a_0}{2z'}\right)^3 d^3R_2$$

$$Now\ r_{12} = r_1 - r_2 = \frac{a_0}{2z'} = (R_1 - R_2)$$

$$I_3 = \left(\frac{z'^3}{\pi a_0^3}\right)^2 \int\int \frac{-2z'}{a_0(R_1 - R_2)}e^{-(R_1+R_2)}\left(\frac{a_0}{2z'}\right)^3 d^3R_1\left(\frac{a_0}{2z'}\right)^3 d^3R_2$$

$$= \left(\frac{z'^3}{\pi a_0^3}\right)^2\left(\frac{2z'}{a_0}\right)\left(\frac{a_0}{2z'}\right)^6 \int\int \frac{e^{-(R_1+R_2)}}{(R_1 - R_2)}d^3R_1 d^3R_2$$

$$= \frac{z'}{32\pi^2 a_0}20\pi^2 \quad \left[\because \frac{e^{-(R_1+R_2)}}{(R_1 - R_2)}d^3R_1 d^3R_2 = 20\pi^2\right]$$

$$I_3 = \frac{5}{8a_0}z' = \frac{5}{8}\frac{2E_H}{e^2}z' = \frac{5E_H}{4e^2}z' \left[\because E_H = -\frac{e^2}{2a_0}\right] \qquad [10]$$

Now using equations [7], [8] and [9] in equation [6]

$$E=\iint \Psi^{\times}\left[-2z'^2E_H+(z'-z)e^2\left(\frac{1}{r_1}+\frac{1}{r_2}\right)+\frac{e^2}{r_{12}}\right]\Psi\, d\tau_1 d\tau_2$$

$$=-2z'^2E_H\iint \Psi^{\times}\Psi\, d\tau_1 d\tau_2+e^2(z'-z)\iint\Psi^{\times}\left(\frac{1}{r_1}+\frac{1}{r_2}\right)\Psi\, d\tau_1 d\tau_2+e^2\iint\Psi^{\times}\frac{1}{r_{12}}\Psi\, d\tau_1 d\tau_2$$

$$=-2z'^2E_H+e^2(z'-z)\frac{4z'}{e^2}E_H+e^2\frac{5E_Hz'}{4e^2}=\left[-2z'^2+4z'(z'-z)+\frac{5}{4}z'\right]E_H \qquad [11]$$

Now we shall choose z' in such a manner that to give minimum energy

$$\frac{dE}{dz'}=0$$

$$\therefore -4z'+8z-4z+\frac{5}{4}=0 \;\; or,\;\; 4z'-4z=-\frac{5}{4} \;\; or,\;\; z'=z-\frac{5}{16}$$

Thus ground state energy of Helium atom

$$E_0=\left[-2z'^2+4z'^2-4z'z+\frac{5}{4}z'\right]E_H=\left[2\left(z-\frac{5}{16}\right)^2 4\left(z-\frac{5}{16}\right)z+\frac{5}{8}\left(z-\frac{5}{16}\right)\right]E_H$$

$$=\left(z-\frac{5}{16}\right)\left[2z-\frac{5}{8}-4z+\frac{5}{4}\right]E_H=\left(z-\frac{5}{16}\right)\left(-2z+\frac{5}{8}\right)E_H=-2\left(z-\frac{5}{16}\right)^2E_H$$

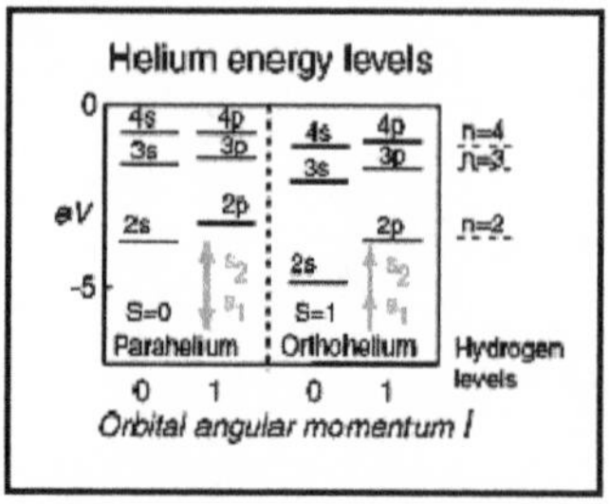

Fig.[15.5]: Shows the energy level of Helium atom

$$Finally,\;\; E_0=-2\left(2-\frac{5}{16}\right)^2\frac{e^2}{2a_0}=-2\left(\frac{27}{16}\right)^2\frac{e^2}{2a_0}$$

$$\therefore\;\; E_0=-2.85\frac{e^2}{2a_0}[For\ Helium\ z=2\ \&\ E_H=\frac{e^2}{2a_0}]$$

15.6 Ground state energy and ground state wave function for simple harmonic oscillator

Given $\quad \Psi(x)=A\exp\left(-\alpha x^2/2\right)$ [1]

First of all we have to find out the Normalization constant A. From the normalization condition

$$\int_{-\infty}^{\infty}\Psi^{\times}\Psi\, d\tau=1 \quad or,\ \int_{-\infty}^{\infty}A^{\times}\exp\left(-\alpha x^2/2\right)A\exp\left(-\alpha x^2/2\right)dx=1$$

$$Or,\quad |A|^2\int_{-\infty}^{\infty}\exp\left(-\alpha x^2\right)dx=1$$

$$Let\quad Z=\alpha x^2\quad or, x=\sqrt{\frac{z}{\alpha}}\quad or,\ dz=2\alpha x dx\quad or,\ dx=\frac{dz}{2\alpha x}\quad or, dx=\frac{dz}{2\alpha}\sqrt{\frac{\alpha}{z}}=\frac{z^{-\frac{1}{2}}dz}{2\sqrt{\alpha}}$$

Therefore,

$$or,\ |A|^2\,2\int_0^{\infty}e^{-z}\frac{z^{-\frac{1}{2}}dz}{2\sqrt{\alpha}}=1\quad or,\frac{|A|^2}{\sqrt{\alpha}}\int_0^{\infty}e^{-z}z^{\frac{1}{2}-1}dz=1$$

$$or,\frac{|A|^2}{\sqrt{\alpha}}\Gamma\left(\frac{1}{2}\right)=1\quad or,\frac{|A|^2}{\sqrt{\alpha}}\sqrt{\pi}=1\quad or, A=\left(\frac{\alpha}{\pi}\right)^{\frac{1}{4}}$$

Thus the normalized wave function is

$$\Psi(x)=\left(\frac{\alpha}{\pi}\right)^{\frac{1}{4}}e^{-\alpha x^2/2} \qquad [2]$$

Now the energy is given by

$$E=\int\Psi^{\times}[H]\Psi\, d\tau$$

For the dimensional harmonic oscillator we consider only the x-term

$$E=\int_{-\infty}^{\infty}\left(\frac{\alpha}{\pi}\right)^{\frac{1}{4}}e^{-\alpha x^2/2}\left[-\frac{\hbar^2}{2m}\frac{d^2}{dx^2}+\frac{1}{2}kx^2\right]\left(\frac{\alpha}{\pi}\right)^{\frac{1}{4}}e^{-\alpha x^2/2}dx$$

$$=\left(\frac{\alpha}{\pi}\right)^{\frac{1}{2}}2\int_0^{\infty}e^{-\alpha x^2/2}\left[-\frac{\hbar^2}{2m}\frac{d^2}{dx^2}\left(e^{-\alpha x^2/2}\right)+\frac{1}{2}kx^2\left(e^{-\alpha x^2/2}\right)\right]dx$$

$$=\left(\frac{\alpha}{\pi}\right)^{\frac{1}{2}}2\int_0^{\infty}e^{-\alpha x^2/2}\left(-\frac{\hbar^2}{2m}\right)\frac{d}{dx}\left[-\alpha x\, e^{-\alpha x^2/2}\right]dx+\left(\frac{\alpha}{\pi}\right)^{\frac{1}{2}}2\int_0^{\infty}e^{-\alpha x^2/2}\,\frac{1}{2}kx^2\,e^{-\alpha x^2/2}dx$$

$$=\left(\frac{\alpha}{\pi}\right)^{\frac{1}{2}}\frac{\alpha\hbar^2}{m}\int_0^{\infty}e^{-\alpha x^2/2}\left[e^{-\alpha x^2/2}-\alpha x^2\,e^{-\alpha x^2/2}\right]dx+\left(\frac{\alpha}{\pi}\right)^{\frac{1}{2}}k\int_0^{\infty}x^2\,e^{-\alpha x^2}dx$$

$$=\frac{\alpha\hbar^2}{m}\left(\frac{\alpha}{\pi}\right)^{\frac{1}{2}}\int_0^{\infty}e^{-\alpha x^2}dx-\frac{\alpha^2\hbar^2}{m}\left(\frac{\alpha}{\pi}\right)^{\frac{1}{2}}\int_0^{\infty}x^2\,e^{-\alpha x^2}dx+\left(\frac{\alpha}{\pi}\right)^{\frac{1}{2}}k\int_0^{\infty}x^2\,e^{-\alpha x^2}dx$$

$$=\frac{\alpha\hbar^2}{m}\left(\frac{\alpha}{\pi}\right)^{\frac{1}{2}}\int_0^{\infty}e^{-\alpha x^2}dx+\left[\left(\frac{\alpha}{\pi}\right)^{\frac{1}{2}}k-\frac{\alpha^2\hbar^2}{m}\left(\frac{\alpha}{\pi}\right)^{\frac{1}{2}}\right]\int_0^{\infty}x^2\,e^{-\alpha x^2}dx$$

$$\text{Let}\quad \alpha x^2=z\ \ or, 2x\,dx=\frac{1}{\alpha}dz\ \ or,\ \ x=\left(\frac{z}{\alpha}\right)^{\frac{1}{2}}\ \ and\ \ dx=\frac{1}{2\alpha}\left(\frac{\alpha}{z}\right)^{\frac{1}{2}}dz\ \ or, dx=\frac{1}{2\sqrt{\alpha}}z^{-\frac{1}{2}}dz$$

$$\therefore\ E=\left(\frac{\alpha\hbar^2}{m}\right)\left(\frac{\alpha}{\pi}\right)^{\frac{1}{2}}\int_0^\infty e^{-z}\,\frac{1}{2\sqrt{\alpha}}z^{-\frac{1}{2}}dz+\left(\frac{\alpha}{\pi}\right)^{\frac{1}{2}}\left[k-\frac{\alpha^2\hbar^2}{m}\right]\int_0^\infty \frac{z}{\alpha}e^{-z}\,\frac{1}{2\sqrt{\alpha}}z^{-\frac{1}{2}}\,dz$$

$$=\frac{1}{2\sqrt{\pi}}\left(\frac{\alpha\hbar^2}{m}\right)\int_0^\infty e^{-z}z^{\frac{1}{2}-1}dz+\frac{1}{2\sqrt{\pi}\alpha}\left[k-\frac{\alpha^2\hbar^2}{m}\right]\int_0^\infty e^{-z}\,z^{\frac{1}{2}}\,dz$$

$$=\frac{1}{2\sqrt{\pi}}\frac{\alpha\hbar^2}{m}\sqrt{\pi}+\frac{1}{2\sqrt{\pi}\alpha}\left(k-\frac{\alpha^2\hbar^2}{m}\right)\frac{\sqrt{\pi}}{2}$$

$$=\frac{\alpha\hbar^2}{2m}+\frac{1}{4\alpha}\left(k-\frac{\alpha^2\hbar^2}{m}\right)$$

$$=\frac{\alpha\hbar^2}{2m}+\frac{k}{4\alpha}-\frac{\alpha\hbar^2}{4m}$$

$$=\frac{\alpha\hbar^2}{4m}+\frac{m\omega^2}{4\alpha}\quad\left[\because\ \omega=\sqrt{\frac{k}{m}}\right] \qquad [3]$$

Now we find out the ground state energy

$$\frac{dE}{d\alpha}=0 \quad or,\ \frac{\hbar^2}{4m}-\frac{m\omega^2}{4\alpha^2}=0\ \ or,\ \alpha^2=\frac{m^2\omega^2}{\hbar^2}\quad or,\ \alpha=\frac{m\omega}{\hbar}$$

Therefore equation [3] becomes I.e., the ground state energy

$$E_0=\frac{\hbar^2}{4m}\frac{m\omega}{\hbar}+\frac{m^2\omega^2}{4}\frac{\hbar}{m\omega}\quad or,\ E_0=\frac{1}{2}\hbar\omega \qquad [4]$$

Equation [4] represents the ground state energy of L.H.O and the ground state wave function of L.H.O is

$$\Psi_0=\left(\frac{\alpha}{\pi}\right)^{\frac{1}{4}}e^{-\alpha x^2/2}=\left(\frac{m\omega}{\pi\hbar}\right)^{\frac{1}{4}}e^{-m\omega x^2/2\hbar}$$

15.7[a] The first excited state energy and wave function for the trial wave function

Given: Normalize wave function $\Psi_1=Bx\,e^{-\beta x^2/2}$

From the normalization condition

$$\int_{-\infty}^{+\infty}\Psi^\times\Psi\,d\tau=1$$

$$Or,\qquad \int_{-\infty}^{+\infty}Bx\,e^{-\beta x^2/2}\,Bx\,e^{-\beta x^2/2}\,dx$$

$$Or,\qquad |B|^2\int_{-\infty}^{+\infty}x^2e^{-\beta x^2}\,dx=1$$

$$Let\ \beta x^2=z\ \ \therefore x=\left(\frac{z}{\beta}\right)^{\frac{1}{2}}\ and\ \ 2\beta x dx=dz\,;\ dx=\frac{1}{2\beta x}dx=\frac{1}{2\beta}\sqrt{\frac{\beta}{z}}\,dz$$

$$Or, |B|^2 2\int_0^\infty \frac{z}{\beta} e^{-z} \frac{1}{2\beta}\sqrt{\frac{\beta}{z}}\, dz = 1 \quad Or, |B|^2 \frac{\sqrt{\beta}}{\beta^2} 2\int_0^\infty e^{-z} z^{-\frac{1}{2}+1} dz = 1$$

$$Or, \frac{|B|^2}{\beta^{\frac{3}{2}}} \Gamma\left(\frac{3}{2}\right) = 1 \quad Or, \frac{|B|^2}{\beta^{\frac{3}{2}}} \frac{\sqrt{\pi}}{2} = 1 \quad Or, |B|^2 = \frac{2\beta^{\frac{3}{2}}}{\sqrt{\pi}} \quad Or, B = \left(\frac{2\beta^{\frac{3}{2}}}{\sqrt{\pi}}\right)^{\frac{1}{2}}$$

The first excited state normalized wave function becomes

$$\Psi_1 = \left(\frac{2\beta^{\frac{3}{2}}}{\sqrt{\pi}}\right)^{\frac{1}{2}} x e^{-\beta x^2/2}$$

[2]

Now the energy of the first excited state is

$$E = \int_{-\infty}^{+\infty} \Psi_1^\times [H] \Psi_1 \, d\tau$$

For one dimensional harmonic oscillator, we consider only the x- component

$$E = \int_{-\infty}^{+\infty} \Psi_1^\times [H] \Psi_1 \, dx$$

$$E = 2\int_0^\infty B^\times x e^{-\beta x^2/2} \left[-\frac{\hbar^2}{2m}\frac{d^2}{dx^2} + \frac{1}{2}kx^2\right] B x e^{-\beta x^2/2} dx$$

$$= 2|B|^2 \int_0^\infty x e^{-\beta x^2/2} \left[-\frac{\hbar^2}{2m}(\beta^2 x^2 - 3\beta) + \frac{1}{2}kx^2\right] x e^{-\beta x^2/2} dx$$

$$[Note: \frac{d}{dx}\left(x e^{-\frac{\beta x^2}{2}}\right) = x\left(-\beta x e^{-\frac{\beta x^2}{2}}\right) + e^{-\frac{\beta x^2}{2}} = -\beta x^2 e^{-\frac{\beta x^2}{2}} + e^{-\frac{\beta x^2}{2}}$$

$$= -\left[\beta x^2\left(-\beta x e^{-\frac{\beta x^2}{2}}\right) + 2\beta x e^{-\frac{\beta x^2}{2}}\right] - \beta x e^{-\frac{\beta x^2}{2}}$$

$$= (\beta^2 x^2 - 3\beta) x e^{-\frac{\beta x^2}{2}} \quad]$$

$$E = 2|B|^2\left[-\frac{\beta^2\hbar^2}{2m}\int_0^\infty x^4 e^{-\beta x^2} dx + 3\beta\frac{\hbar^2}{2m}\int_0^\infty x^2 e^{-\beta x^2} dx + \frac{k}{2}\int_0^\infty x^4 e^{-\beta x^2} dx\right]$$

$$or, E = 2|B|^2\left[3\beta\frac{\hbar^2}{2m}\int_0^\infty x^2 e^{-\beta x^2} dx + \left(\frac{k}{2} - \frac{\beta^2\hbar^2}{2m}\right)\int_0^\infty x^4 e^{-\beta x^2} dx\right]$$

$$Let\ \beta x^2 = z \quad \therefore x = \left(\frac{z}{\beta}\right)^{\frac{1}{2}} \quad Therefore,\ 2\beta x dx = dz;\ dx = \frac{1}{2\beta x}\, dz = \frac{1}{2\beta}\sqrt{\frac{\beta}{z}}\, dz$$

$$\therefore\ E = 2|B|^2\left[\ 3\beta\frac{\hbar^2}{2m}\int_0^\infty \frac{z}{\beta}\frac{1}{2\beta}\sqrt{\frac{\beta}{z}}\,e^{-z}\,dz + \left(\frac{k}{2}-\frac{\beta^2\hbar^2}{2m}\right)\int_0^\infty \frac{z^2}{\beta^2}\frac{1}{2\beta}\sqrt{\frac{\beta}{z}}\,e^{-z}\,dz\ \right]$$

$$= 2|\beta|^2\left[\ \frac{3\hbar^2}{4m\sqrt{\beta}}\int_0^\infty e^{-z}\,z^{\frac{1}{2}}\,dz + \left(\frac{k}{2}-\frac{\beta^2\hbar^2}{2m}\right)\int_0^\infty \frac{1}{2\beta^2\sqrt{\beta}}\,e^{-z}\,z^{\frac{3}{2}}dz\ \right]$$

$$= 2|B|^2\left[\ \frac{3\hbar^2}{4m\sqrt{\beta}}\int_0^\infty e^{-z}\,z^{\frac{3}{2}-1}\,dz + \left(\frac{k}{2}-\frac{\beta^2\hbar^2}{2m}\right)\frac{1}{2\beta^2\sqrt{\beta}}\int_0^\infty e^{-z}\,z^{\frac{5}{2}-1}dz\ \right]$$

$$= 2|B|^2\left[\ \frac{3\hbar^2}{4m\sqrt{\beta}}\Gamma\left(\frac{3}{2}\right) + \left(\frac{k}{2}-\frac{\beta^2\hbar^2}{2m}\right)\frac{1}{2\beta^2\sqrt{\beta}}\Gamma\left(\frac{5}{2}\right)\right]$$

$$= 2\frac{2}{\sqrt{\pi}}\beta^{3/2}\,\frac{3\hbar^2}{4m\sqrt{\beta}}\frac{\sqrt{\pi}}{2} + \left(\frac{k}{2}-\frac{\beta^2\hbar^2}{2m}\right)2\frac{2}{\sqrt{\pi}}\beta^{3/2}\,\frac{1}{2\beta^2\sqrt{\beta}}\,\frac{3}{2}\frac{\sqrt{\pi}}{2}$$

$$= \frac{3}{2m}\hbar^2\beta + \left(\frac{k}{2}-\frac{\beta^2\hbar^2}{2m}\right)\frac{3}{2}\frac{\beta^{3/2}}{\beta^2\sqrt{\beta}}$$

$$= \frac{3}{2m}\hbar^2\beta + \left(\frac{k}{2}-\frac{\beta^2\hbar^2}{2m}\right)\frac{3}{2\beta}$$

$$= \frac{3}{2m}\hbar^2\beta + \frac{3k}{4\beta} - \frac{3\beta^2\hbar^2}{4m\beta}$$

$$= \frac{3}{2m}\hbar^2\beta + \frac{3m\omega^2}{4\beta} - \frac{3\beta\hbar^2}{4m} \qquad \left[\because\ \omega = \sqrt{\frac{k}{m}}\ \right]$$

$$= \frac{3}{4m}\hbar^2\beta + \frac{3m}{4\beta}\omega^2$$

$$= \frac{3}{4}\left(\frac{\beta\hbar^2}{m} + \frac{m\omega^2}{\beta}\right)$$

This is minimum at

$$\frac{dE}{d\beta} = 0$$

$$or,\qquad \frac{3}{4}\left(\frac{\hbar^2}{m} - \frac{m\,\omega^2}{\beta^2}\right) = 0 \quad or,\ \frac{\hbar^2}{m} = \frac{m\,\omega^2}{\beta^2}$$

$$or,\qquad \beta^2 = \frac{m^2\omega^2}{\hbar^2} \quad or,\qquad \beta = \frac{m\,\omega}{\hbar}$$

Thus, the energy of first excited state is

$$E_1 = \frac{3}{4}\left(\frac{m\omega}{\hbar}\frac{\hbar^2}{m} + m\omega^2\frac{\hbar}{m\omega}\right) = \frac{3}{4}(\hbar\omega + \hbar\omega) = \frac{3}{4}\ 2\hbar\omega$$

Or, $\quad Or,\ E_1 = \frac{3}{2}\hbar\omega$ [3]

∴ The 1st excited state wave function

$$\Psi_1 = B\,x\,e^{-\beta x^2/2} = \left(\frac{2\beta^{3/2}}{\sqrt{\pi}}\right)^{\frac{1}{2}} x\,e^{-\frac{m\omega x^2}{2\hbar}} = \left[\frac{2}{\sqrt{\pi}}\left(\frac{m\omega}{\hbar}\right)^{\frac{3}{2}}\right]^{\frac{1}{2}} x\,e^{-\frac{m\omega x^2}{2\hbar}}$$

15.7[b] The first excited state energy and wave function for the given trial wave function

For the 1st excited state, we take the trial wave function

$$\Psi = \left(B(1 + C\frac{r}{r_0}\right)e^{-\alpha r/r_0}$$

But we know, the ground state wave function for hydrogen atom

$$\Psi_0 = \frac{1}{\sqrt{\pi r_0^3}} e^{-r/r_0}$$

From the orthogonality condition

$$\left(\Psi_{0,}\Psi_1\right) = \int_{-\infty}^{\infty} \Psi_0^{\times}\ \Psi_1 d\tau = 0$$

$$Or, \quad \int_0^{\infty}\int_0^{\pi}\int_0^{2\pi} \frac{1}{\sqrt{\pi r_0^3}} e^{-r/r_0}\ B\left(1 + C\frac{r}{r_0}\right)e^{-\alpha r/r_0}\ r^2 \sin\theta\, d\theta\, d\varphi\, dr = 0$$

$$Or, \quad \frac{B}{\sqrt{\pi r_0^3}} 4\pi \int_0^{\infty}\left(1 + C\frac{r}{r_0}\right)e^{-(1+\alpha)\, r/r_0}\quad r^2 dr = 0$$

$$Or, \quad \frac{\beta}{\sqrt{\pi r_0^3}} 4\pi\left[\int_0^{\infty} e^{-(1+\alpha)\, r/r_0} r^2 dr + \frac{c}{r_0}\int_0^{\infty} e^{-(1+\alpha)\, r/r_0} r^3 dr\right] = 0$$

$$Let \quad \theta = (1+\alpha)\frac{r}{r_0} \quad Or, \quad r = \frac{r_0}{1+\alpha}\theta \quad \therefore\ dr = \frac{r_0}{1+\alpha}d\theta$$

$$when \quad r = \infty \ then\ \theta = \infty \quad when \quad r = 0 \ then\ \theta = 0$$

$$Or, \left[\int_0^{\infty} e^{-\theta}\left(\frac{r_0}{1+\alpha}\theta\right)^2\left(\frac{r_0}{1+\alpha}\right)d\theta + \frac{c}{r_0}\int_0^{\infty} e^{-\theta}\left(\frac{r_0}{1+\alpha}\theta\right)^3\left(\frac{r_0}{1+\alpha}\right)\ d\theta\right] = 0$$

$$Or, \left[\left(\frac{r_0}{1+\alpha}\right)^2\left(\frac{r_0}{1+\alpha}\right)\int_0^{\infty} e^{-\theta}\,\theta^{3-1}\, d\theta + \frac{c}{r_0}\left(\frac{r_0}{1+\alpha}\theta\right)^3\left(\frac{r_0}{1+\alpha}\right)\int_0^{\infty} e^{-\theta}\,\theta^{4-1}\ d\theta\right] = 0$$

$$Or, \left[\left(\frac{r_0}{1+\alpha}\right)^3 2 + \frac{c\, r_0^3}{(1+\alpha)^4}6 = 0 \quad Or, \quad \frac{2r_0^3}{(1+\alpha)^4}(1+\alpha) + \frac{2\, r_0^3}{(1+\alpha)^4}3c\right] = 0$$

$$Or,\ \frac{2r_0^3}{(1+\alpha)^4}[1+\alpha+3c]=0 \quad or,\ 1+\alpha+3c=0 \quad \therefore c=-\frac{1}{3}(1+\alpha)$$

Again, $\int_{-\infty}^{\infty}\Psi_1^\times \Psi_1 d\tau=1$

$$\int_0^\infty\int_0^\pi\int_0^{2\pi} B^\times\left(1+C\frac{r}{r_0}\right)e^{-\alpha r/r_0} B\left(1+C\frac{r}{r_0}\right)e^{-\alpha r/r_0} r^2 \sin\theta d\theta d\varphi dr=1$$

$$or, 4\pi|B|^2\int_0^\beta\left(1+C\frac{r}{r_0}\right)^2 e^{-2\alpha r/r_0} r^2\,dr=1$$

$$or, 4\pi B^2\left[\int_0^\beta e^{-2\alpha r/r_0} r^2\,dr+\frac{2c}{r_0}\int_0^\infty e^{-2\alpha r/r_0} r^3\,dr+\frac{c^2}{r_0^2}\int_0^\infty e^{-2\alpha r/r_0} r^4\,dr\right]=1$$

$$Let\ \frac{2\alpha r}{r_0}=\theta \ \ or,\ r=\frac{r_0}{2\alpha}\theta \ \ \therefore dr=\frac{r_0}{2\alpha}d\theta$$

$$or, 4\pi B^2\left[\left(\frac{r_0}{2\alpha}\right)^2\frac{r_0}{2\alpha}\int_0^\beta e^{-\theta}\theta^{3-1}d\theta+\frac{2c}{r_0}\left(\frac{r_0}{2\alpha}\right)^3\frac{r_0}{2\alpha}\int_0^\infty e^{-\theta}\theta^{4-1}\,d\theta\right]$$

$$+4\pi B^2\left[\frac{c^2}{r_0^2}\left(\frac{r_0}{2\alpha}\right)^4\frac{r_0}{2\alpha}\int_0^\infty e^{-\theta}\theta^{5-1}\,d\theta\right]=1$$

$$Or, 4\pi B^2\left[\frac{r_0^3}{8\alpha^3}2+\frac{cr_0^3}{8\alpha^4}6+\frac{c^2r_0^3}{32\alpha^5}24\right]=1$$

$$Or,\ \pi B^2 r_0^3\left[\frac{1}{\alpha^3}+\frac{3c}{\alpha^4}+\frac{3}{\alpha^5}c^2\right]=0$$

$$Or, \pi B^2 r_0^3\left[\frac{1}{\alpha^3}+\frac{3}{\alpha^4}\left\{-\frac{1}{3}(1+\alpha)\right\}+\frac{3}{\alpha^5}\left\{\frac{1}{9}(1+\alpha)^2\right\}\right]=0$$

$$Or, \pi B^2 r_0^3\left[\frac{1}{\alpha^3}-\frac{1+\alpha}{\alpha^4}+\frac{(1+\alpha)}{3\alpha^5}\right]=1$$

$$Or,\ \pi B^2 r_0^3\left[\frac{3\alpha^2-3\alpha(1+\alpha)+(1+\alpha)^2}{3\alpha^5}\right]=1 \ \ Or,\ B^2\left[3\alpha^2-3\alpha-3\alpha^2+1+2\alpha+\alpha^2\right]=\frac{3\alpha^5}{\pi r_0^3}$$

$$Or,\ B^2\left[\alpha^2-\alpha+1\right]=\frac{3\alpha^5}{\pi r_0^3} \quad or,\ B=\left[\frac{3\alpha^5}{\pi r_0^3}\left(\alpha^2-\alpha+1\right)\right]^{\frac{1}{2}}$$

Hence the normalized wave function

$$\Psi_1=\left[\frac{3\alpha^5}{\pi r_0^3\left(\alpha^2-\alpha+1\right)}\right]^{\frac{1}{2}}\left[1-\frac{1}{3}(1+\alpha)\frac{r}{r_0}\right]e^{-\frac{\alpha r}{r_0}}$$

$$Therefore,\ E=\int_{-\infty}^{\infty}\Psi_1^{\times}H\,\Psi_1\,d\tau$$

$$or,\ E=\frac{12\alpha^5}{r_0^3\left(\alpha^2-\alpha+1\right)}\left[\int_0^{\infty}\left(-\frac{r}{3r_0}(1+\alpha)\right)e^{-\alpha r/r_0}\right]$$

$$\times\left[-\frac{\hbar^2}{2m}\frac{1}{r^2}\frac{d}{dr}\left(r^2\frac{d}{dr}\right)\left\{\left(1-\frac{r}{3r_0}\right)(1+\alpha)\right\}e^{-\alpha r/r_0}\,r^2dr-e^2\int_0^{\infty}\left(1-\frac{r}{3r_0}(1+\alpha)\right)^2e^{-2\alpha r/r_0}\,rdr\right]$$

$$or,\ E=\frac{12\alpha^5}{r_0^3\left(\alpha^2-\alpha+1\right)}\left[\frac{e^2r_0^2}{24\alpha^4}\frac{7\alpha^3-\alpha^2-\alpha}{3}-\frac{e^2r_0^2}{24\alpha^4}\left(3\alpha^2-2\alpha+1\right)\right]$$

$$Or,\ E=\frac{12\alpha^5}{r_0^3\left(\alpha^2-\alpha+1\right)}\frac{e^2r_0^2}{72\alpha^4}\left[7\alpha^3-\alpha^2+\alpha-9\alpha^2+6\alpha-3\right]$$

$$Or,\ E=\frac{e^2\alpha}{6r_0\left(\alpha^2-\alpha+1\right)}\left[7\alpha^3-10\alpha^2+7\alpha-3\right]$$

$$=\frac{e^2\alpha}{6r_0\left(\alpha^2-\alpha+1\right)}\left[7\alpha\left(\alpha^2+\alpha-1\right)-3\left(\alpha^2+\alpha-1\right)-3\alpha\right]$$

$$=\frac{e^2}{r_0}\left[\frac{7}{6}\alpha^2-\frac{1}{2}\alpha-\frac{\alpha^2}{2\left(\alpha^2+\alpha-1\right)}\right]$$

$$\therefore\quad \frac{dE}{d\alpha}=0$$

$$Or,\ 14\alpha(\alpha^2-\alpha+1)^2-3(\alpha^2-\alpha+1)^2-6\alpha(\alpha^2-\alpha+1)+3\alpha^2(2\alpha-1)=0$$

$$Or,\ 2\alpha-1=0\ \ Or,\ \alpha^2-\alpha+1=0\ \ Or,\ \alpha=\frac{1}{2}\ \ Or,\ \alpha=\frac{1}{2}+i\frac{\sqrt{3}}{2}$$

So, Physically acceptable value $\alpha=\frac{1}{2}$ $\therefore$ *First excited state energy*

$$or,\quad E_1=\frac{e^2}{r_0}\left[\frac{7}{6}\left(\frac{1}{2}\right)^2-\frac{1}{2}\frac{1}{2}-\frac{\frac{1}{4}}{2\left(\frac{1}{4}-\frac{1}{2}+1\right)}\right]=\frac{e^2}{r_0}\left[\frac{7}{24}-\frac{1}{4}-\frac{1}{6}\right]=-\frac{3e^2}{24r_0}=-\frac{3e^2}{8a_0}$$

Therefore 1st excited state wave function

$$\Psi_1=\left(\frac{3}{\pi r_0^3}\frac{\left(\frac{1}{2}\right)^5}{\frac{1}{4}-\frac{1}{2}+1}\right)\left[1-\frac{1}{3}\left(1+\frac{1}{2}\right)\frac{r}{r_0}\right]e^{-r/2r_0}=\left(\frac{1}{8\pi r_0^3}\right)^{\frac{1}{2}}\left[1-\frac{r}{2r_0}\right]e^{-r/2r_0}$$

15.8 Ground state energy for the given trial wave function $\Psi = e^{-\frac{\alpha r}{a_0}}$

The trial wave function $\Psi = e^{-\frac{\alpha r}{a_0}}$

Let, the normalized trial wave function $\Psi = Ne^{-\frac{\alpha r}{a_0}}$. we also consider $\beta = \frac{\alpha}{a_0}$ then,

$$\Psi = Ne^{-\beta r}$$

To find out N we shall use the normalization condition

$$\int_{-\infty}^{\infty} \Psi^{\times}\Psi \, dy = 1$$

Or, $\int_{-\infty}^{\infty} N^{\times}e^{-\beta r}Ne^{-\beta r} r^2 \sin\theta \, d\theta \, d\phi \, dr = 1$

Or, $|N|^2 \int_0^{\infty}\int_0^{\pi}\int_0^{2\pi} e^{-2\beta r} r^2 \sin\theta \, d\theta \, d\phi \, dr = 1$

Or, $4\pi|N|^2 \int_0^{\infty} e^{-2\beta r} r^2 \, dr = 1$; Let, $x = 2\beta r$ Or, $r = \frac{x}{2\beta}$ Or, $dr = \frac{1}{2\beta}dx$

Or, $4\pi|N|^2 \int_0^{\infty} e^{-x} \frac{x^2}{4\beta^2}\frac{1}{2\beta} dx = 1$

Or, $\frac{\pi|N|^2}{2\beta^3} \int_0^{\infty} e^{-x} x^2 \, dx = 1$

Or, $\frac{\pi|N|^2}{2\beta^3} 2! = 1$

Or, $|N|^2 = \frac{\beta^3}{\pi}$ Or, $N = \left(\frac{\beta^3}{\pi}\right)^{1/2}$

Therefore, the normalized wave function be

$$\Psi = Ne^{-\frac{\alpha r}{a_0}}$$

$$\Psi = \left(\frac{\beta^3}{\pi}\right)^{1/2} e^{-\beta r}$$

Now the energy is $E = \int \Psi^{\times} [H]\Psi \, d\tau$

Or, $$E = \int \left(\frac{\beta^3}{\pi}\right)^{1/2} e^{-\beta r} \left[-\frac{\hbar^2}{2m}\nabla^2 + V(r)\right] \left(\frac{\beta^3}{\pi}\right)^{1/2} e^{-\beta r} d\tau \quad [1]$$

We know, the Laplacian operator $\nabla^2 = \frac{1}{r^2}\frac{\partial}{\partial r}\left(r^2 \frac{\partial}{\partial r}\right) + \frac{1}{r^2 \sin\theta}\frac{\partial}{\partial\theta}\left(\sin\theta \frac{\partial}{\partial\theta}\right) + \frac{1}{r^2 sin^2\varphi}\frac{\partial^2}{\partial\varphi^2}$

As the energy occurs in the radial part of the Schrödinger equation so we consider the radial part of the Laplacian operator ∇^2 [Since, ∇^2 will operate only on r.]

$$\nabla^2 = \frac{1}{r^2}\frac{\partial}{\partial r}(r^2 \frac{\partial}{\partial r}) = \frac{1}{r^2}(2r\frac{\partial}{\partial r} + r^2 \frac{\partial^2}{\partial r^2}) = \frac{2}{r}\frac{\partial}{\partial r} + \frac{\partial^2}{\partial r^2}$$

The potential energy of Hydrogen-atom $V(r) = -\frac{e^2}{r}$

From equation [1]

$$E = \int \left(\frac{\beta^3}{\pi}\right)^{1/2} e^{-\beta r}\left[-\frac{\hbar^2}{2m}\nabla^2 + V(r)\right]\left(\frac{\beta^3}{\pi}\right)^{1/2} e^{-\beta r} d\tau$$

$$E = \frac{\beta^3}{\pi}\int_0^\infty \int_0^\pi \int_0^{2\pi} e^{-\beta r}\left[-\frac{\hbar^2}{2m}\frac{1}{r^2}\frac{\partial}{\partial r}\left(r^2\frac{\partial}{\partial r}e^{-\beta r}\right) - \frac{e^2}{r}e^{-\beta r}\right] r^2 \sin\theta \, d\theta \, d\phi dr$$

$$E = \frac{\beta^3}{\pi} 4\pi \int_0^\infty e^{-\beta r}\left[-\frac{\hbar^2}{2m}\frac{1}{r^2}\frac{\partial}{\partial r}\left(-\beta r^2 e^{-\beta r}\right) - \frac{e^2}{r}e^{-\beta r}\right] r^2 dr$$

$$E = 4\beta^3 \int_0^\infty e^{-\beta r}\left[\frac{\beta\hbar^2}{2mr^2}\left(2re^{-\beta r} - \beta r^2 e^{-\beta r}\right) - \frac{e^2}{r}e^{-\beta r}\right] r^2 dr$$

$$E = 4\beta^3 \int_0^\infty \left[\frac{\beta\hbar^2}{mr^2}e^{-2\beta r} - \frac{\beta^2\hbar^2}{2m}e^{-2\beta r} - \frac{e^2}{r}e^{-2\beta r}\right] r^2 dr$$

$$E = \frac{4\beta^4\hbar^2}{m}\int_0^\infty r\, e^{-2\beta r}\, dr - \frac{2\beta^5\hbar^2}{m}\int_0^\infty r^2 e^{-2\beta r}\, dr - 4\beta^3 e^2 \int_0^\infty r e^{-2\beta r}\, dr \qquad [1]$$

Let, $I_1 = \int_0^\infty r\, e^{-2\beta r}\, dr$ & $I_2 = \int_0^\infty r^2 e^{-2\beta r}\, dr$; Suppose, $x = 2\beta r$; $dr = \frac{dx}{2\beta}$; then

$$I_1 = \int_0^\infty \frac{x}{2\beta}\, e^{-x}\frac{dx}{2\beta} = \frac{1}{4\beta^2}$$

$$I_2 = \int_0^\infty \frac{x^2}{4\beta^2}\, e^{-x}\, \frac{dx}{2\beta} = \frac{1}{8\beta^3}\, 2 = \frac{1}{4\beta^3}$$

Using these in [1]

$$E = \frac{4\beta^4\hbar^2}{m}\frac{1}{4\beta^2} - \frac{2\beta^5\hbar^2}{m}\frac{1}{4\beta^3} - 4\beta^3 e^2\, \frac{1}{4\beta^2}$$

$$E = \frac{\beta^2\hbar^2}{m} - \frac{\beta^2\hbar^2}{2m} - e^2\beta$$

$$E = \frac{\beta^2\hbar^2}{2m} - \beta e^2 \qquad [2]$$

To find out the ground state energy

$$\frac{dE}{d\beta} = 0$$

Or, $$\frac{\beta\hbar^2}{m} - e^2 = 0$$

$$\therefore \beta = \frac{me^2}{\hbar^2}$$

Using the value of β in [2] the ground state energy will be

$$E_0 = \frac{\hbar^2}{2m} \times \frac{m^2 e^4}{\hbar^4} - \frac{m\, e^2}{\hbar^2}e^2$$

Or, $$E_0 = \frac{m\, e^4}{2\hbar^2} - \frac{m\, e^4}{\hbar^2} = -\frac{m\, e^4}{2\hbar^2}$$

$$\boxed{E_0 = -\frac{m\, e^4}{2\hbar^2}}$$; Which is the ground state energy of the given trial wave function.

Exercise

1. Write the significance of an approximation method and Importance of variation method.
2. Explain the Principle of variational method.
3. Show that the expectation value of the Hamiltonian in any state ψ gives the upper bound to the ground state energy.
4. Find out the ground state energy of hydrogen atom by using variation method.
5. Show that the ground state energy of hydrogen atom

   $E_{\text{minimum}} = -\frac{e^2}{2r_0} = -\frac{e^2}{2}\frac{me^2}{\hbar^2} = -\frac{me^4}{2\hbar^2}$; by using variation method
6. Find the ground state wave function

   $$\Psi = \left(\frac{\alpha^3}{\pi r_0^3}\right)^{\frac{1}{2}} \exp(-\alpha r / r_0) \quad Or, \Psi_0 = \frac{1}{\sqrt{\pi r_0^3}} \exp(-r / r_0) \, [\because \alpha = 1]$$
7. What is the Importance of Many-electron atom in variational method?
8. Find out the ground state energy of Helium atom by using variation method.
9. Show that the ground state energy of Helium atom

   $E_0 = -2.85 \frac{e^2}{2a_0} [For\ Helium\ z = 2\ \&\ E_H = \frac{e^2}{2a_0}]$; by using variation method
10. Find out the ground state energy and ground state wave function for simple harmonic oscillator by using variation method.
11. By using variation method find the first excited state energy and wave function for the trial wave function

    $$\Psi_1 = Bx\, e^{-\beta x^2/2}$$
12. By using variation method find he first excited state energy and wave function for the given trial wave function

    $$\Psi = \left(B(1 + C\frac{r}{r_0} \right) e^{-\alpha r / r_0}$$
13. By using variation method find the ground state energy for the given trial wave function

    $$\Psi = e^{-\frac{\alpha r}{a_0}}$$
14. By using variation method find Show that the ground state energy for the given trial wave function $\Psi = e^{-\frac{\alpha r}{a_0}}$ is $E_0 = -\frac{m e^4}{2\hbar^2}$
15. Using the variation method to estimate the ground energy of a particle in the potential

    $V(x) = \infty$ for $x < 0$

    $V(x) = -kx$ for $x > a$

16. Using the variational principle, calculate the binding energy of deuteron with the nuclear potential

$V(r) = V_0 \frac{e^{-r/a}}{r/a}$; Where the range of the potential $a = 2.8 \times 10^{-13} cm$ and experimental binding energy $E_B = -2.23\ MeV$ which may be used to estimate V_0.

17. Using the variational method with the trial wave function

$\psi = \frac{1}{\pi a^3} e^{-(r_1+r_2)}/a$; where a is the variational parameter, to find the energy of the helium atom.

Chapter 16 Approximation Method: Perturbation Theory

16.1 Need for perturbation theory in Quantum Mechanics

Perturbation means small disturbance. Remember that the Hamiltonian of a system is nothing but the total energy of that system. Some external factors can always affect the energy of the system and its behavior. To analyze a system's energy, if we don't know the exact way of solution, then we can study the effects of external factors (perturbation) on the Hamiltonian.

Perturbation Theory is an extremely important method of seeing how a Quantum System will be affected by a small change in the potential. It allows us to get good approximations for systems where the Eigen states are not all easily findable.

So far we have concentrated on systems for which we could find exactly the Eigen values and Eigen functions of the Hamiltonian, like e.g. the harmonic oscillator, the quantum rotator, or the hydrogen atom. However the vast majority of systems in nature cannot be solved exactly, and we need develop appropriate tools to deal with them. *Perturbation theory is extremely successful in dealing with those cases that can be modelled as a "small deformation" of a system that we can solve exactly.* Let us translate the above statement into a precise mathematical frame work. We are going to consider systems that have an Hamiltonian:

$$\hat{H} = \hat{H}_0 = \lambda\hat{V}$$

Where $\hat{H}_0$ the Hamiltonian of the unperturbed system and λ is is a small parameter. $\hat{V}$ is the potential describing the perturbation. We shall assume that the perturbation $\hat{V}$ is independent of time.

Perturbation applied to a system is of two types: time dependent and time independent .We have to split the Hamiltonian into two parts. One part is a Hamiltonian whose solution we know exactly and the other part is the perturbation term. [Hamiltonian can be broken up into two parts. One of, which is large and characterizes a system for which the Schrödinger equation can be solved exactly, while the other part is small and can be treated as a perturbation] By this way we can solve the problems with a very good approximation. For an example of this method in quantum mechanics, we can use the Hamiltonian of the hydrogen atom to solve the problem of helium ion.

When the Hamiltonian is changed *suddenly,* the state cannot catch up with the change and basically remains unchanged. This is the basis of the **sudden approximation.**

When the Hamiltonian is changed *slowly,* the state does not realize that it is subject to a time-dependent Hamiltonian, and simply tracks the *instantaneous Eigen states.* This is the basis of the **adiabatic approximation.**

16.2 Time independent perturbation theory: Non-degenerate case

The time-independent perturbation theory is very successful when the system posses a small dimensionless parameter. It allows us to work out corrections to the energy Eigen values and Eigen states. However, it is not capable of working out consequences of a perturbation that depends on time.

A small disturbance on a pure system is called perturbation. We consider a perturbed Hamiltonian. In many cases the Hamiltonian can be written as a sum of two parts

$$H = H_0 + H' \qquad [1]$$

Here H_0 & H' are unperturbed and perturbed Hamiltonian. The Eigen value & the Eigen function of the unperturbed Hamiltonian H_0 is known. Thus the Eigen value equation for H_0 is,

$$H_0 u_n = E_n u_n \qquad [2]$$

Where E_n & u_n is the Eigen value and the orthogonal Eigen function of H_0 respectively. Our aim is to develop a perturbation approach to solve the Schrödinger equation.

$$H\Psi_n = W_n \Psi_n \qquad [3]$$

We shall obtain the approximate values of W_n & Ψ_n. Here we consider the non-degenerate case. We introduce a parameter λ & write

$$H = H_0 + \lambda H' \qquad [4]$$

Where, 'λ' lies between 0 and 1. We expand W_n & Ψ_n as a power series in λ I.e.,

$$\Psi_n = \Psi_n^{(0)} + \lambda\Psi_n^{(1)} + \lambda\Psi_n^{(2)} + \cdots\cdots \qquad [5]$$

$$W_n = W_n^{(0)} + \lambda W_n^{(1)} + \lambda W_n^{(2)} + \cdots\cdots \qquad [6]$$

Here 0, 1, 2, ----------- refer to zeroth order, 1st order , 2nd order-----respectively. If we put λ =1. We get the solution for [1]. Using [4], [5], & [6] in [3] :

$$H\Psi_n = W_n \Psi_n$$

$$(H_0 + \lambda H')\left(\Psi_n^{(0)} + \lambda\Psi_n^{(1)} + \lambda^2\Psi_n^{(2)} + ----------------\right)$$
$$= \left(W_n^{(0)} + \lambda W_n^{(1)} + \lambda^2 W_n^{(2)} + \;-------\right)\left(\Psi_n^{(0)} + \lambda\Psi_n^{(1)} + \lambda^2\Psi_n^{(2)} + -------\right)$$

Now equality the coefficient of λ^0, λ^1, & λ^2 :

$$For\ \lambda^0 : \quad H_0\Psi_n^{(0)} = W_n^{(0)}\Psi_n^{(0)} \qquad [7]$$

$$For\ \lambda^1 : \quad H'\Psi_n^{(0)} + H_0\Psi_n^{(0)} = W_n^{(0)}\Psi_n^{(1)} + W_n^{(1)}\Psi_n^{(0)} \qquad [8]$$

$$For\ \lambda^2:\ H'\Psi_n^{(1)} + H_0\Psi_n^{(2)} = W_n^{(0)}\Psi_n^{(2)} + W_n^{(1)}\Psi_n^{(1)} + W_n^{(2)}\Psi_n^{(0)} \qquad [9]$$

Equation [7] represents the Eigen value equation for the unperturbed Hamiltonian

$$\Psi_n^{(0)} = u_n \ \ and\ W_n^{(0)} = E_n$$

Now in ket notation the Eigen value equation [2] for the unperturbed Hamiltonian can be written as

$$H_0|n\rangle = E_n|n\rangle \qquad [10]$$

16.3(a) First order perturbation to the energy Eigen value and Eigen function

Now we shall calculate the 1st order perturbation: In this case we can expand $\Psi_n^{(1)}$ as a linear combination of u_n . Thus

$$\Psi_n^{(1)} = \sum_m a_m^{(1)} u_m \qquad [11]$$

Here the subscript 1 represents the 1st order perturbation. Putting this is [8]

$$H'u_n + H_0\sum_m a_m^{(1)}u_m = E_n\sum_m a_m^{(1)}u_m + W_n^{(1)}u_n$$

In ket notation :

$$H'|n\rangle + \sum_m a_m^{(1)}H_0|m\rangle = E_n\sum_m a_m^{(1)}|m\rangle + \omega_n^{(1)}|n\rangle$$

Taking dot product of $\langle k|$ *to the left on both sides of the above equation*

$$or,\ \langle k|H'|n\rangle + \sum_m a_m^{(1)}\langle k|E_m|m\rangle = E_n\sum_m a_m^{(1)}\langle k|m\rangle + W_n^{(1)}\langle k|n\rangle$$

$$or,\ \langle k|H'|n\rangle + \sum_m a_m^{(1)}E_m\langle k|m\rangle = E_n\sum_m a_m^{(1)}\delta_{km} + W_n^{(1)}\delta_{kn}$$

$$\langle k|H'|n\rangle + \sum_m a_m^{(1)}E_m\delta_{km} = E_n\sum_m a_m^{(1)}\delta_{km} + W_n^{(1)}\delta_{kn} \quad [when, \delta_{ij}=1, if\ i=j\ \&\ \delta_{ij}=0, if\ i\neq j]$$

$$\langle k|H'|n\rangle + a_k^{(1)}E_k = E_n a_k^{(1)} + W_n^{(1)}\delta_{kn}$$

$$or,\ a_k^{(1)}(E_n - E_k) + W_n^{(1)}\delta_{kn} = \langle k|H'|n\rangle \qquad [12]$$

If we put $k = n$

$$\therefore W_n^1 = \langle n|H'|n\rangle = \int u_n^{\times} H' u_n\, d\tau = H'_{nn} \qquad [13]$$

[*this gives the first order perturbation to the energy eigen value*]

This is the 1st order correction for Eigen value. Again putting $k = m$ in [12]

$$a_m^{(1)}(E_n - E_m) = \langle m|H'|n\rangle = \int u_m^{\times} H' u_n = H'_{mn}$$

$$or,\ a_m^{(1)}(E_n - E_m) = H'_{mn}$$

$$or,\ a_m^{(1)} = \frac{H'_{mn}}{E_n - E_m}$$

Using this in Equation [11], the 1st order correction to Eigen function is

$$or,\ \Psi_n^{(1)} = \sum_m \frac{H'_{mn}}{(E_n - E_m)} u_m \qquad [14]$$

16.3(b) Second order perturbation to the energy Eigen value and Eigen function

Now we shall calculate the 2nd order correction: In this case, again we can expand $\Psi_n^{(2)}$ as a linear combination of u_n, thus,

$$\Psi_n^{(2)} = \sum_m a_m^{(2)}\ u_m \qquad [15]$$

Here 2 refer to the 2nd order perturbation putting [15] & [11] in equation [9]

$$H'\sum_m a_m^{(1)} u_m + H_0 \sum_m a_m^{(2)} u_m = W_n^{(0)} \sum_m a_m^{(2)} u_m + \omega_n^{(1)} \sum_m a_m^{(1)} u_m + \omega_n^{(2)} \Psi_n^{(0)}$$

$$or,\quad H'\sum_m a_m^{(1)} u_m + \sum_m a_m^{(2)} E_m u_m = W_n^{(0)} \sum_m a_m^{(2)} u_m + \omega_n^{(1)} \sum_m a_m^{(1)} u_m + \omega_n^{(2)} u_m$$

In ket notation:

$$H'\sum_m a_m^{(1)} |m\rangle + \sum_m a_m^{(2)} E_m |m\rangle = W_n^{(0)} \sum_m a_m^{(2)} |m\rangle + W_n^{(1)} \sum_m a_m^{(1)} |m\rangle + W_n^{(2)} |n\rangle$$

Taking scalar product of bra vector $\langle k|$ to the left on both sides:

$$\sum_m a_m^{(1)} \langle k|\mathrm{H}'|m\rangle + \sum_m a_m^{(2)} E_m \langle k|m\rangle = W_n^{(0)} \sum_m a_m^{(2)} \langle k|m\rangle + W_n^{(1)} \sum_m a_m^{(1)} \langle k|m\rangle + W_n^{(2)} \langle k|n\rangle$$

$$\sum_m a_m^{(1)} \langle k|\mathrm{H}'|m\rangle + \sum_m a_m^{(2)} E_m \delta_{km} = W_n^{(0)} \sum_m a_m^{(2)} \delta_{km} + W_n^{(1)} \sum_m a_m^{(1)}\ \delta_{km} + W_n^{(2)}\ \delta_{kn}$$

$$Here\quad \delta_{ij} = 1\ \ if\ \ i = j\quad or, \delta_{ij} = 0\ \ if\ \ i \neq j$$

$$or,\ \sum_m a_m^{(1)} \langle k|H'|m\rangle + a_K^{(2)} E_k = E_n a_K^{(2)} + W_n^{(1)} a_K^{(1)} + W_n^{(2)}\ \delta_{kn}$$

$$or,\ a_K^{(2)}(E_n - E_k) + W_n^{(2)}\ \delta_{kn} = \sum_m a_m^{(1)} \langle k|H'|m\rangle - a_K^{(1)} W_n^{(1)} \qquad [16]$$

When k = n :

$$W_n^{(2)} = \sum_m a_m^{(1)} \langle k|\mathrm{H}'|m\rangle - a_n^{(1)} W_n^{(1)}$$

$$= \sum_m a_m^{(1)} \langle k|\mathrm{H}'|m\rangle - a_n^{(1)} \langle k|\mathrm{H}'|m\rangle$$

$$W_n^{(2)} = \sum_{m\neq n} a_m^{(1)} \langle k|\mathrm{H}'|m\rangle - a_n^{(1)} \langle k|\mathrm{H}'|m\rangle \quad \left[\mathrm{since}\ a_n^{(1)} = 0\right] \qquad [17]$$

$$W_n^{(2)} = \sum_{m\neq n} \frac{H'_{mn}}{(E_n - E_m)} H'_{mn} = \sum_{m\neq n} \frac{|H'_{mn}|^2}{(E_n - E_m)}$$

This is the 2nd order correction for Eigen value. [This gives the 2nd order perturbation to the energy Eigen value]

Again from [16] for $k = m$:

$$a_m^{(2)}\left(E_n - E_m\right) = \sum_m a_m^{(1)} \langle m|\mathrm{H}'|m\rangle - a_m^{(1)} W_n^{(1)}$$

$$= \sum_m a_m^{(1)} \mathrm{H}'_{mm} - a_m^{(1)} \mathrm{H}'_{nn}$$

$$= \sum_m \frac{\mathrm{H}'_{mn}}{E_n - E_m} \mathrm{H}'_{mm} - \frac{\mathrm{H}'_{mn}}{E_n - E_m} \mathrm{H}'_{nn}$$

$$a_m^{(2)} = \sum_m \frac{\mathrm{H}'_{mn}\mathrm{H}'_{mm}}{\left(E_n - E_m\right)^2} - \frac{\mathrm{H}'_{mn}\mathrm{H}'_{nn}}{\left(E_n - E_m\right)^2}$$

Thus the second order correction for Eigen function is

$$\Psi_n^{(2)} = \sum_m a_m^{(2)} u_m$$

$$= \sum_m \left[\sum_m \frac{\mathrm{H}'_{mn}\mathrm{H}'_{mn}}{\left(E_n - E_m\right)^2} - \frac{\mathrm{H}'_{mn}\mathrm{H}'_{nn}}{\left(E_n - E_m\right)^2} \right] u_m$$

Therefore, the total Eigen value taking up to and order correction is

$$W_n = W_n^{(0)} + W_n^{(1)} + W_n^{(2)}$$

$$W_n = E_n + H'_{nn} + \sum_{m \neq n} \frac{\left|\mathrm{H}_{mn}\right|^2}{\left(E_n - E_m\right)}$$

The total Eigen function taking up to second order correction is

$$\Psi_n = \Psi_n^{(0)} + \Psi_n^{(1)} + \Psi_n^{(2)}$$

$$= u_m + \sum_m \frac{\mathrm{H}'_{mn}}{\left(E_n - E_m\right)} u_m + \sum_m \left[\sum_m \frac{\mathrm{H}'_{mn}\mathrm{H}'_{mm}}{\left(E_n - E_m\right)^2} - \frac{\mathrm{H}'_{mm}\mathrm{H}'_{nn}}{\left(E_n - E_m\right)^2} \right] u_m$$

16.4 Dirac's time dependent perturbation theory

We know the dependent Schrödinger equation is

$$H\Psi = E\Psi$$

Or, $$H\Psi(\bar{r}, t) = i\hbar \frac{\partial}{\partial t} \Psi(\bar{r}, t) \qquad [1]$$

Since the perturbation method is applicable when the Hamiltonian can be split into two parts. One is time independent & other is time dependent. I.e.,

$$H = H_0 + H'(t) \qquad [2]$$

Here H_0, is usually referred to as the unperturbed Hamiltonian and $H'(t)$is known as the perturbation Hamiltonian. Hence [1] becomes

$$i\hbar\frac{\partial}{\partial t}\Psi(\bar{r},t) = [H_0 + H'(t)]\Psi(\bar{r},t)$$

Or, $$\frac{\hbar}{i}\frac{\partial}{\partial t}\Psi(\bar{r},t) = -[H_0 + H'(t)]\Psi(\bar{r},t)$$

Or, $$\frac{\hbar}{i}\frac{\partial}{\partial t}\Psi(\bar{r},t) + [H_0 + H'(t)]\Psi(\bar{r},t) = 0 \qquad [3]$$

Let $\phi(\bar{r},t)$be the state when H reduced to H_0 . Then [1] becomes

$$i\hbar\frac{\partial}{\partial t}\phi(\bar{r},t) = H_0\phi(\bar{r},t) \qquad [3a]$$

$$H_0\phi(\bar{r},t) = E_n\phi(\bar{r},t) \qquad [4]$$

Now the general solution of equation [3a]

$$\phi(\bar{r},t) = \textstyle\sum_n C_n\phi_n(\bar{r})\, exp\left(-\frac{i}{\hbar}E_n t\right) \qquad [5]$$

On the other hand, we seek the general solution of the complete problem [3] in the form,

$$\Psi(\bar{r},t) = \textstyle\sum_n C_n(t)\phi_n(\bar{r})\, exp\left(-\frac{i}{\hbar}E_n t\right) \qquad [6]$$

Substituting [6] in [3],

$$\frac{\hbar}{i}\frac{\partial}{\partial t}\Psi(\bar{r},t) + [H_0 + H'(t)]\Psi(\bar{r},t) = 0$$

$$\frac{\hbar}{i}\left[\textstyle\sum_n \dot{c}_n(t)\phi_n(\bar{r})\, exp\left(-\frac{i}{\hbar}E_n t\right) + \sum_n C_n(t)\phi_n(\bar{r})\left(-\frac{i}{\hbar}E_n\right) exp\left(-\frac{i}{\hbar}E_n t\right)\right]$$

$$+H_0\textstyle\sum_n C_n(t)\phi_n(\bar{r})\, exp\left(-\frac{i}{\hbar}E_n t\right) + H'(t)\sum_n C_n(t)\phi_n(\bar{r})\, exp\left(-\frac{i}{\hbar}E_n t\right) = 0$$

$$\frac{\hbar}{i}\left[\textstyle\sum_n \dot{c}_n(t)\phi_n(\bar{r})\, exp\left(-\frac{i}{\hbar}E_n t\right) - \sum_n C_n(t)\phi_n(\bar{r})E_n\, exp\left(-\frac{i}{\hbar}E_n t\right)\right]$$

$$+H_0\textstyle\sum_n C_n(t)\phi_n(\bar{r})\, exp\left(-\frac{i}{\hbar}E_n t\right) + H'(t)\sum_n C_n(t)\phi_n(\bar{r})\, exp\left(-\frac{i}{\hbar}E_n t\right) = 0$$

$$\frac{\hbar}{i}\textstyle\sum_n \dot{c}_n(t)\phi_n(\bar{r})\, exp\left(-\frac{i}{\hbar}E_n t\right) - \sum_n C_n(t)\phi_n(\bar{r})E_n\, exp\left(-\frac{i}{\hbar}E_n t\right)$$

$$+\textstyle\sum_n C_n(t)\phi_n(\bar{r})E_n\, exp\left(-\frac{i}{\hbar}E_n t\right) + H'(t)\sum_n C_n(t)\phi_n(\bar{r})\, exp\left(-\frac{i}{\hbar}E_n t\right) = 0$$

$$[\because H_0\phi(\bar{r},t) = E_n\phi(\bar{r},t)]$$

$$\frac{\hbar}{i}\textstyle\sum_n \dot{c}_n(t)\phi_n(\bar{r})\, exp\left(-\frac{i}{\hbar}E_n t\right) + H'(t)\sum_n C_n(t)\phi_n(\bar{r})\, exp\left(-\frac{i}{\hbar}E_n t\right) = 0$$

Now multiply the above equation by $\phi_k(\bar{r})$

$$\frac{\hbar}{i}\textstyle\sum_n \dot{c}_n(t)\, exp\left(-\frac{i}{\hbar}E_n t\right)\left(\phi_k(\bar{r}),\phi_n(\bar{r})\right) + H'(t)\sum_n C_n(t)\, exp\left(-\frac{i}{\hbar}E_n t\right)\left(\phi_k(\bar{r}),\phi_n(\bar{r})\right) = 0$$

$\frac{\hbar}{i}\sum_n \dot{c}_n(t)\, exp\left(-\frac{i}{\hbar}E_n t\right)\delta_{kn} = -\sum_n C_n(t)\, exp\left(-\frac{i}{\hbar}E_n t\right) H'_{kn}(t)$

Where $(\phi_k(\bar{r}), \phi_n(\bar{r})) = \delta_{kn}$ and $(\phi_k(\bar{r}), H'(t)\phi_n(\bar{r})) = H'_{kn}(t)$; When $n = k$;

$\frac{\hbar}{i}\dot{c}_k(t)\, exp\left(-\frac{i}{\hbar}E_k t\right) = -\sum_n C_n(t)\, exp\left(-\frac{i}{\hbar}E_n t\right) H'_{kn}(t)$ $[\because \delta_{nk} = 1, when\; n = k]$

$$\dot{c}_k(t) = -\frac{i}{\hbar}\sum_n C_n(t)\, exp\left(\frac{i}{\hbar}(E_k - E_n)t\right) H'_{kn}(t)$$

$\dot{c}_k(t) = -\frac{i}{\hbar}\sum_n C_n(t)\, exp(i\omega_{kn}t)\, H'_{kn}(t)$ [7]

Where $\omega_{kn} = \frac{E_k - E_n}{\hbar}$ is the Bohr angular frequency.

If the time dependent H' is small, we can attempt to solve [7] by the method successive approximation. In the zeroth approximation we set $C_n^0(t) = C_n$. We insert this value on the R.H.S of [7] and integrated from $t = 0$ to $t = t$. We obtain immediately the first Born approximation:

$$C_k^1(t) = -\frac{i}{\hbar}\sum_n C_n \int_0^t H'_{kn}(t)\, exp(i\omega_{kn}t')\, dt' + C_k^1(0) \qquad [8]$$

The simplest and one of the most common situations is, the interaction H' is switched on at $t = 0$. I.e., at $t = 0$, the unperturbed and the perturbed generation solutions are coincide. I.e., $\phi(r, 0) = \psi(r, 0)$. In view of [5] and [6] the unperturbed and the perturbed generation solutions mean that $C_n(0) = C_n$ for all n. Furthermore, in most cases at $t = 0$, the system is in a definite stationary energy Eigen state.

I.e., $\phi(\bar{r}, 0) = \phi_i(\bar{r})$. This means that $C_n = \delta_{in}$ So, $C_n = 1$ for $i = n$ and $C_n = 0$ for $i \neq n$

Hence [8] becomes

$$C_k^1(t) = -\frac{i}{\hbar}\int_0^t H'_{ki}(t')\, exp(i\omega_{ki}t')dt' + \delta_{ik}$$

For k$\neq i$, then the above equation becomes

$$C_k^1(t) = -\frac{i}{\hbar}\int_0^t H'_{ki}(t')\, exp(i\omega_{ki}t')dt' \qquad [9]$$

In many applications it so happens that the perturbation is of short duration and H'varies slowly in time. Another case is that $H'$is switched on at $t = 0$ stay constant until some later instant t and then suddenly switched off. In both cases we can take out H'_{ki} in front of the integral.

$$C_k^1(t) = -\frac{i}{\hbar}H'_{ki}\int_0^t exp(i\omega_{ki}t')\, dt' = -\frac{i}{\hbar}H'_{ki}\left[\frac{exp(i\omega_{ki}t')}{i\omega_{ki}}\right]_0^t = -\frac{H'_{ki}}{\hbar}\left[\frac{exp(i\omega_{ki}t') - exp(0)}{\omega_{ki}}\right]$$

$$C_k^1(t) = -\frac{H'_{ki}}{\hbar}\left[\frac{exp(i\omega_{ki}t') - 1}{\omega_{ki}}\right] \qquad [10]$$

Equation [10] is known as 1st order perturbation theory. At, $k = n$,

$$C_n^1(t) = -\frac{H'_{ni}}{\hbar}\left[\frac{exp(i\omega_{ni}t')-1}{\omega_{ni}}\right] \quad [11]$$

Substituting [11] in [7]

$$\dot{c}_k(t) = -\frac{i}{\hbar}\sum_n C_n(t)\, exp(i\omega_{kn}t)\, H'_{kn}(t)$$

$$\dot{c}_k(t) = \frac{i}{\hbar^2}\sum_n \frac{H'_{ni}H'_{kn}}{\omega_{ni}}[exp(i\omega_{ni}t)\, exp(i\omega_{kn}t) - exp(i\omega_{nk}t)]$$

$$= \frac{i}{\hbar^2}\sum_n \frac{H'_{ni}H'_{kn}}{\omega_{ni}}[exp(i(\omega_{ni}+\omega_{kn})t) \; - exp(i\omega_{kn}t)]$$

Where $\omega_{ni} + \omega_{kn} = \frac{E_n - E_i}{\hbar} + \frac{E_k - E_n}{\hbar} = \frac{E_k - E_i}{\hbar} = \omega_{ki}$

Therefore,

$$\dot{c}_k(t) = \frac{i}{\hbar^2}\sum_n \frac{H'_{ni}H'_{kn}}{\omega_{ni}}[exp(i\omega_{ki}t) \; - exp(i\omega_{kn}t)]$$

16.5 Fermi Golden rule

The probability of finding the particle in k[th] state can be calculated from 1st order perturbation theory

$$C_k^1(t) = -\frac{H'_{ki}}{\hbar}\left[\frac{exp(i\omega_{ki}t')-1}{\omega_{ki}}\right] \quad [1]$$

Hence the probability of finding the particle

$$C_k^1(t)C_k^1(t)^{\times} = |C_k^1(t)|^2 = -\frac{H'_{ki}}{\hbar}\left[\frac{exp(i\omega_{ki}t')-1}{\omega_{ki}}\right] - \frac{H'_{ki}}{\hbar}\left[\frac{exp(-i\omega_{ki}t')-1}{\omega_{ki}}\right]$$

$$= \frac{1}{\omega_{ki}^2}[\{exp(i\omega_{ki}t') - 1\} \;\times\; \{exp(-i\omega_{ki}t') - 1\}]\frac{|H'_{ki}|^2}{\hbar^2}$$

$$= \frac{1}{\omega_{ki}^2}\left[1 + 1 - 2\frac{\left(exp(i\omega_{ki}t')+exp(-i\omega_{ki}t')\right)}{2}\right]\frac{|H'_{ki}|^2}{\hbar^2}$$

$$= \frac{1}{\omega_{ki}^2}[2(1 - cos\omega_{ki}t')]\frac{|H'_{ki}|^2}{\hbar^2}$$

$$|C_k^1(t)|^2 = \frac{4}{\omega_{ki}^2}sin^2\frac{\omega_{ki}t'}{2}\frac{|H'_{ki}|^2}{\hbar^2} = \frac{4|H'_{ki}|^2\left(sin^2\frac{\omega_{ki}t'}{2}\right)}{\hbar^2\omega_{ki}^2} \quad [2]$$

The factor $\left[\frac{sin^2\frac{\omega_{ki}t'}{2}}{\omega_{ki}^2}\right]$ is plotted as a function of ω_{ki} as shown in the following figure [16.4]

Physical interpretation of the curve:

The maximum value of $\left[\frac{sin^2\frac{\omega_{ki}t'}{2}}{\omega_{ki}^2}\right]$ occurs when ω_{ki} is zero. Let $\omega_{ki} = x$. Therefore,

$$\frac{sin^2\frac{xt'}{2}}{x^2} = \frac{1}{x^2}\left[\frac{xt'}{2} - \frac{(xt')^2}{3.2^3} + \dots\dots\dots\dots\dots\dots\dots\right]^2$$

$$= \frac{1}{x^2}\left[\frac{xt'}{2}\right]^2; \text{ neglecting the higher power of x}$$

$$\frac{sin^2\frac{xt'}{2}}{x^2} = \frac{t'^2}{4} \quad [3]$$

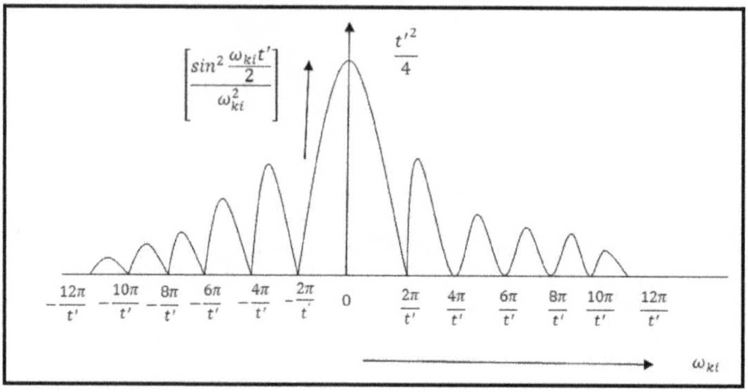

Fig.[16.5]: Shows Fermi's Golden rule and states that the transition probability per unit

So the highest peak value is $\frac{t'^2}{4}$, which can be seen in the graph. The value of $\frac{sin^2\frac{xt'}{2}}{x^2}$ will be zero when

$$\frac{sin^2\frac{xt'}{2}}{x^2} = 0$$

Or, $\frac{xt'}{2} = \pm n\pi \ or, \ x = \pm\frac{2n\pi}{t'}$

Therefore, $\omega_{ki} = x = \pm\frac{2n\pi}{t'}$ [4]

For $n = 1$ $\omega_{ki}[= x] = \pm\frac{2\pi}{t'}$

For $n = 2$ $\omega_{ki}[= x] = \pm\frac{4\pi}{t'}$

For $n = 3$ $\omega_{ki}[= x] = \pm\frac{6\pi}{t'}$

It should be noted that the height of the main peak increases in proportion to t'^2[see equation-3] while its breadth decreases inversely as t' [see equation-4]. The area under the curve is proportional to t'.The probability of finding the particle in k[th] state or The probability of transition of kth state is

$$|C_k^1(t)|^2 = \frac{4|H'_{ki}|^2\left(sin^2\frac{\omega_{ki}t'}{2}\right)}{\hbar^2\omega_{ki}^2}$$

Because of the factor $\frac{1}{\omega_{ki}^2}$, the probability is largest for those states in which the unperturbed energies $E_k^{(0)}$are closet to $E_i^{(0)}$. If $E_n^{(0)}$ is far away from $E_i^{(0)}$, the probability is small. Since the levels are very close to one another, they form a cluster around $E_k^{(0)} = E_i^{(0)}$. Since all levels in the cluster represent nearly the same physical properties, we may add up the probabilities of all levels in the cluster and consider them as a whole.

Total probability of transition $= \sum |C_k^1(t)|^2$. Now we define the density of final states $\rho(k)$; such that $\rho(k)dE$ is the number of such states in the energy range E_k and $E_k + dE$. The total probability for transition into these states is obtained multiplying equation [2] by $\rho(k)dE$ and integrating with respect to E. So, we can write

$$T = \sum |C_k^1(t)|^2 = \int |C_k^1(t)|^2 \rho(k)dE \qquad [5]$$

If the central peak of $\left[\frac{sin^2\frac{\omega_{ki}t'}{2}}{\omega_{ki}^2}\right]$ in the domain of integration [I.e., energy conserving transition]the main contribution comes from this peak and only a small error is involved in extending the limit of this integral to $-\infty \; to + \infty$.

$$T = \int_{+\infty}^{-\infty} \frac{4|H'_{ki}|^2\left(sin^2\frac{\omega_{ki}t'}{2}\right)}{\hbar^2\omega_{ki}^2}\rho(k)dE \qquad [5]$$

But $\qquad dE = \hbar d\omega \;\; as \; E_k - E_i = \hbar\omega_k$

$$T = \int_{+\infty}^{-\infty} \frac{4|H'_{ki}|^2\left(sin^2\frac{\omega_{ki}t'}{2}\right)}{\hbar^2\omega_{ki}^2}\rho(k)dE = \int_{+\infty}^{-\infty} \frac{4|H'_{ki}|^2\left(sin^2\frac{\omega_{ki}t'}{2}\right)}{\hbar^2\omega_{ki}^2}\rho(k)\hbar d\omega$$

$$= \frac{1}{\hbar}\; |H'_{ki}|^2\rho(k)\int_{+\infty}^{-\infty}\frac{4\left(sin^2\frac{\omega_{ki}t'}{2}\right)}{\omega_{ki}^2}d\omega \qquad [6]$$

Consider the integral $\int_{+\infty}^{-\infty}\frac{4\left(sin^2\frac{\omega_{ki}t'}{2}\right)}{\omega_{ki}^2}d\omega$; put $x = \frac{\omega_{ki}t'}{2} \;\; or, dx = \frac{t'}{2}d\omega_{ki} \;\; or, \omega_{ki}^2 = \frac{4x^2}{t'^2}$

$$4\int_{+\infty}^{-\infty}\frac{sin^2x}{\frac{4x^2}{t'^2}}\frac{2}{t'}dx = 2\,t'\int_{-\infty}^{+\infty}\frac{sin^2x}{x^2}dx = 2t'\pi \;;\; \text{where} \int_{-\infty}^{+\infty}\frac{sin^2x}{x^2}dx = \pi$$

Finally equation [6] becomes

$$T = \frac{1}{\hbar}\; |H'_{ki}|^2\rho(k)2t'\pi \qquad [7]$$

So the transition probability per unit time

$$T = \frac{2\pi}{\hbar}\; |H'_{ki}|^2\rho(k) \qquad [8]$$

The above equation [8] has wide applications in Quantum mechanics. Quantum theory of scattering can be explained with the help of this formula. *Equation [8] is Fermi's Golden rule and states that the transition probability per unit time.*

Significance of Fermi's Golden rule: *the transition probability per unit time is*

- Non zero only between continuum states of the same energy
- Proportional to the square of H'_{ki} of the perturbation connecting the states
- Proportional to the density of final states
- Expressions for Fermi golden rule are different for different forms of perturbation $H'(t)$.

16.6 Harmonic Perturbation

The harmonic perturbation is a time-dependent perturbation

$$V(t) = 2V_0 \cos\omega t \tag{1}$$

where V_0 is in general an operator. If the perturbation is time dependent $V(t) = V$, turned on at $t = 0$, the lowest order term in the expansion of the transition amplitude for $i \neq f$ is the transition probability at the first order in perturbation,

$$\frac{-i}{\hbar}\int_0^t dt'\langle f|V_I(t')|i\rangle = \frac{-i}{\hbar}\int_0^t dt'\left\langle f\left|e^{iH_0t'/\hbar}V_0\cos\omega t\, e^{-iH_0t'/\hbar}\right|i\right\rangle \tag{2}$$

$$= \frac{-i}{\hbar}\int_0^t dt' e^{-i(E_i-E_f)t'/\hbar} 2\cos\omega t\,\langle f|V_0|i\rangle$$

$$= \left(\frac{e^{-i(E_i-E_f+\hbar\omega)t/\hbar}-1}{E_i-E_f+\hbar\omega} + \frac{e^{-i(E_i-E_f-\hbar\omega)t/\hbar}-1}{E_i-E_f+\hbar\omega}\right)V_{fi} \tag{3}$$

Transition probability per unit time

$$\Gamma(i\to f) = lim_{t\to\infty}\frac{P(i\to f,t)}{t} = lim_{t\to\infty}\, 4\frac{sin^2\Delta Et/2\hbar}{t(\Delta E)^2}\left|V_{fi}\right|^2$$

$$= lim_{t\to\infty}\, 4\frac{sin^2\Delta Et/2\hbar}{t(\Delta E)^2}\left|V_{fi}\right|^2 = 4\, lim_{t\to\infty}\frac{sin^2(E_f-E_i)t/2\hbar}{t(E_f-E_i)^2} = 2\pi\delta(E_f-E_i)\frac{1}{\hbar} \tag{4}$$

The Transition rate is

$$\Gamma(i\to f) = \frac{2\pi}{\hbar}\left[\delta(E_i-E_f+\hbar\omega)+\delta(E_i-E_f-\hbar\omega)\right]\left|V_{fi}\right|^2 \tag{5}$$

Therefore, the "energy conservation" is now changed to

$$E_f = E_i \pm \hbar\omega \tag{6}$$

It was expected that the energy is strictly not conserved in the presence of a harmonic perturbation, because there is no time-translation invariance. However, there is a discrete time translational invariance by $t \to t' + \frac{2\pi}{\omega}$. Similarly to the discrete spatial translational invariance $x \to x + a$ that led to the Bloch wave, which "conserves the momentum modulo $\frac{2\pi\hbar}{a}$, the energy is "conserved modulo $\hbar\omega$," consistent with the above result. In fact, the second-order contribution can be seen to "conserve" the energy as

$$\boldsymbol{E_f = E_i \pm 2\hbar\omega \;\; or, E_f = E_i} \qquad [7]$$

The third order term

$$\boldsymbol{E_f = E_i \pm 3\hbar\omega \;\; or, E_f = E_i \pm \hbar\omega \;\; etc.} \qquad [8]$$

Sakurai discusses the photoelectric effect using this formalism. Strictly speaking, the photoelectric effect must be discussed only after properly quantizating the electromagnetic field to obtain "photons." We will not go into this discussion further here.

16.7 Periodic Time Dependent Perturbation

Let us consider the time-dependent perturbation $H'(t) = V(\vec{r})\cos\omega t$.

We can rewrite it as $H'(t) = V(\vec{r})\frac{(e^{i\omega t}+e^{-i\omega t})}{2} = \frac{1}{2}V(\vec{r})(e^{i\omega t} + e^{-i\omega t})$ $[\cos\omega t = \frac{(e^{i\omega t}+e^{-i\omega t})}{2}]$

$$H'_{mn} = \left(u_m^{(0)}, H'u_n^{(0)}\right) = \left(u_m^{(0)}, \frac{1}{2}V(\vec{r})(e^{i\omega t} + e^{-i\omega t})u_n^{(0)}\right)$$

$$= \frac{1}{2}\left(u_m^{(0)}, V(\vec{r})u_n^{(0)}\right)(e^{i\omega t} + e^{-i\omega t})$$

$$= \frac{1}{2}V_{mn}(e^{i\omega t} + e^{-i\omega t})$$

Because, scalar products are definite integrals with respect to space coordinates only.

Let unperturbed system be in the Eigen state $u_n^{(0)}$ only. After application of a time-dependent perturbation H', probability of transition of the system from $u_n^{(0)}$ to $u_m^{(0)}$, i.e., from energy level $E_n^{(0)}$ to $E_m^{(0)}$ within 1st order is given by

$$P_{n\to m} = \left|C_m^{(1)}\right|^2 = \left|\frac{1}{i\hbar}\int_0^1 H'_{mn}e^{i\omega_{mn}t}dt\right|^2 = \left|\frac{1}{i\hbar}\int_0^1 \frac{1}{2}V_{mn}(e^{i\omega t} + e^{-i\omega t})e^{i\omega_{mn}t}dt\right|^2$$

$$P_{n\to m} = \left|\frac{1}{i\hbar}\frac{1}{2}V_{mn}\left(\int_0^t e^{i(\omega+\omega_{mn})t}dt + \int_0^t e^{-i(\omega-\omega_{mn})t}dt\right)\right|^2$$

$$P_{n\to m} = \left|\frac{V_{mn}}{i2\hbar}\left(\frac{e^{i(\omega+\omega_{mn})t}}{i(\omega+\omega_{mn})}\bigg|_0^t + \frac{e^{-i(\omega-\omega_{mn})t}}{-i(\omega-\omega_{mn})}\bigg|_0^t\right)\right|^2$$

$$P_{n\to m} = \left|\frac{V_{mn}}{i2\hbar}\left(\frac{e^{i(\omega+\omega_{mn})t}-1}{i(\omega+\omega_{mn})} + \frac{e^{-i(\omega-\omega_{mn})t}-1}{-i(\omega-\omega_{mn})}\right)\right|^2$$

$$P_{n\to m} = \left|\frac{V_{mn}}{2i^2}\left(\frac{e^{\frac{i}{\hbar}\left(\hbar\omega+E_m^{(0)}-E_n^{(0)}\right)t}-1}{\hbar\omega+E_m^{(0)}-E_n^{(0)}} + \frac{e^{-\frac{i}{\hbar}\left(\hbar\omega-E_m^{(0)}+E_a^{(0)}\right)t}-1}{E_m^{(0)}-E_n^{(0)}-\hbar\omega}\right)\right|^2 \quad [\because \quad \omega_{mn} = \frac{E_m^{(0)}-E_n^{(0)}}{\hbar}]$$

Thus $\quad P_{n\to m} = \left|C_m^{(1)}\right|^2$

$$= \left|\frac{V_{mn}}{2}\left(\frac{1-e^{\frac{i}{\hbar}\left(\hbar\omega+E_m^{(0)}-E_n^{(0)}\right)t}}{E_m^{(0)}-\left(E_n^{(0)}-\hbar\omega\right)}+\frac{1-e^{-\frac{i}{\hbar}\left(\hbar\omega-E_m^{(0)}+E_n^{(0)}\right)t}-1}{E_m^{(0)}-\left(E_n^{(0)}+\hbar\omega\right)}\right)\right|^2 \qquad [1]$$

For $E_m^{(0)} \approx E_n^{(0)} + \hbar\omega \approx E_n^{(0)} + h\nu$, second term on R.H.S of equation (1) gets large amplitude making $P_{n\to m}$ appreciable, and hence we have appreciable transition of the system from energy level $E_n^{(0)}$ to higher energy level $E_m^{(0)}$. This is resonant absorption of energy $h\nu \approx E_m^{(0)} - E_n^{(0)}$ from perturbation of frequency $\nu = \frac{\omega}{2\pi} \approx \frac{E_m^{(0)}-E_n^{(0)}}{h}$.
Again, for $E_m^{(0)} \approx \left(E_n^{(0)} - \hbar\omega\right) \approx E_n^{(0)} - h\nu$, 1st term of R.H.S. of equation [1] gets large amplitude making $P_{n\to m}$ appreciable, and hence we have appreciable transition of the system from energy level $E_n^{(0)}$ to lower energy level $E_m^{(0)}$, and we have associated stimulated emission of energy $h\nu = \hbar\omega \approx E_n^{(0)} - E_m^{(0)}$ from the system.

The resonant absorption and stimulated emission are applicable to any quantum mechanical system. In case of optical phenomena, we have resonant absorption and stimulated emission of photon. This stimulated emission explains Einstein's stimulated emission associated with LASER action. The absorption and emission of photon of energy$h\nu = \hbar\omega \approx E_n^{(0)} - E_m^{(0)}$ explains Bohr frequency rule for hydrogen atom. (Spontaneous emission is also stimulated emission with help of residual or background radiation rather than by intentionally applied radiation. There is nothing called spontaneous emission.)

16.8 The Stark effect: A brief description

The effect is named after the German physicist Johannes Stark, who discovered it in 1913. It was independently discovered in the same year by the Italian physicist Antonino Lo Surdo, and in Italy it is thus sometimes called the **Stark–Lo Surdo effect**. The discovery of this effect contributed importantly to the development of quantum theory and was rewarded with the Nobel Prize in Physics for Johannes Stark in the year 1919.

The **Stark effect** is the shifting and splitting of spectral lines of atoms and molecules due to the presence of an external electric field. The amount of splitting or shifting is called the stark splitting or stark shift. There are two types of Stark Effect.

- First order Stark effect [weak field Stark effect]
- Second order Stark effect [strong field Stark effect]

It is the electric-field analogue of the Zeeman effect, where a spectral line is split into several components due to the presence of the magnetic field. Although initially coined for the static case, it is also used in the wider context to describe effect of time-dependent electric fields. In particular, the Stark effect is responsible for the pressure broadening (Stark broadening) of spectral lines by charged particles in plasmas. For majority of spectral lines, the Stark effect is either linear (proportional to the applied electric field) or quadratic with a high accuracy.

The Stark effect can be observed both for emission and absorption lines. The latter is sometimes called the **inverse Stark effect**, but this term is no longer used in the modern literature. In a semiconductor heterostructure, where a small band gap material is sandwiched between two layers of a larger band gap material, the Stark effect can be dramatically enhanced by bound excitons. This is because the electron and hole which form the exciton are pulled in opposite directions by the applied electric field, but they remain confined in the smaller band gap material, so the exciton is not merely pulled apart by the field. The quantum-confined Stark effect is widely used for semiconductor-based optical modulators, particularly for optical fiber communications.

16.9 Stark effect: perturbation theory for degenerate state

The splitting of spectral lines due to an electric field is known as stark effect. Let us consider the 1st order change in energy level of a hydrogen atom due to an external electric field of strength $\mathcal{E}$ directed along the Z- axis. This is known as stark effect in hydrogen[22,23]. Now we shall study the perturbation for degenerate state for H_2 atom with $n = 2$. We know, for the unperturbed Hamiltonian, the Eigen value equation is

$$H_0 \Psi_n^{(0)} = W_n^{(0)} \Psi_n^{(0)} \qquad [1]$$

$$Here\ \Psi_n^{(0)} = u_n \ and \ W_n^{(0)} = E_n;$$

The Eigen function u_n & the Eigen value E_n are known. Since $n = 2$ therefore $l = 0$ or $l = 1$ and m may have the values I.e., (0, 1, -1)

$$\left.\begin{aligned} For\ n = 2,\ l &= 0, & m &= 0 \\ &= 1, & m &= 0 \\ & & &= 1 \\ & & &= -1 \end{aligned}\right\} 4-states$$

Thus the quantum numbers $l\ and\ m$ have the following combination

$$(0,0); (1,0); (1,1); (1,-1)$$

Therefore, $n = 2$, the state of Hydrogen atom is 4- fold degenerate. The Eigen function for these states

$$u = R_{nl}(r)Y_{lm}(\theta,\varphi)$$

$$u_1 = R_{20}Y_{00} = \frac{1}{\sqrt{4\pi}}R_{20}(r)\ ; \qquad n = 2,\ l = 0,\ m = 0$$

$$u_2 = R_{21}Y_{10} = \sqrt{\frac{3}{4\pi}}R_{21}(r)\cos\theta\ ; \qquad n = 2,\ l = 1,\ m = 0$$

$$u_3 = R_{21}Y_{11} = -\sqrt{\frac{3}{8\pi}}R_{21}(r)\sin\theta\, e^{j\varphi}; \qquad n=2,\ l=1,\ m=1$$

$$u_4 = R_{21}Y_{1,-1} = \sqrt{\frac{3}{8\pi}}R_{21}(r)\sin\theta\, e^{-j\varphi}; \qquad n=2,\ l=1,\ m=-1 \qquad [2]$$

Here $R_{20}(r) = \frac{1}{\sqrt{2}}\frac{1}{a_0^{3/2}}\left(1-\frac{r}{2a_0}\right)e^{-r/2a_0}$ & $R_{21}(r) = \frac{1}{2\sqrt{6}}\frac{1}{a_0^{3/2}}\frac{r}{a_0}e^{-r/2a_0}$ [3]

For $n = 2$, the energy Eigen value for Hydrogen atom is

$$E_2 = W_2^{(0)} = -\frac{\mu e^4}{8\hbar^2} \qquad [\because E_n = -\frac{\mu e^4}{2\hbar^2}\left(\frac{1}{n^2}\right)] \qquad [4]$$

The Eigen function $\Psi^{(0)}$ can be expanded as a linear combination of u_1, u_2, u_3, u_4. Thus the Eigen function

$$\Psi^{(0)} = c_1u_1 + c_2u_2 + c_3u_3 + c_4u_4 \qquad [5]$$

We know, the equation of 1st order perturbation is,

$$H_0\Psi^{(1)} + H'\Psi^{(0)} = W_2^{(0)}\Psi^{(1)} + W_2^{(1)}\Psi^{(0)} \qquad [6]$$

We can also expand $\Psi^{(1)}$ as a linear combination of u_m. Let

$$\Psi^{(1)} = \sum_m a_m^{(1)}u_m \qquad [7]$$

Using [7] & [5] in [6] we have

$$H_0\sum_m a_m^{(1)}u_m + H'[c_1u_1 + c_2u_2 + c_3u_3 + c_4u_4]$$
$$= W_2^{(0)}\sum_m a_m^{(1)}u_m + W_2^{(1)}[c_1u_1 + c_2u_2 + c_3u_3 + c_4u_4]$$

$$\sum_m a_m^{(1)}E_mu_m + H'[c_1u_1 + c_2u_2 + c_3u_3 + c_4u_4]$$
$$= W_2^{(0)}\sum_m a_m^{(1)}u_m + W_2^{(1)}[c_1u_1 + c_2u_2 + c_3u_3 + c_4u_4]$$

In ket notation the above equation can be written

$$\sum_m a_m^{(1)}E_m|m\rangle + H'[c_1|1\rangle + c_2|2\rangle + c_3|3\rangle + c_4|4\rangle]$$
$$= W_2^{(0)}\sum_m a_m^{(1)}|m\rangle + w_2^{(1)}[c_1|1\rangle + c_2|2\rangle + c_3|3\rangle + c_4|4\rangle]$$

Taking scalar product of $\langle 1|$ to the left on both sides

$$\langle 1|\sum_m a_m^{(1)}E_m|m\rangle + c_1\langle 1|H'|1\rangle[+c_2\langle 1|H'|2\rangle + c_3\langle 1|H'|3\rangle + c_4\langle 1|H'|4\rangle]$$
$$= W_2^{(0)}\sum_m a_m^{(1)}\langle 1|m\rangle + W_2^{(1)}[c_1\langle 1|1\rangle + c_2\langle 1|2\rangle + c_3\langle 1|3\rangle + c_4\langle 1|4\rangle]$$

Or, $\sum_m a_m^{(1)} E_m \delta_{1m} + c_1 H'_{11} + c_2 H'_{12} + c_3 H'_{13} + c_4 H'_{14}$

$$= W_2^{(0)} \sum_m a_m^{(1)} \delta_{1m} + W_2^{(1)} [c_1 \delta_{11} + c_2 \delta_{12} + c_3 \delta_{13} + c_4 \delta_{14}]$$

Or, $a_1^{(1)} E_1 + c_1 H'_{11} + c_2 H'_{12} + c_3 H'_{13} + c_4 H'_{14} = W_2^{(0)} a_1^{(1)} + W_2^{(1)} c_1$

Or, $a_1^{(1)} W_2^{(0)} + c_1 H'_{11} + c_2 H'_{12} + c_3 H'_{13} + c_4 H'_{14} = W_2^{(0)} a_1^{(1)} + W_2^{(1)} c_1 \; [\because E_1 = W_2^{(0)}]$

Or, $c_1 \left(H'_{11} - W_2^{(1)} \right) + c_2 H'_{12} + c_3 H'_{13} + c_4 H'_{14} = 0$

Similarly,

$$c_1 H'_{21} + c_2 \left(H'_{22} - W_2^{(1)} \right) + c_3 H'_{23} + c_4 H'_{24} = 0$$

$$c_1 H'_{31} + c_2 H'_{32} + c_3 \left(H'_{33} - W_2^{(1)} \right) + c_4 H'_{34} = 0$$

$$c_1 H'_{41} + c_2 H'_{42} + c_3 H'_{43} + c_4 \left(H'_{44} - W_2^{(1)} \right) = 0 \qquad [9]$$

For non-trivial solution:

$$\begin{vmatrix} H'_{11} - W_2^{(1)} & H'_{12} & H'_{13} & H'_{14} \\ H'_{21} & H'_{22} - W_2^{(1)} & H'_{23} & H'_{24} \\ H'_{31} & H'_{32} & H'_{33} - W_2^{(1)} & H'_{34} \\ H'_{41} & H'_{42} & H'_{43} & H'_{44} - W_2^{(1)} \end{vmatrix} = 0 \qquad [10]$$

This is called the secular equation for 4-fold degenerate system. Thus, for g–fold degenerate system, the secular equation becomes

$$\begin{vmatrix} H'_{11} - W_2^{(1)} & H'_{12} & \cdots & H'_{g1} \\ H'_{21} & H'_{22} - W_2^{(1)} & \cdots & H'_{g2} \\ \vdots & \vdots & \cdots & \vdots \\ H'_{g1} & H'_{g2} & \cdots & H'_{gg} - W_2^{(1)} \end{vmatrix} = 0$$

Solving this determinant, one can get $W_2^{(1)}$ for g-fold degenerate state. Now we study the Stark effect. We consider the perturbation H'; which is the extra energy of nucleus & electron due to external energy $\mathcal{E}$ is

$$H' = -e\mathcal{E}z = e\mathcal{E}r \cos\theta \qquad [11]$$

Where, e is the electronic charge and $z = r\cos\theta$ in polar coordinates. In this case to calculate the 1st order correction for energy Eigen value $W_2^{(1)}$, we also have to calculate the matrix elements. Here

$$H'_{11} = \langle 1|H'|1\rangle = \int u_1^{\times} H' u_1 d\tau$$

Or, $H'_{11} = \int_{r=0}^{\infty}\int_{\theta=0}^{\pi}\int_{\varphi=0}^{2\pi}\frac{1}{\sqrt{4\pi}}R_{20}(r)[-e\mathcal{E}r\cos\theta]\frac{1}{\sqrt{4\pi}}R_{20}(r)r^2dr\sin\theta\, d\theta d\varphi$

Or, $H'_{11} = \frac{1}{4\pi}(-e\mathcal{E})\int_{r=0}^{\infty}R_{20}^2(r)\,r^3dr\int_{\varphi=0}^{2\pi}d\varphi\int_{\theta=0}^{\pi}\sin\theta\cos\theta\, d\theta$

Or, $H'_{11} = 0$; Because $\int_{\theta=0}^{\pi}\sin\theta\cos\theta\, d\theta = 0$

And $H'_{23} = \langle 2|H'|3\rangle = \int u_2^{\times}H'u_3d\tau$

Or, $H'_{23} = \int_{r=0}^{\infty}\int_{\theta=0}^{\pi}\int_{\varphi=0}^{2\pi}\sqrt{\frac{3}{4\pi}}R_{21}(r)\cos\theta\,[-e\mathcal{E}r\cos\theta]\sqrt{\frac{3}{4\pi}}R_{21}(r)\sin\theta\, e^{j\varphi}r^2dr\sin\theta\, d\theta d\varphi$

Here $\int_{\varphi=0}^{2\pi}e^{j\varphi}\,d\varphi = \frac{1}{j\varphi}\left[e^{j\varphi}\right]_0^{2\pi} = \frac{1}{j\varphi}[\cos\varphi + j\sin\varphi]_0^{2\pi} = \frac{1}{j\varphi}[\cos 2\pi + j\sin 2\pi - e^0]$

$= \frac{1}{j\varphi}[1 + j.0 - 1] = 0$. Thus $H'_{23} = 0$

Similarly,

$$H'_{11} = H'_{13} = H'_{14} = H'_{22} = H'_{23} = H'_{24} = H'_{31} = H'_{32} = H'_{33} = H'_{34} = H'_{41}$$
$$= H'_{42} = H'_{43} = H'_{44} = 0$$

Here

$$H'_{12} = \langle 1|H'|2\rangle = \int u_1^{\times}H'u_2d\tau$$

$$H'_{12} = \int_{r=0}^{\infty}\int_{\theta=0}^{\pi}\int_{\varphi=0}^{2\pi}\frac{1}{\sqrt{4\pi}}R_{20}(r)[-e\mathcal{E}r\cos\theta]\sqrt{\frac{3}{4\pi}}R_{21}(r)\cos\theta\, r^2dr\sin\theta\, d\theta d\varphi$$

$$= -\frac{e\mathcal{E}\sqrt{3}}{4\pi}\int_{r=0}^{\infty}\int_{\theta=0}^{\pi}\int_{\varphi=0}^{2\pi}\frac{1}{\sqrt{2}}\frac{1}{a_0^{3/2}}\left(1-\frac{r}{2a_0}\right)^{-r/2a_0} r\cos\theta\frac{1}{2\sqrt{b}}\frac{1}{a_0^{3/2}}\frac{r}{a_0}e^{-r/2a_0}\cos\theta\, r^2dr\sin\theta\, d\theta d\varphi$$

$$H'_{12} = -\frac{e\mathcal{E}}{16\pi a_0^4}\int_{r=0}^{\infty}\left(1-\frac{x}{2}\right)x^4a_0^4e^{-x}a_0\,dx\int_{+1}^{-1}y^2(-dy)2\pi$$

Let $x = \frac{r}{a_0}$ $or,\ dx = \frac{1}{a_0}dr$ $or, dr = a_0dx$ and $y = \cos\theta$ $or, dy = -\sin\theta\, d\theta$

If $\theta = 0, y = 1$ and If $\theta = \pi, y = -1$

$$H'_{12} = -\frac{e\mathcal{E}}{16\pi a_0^4}a_0^5\,2\pi\int_0^{\infty}\left(1-\frac{x}{2}\right)x^4e^{-x}\,dx\int_{-1}^{+1}y^2dy$$

$$H'_{12} = -\frac{a_0e\mathcal{E}}{8}\left[\left(\int_0^{\infty}x^4e^{-x}dx - \frac{1}{2}\int_0^{\infty}x^5e^{-x}dx\right)\int_{-1}^{+1}y^2dy\right]$$

$$H'_{12} = -\frac{a_0e\mathcal{E}}{8}\left[\left(4! - \frac{1}{2}5!\right)\frac{2}{3}\right]$$

$$H'_{12} = -\frac{e\mathcal{E}a_0}{8}\left[(24-60)\frac{2}{3}\right] = -\frac{e\mathcal{E}a_0}{8}(-36)\frac{2}{3} = 3e\mathcal{E}a_0$$

$H'_{12} = -g$; where $g = 3|e|\mathcal{E}a_0$

Similarly, $H'_{21} = -g$.

Putting these matrix elements in [10]

$$\begin{vmatrix} -w_2^{(1)} & -g & 0 & 0 \\ -g & -w_2^{(1)} & 0 & 0 \\ 0 & 0 & -w_2^{(1)} & 0 \\ 0 & 0 & 0 & -w_2^{(1)} \end{vmatrix} = 0$$

or, $$-w_2^{(1)} \begin{vmatrix} -w_2^{(1)} & 0 & 0 \\ 0 & -w_2^{(1)} & 0 \\ 0 & 0 & -w_2^{(1)} \end{vmatrix} + g \begin{vmatrix} -g & 0 & 0 \\ 0 & -w_2^{(1)} & 0 \\ 0 & 0 & -w_2^{(1)} \end{vmatrix} = 0$$

Or, $$-W_2^{(1)}\left[-W_2^{(1)}\left(W_2^{(1)}\right)^2\right] + g\left[-g\left(W_2^{(1)}\right)^2\right] = 0$$

Or, $$\left(W_2^{(1)}\right)^4 - g^2\left(W_2^{(1)}\right)^2 = 0$$

Or, $$\left(W_2^{(1)}\right)^2\left[\left(W_2^{(1)}\right)^2 - g^2\right] = 0$$

Either $\left(W_2^{(1)}\right)^2 = 0 \quad \therefore W_2^{(1)} = 0,0$

And $\left(W_2^{(1)}\right)^2 - g^2 = 0 \quad \therefore W_2^{(1)} = \pm g$

Thus $W_2^{(1)} = -g, \ +g, \ 0, \ 0$ [roots]

Thus, the energy level levels with 1st order correction is

$$W_2 = W_2^{(0)} + W_2^{(1)} = -\frac{\mu e^4}{8\hbar^2} + g, \quad -\frac{\mu e^4}{8\hbar^2} - g, \quad -\frac{\mu e^4}{\hbar^2}, \quad -\frac{\mu e^4}{\hbar^2}$$

Thus, the splitting of energy level in stark effect is shown in the following:

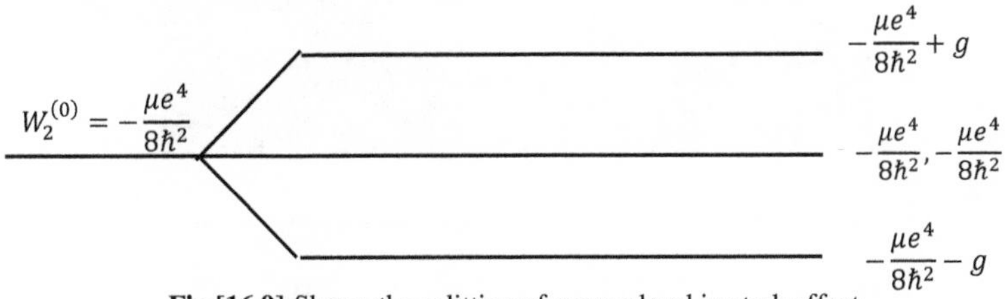

Fig.[16.9]: Shows the splitting of energy level in stark effect

Hence the zero –order wave function

$$u_1 = c_1(\Psi_{200} + \Psi_{210})$$

This wave function corresponds to the energy level

$$W_2 = W_2^{(0)} + W_2^{(1)} \quad [E_2 = E_2^{(0)} + E_2^{(1)}]$$

$$= -\frac{\mu e^4}{8\hbar^2} - 3e\mathcal{E}a_0 = -\frac{e^2}{8a_0} - 3e\mathcal{E}a_0$$; Where, a_0 is Bohr magneton

Similarly, for the second root, the wave function

$$u_2 = c_1(\Psi_{200} + \Psi_{210})$$

And it correspondence to energy

$$W_2 = -\frac{e^2}{8a_0} + 3e\mathcal{E}a_0$$

The other two cases correspond to unchanged energy levels and their wave functions may be taken as Ψ_{211} and $\Psi_{21,-1}$. Thus for $\mu[= m] \pm 1$; there is no splitting of the energy levels in the electric field.

16.10 Zeeman Effect

The atomic energy levels, the transitions between these levels, and the associated spectral lines discussed to this point have implicitly assumed that there are no magnetic fields influencing the atom. If there are magnetic fields present, the atomic energy levels are split into a larger number of levels and the spectral lines are also split. This splitting is called the *Zeeman Effect*. The following figure illustrates the Zeeman Effect. The Zeeman Effect can be interpreted in terms of the precession of the orbital angular momentum vector in the magnetic field, similar to the precession of the axis of a spinning top in a gravitational field.

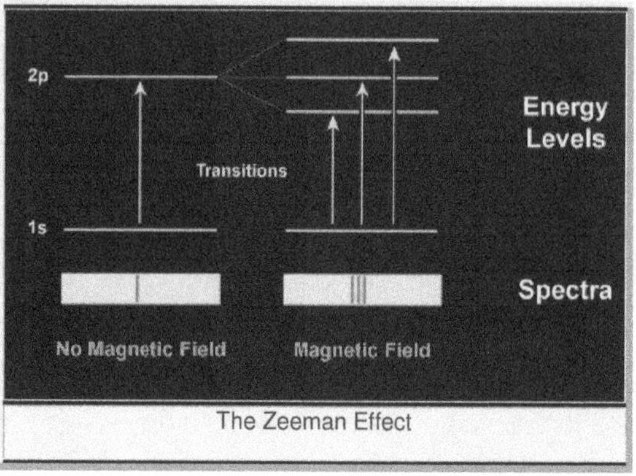

The Zeeman Effect

[Fig.16.10]: illustrates the Zeeman Effect

16.11 Zeeman Effect: Perturbation energy in case of hydrogen atom

The change in the energy levels of an atom caused by a uniform external magnetic field is called the Zeeman Effect. The unperturbed wave functions of the electron are determined by the

Hamiltonian $H^{(0)}$, given by

$$H^{(0)} = -\frac{\hbar^2}{8\pi^2\mu}\nabla^2 - \frac{e^2}{r} \quad [1]$$

The wave functions can be written in the following form due to the central force

$$\psi_{nlm}^{(0)} = Y_{nlm}\rho'\, L_{n+l}^{2l+1}(\rho)P_l^m(cos\theta)e^{im\phi} \quad [2]$$

Thus the Quantum mechanical Hamiltonian in case of hydrogen atom is given by

$$H = -\frac{\hbar^2}{8\pi^2\mu}\nabla^2 - \frac{ieh}{2\pi\mu c}\frac{\mathcal{H}}{2}\frac{\partial}{\partial\phi} - \frac{e^2}{r} - \frac{e^2}{2\mu c^2}\frac{1}{2}\mathcal{H}^2 r^2 sin^2\theta$$

Where $\mathcal{H} = curlA = \overline{\nabla} \times \bar{A}$; A *is the vector potential* [3]

The first-order perturbed Hamiltonian is given by setting $\lambda = 1$;

$$H^{(1)} = -\frac{ieh}{2\pi\mu c}\frac{\mathcal{H}}{2}\frac{\partial}{\partial\phi} \quad [4]$$

The first-order perturbed Eigen values

$$E_{nlm} = E_{nlm}^{(0)} + \lambda E_{nlm}^{(1)} \quad [5]$$

The first-order perturbed energy

$$E^{(1)} = \int \psi_{nlm}^{(0)\times} H^{(1)}\psi_{nlm}^{(0)} d\tau \quad [6]$$

$$E^{(1)} = \int \psi_{nlm}^{(0)\times} - \frac{ieh}{2\pi\mu c}\frac{\mathcal{H}}{2}\frac{\partial}{\partial\phi}[Y_{nlm}\rho'\, L_{n+l}^{2l+1}(\rho)P_l^m(cos\theta)e^{im\phi}]d\tau$$

$$= -\frac{ieh\mathcal{H}}{4\pi\mu c}\, im \int \psi_{nlm}^{(0)\times}\, \psi_{nlm}^{(0)}\, d\tau$$

$$E^{(1)} = -\frac{ieh\mathcal{H}}{4\pi\mu c}\, im = \frac{eh\mathcal{H}}{4\pi\mu c}m$$

The perturbed Eigen values

$$E = E^{(0)} + \frac{eh\mathcal{H}}{4\pi\mu c}m \quad [7]$$

Let the transition take place from an energy state to another energy state

$$E_1 = E_1^{(0)} + \frac{eh\mathcal{H}}{4\pi\mu c}m_1 \quad [8]$$

$$E_2 = E_2^{(0)} + \frac{eh\mathcal{H}}{4\pi\mu c}m_2 \quad [9]$$

The frequencies of lines emitted in transition between two states of the system are given by

$$\upsilon = \frac{E_2 - E_1}{h} = \frac{E_2^{(0)} - E_1^{(0)}}{h} + \frac{e\mathcal{H}}{4\pi\mu c}\Delta m \quad [10]$$

We know that only those transitions are possible for which

$$\Delta m = 0 \quad or, \Delta m = \pm 1 \qquad [11]$$

Therefore, $\upsilon_1 = \upsilon_0$

$$\upsilon_2 = \upsilon_0 + \frac{e\mathcal{H}}{4\pi\mu c} \qquad [12]$$

$$\upsilon_3 = \upsilon_0 - \frac{e\mathcal{H}}{4\pi\mu c}$$

which are identical with those obtained classically by means of the Larmour theorem. A schematic diagram of the possible transitions between two levels with $l_1 = 2$ and $l_2 = 1$ as shown in figure [16.9]

The splitting of energy levels derived above gives rise to change in the pattern of spectral lines emitted by an atom. *This change is known as Zeeman effect.* Equation [7] may be interpreted to mean that ψ_{nlm} has a magnetic moment

$$M = \frac{eh}{4\pi\mu c} \qquad [13]$$

This value suggests that the Z- component of the magnetic moment is quantized to integral multiples of the Bohr magneton.

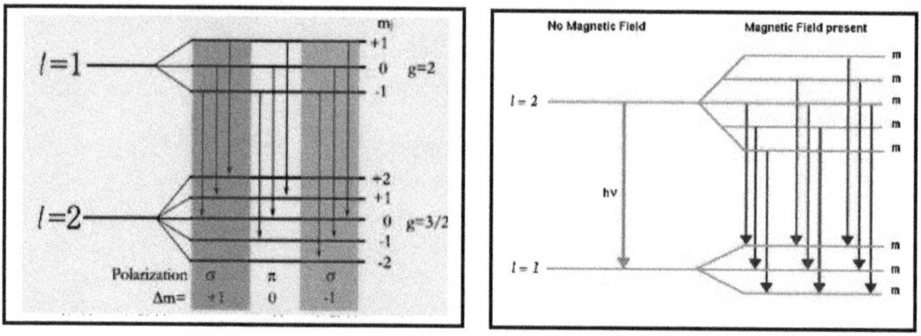

[Fig.16.11]: Shows the possible transitions between two levels

16.12 First order change in energy of the oscillator for the perturbation $ax^3 + bx^4$

The 1st order correction to the energy Eigen value

$$E_n^{(1)} = \int \left(\Psi_n^{(0)}\right)^{\times} H' \Psi_n^{(0)}\, d\tau \qquad [1]$$

$$= \int_{-\infty}^{+\infty} \left(\Psi_n^{(0)}\right)^{\times} [ax^3 + bx^4] \Psi_n^{(0)}\, dx$$

$$= a\int_{-\infty}^{+\infty}\left(\Psi_n^{(0)}\right)^{\times} x^3\Psi_n^{(0)}\,dx + b\int_{-\infty}^{+\infty}\left(\Psi_n^{(0)}\right)^{\times} x^4\Psi_n^{(0)}\,dx \qquad [2]$$

Since x^3 is an odd function and $\left(\Psi_n^{(0)}\right)^{\times}\Psi_n^{(0)}$ is an Eigen function, hence the value of the 1st integral is zero I.e., 1st order perturbation due to $x^3$is zero. Thus

$$E_n^{(1)} = b\int_{-\infty}^{+\infty}\left(\Psi_n^{(0)}\right)^{\times} x^4\Psi_n^{(0)}\,dx \qquad [3]$$

The ground state wave function

$$\Psi_n^{(0)} = \left(\frac{k}{\pi\hbar\omega}\right)^{\frac{1}{4}} exp\left(-\frac{kx^2}{2\hbar\omega}\right)$$

$$E_n^{(1)} = b\int_{-\infty}^{+\infty}\left(\frac{k}{\pi\hbar\omega}\right)^{\frac{1}{4}} exp\left(-\frac{kx^2}{2\hbar\omega}\right)x^4\left(\frac{k}{\pi\hbar\omega}\right)^{\frac{1}{4}} exp\left(-\frac{kx^2}{2\hbar\omega}\right)dx$$

$$= b\left(\frac{k}{\pi\hbar\omega}\right)^{\frac{1}{2}} 2\int_{0}^{+\infty} x^4\, exp\left(\frac{-kx^2}{\hbar\omega}\right)dx$$

But $$\int_0^{\infty} x^m\, exp(-ax^n)dx = \frac{1}{n}\frac{\Gamma\left(\frac{m+1}{n}\right)}{a^{\frac{m+1}{n}}}$$

$$= b\left(\frac{k}{\pi\hbar\omega}\right)^{\frac{1}{2}}\frac{\Gamma\frac{5}{2}}{\left(\frac{k}{\hbar\omega}\right)^{\frac{5}{2}}} = b\frac{k^{1/2}}{\pi^{\frac{1}{2}}(\hbar\omega)^{\frac{1}{2}}}\ \frac{3}{2}\ \frac{1}{2}\sqrt{\pi}\ \frac{(\hbar\omega)^{\frac{5}{2}}}{k^{\frac{5}{2}}}$$

$$= \frac{3b}{4}k^{-1}(\hbar\omega)^2 = \frac{3b}{4}\frac{(\hbar\omega)^2}{k}$$

Therefore, the total energy of the L.H.S with second order correction is

$$E_n = E_n^{(0)} + E_n^{(1)}$$

But we know the unperturbed energy

$$E_n^{(0)} = \left(n+\frac{1}{2}\right)\hbar\omega$$

$$\therefore\ E_n = \left(n+\frac{1}{2}\right)\hbar\omega + \frac{3b}{4}\frac{(\hbar\omega)^2}{k}$$

16.13 First order energy correction for one dimensional Harmonic Oscillator for the perturbing Hamiltonian ax^4

The 1st order correction to the energy Eigen value

$$E_n^{(1)} = \int\left(\Psi_n^{(0)}\right)^{\times} H'\Psi_n^{(0)}\,d\tau \qquad [1]$$

$$= \int_{-\infty}^{+\infty}\left(\Psi_n^{(0)}\right)^{\times}[ax^4]\Psi_n^{(0)}\,dx \qquad [2]$$

Where the perturbing Hamiltonian $H' = ax^4$ [3]

And the ground state wave function $\Psi_n^{(0)} = \left(\frac{k}{\pi\hbar\omega}\right)^{\frac{1}{4}} exp\left(-\frac{kx^2}{2\hbar\omega}\right)$ [4]

So, equation [2] becomes

$$E_n^{(1)} = \int_{-\infty}^{+\infty} \left(\frac{k}{\pi\hbar\omega}\right)^{\frac{1}{4}} exp\left(-\frac{kx^2}{2\hbar\omega}\right) [ax^4] \left(\frac{k}{\pi\hbar\omega}\right)^{\frac{1}{4}} exp\left(-\frac{kx^2}{2\hbar\omega}\right) dx$$

$$= 2a\left(\frac{k}{\pi\hbar\omega}\right)^{\frac{1}{2}} \int_0^{+\infty} exp\left(-\frac{kx^2}{\hbar\omega}\right) x^4\, dx \quad [5]$$

From standard integral

$$\int_0^{\infty} x^m exp(-\alpha x^n) = \frac{1}{n}\frac{\Gamma(m+1/n)}{\alpha^{(m+1/n)}} \quad [6]$$

Or,

$$E_n^{(1)} = 2a\left(\frac{k}{\pi\hbar\omega}\right)^{\frac{1}{2}} \frac{1}{2}\frac{\Gamma(4+1/2)}{\left(\frac{k}{\hbar\omega}\right)^{(4+1/2)}}$$

$$E_n^{(1)} = 2a\left(\frac{k}{\pi\hbar\omega}\right)^{\frac{1}{2}} \frac{1}{2}\frac{\Gamma(5/2)}{\left(\frac{k}{\hbar\omega}\right)^{(5/2)}} = 2a\left(\frac{\hbar\omega}{k}\right)^2 \frac{1}{\pi^{1/2}}\frac{1}{2}\frac{3}{2}\frac{1}{2}\pi^{1/2} = \frac{3a}{4}\left(\frac{\hbar\omega}{k}\right)^2 \quad [7]$$

Therefore, the total energy of the L.H.S with second order correction is

$$E_n = E_n^{(0)} + E_n^{(1)}$$

But we know the unperturbed energy

$$E_n^{(0)} = \left(n+\frac{1}{2}\right)\hbar\omega = \frac{1}{2}\hbar\omega \;\; ; \text{when } n = 0$$

$$\therefore \;\; E_n = \frac{1}{2}\hbar\omega + \frac{3a}{4}\frac{(\hbar\omega)^2}{k} = \frac{1}{2}\hbar\omega\left[1+\frac{3a}{2k^2}\right]$$

16.14 Selection Rule: Electrostatic polarization and the dipole moment

We consider an electron bound in an atom and placed in a weak uniform external electric field $\vec{E}$. The field can be derived from an electrostatic potential. Suppose the atom is hydrogen like having electric dipole moment

$\vec{p} = -q\vec{r}$; where $\vec{r}$ is the location of electron with respect to nucleus.The electrostatic potential of the electric dipole given by

$\varphi(r) = -\vec{E}.\bar{r}$; Where the coordinate origin is most conveniently chosen at the position of the nucleus. And the perturbation potential

$$gV = -e\varphi = e\vec{E}.\bar{r} \quad [1]$$

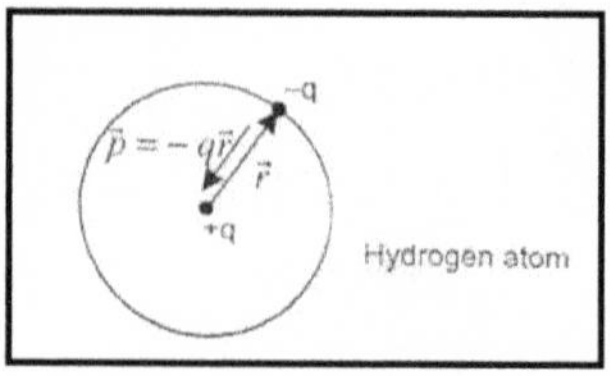

Fig.[16.14a]: Shows two body systems like hydrogen atom

The energy of the system to second order is given by

$$E_n = E_n^{(0)} + e\vec{E}.\bar{r}_{nn} + e^2 \sum_{k\neq n} \frac{(\vec{E}.\bar{r}_{nk})(\vec{E}.\bar{r}_{kn})}{E_n^{(0)} - E_k^{(0)}} \quad [2]$$

where all matrix elements are to be taken with respect to the unperturbed Eigen states. *The shift of energy levels in an electric field $\vec{E}$ is known as the Stark effect.* The first two terms of the perturbation expansion give accurate results for applied fields which are small compared to the internal electric field of the atom.

The latter is in order of magnitude given by $\frac{E^{(0)}}{ea} \approx 10^{10} volts/meter$. In practice, this condition is always well satisfied and successive terms in the perturbation expansion decrease rapidly and uniformly, except that some terms may vanish owing to certain symmetry properties of the system. The most important example of this is the absence of the first- order term in almost all atomic states, with the important exception of hydrogenic atoms.

The particular property which causes the expectation value $\langle\bar{r}\rangle_n = \bar{r}_{nn}$ of the position operator in the nth Eigen state to vanish is symmetry of the unperturbed system with respect to coordinate reflection. For, assuming that the parity operator U_p commutes with unperturbed Hamiltonian H_0 and that $E_n^{(0)}$ belongs to a non-degenerate Eigen states of H_0, this state $\psi_n^{(0)}$ must be an Eigen state of U_p with Eigen value $+1\ or\ -1$.

$$U_p\psi_n^{(0)} = \pm 1\psi_n^{(0)}$$

The expectation value of the position operator in such a state is

$$\langle\bar{r}\rangle_n = \left(\psi_n^{(0)}, \bar{r}\psi_n^{(0)}\right) = -\left(\psi_n^{(0)}, U_p^\dagger \bar{r} U_p \psi_n^{(0)}\right)$$

$$= -\left(U_p\psi_n^{(0)}, \bar{r}U_p\psi_n^{(0)}\right) = -\left(\psi_n^{(0)}, \bar{r}\psi_n^{(0)}\right) = -\langle\bar{r}\rangle_n \quad [\because U_p^{\times}U_p = 1]$$

Hence $\langle\bar{r}\rangle_n = 0$. In terms of ordinary wave function we may say that if $\psi(r)$ has definite parity, then the probability density $\rho = |\psi(r)|^2$ is an even function of the coordinates and the center of probability must be at the coordinate origin. A similar argument shows that the matrix element of $\bar{r}$ between any two states of the same parity vanishes. If symmetry properties cause certain matrix elements of an operator to be zero, we say that a selection rule is in effect. On the other hand if $\psi_{na}^{(0)}$ and $\psi_{nb}^{(0)}$ are any two states of the different parity, then

$$\left(\psi_{na}^{(0)}, \bar{r}\psi_{nb}^{(0)}\right) = \left(\psi_{na}^{(0)}, x\psi_{nb}^{(0)}\right) \neq 0,\ \left(\psi_{na}^{(0)}, y\psi_{nb}^{(0)}\right) \neq 0 \text{ and } \left(\psi_{na}^{(0)}, z\psi_{nb}^{(0)}\right) \neq 0$$

i.e., $\int \psi_{na}^{(0)\times} x\psi_{nb}^{(0)} d\tau \neq 0,\ \int \psi_{nb}^{(0)\times} y\psi_{nb}^{(0)} d\tau \neq 0,\ \int \psi_{na}^{(0)\times} z\psi_{nb}^{(0)} d\tau \neq 0.$

This can happen if any one of $\psi_{na}^{(0)}$ and $\psi_{nb}^{(0)}$ is odd function and the other is even function. If the integrands are odd functions, the integrals become zero, i.e.,

$$\int \psi_{na}^{(0)\times} x\psi_{nb}^{(0)} d\tau = 0,\ \int \psi_{nb}^{(0)\times} y\psi_{nb}^{(0)} d\tau = 0,\ \int \psi_{na}^{(0)\times} z\psi_{nb}^{(0)} d\tau = 0$$

This can happen if both $\psi_{na}^{(0)}$ and $\psi_{nb}^{(0)}$ are even functions or both $\psi_{na}^{(0)}$ and $\psi_{nb}^{(0)}$ are odd functions. In case of hydrogen atom, $\psi_{na}^{(0)}$ and $\psi_{nb}^{(0)}$ depend on quantum numbers n, ℓ and m_ℓ. Thus for the integrals to be non-zero, we get some restrictions or selection rules $\Delta\ell = \pm 1$, $\Delta m_\ell = 0, \pm 1$. $\Delta n =$ Anything.

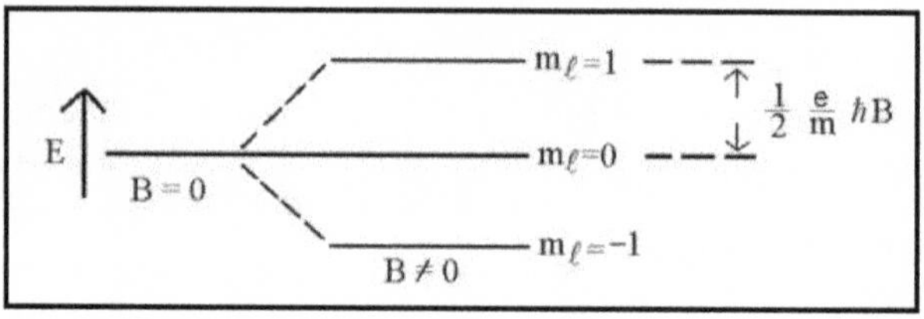

Fig.[16.14b]: Shows selection rules $\Delta\ell = \pm 1$, $\Delta m_\ell = 0, \pm 1$.

Exercise

1. Explain why we need the perturbation theory in Quantum Mechanics.
2. Derive the time independent perturbation theory for non-degenerate case.
3. Calculate the first order perturbation to the energy Eigen value and Eigen function.
4. Calculate the Second order perturbation to the energy Eigen value and Eigen function.
5. Establish Dirac's time dependent perturbation theory.
6. What is the significance of Fermi's Golden rule?
7. Show that the transition probability per unit time in perturbation theory.
8. Give a brief discussion about Fermi Golden rule.
9. Explain in detail the Harmonic Perturbation.
10. Discuss the Periodic time Dependent Perturbation theory.
11. Give a brief description of Stark effect. Calculate the first order Stark effect in Hydrogen atom.
12. Use the Stark effect in perturbation theory for degenerate state.
13. What is Zeeman Effect? Find Perturbation energy in case of hydrogen atom by the Zeeman Effect.
14. Find the first order change in energy of the oscillator for the perturbation $ax^3 + bx^4$.
15. Find the first order energy correction for one dimensional Harmonic Oscillator for the perturbing Hamiltonian ax^4.
16. To determine the Electrostatic polarization and the dipole moment by using Selection Rule.
17. Work out the first- order Stark effect in the $n = 3$ state of the hydrogen atom.
18. Calculate the classical and Quantum mechanical Hamiltonian when electromagnetic field is applied.

Chapter 17 WKB approximation method

17.1 WKB approximation[Wentzel–Kramers–Brillouin]

The name is initialism for **Wentzel–Kramers–Brillouin**. It is also known as the **LG** or **Liouville–Green method**. Other often-used letter combinations include **JWKB** and **WKBJ**, where the "J" stands for Jeffreys.

In mathematical physics, the **WKB approximation** or **WKB method** is a method for finding approximate solutions to linear differential equations with spatially varying coefficients. It is typically used for a semi classical calculation in quantum mechanics in which the wave function is recast as an exponential function, semi classically expanded, and then either the amplitude or the phase is taken to be changing slowly.

17.2 Importance of WKB [Wentzel, Kramer's and Brillouin] approximation method

WKB approximation method is applied to those Quantum mechanical problems where the potential function is a slowly varying function of the position. *A slowly changing potential means the variation of V(r) is slightly over several wave lengths (de-Broglie waves) of the particle.* Thus when the potential energy function *V(x)* is not a simple form and when it is slowly varying function of position then the solution of the Schrödinger equation even in one dimension is usually a complicated problem, which requires the use of approximate methods[17]. A particular method however is found is known as WKB approximate method, which can be solved the problems exactly. Problem in one dimension and also of three dimensions reducible in one dimensional problem are solved by this method.

This is the approximate method of solution of the wave function based on the expansion, *the first term of which leads to the classical results. The second term leads to the old Quantum theory results and the higher terms to the characteristics of the new mechanics.* This suggests that if *V(x)*, while no longer constant varies slowly with x, we might try a solution of the form

$$\Psi(x) = exp\left(\frac{i}{\hbar}S_0(x) + S_1(x)\right) = exp\left(\frac{i}{\hbar}S_0(x)\right)exp\big(S_1(x)\big) \qquad [1]$$

$$\Psi(x) = exp\left(\pm\frac{i}{\hbar}\int p(x)\,dx\right)\frac{C}{\sqrt{|p(x)|}} = \frac{C}{\sqrt{|p(x)|}}exp\left(\pm\frac{i}{\hbar}\int p(x)\,dx\right) \qquad [2]$$

17.3 WKB approximation is a linear combination of two solutions

From one dimensional Schrodinger equation

$$\frac{\partial^2\Psi(x)}{\partial x^2} + \frac{2m}{\hbar^2}\big(E - V(x)\big)\Psi(x) = 0 \qquad [1]$$

For free particle $V(x) = V_0 = constant$. The solution is

$$\Psi(x) = exp\left(\pm\, i\, p_0 \frac{x}{\hbar}\right) \qquad [2]$$

Where $p_0 = \sqrt{2m(E - V_0)} = const.$ [3]

This suggests that when $V(x)$ is slowly varying, we try a solution of the form

$$\Psi(x) = exp\left(i\, \frac{S(x)}{\hbar}\right) \qquad [4]$$

$$\frac{\partial\Psi(x)}{\partial x} = \frac{i}{\hbar}\frac{\partial S(x)}{\partial x}\, exp\left(i\, \frac{S(x)}{\hbar}\right)$$

$$\frac{\partial^2\Psi(x)}{\partial x^2} = \frac{i}{\hbar}\frac{\partial^2 S(x)}{\partial x^2}\, exp\left(i\, \frac{S(x)}{\hbar}\right) - \frac{1}{\hbar^2}\left(\frac{\partial S(x)}{\partial x}\right)^2 exp\left(i\, \frac{S(x)}{\hbar}\right) \qquad [5]$$

Substituting [4] and [5] in equation [1] and cancelling the exponential factor we get

$$-\frac{1}{\hbar^2}\left(\frac{\partial S(x)}{\partial x}\right)^2 + \frac{i}{\hbar}\frac{\partial^2 S(x)}{\partial x^2} + \frac{2m}{\hbar^2}(E - V(x)) = 0 \qquad [6]$$

$$\left(\frac{\partial S(x)}{\partial x}\right)^2 - i\hbar\,\frac{\partial^2 S(x)}{\partial x^2} - 2m\big(E - V(x)\big) = 0 \qquad [7]$$

Defining $p(x) = \sqrt{2m\big(E - V(x)\big)}$ [8]

The above equation can be written as

$$\left(\frac{\partial S(x)}{\partial x}\right)^2 - p^2(x) - i\hbar\,\frac{\partial^2 S(x)}{\partial x^2} = 0 \qquad [9]$$

Note that for free particle $\frac{\partial^2 S(x)}{\partial x^2} = 0$ Since $S = S_0 = constant$. This suggests that for a slowly varying potential Since S= S_0 = constant. This suggests that for a slowly varying potential $\frac{\partial^2 S(x)}{\partial x^2}$ is small or, we are led to the suspect that this second derivative remains relatively small if the potential does not vary too violently. The zeroth-order term, in other words, the dominating term of S is obtained by neglecting the third term in equation [9] I.e., by setting $\hbar = 0$. The correction is of the order of $\hbar$ Thus we write

$$S(x) = S_0(x) + \frac{\hbar}{i}S_1(x) + \left(\frac{\hbar}{i}\right)^2 S_2(x) + \ldots\ldots\ldots \qquad [10]$$

The WKB approximation consists of retaining only the first two terms in equation [9]. Now Substituting [10] in [9] and equating the coefficient of the successive powers of $\hbar$ to zero, we get the following systems of coupled equations:

Zeroth order: $\left(\frac{\partial S_0}{\partial x}\right)^2 - p^2(x) = 0$ Or, $\left(\frac{\partial S_0}{\partial x}\right)^2 = p^2(x)$ [11]

First order:

$$\left(\frac{\partial S(x)}{\partial x}\right)^2 - p^2(x) - i\hbar\,\frac{\partial^2 S(x)}{\partial x^2} = 0$$

$$\left[\frac{\partial}{\partial x}\left(S_0 + \frac{\hbar}{i}S_1(x)\right)\right]^2 - p^2(x) - i\hbar\,\frac{\partial^2}{\partial x^2}\left(S_0 + \frac{\hbar}{i}S_1(x)\right) = 0$$

$$\left[\frac{\partial S_0}{\partial x} + \frac{\hbar}{i}\frac{\partial S_1(x)}{\partial x}\right]^2 - p^2(x) - i\hbar\,\frac{\partial^2 S_0}{\partial x^2} - \hbar^2\frac{\partial^2 S_1(x)}{\partial x^2} = 0$$

$$\frac{\partial^2 S_0}{\partial x^2} + 2\,\frac{\partial S_0}{\partial x}\,\frac{\hbar}{i}\,\frac{\partial S_1(x)}{\partial x} + \left(\frac{\hbar}{i}\right)^2\frac{\partial^2 S_1(x)}{\partial x^2} - p^2(x) - i\hbar\,\frac{\partial^2 S_0}{\partial x^2} - \hbar^2\frac{\partial^2 S_1(x)}{\partial x^2} = 0$$

Now equating the coefficients of successive power of $\hbar$ to zero.

$$\frac{2\hbar}{i}\,\frac{\partial S_0}{\partial x}\frac{\partial S_1(x)}{\partial x} - i\hbar\,\frac{\partial^2 S_0}{\partial x^2} = 0 \qquad [12]$$

$$\frac{\partial^2 S_0}{\partial x^2} + 2\,\frac{\partial S_0}{\partial x}\frac{\partial S_1}{\partial x} = 0 \qquad [13]$$

From equation [11]

$$S_0(x) = \pm\int p(x)\,\partial x \qquad [14]$$

Next we rewrite equation [12] as

$$\frac{\partial S_1}{\partial x} = -\frac{1}{2}\,\frac{S_0''(x)}{S_0'(x)} \qquad [15]$$

Or, $S_1(x) = -\frac{1}{2}\ln|S_0'(x)| + \ln|C|$ [16]

Where, C is an arbitrary constant.

$$S_1(x) = ln\frac{c}{\sqrt{|S_0'(x)|}}$$

$$S_1(x) = \ln\frac{C}{\sqrt{|p(x)|}} \qquad [17]$$

The approximation wave function can be written as

$$\Psi(x) = exp\left(\frac{i}{\hbar}\left[S_0(x) + \frac{\hbar}{i}S_1(x)\right]\right)$$

$$\Psi(x) = exp\left(\frac{i}{\hbar}S_0(x) + S_1(x)\right) = exp\left(\frac{i}{\hbar}S_0(x)\right)exp(S_1(x))$$

$$\Psi(x) = exp\left(\pm\frac{i}{\hbar}\int p(x)\,dx\right)\frac{C}{\sqrt{|p(x)|}} = \frac{C}{\sqrt{|p(x)|}}exp\left(\pm\frac{i}{\hbar}\int p(x)\,dx\right) \qquad [18]$$

Thus there are two linearly independent WKB solutions are

$$\Psi_+(x) = \frac{C_+}{\sqrt{|p(x)|}}\,exp\left(+\frac{i}{\hbar}\int p(x)\,dx\right) \qquad [19]$$

And $\Psi_-(x) = \frac{C_-}{\sqrt{|p(x)|}}\,exp\left(-\frac{i}{\hbar}\int p(x)\,dx\right)$ [20]

A general solution in the WKB approximation is linear combination of these two solutions.

17.4 Classical turning point

Classical turning point can be described by the following figure if V is an increasing function of x. When $x > x_1$ is classically forbidden region, $x < x_1$ is classically allowed region and $x = x_1$ is called classical turning point. *WKB solutions are invalid near* $x = x_1$. Because E is close to V.

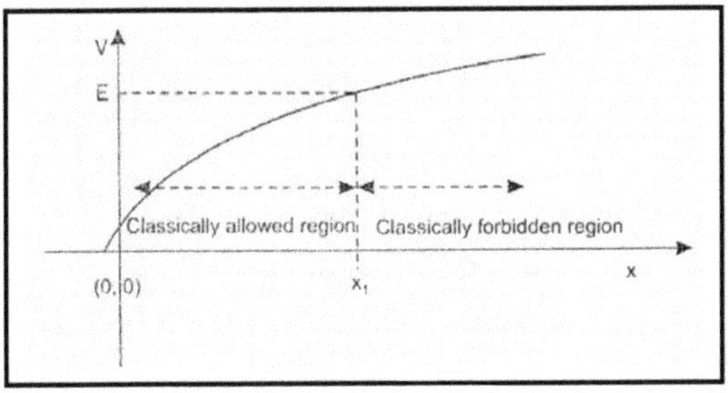

Fig. [17.4a]: Shows the classical turning point if $V$is a increasing function of x.

Classical turning point can be described by the following figure if $V$is a decreasing function of x. *WKB solutions are invalid near* $x = x_2$. Because E is close to V.

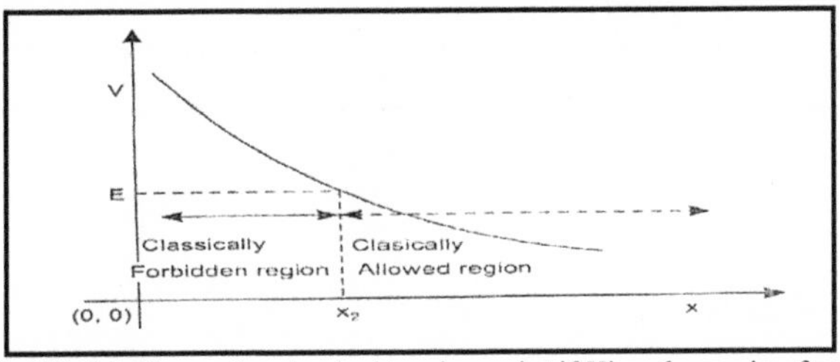

Fig.[17.4b]: Shows the classical turning point if $V$is a decreasing function of x.

Classical turning points (a and b) for a particle of total energy E encountering a single potential barrier of general shape as shown in the following figure. *WKB solutions are valid in the three regions because the regions are away from the classical turning points.*

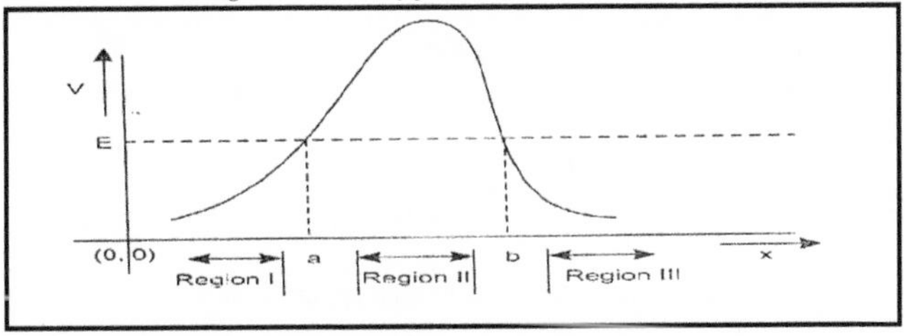

Fig.[17.4c]: Shows classical turning points ($a$and b) for a particle of total energy E encountering a single potential barrier of general shape.

Classical turning points (a and b) for a particle of total energy E residing in the quantum well of general shape as shown in the following figure. *WKB solutions are valid in the three regions because the regions are away from the classical turning points* [24].

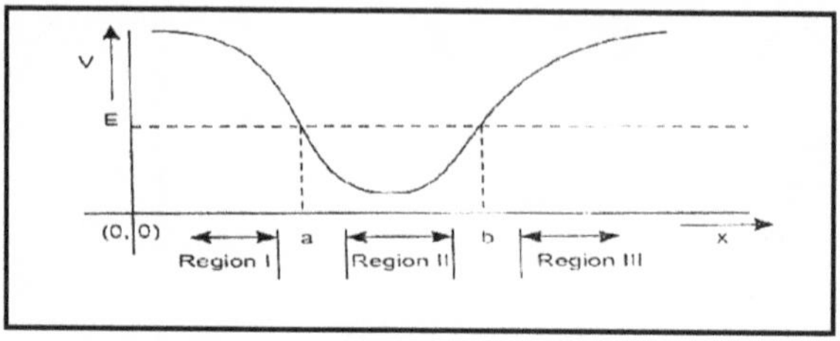

Fig.[17.4d]: Shows classical turning points ($a$and b) for a particle of total energy E residing in the quantum well of general shape.

17.5 Derivation of WKB connection formula [case-I]: For Rising potential

The WKB solutions $\Psi(x) = \frac{C}{\sqrt{|p(x)|}} exp\left(\pm\frac{i}{\hbar}\int p(x)\,dx\right)$ are break down near the classical turning points. At a turning point *E* = *V(x)*. Therefore *P(x)* = *0* and The WKB solutions become infinity.

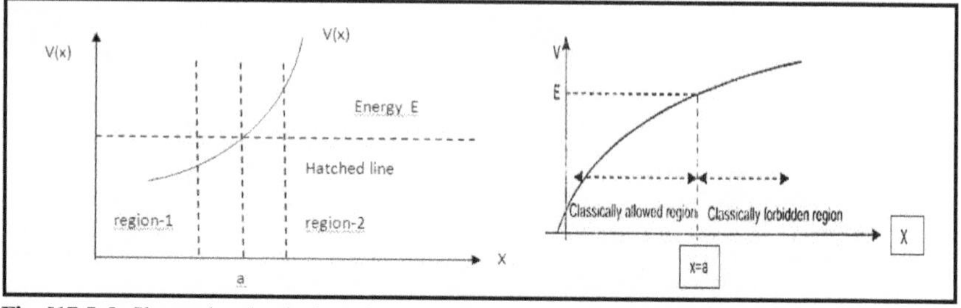

Fig. [17.5a]: Shows the classical turning point if $V$is a increasing function of x

The figure shows that a turning point at $x = a$. Further $V(x)$ is increasing at $x = a$. In region 1, I.e., far left of the turning point $E > V(x)$, $p(x)$) is real and positive. So the WKB solutions are oscillatory. Thus we can write

$$\Psi_{1,\ WKB}(x) = \frac{A_1}{\sqrt{|p(x)|}} Sin\left[\frac{1}{\hbar}\int_x^a p(x)\,dx + \frac{\pi}{4}\right] + \frac{B_1}{\sqrt{|p(x)|}} Cos\left[\frac{1}{\hbar}\int_x^a p(x)\,dx + \frac{\pi}{4}\right] \qquad [1]$$

Where the phase $\frac{\pi}{4}$ is chosen purely for convenience as will be apparent later. To the far right of the turning point, I.e., in region 2, we have E < V(x), therefore *p(x)* is purely imaginary. Thus we have

$$p(x) = i\,|p(x)| \quad [2]$$

The WKB solution in this region is exponential. We write

$$\Psi_{2,\ WKB}(x) = \frac{A_2}{\sqrt{|p(x)|}}\, exp\left(-\frac{1}{\hbar}\int_a^x |p(x)|\, dx\right) + \frac{B_2}{\sqrt{|p(x)|}}\, exp\left(+\frac{1}{\hbar}\int_a^x |p(x)|\, dx\right) \quad [3]$$

Equation [1] and [3] are expressions of the same wave function in different regions. The constants A_1, B_1 and A_2, B_2 cannot be arbitrarily chosen. Thus there must be a connection between the WKB solutions in regions 1 and 2.

In order to discover this connection, we must follow the variation of the wave function from region 1 to region 2. In doing so, it is necessary to pass through the hatched region of figure near the turning point, where the WKB solution is not valid. In order to find the behavior of $\Psi(x)$ in this region, an exact solution of the Schrodinger equation must be considered. This exact solution will be used as a link between the WKB solutions in regions 1 and 2.
To find a solution of the turning point 'a', we assume $V(x)$ to be linear at 'a'. So we can write

$$V(x) = V(a) + (x-a)\left(\frac{\partial V}{\partial x}\right)_{x=a} = E + A(x-a) \ [y = c + mx] \quad [4]$$

Where A is the slope and which is positive in the present case. The Schrodinger equation becomes

$$\frac{\partial^2 \Psi}{\partial x^2} - \frac{2mA}{\hbar}(x-a)\Psi = 0 \quad [5]$$

Now making a change of variable

$$Z = \left(\frac{2mA}{\hbar^2}\right)^{\frac{1}{3}}(x-a) \quad [6]$$

Therefore equation [5] transform to

$$\frac{\partial \Psi}{\partial x} = \frac{\partial \Psi}{\partial Z}\frac{\partial Z}{\partial x} = \frac{\partial \Psi}{\partial Z}\left(\frac{2mA}{\hbar^2}\right)^{\frac{1}{3}}$$

$$\frac{\partial^2 \Psi}{\partial x^2} = \frac{\partial}{\partial x}\left[\left(\frac{2mA}{\hbar^2}\right)^{\frac{1}{3}}\frac{\partial \Psi}{\partial Z}\right] = \frac{\partial}{\partial Z}\left[\left(\frac{2mA}{\hbar^2}\right)^{\frac{1}{3}}\frac{\partial \Psi}{\partial Z}\right]\frac{\partial Z}{\partial x}$$

$$\frac{\partial^2 \Psi}{\partial x^2} = \left(\frac{2mA}{\hbar^2}\right)^{\frac{1}{3}}\frac{\partial^2 \Psi}{\partial Z^2}\left(\frac{2mA}{\hbar^2}\right)^{\frac{1}{3}} = \left(\frac{2mA}{\hbar^2}\right)^{\frac{2}{3}}\frac{\partial^2 \Psi}{\partial Z^2}$$

Equation [5] becomes

$$\frac{\partial^2 \Psi}{\partial x^2} - \frac{2mA}{\hbar^2}(x-a)\Psi = 0$$

$$\left(\frac{2mA}{\hbar^2}\right)^{\frac{2}{3}}\frac{\partial^2 \Psi}{\partial Z^2} - \left(\frac{2mA}{\hbar^2}\right) Z \left(\frac{2mA}{\hbar^2}\right)^{-\frac{1}{3}}\Psi = 0$$

$$\left(\frac{2mA}{\hbar^2}\right)^{\frac{2}{3}}\frac{\partial^2 \Psi}{\partial Z^2} - \left(\frac{2mA}{\hbar^2}\right)^{\frac{2}{3}} Z\Psi = 0$$

$$\therefore \quad \frac{\partial^2 \Psi}{\partial Z^2} - Z\Psi = 0 \quad [7]$$

Let the solution of equation [7] are [Since equation [7] is standard in mathematical Physics]

$$\Psi = CA_i(Z) + DB_i(Z) \qquad [8]$$

Where $A_i(Z)$ and $B_i(Z)$ are known as Airy functions and are given by

$$A_i(Z) = \frac{1}{\pi}\int_0^\infty Cos\left(\frac{S^3}{3} + SZ\right)\partial S \qquad [9a]$$

$$B_i(Z) = \frac{1}{\pi}\int_0^\infty \left[exp\left(SZ - \frac{S^3}{3}\right) + Sin\left(SZ + \frac{S^3}{3}\right)\right]\partial S \qquad [9b]$$

We are only interested in the asymptotic form of the airy functions. Defining

$$\xi = \frac{2}{3}|Z|^{\frac{2}{3}} \qquad [10]$$

The asymptotic forms are

$$A_i(Z) \xrightarrow[Z\to\infty]{} \frac{1}{2\sqrt{\pi}} Z^{\frac{1}{4}}\exp(-\xi) \xrightarrow[Z\to-\infty]{} \frac{1}{\sqrt{\pi}}|Z|^{-\frac{1}{4}}\text{Sin}\left(\xi + \frac{\pi}{4}\right) \qquad [11a]$$

And $$B_i(Z) \xrightarrow[Z\to\infty]{} \frac{1}{\sqrt{\pi}} Z^{-\frac{1}{4}}\exp(\xi) \xrightarrow[Z\to-\infty]{} \frac{1}{\sqrt{\pi}}|Z|^{-\frac{1}{4}} Cos\left(\xi + \frac{\pi}{4}\right) \qquad [11b]$$

Now to the left of the turning point Z is negative and $p(x)dx = \sqrt{2m[E - V(x)]}$ is Positive. [refer to fig.1] Furthermore,

$$p(x)dx = \sqrt{2m[E - V(x)]}$$

$$\int_x^a p(x)\,dx = \int_x^a \sqrt{2m[E - V(x)]}\,dx \qquad [12]$$

From equation [4]

$$V(x) = V(a) + A(x-a) = E + A(x-a)$$
$$E - V(x) = -A(x-a) = A(a-x)$$

And $$Z = \left(\frac{2mA}{\hbar^2}\right)^{\frac{1}{3}}(x-a) \quad \text{or,} \quad |Z|^{\frac{2}{3}} = \sqrt{\left(\frac{2mA}{\hbar^2}\right)}\,(x-a)^{\frac{2}{3}}$$

∴ Equation [12] becomes

$$\int_x^a p(x)\,dx = \int_x^a \sqrt{2m[E - V(x)]}\,dx$$
$$= \sqrt{2mA}\int_x^a (a-x)^{\frac{1}{2}}\,dx$$
$$= \sqrt{2mA}\,\frac{2}{3}(a-x)^{\frac{3}{2}}$$
$$= \frac{2}{3}\hbar\sqrt{\frac{2mA}{\hbar^2}(a-x)^{\frac{3}{2}}}$$

$$\int_x^a p(x)\,dx = \frac{2}{3}\hbar|Z|^{\frac{3}{2}} = \hbar\xi \quad \left[\because \xi = \frac{2}{3}|Z|^{\frac{3}{2}}\right]$$

∴ $$\int_x^a p(x)\,dx = \hbar\xi \qquad [13]$$

Also to the left of the turning point, we have

$$Z = \left(\frac{2mA}{\hbar^2}\right)^{\frac{1}{3}}(x-a) \;\; or, \;\; |Z|^{-\frac{1}{4}} = \left(\frac{2mA}{\hbar^2}\right)^{-\frac{1}{12}}(x-a)^{-\frac{1}{4}}$$

But $\quad p(x) = \sqrt{2mA}\,(a-x)^{\frac{1}{2}}$

$$\sqrt{p(x)} = (2mA)^{\frac{1}{4}}(a-x)^{\frac{1}{4}} \;\; or, \;\; \frac{1}{\sqrt{p(x)}} = (2mA)^{-\frac{1}{4}}(a-x)^{-\frac{1}{4}}$$

$$|Z|^{-\frac{1}{4}} = \left(\frac{2mA}{\hbar^2}\right) = \frac{\alpha}{\sqrt{p(x)}} \qquad [14]$$

Where α is constant which can be determined. Next referring to figure-1, we see that to the right of the turning point $E < V(x)$. Therefore

$$p(x) = \sqrt{2m[E-V(x)]} = \sqrt{-2m[V(x)-E]} = i|p(x)|$$

And $\quad p(x) = \sqrt{2m[V(x)-E]} = \sqrt{2mA(x-a)}$

$V(x) = V(a) + A(x-a)$ $= E + A(x-a)$ $V(x) - E = A(x-a)$

$$\int_a^x |p(x)|\,dx = (2mA)^{\frac{1}{2}}\int_a^x (x-a)^{\frac{1}{2}}\,dx = \sqrt{2mA}\,\frac{2}{3}(x-a)^{\frac{3}{2}} = \hbar\frac{2}{3}\sqrt{\frac{2mA}{\hbar^2}}(x-a)^{\frac{3}{2}} = \hbar\xi \qquad [15]$$

And $\quad |Z|^{-\frac{1}{4}} = \left(\frac{2mA}{\hbar^2}\right) = \frac{\alpha}{\sqrt{p(x)}} \qquad [16]$

Thus the asymptotic forms of $A_i(Z)$ and $B_i(Z)$ can be written as

$$A_i(Z) \xrightarrow[Z\to-\infty]{} \frac{1}{\sqrt{\pi}} Z^{-\frac{1}{4}}\exp(-\xi)$$

$$\xrightarrow[Z\to-\infty]{} \frac{1}{\sqrt{\pi}} |Z|^{-\frac{1}{4}}\,\mathrm{Sin}\left(\xi + \frac{\pi}{4}\right)$$

$$\xrightarrow[Z\to-\infty]{} \frac{1}{\sqrt{\pi}}\frac{\alpha}{\sqrt{p(x)}}\,Sin\left[\frac{1}{\hbar}\int_x^a p(x)dx + \frac{\pi}{4}\right] \qquad [17]$$

$$\xrightarrow[Z\to+\infty]{} \frac{1}{2\sqrt{\pi}}\frac{\alpha}{\sqrt{p(x)}}\exp\left[-\frac{1}{\hbar}\int_a^x |p(x)|dx\right]$$

$$B_i(Z) \xrightarrow[Z\to-\infty]{} \frac{1}{\sqrt{\pi}} |Z|^{-\frac{1}{4}}\,\mathrm{Cos}\left(\xi + \frac{\pi}{4}\right)$$

$$\xrightarrow[Z\to-\infty]{} \frac{1}{\sqrt{\pi}}\frac{\alpha}{\sqrt{p(x)}}\,\mathrm{Cos}\left[\frac{1}{\hbar}\int_x^a p(x)dx + \frac{\pi}{4}\right] \qquad [18]$$

$$\xrightarrow[Z\to+\infty]{} \frac{1}{\sqrt{\pi}}\frac{\alpha}{\sqrt{p(x)}}\exp\left[\frac{1}{\hbar}\int_a^x |p(x)|dx\right]$$

Now suppose that the exact wave function near the turning point is taken as

$$\Psi(x) = CA_i(Z) \qquad [19]$$

This wave function, when extrapolated to the far left of the turning point in region 1, must watch smoothly with the WKB wave function $\Psi_{1,WKB}(x)$ is given in equation [1]. The extrapolation of $A_i(Z)$ to region 1 is given by the first of equation [17]. Hence we must have $B_1 = 0$.

And $\quad A_1 = C\frac{\alpha}{\sqrt{\pi}} \qquad [20]$

Thus the corresponding WKB solution in region 1 is

$$\Psi_{1,\ WKB}(x) = C\frac{\alpha}{\sqrt{\pi}}\frac{1}{\sqrt{|p(x)|}}\ Sin\left[\frac{1}{\hbar}\int_x^a p(x)dx + \frac{\pi}{4}\right] \qquad [21]$$

Similarly, for the wave function [21] to match smooth with the $\Psi_{2,WKB}(x)$, we must choose in equation [3] $B_2 = 0$

Therefore, $A_2 = C\frac{\alpha}{2\sqrt{\pi}}$ [22]

Thus the corresponding $\Psi_{2,WKB}(x)$ is

$$\Psi_{2,\ WKB}(x) = C\frac{\alpha}{2\sqrt{\pi}}\frac{1}{\sqrt{|p(x)|}}\exp\left[-\frac{1}{\hbar}\int_a^x p(x)dx\right] \qquad [23]$$

Thus the functions [19], [21] and [23] match smoothly. Hence the connection formulas are

$$\frac{1}{\sqrt{|p(x)|}}\ Sin\left[\frac{1}{\hbar}\int_x^a p(x)dx + \frac{\pi}{4}\right] \Leftrightarrow \frac{1}{2\sqrt{|p(x)|}}\exp\left[-\frac{1}{\hbar}\int_a^x |p(x)|dx\right] \qquad [24]$$

Next let us take $B_i(Z)$ as the exact wave function near the turning point. Again it must match with the WKB solutions in region 1[I.e. $Z \to -\infty$ and in region 2[I.e., $Z \to +\infty$]. Thus the WKB solution in region 1 is given by the first of equation [18] and in region 2 is given by the second of equation [18]. Hence the connection formula is

$$\frac{1}{\sqrt{|p(x)|}}\ Cos\left[\frac{1}{\hbar}\int_x^a p(x)dx + \frac{\pi}{4}\right] \Leftrightarrow \frac{1}{\sqrt{|p(x)|}}\exp\left[\frac{1}{\hbar}\int_a^x p(x)dx\right] \qquad [25]$$

Now $k(x) = \frac{p(x)}{\hbar} = \sqrt{2m[E - V(x)]}$ [26]

Therefore, the connection formulas can be written as

$$\frac{2}{\sqrt{\hbar k(x)}}Sin\left[\frac{1}{\hbar}\int_x^a \hbar k(x)dx + \frac{\pi}{4}\right] \Leftrightarrow \frac{1}{\sqrt{\hbar|k(x)|}}exp\left[-\frac{1}{\hbar}\int_a^x \hbar|k(x)|dx\right]$$

$$\left.\begin{aligned} &\frac{2}{\sqrt{k(x)}}Sin\left[\int_x^a k(x)dx + \frac{\pi}{4}\right] \Leftrightarrow \frac{1}{\sqrt{|k(x)|}}\ exp\left[-\int_a^x |k(x)|dx\right] \\ &\frac{1}{\sqrt{k(x)}}\ Cos\left[\int_x^a k(x)dx + \frac{\pi}{4}\right] \Leftrightarrow \frac{1}{\sqrt{|k(x)|}}\ exp\left[\int_a^x |k(x)|dx\right] \end{aligned}\right\} \qquad [27]$$

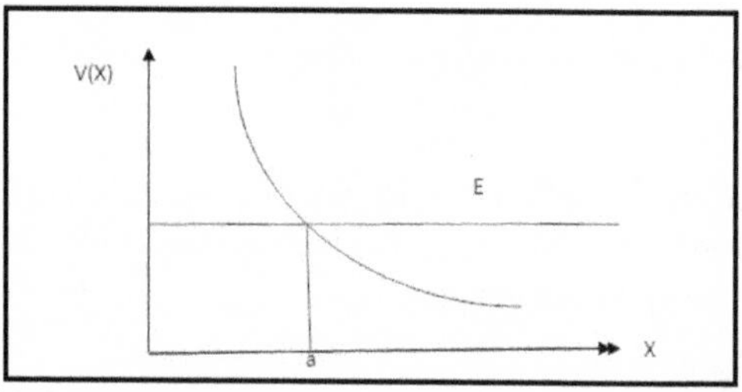

Fig.[17.5b]: Shows WKB connection formula for falling potential I.e., the classical turning point if $V$is a decreasing function of x

We have derived the connection formulas about a turning point ***when the potential energy is a rising function at the turning point as shown in fig.* [17.5a].** Using the same procedure we can obtain the connection formulas *when the potential energy is a falling function at the turning point as shown in fig.* [17.5b]. We have

$$\left.\begin{array}{l} \frac{1}{\sqrt{k(x)}}\, exp\left[-\int_x^a |k(x)|dx\right] \Leftrightarrow \frac{2}{\sqrt{k(x)}}\, Sin\left[\int_a^x |k(x)|dx + \frac{\pi}{4}\right] \\ \text{And} \quad \frac{1}{\sqrt{k(x)}}\, exp\left[\int_x^a |k(x)|dx\right] \Leftrightarrow \frac{1}{\sqrt{k(x)}}\, Cos\left[\int_a^x |k(x)|dx + \frac{\pi}{4}\right] \end{array}\right\} \quad [28]$$

17.6 Derivation of WKB connection formula [case II]: For Falling potential

Using the same procedure [see section 17.5] we can obtain the connection formulas ***when the potential energy is a falling function at the turning point as shown in fig.* [17.6].**We consider $V(x)$ as a decreasing function of x in the region II as shown in the right figure. x_2 is the classical turning point for a particle of total energy E approaching the potential from right.

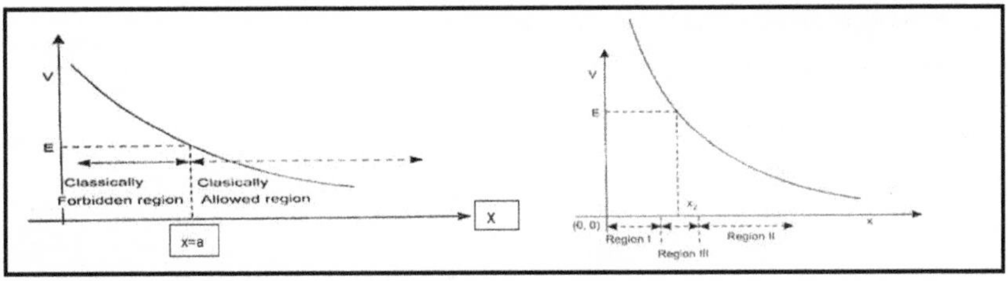

Fig. [17.6]: Shows WKB connection formula for falling potential I.e., the classical turning point if $V$is a decreasing function of x

WKB solutions for the region I

$$\Psi_I(x) = \frac{P_1}{\sqrt{k'(x)}} e^{+\int_{x_2}^{x} k'(x)\,dx} + \frac{P_2}{\sqrt{k'(x)}} e^{-\int_{x_2}^{x} k'(x)\,dx} \quad [1]$$

WKB solutions for the region for the region II

$$\Psi_{II}(x) = \frac{Q_1}{\sqrt{k(x)}} e^{+i\int_{x_2}^{x} k(x)\,dx} + \frac{Q_2}{\sqrt{k(x)}} e^{-i\int_{x_2}^{x} k(x)\,dx} \quad [2]$$

Ψ_I becomes Ψ_{II} as we go from region I to region II.

In region III: Schrödinger equation is

$$\frac{d^2\Psi_{III}}{dx^2} + \frac{2m}{\hbar^2}\big(E - V(x)\big)\Psi_{III} = 0 \quad [3]$$

In region III, let $V(x)$ be a linear function of x. Therefore, $\frac{2m}{\hbar^2}\big(E - V(x)\big) \approx \alpha'(x - x_2)$

Where α' positive. And $\frac{2m}{\hbar^2}\big(E - V(x)\big) \approx \alpha'(x - x_2) \approx -\alpha(x - x_2)$ [3a]

Finally [3] becomes

$$\frac{d^2\Psi_{III}}{dx^2} + \alpha'(x - x_2)\Psi_{III} \approx 0$$

Or, $\frac{d^2\Psi_{III}}{dx^2} - \alpha(x - x_2)\Psi_{III} \approx 0$ [$\approx -\alpha(x - x_2)$; Where α is negative, $\alpha = -\alpha'$] [3b]

Or, $\frac{d^2\Psi_{III}}{dz^2} - z\Psi_{III} \approx 0$ [4]

Where $z = \alpha^{\frac{1}{3}}(x - x_2) = (-1)^{\frac{1}{3}}|\alpha|^{\frac{1}{3}}(x - x_2) = -|\alpha|^{\frac{1}{3}}(x - x_2)$ [5]

Equation [4] is Airy equation having two solutions $A_i(z)$ and $B_i(z)$.

We can take $A_i(\text{large negative } z) = \Psi_{II}$

$A_i(\text{large positive } z) = \Psi_I$

We can also take $B_i(\text{large negative } z) = \Psi_{II}$

$B_i(\text{large positive } z) = \Psi_I$

Because equation [5] shows that z is negative for $x > x_2$.

Ψ_I becomes Ψ_{II} as we go from region I to region II, I.e.,

$$A_i(\text{large positive } z) \overset{I-II}{\longleftrightarrow} A_i(\text{large negative } z)$$

$$B_i(\text{large positive } z) \overset{I-II}{\longleftrightarrow} B_i(\text{large negative } z)$$

I.e., $\frac{1}{2\sqrt{\pi}} z^{-\frac{1}{4}} e^{-\xi} \overset{I-II}{\longleftrightarrow} \frac{1}{\sqrt{\pi}} |z|^{-\frac{1}{4}} \sin\left(\xi + \frac{\pi}{4}\right)$ [6]

and $\frac{1}{\sqrt{\pi}} z^{-\frac{1}{4}} e^{+\xi} \overset{I-II}{\longleftrightarrow} \frac{1}{\sqrt{\pi}} |z|^{-\frac{1}{4}} \cos\left(\xi + \frac{\pi}{4}\right)$ [7]

where $\xi = \frac{2}{3}|z|^{3/2}$ and $z = -|\alpha|^{\frac{1}{3}}(x - x_2)$

Equations [6] and [7] are connection formula. We now give the useful forms of the two formula as follows.

For region II: $E > V(x)$, $x \gg x_2$ and z is large negative

$k^2(x) = \frac{2m}{\hbar^2}(E - V(x)) \approx -\alpha(x - x_2) = |\alpha|(x - x_2) = |\alpha|^{\frac{2}{3}}|z|$ [8]

where α is negative and hence $\alpha = -|\alpha|$. We used equations [3] and [5]

$$z = -|\alpha|^{\frac{1}{3}}(x - x_2)$$

Or, $|z| = |\alpha|^{\frac{1}{3}}(x - x_2)$

Or, $|z| = |\alpha|^{-\frac{2}{3}} k^2(x)$

Or, $|z|^{-\frac{1}{4}} = \frac{|\alpha|^{\frac{1}{6}}}{\sqrt{k(x)}}$ [9]

Again, $\int_{x_2}^{x} k(x)\,dx = \int_{x_2}^{x} |\alpha|^{\frac{1}{3}} |z|^{\frac{1}{2}}\,dx = \int_0^{|z|} |\alpha|^{\frac{1}{3}} |z|^{\frac{1}{2}} |\alpha|^{-\frac{1}{3}} d|z| = \frac{2}{3}|z|^{\frac{3}{2}} = \xi$

Using equations [8] and [5]:

$$z = -|\alpha|^{\frac{1}{3}}(x - x_2),\ dz = -|\alpha|^{\frac{1}{3}}\,dx,\ \ dx = -|\alpha|^{-\frac{1}{3}}\ ,\ \ dz = |\alpha|^{-\frac{1}{3}} d|z|$$

Thus $\xi = \int_{x_2}^{x} k(x)\,dx$ [10]

For region I: we have $E < V(x)$ and $x \ll x_2$, equation [3a] gives

$$k'^2(x) = \frac{2m}{\hbar^2}(V(x) - E) \approx \alpha(x - x_2) \approx -|\alpha|(-1)(x_2 - x) \approx |\alpha|(x_2 - x)$$

$$\approx |\alpha|^{\frac{2}{3}}|\alpha|^{\frac{1}{3}}(x_2 - x) \approx |\alpha|^{\frac{2}{3}} z \qquad [11]$$

where α is negative, using equation [5]

Thus $z = |\alpha|^{-\frac{2}{3}} k'^2(x)$

Or, $z^{-\frac{1}{4}} = \frac{|\alpha|^{\frac{1}{6}}}{\sqrt{k'(x)}}$ [12]

Again, $\int_x^{x_2} k'(x)\,dx = \int_x^{x_2} |\alpha|^{\frac{1}{3}}\, z^{\frac{1}{2}}\,dx = -\int_z^{0} |\alpha|^{\frac{1}{3}}\, z^{\frac{1}{2}} |\alpha|^{-\frac{1}{3}}\,dz = \int_0^{z} z^{\frac{1}{2}}\,dz = \frac{2}{3} z^{\frac{3}{2}} = \xi$ [13]

We have used equation [11] and [5]

$$z = -|\alpha|^{\frac{1}{3}}(x - x_2) = |\alpha|^{\frac{1}{3}}(x_2 - x)$$

Or, $dz = |\alpha|^{\frac{1}{3}}(-\,dx)$

Or, $dx = -|\alpha|^{-\frac{1}{3}}\,dz$

We now use equations [12] and [13] on L.H.S. of equations [6] and [7] and we use equations [9] and [10] on R.H.S. of equations [6] and [7] to get

$$\frac{1}{2\sqrt{\pi}} \frac{|\alpha|^{\frac{1}{6}}}{\sqrt{k'(x)}} e^{-\int_x^{x_2} k'(x)\,dx} \overset{I-II}{\longleftrightarrow} \frac{1}{\sqrt{\pi}} \frac{|\alpha|^{\frac{1}{6}}}{\sqrt{k(x)}} \sin\left(\int_{x_2}^{x} k(x)\,dx + \frac{\pi}{4}\right)$$

$$\frac{1}{\sqrt{k'(x)}} e^{-\int_x^{x_2} k'(x)\,dx} \overset{I-II}{\longleftrightarrow} \frac{2}{\sqrt{k(x)}} \sin\left(\int_{x_2}^{x} k(x)\,dx + \frac{\pi}{4}\right) \qquad [14]$$

And $\frac{1}{\sqrt{\pi}} \frac{|\alpha|^{\frac{1}{6}}}{\sqrt{k'(x)}} e^{+\int_x^{x_2} k'(x)\,dx} \overset{I-II}{\longleftrightarrow} \frac{1}{\sqrt{\pi}} \frac{|\alpha|^{\frac{1}{6}}}{\sqrt{k(x)}} \cos\left(\int_{x_2}^{x} k(x)\,dx + \frac{\pi}{4}\right)$

Or, $\frac{1}{\sqrt{k'(x)}} e^{+\int_x^{x_2} k'(x)\,dx} \overset{I-II}{\longleftrightarrow} \frac{1}{\sqrt{k(x)}} \cos\left(\int_{x_2}^{x} k(x)\,dx + \frac{\pi}{4}\right)$ [15]

Finally,

$$\frac{1}{\sqrt{k'(x)}} e^{-\int_x^{x_2} k'(x)\,dx} \overset{I-II}{\longleftrightarrow} \frac{2}{\sqrt{k(x)}} \sin\left(\int_{x_2}^{x} k(x)\,dx + \frac{\pi}{4}\right) \quad [14]$$

And $$\frac{1}{\sqrt{k'(x)}} e^{+\int_x^{x_2} k'(x)\,dx} \overset{I-II}{\longleftrightarrow} \frac{1}{\sqrt{k(x)}} \cos\left(\int_{x_2}^{x} k(x)\,dx + \frac{\pi}{4}\right) \quad [15]$$

Relations [14] and [15] are WKB connection formula.

17.7 Somerfield-Wilson quantization condition

The WKB approximation can be applied to derive an equation for energies of bound states. Consider a simple well-shaped potential with two classical turning points as shown in figure. Turning points mean where $E = V$ and WKB formula is not valid at these points.

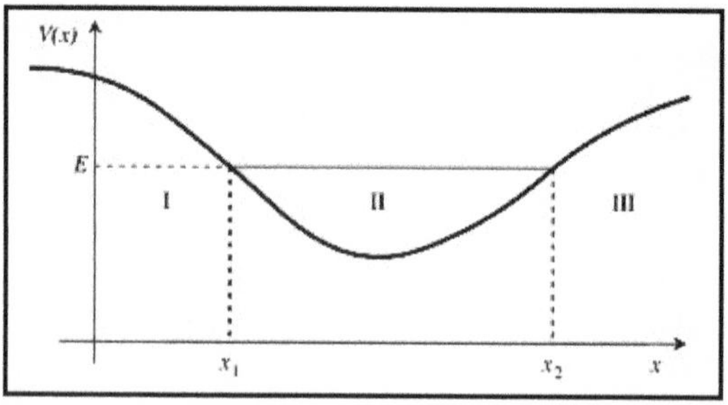

Fig.[17.7a]: Shows a simple well-shaped potential with two classical turning points

The WKB approximation will be used in regions 1, 2 and 3 away from the turning points and the Connection formulas will serve near $x = a$ and $x = b$. The usual requirement that Ψ must be finite dictates that the solutions which increase exponentially as one move outward from the turning points must vanish rigorously[17].

In region 1; I.e., $< a$, the wave function exponentially decaying.

$$\Psi_1(x) = \frac{C_+}{\sqrt{|k(x)|}} \exp\left(\int_x^a k(x)\,dx\right) + \frac{C_-}{\sqrt{|k(x)|}} \exp\left(-\int_x^a k(x)\,dx\right) \quad [1]$$

Since $\int_x^a k(x)dx \to \infty \; as\, x \to -\infty$, we must choose $C_+ = 0$

Thus to the left of the first turning point we have

$$\Psi_1(x) = \frac{C}{\sqrt{|k(x)|}} \exp\left(-\int_x^a k(x)\,dx\right) \quad [2]$$

In region 2; Using the Connection formula, the wave function

$$\Psi_2(x) = C\frac{2}{\sqrt{|k(x)|}} \operatorname{Sin}\left[\int_a^x k(x)\,dx + \frac{\pi}{4}\right]$$

$$= C\frac{2}{\sqrt{|k(x)|}}\,\text{Cos}\left[\int_a^x k(x)\,dx - \frac{\pi}{4}\right] \quad ; a \le x \le b \quad \because Sin\theta = \text{Cos}\left(\theta - \frac{\pi}{2}\right)$$

Rewriting

$$= C\frac{2}{\sqrt{|k(x)|}}\,\text{Cos}\left[\int_a^b k(x)\,dx - \int_x^b k(x)\,dx - \frac{\pi}{4}\right] \quad ; a \le x \le b$$

$$\Psi_2(x) = C\frac{2}{\sqrt{|k(x)|}}\,\text{Cos}\left[\theta - \int_x^b k(x)\,dx - \frac{\pi}{4}\right]$$

Where, $\theta = \int_a^b k(x)\,dx$ [3]

We can expand the cosine in $\Psi_2(x)$ to get

$$\Psi_2(x) = C\frac{2}{\sqrt{|k(x)|}}\,\text{Cos}\theta\,\text{Cos}\left[\int_x^b k(x)\,dx + \frac{\pi}{4}\right] + C\frac{2}{\sqrt{|k(x)|}}\,\text{Sin}\theta\,\text{Sin}\left[\int_x^b k(x)\,dx + \frac{\pi}{4}\right] \quad [4]$$

In region 3; The wave function exponentially decaying. But the $Cos[\int_x^b k(x)\,dx + \frac{\pi}{4}]$ terms in $\Psi_2(x)$ goes over to an exponentially increasing function in region 3. Hence the term in $\Psi_2(x)$which involves the Cos function must be made exactly zero. We therefore put

$$\text{Cos}\theta = 0 \qquad or,\ \theta = \left(n + \frac{1}{2}\right)\pi\ ; \qquad \text{n} = 0, 1, 2 \ldots\ldots\ldots \quad [5]$$

$$\int_a^b k(x)\,dx = \left(n + \frac{1}{2}\right)\pi; \qquad \text{n} = 0, 1, 2 \ldots\ldots\ldots \quad [6]$$

Equation [6] gives the possible discrete values of E ; which is the well-known Somerfield-Wilson quantization condition. E appears in the integrand as well as in the limits of integration, since the turning points a and b are determined such that $V(a) = V(b) = E$. If we introduce the classical momentum $p(x) = \pm\hbar k(x)$ and plot $p(x)$ versus x in phase space, the bounded motion in a potential well can be pictured by a closed curve as shown in figure [17.5b]. It is evident that the condition [6] may be written as

$$J = \oint p(x)dx = \pi\left(n + \frac{1}{2}\right)\hbar \quad [7]$$

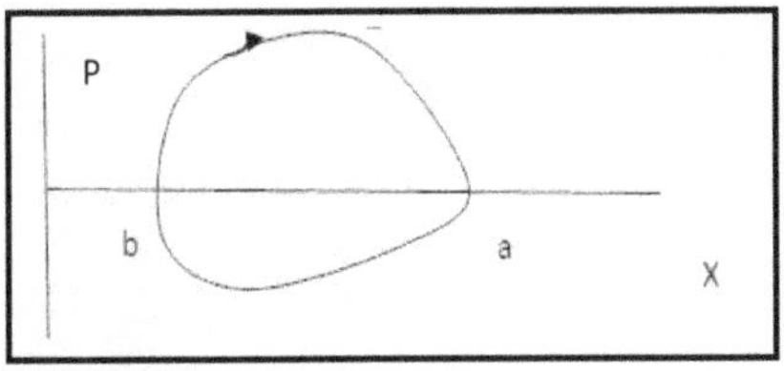

Fig. [17.7b]Shows: Phase space representation of the periodic motion of a particle confined between the classical turning points $x = a$ and $x = b$

The L.H.S of equation [7] equals the area enclosed by the curve representing the motion in phase space, is called the phase integral J in classical terminology.

17.8 Energy levels of the potential $V(x) = \frac{1}{2}kx^2$ for one dimensional harmonic oscillator by using WKB approximation method

The potential for one dimensional harmonic oscillator[30]

$$V(x) = \frac{1}{2}kx^2 = \frac{1}{2}m\omega^2 x^2 \qquad [1]$$

Equation [1] is a parabolic potential well as shown in figure.

From Somerfield-Wilson quantization condition

$$\int_{-a}^{a} k(x)\,dx = \left(n + \frac{1}{2}\right)\pi \qquad [2]$$

Where $k(x) = \sqrt{\frac{2m}{\hbar^2}\left(E - V(x)\right)}$ [3]

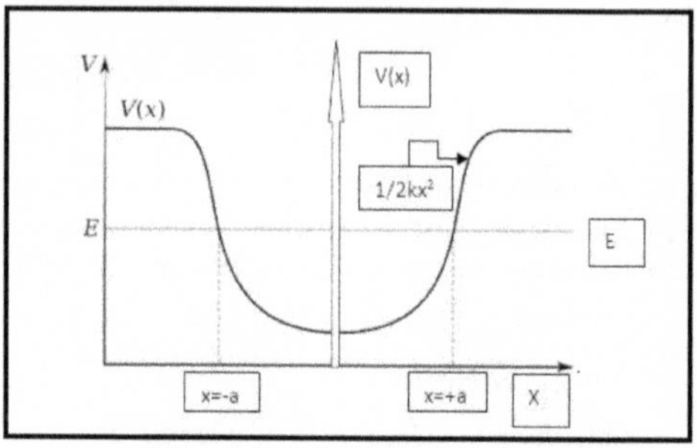

Fig.[17.8]: Shows is a parabolic potential well

$$\therefore \quad E[= V(x)] = \frac{1}{2}kx^2 = \frac{1}{2}m\omega^2 x^2 = \frac{1}{2}m\omega^2 a^2 \qquad [4]$$

$$a^2 = \frac{2E}{m\omega^2} \qquad \text{or, } a = \pm\sqrt{\frac{2E}{m\omega^2}} \qquad [5]$$

And the turning points x ± a are obtained in terms energy E from the equation

$$\therefore \qquad \frac{\sqrt{2m}}{\hbar}\int_{-a}^{a}\left(E - \frac{1}{2}m\omega^2 x^2\right)^{\frac{1}{2}} dx = \left(n + \frac{1}{2}\right)\pi$$

Or, $$\frac{\sqrt{2mE}}{\hbar}\int_{-a}^{a}\left(1 - \frac{1}{2E}m\omega^2 x^2\right)^{\frac{1}{2}} dx = \left(n + \frac{1}{2}\right)\pi \qquad [6]$$

Let $y = \sqrt{\frac{m\omega^2}{2E}}\,x$ Or, $y^2 = \frac{m\omega^2 x^2}{2E}$ or, $x = \sqrt{\frac{2E}{m\omega^2}}\,y$ or, $dx = \sqrt{\frac{2E}{m\omega^2}}$ dy

When $x = -a = -\sqrt{\frac{2E}{m\omega^2}}$; $y = -1$, $y = -1$, When $x = a = \sqrt{\frac{2E}{m\omega^2}}$; $y = +1$

$$\therefore \quad \frac{\sqrt{2mE}}{\hbar}\int_{-1}^{+1}(1 - y^2)^{\frac{1}{2}}\sqrt{\frac{2E}{m\omega^2}}\,\mathrm{dy} = \left(n + \frac{1}{2}\right)\pi$$

Or, $$\frac{\sqrt{2mE}}{\hbar}\sqrt{\frac{2E}{m\omega^2}}\int_{-1}^{+1}(1 - y^2)\,\mathrm{dy} = \left(n + \frac{1}{2}\right)\pi$$

$$\frac{2E}{\hbar\omega}\int_{-1}^{+1}(1 - y^2)\,\mathrm{dy} = \left(n + \frac{1}{2}\right)\pi \qquad [7]$$

$When\ y = -1,\ \ \theta = \pi\ \ and\ when\ y = 1,\ \ \theta = 0\ ;\ \ put\ y = cos\theta$

$\therefore\quad \int_{-1}^{1}\sqrt{(1-y^2)}\,\mathrm{dy} = \int_{\pi}^{0}\sqrt{(1-Cos^2\theta)}d(cos\theta) = \int_{\pi}^{0} Sin\theta\,(-Sin\theta\; d\theta) = -\int_{\pi}^{0} Sin\theta^2\, d\theta$

Or, $\int_{-1}^{1}\sqrt{(1-y^2)}\,\mathrm{dy} = \frac{1}{2}\int_{0}^{\pi} 2Sin^2\theta d\,\theta = \int_{0}^{\pi}(1-Cos2\theta\,)d\theta = \frac{\pi}{2}$

Hence equation [7] becomes

$$\frac{2E}{\hbar\omega}\,\frac{\pi}{2} = \left(n+\frac{1}{2}\right)\pi \qquad \text{Or,}\ \ E = \left(n+\frac{1}{2}\right)\hbar\,\omega \qquad [8]$$

Where $n = 0, 1, 2, 3 \ldots\ldots\ldots\ldots\ldots$

Equation [8] represents the energy levels for one dimensional harmonic oscillator by using WKB approximation method. Also it gives us discrete or quantized allowed values of energies of a particle in parabolic potential well. I.e., the system is completely non- degenerate and equally spaced[26].

17.9 Energy levels of the potential $V(x) = \lambda x^4$ by using WKB approximation method

The energy levels of the potential

$$V(x) = \lambda x^4 \qquad [1]$$

From Somerfield-Wilson quantization condition

$$\int_{-a}^{a} k(x)\,dx = \left(n+\frac{1}{2}\right)\pi \qquad [2]$$

Where $$k(x) = \sqrt{\frac{2m}{\hbar^2}\left(E - V(x)\right)} \qquad [3]$$

And the turning points x = ± a are obtained in terms energy E from the equation

$$E[= V(x)] = \lambda x^4 = \lambda a^4 \ \text{Or,}\ a = \left(\frac{E}{\lambda}\right)^{\frac{1}{4}} \qquad [4]$$

Or, $\int_{-a}^{a}\sqrt{\frac{2m}{\hbar^2}\left(E - V(x)\right)}\,dx = \left(n+\frac{1}{2}\right)\pi\,; \qquad n = 0, 1, 2, \ldots\ldots\ldots\ldots\ldots\ldots\ldots\ldots$

Or, $\sqrt{\frac{2mE}{\hbar^2}}\int_{-a}^{a}\left(1-\frac{V(x)}{E}\right)^{\frac{1}{2}} dx = \left(n+\frac{1}{2}\right)\pi$

Or, $\sqrt{\frac{2mE}{\hbar^2}}\int_{-a}^{a}\left(1-\frac{\lambda x^4}{\lambda a^4}\right)^{\frac{1}{2}} dx = \left(n+\frac{1}{2}\right)\pi$

Or, $\sqrt{\frac{2mE}{\hbar^2}}\int_{-a}^{a}\left(1-\frac{x^4}{a^4}\right)^{\frac{1}{2}} dx = \left(n+\frac{1}{2}\right)\pi$

Let, $y = \frac{x}{a}$, $x = ay$, $\therefore dx = a\,dy$

$\therefore$ $\sqrt{\frac{2mE}{\hbar^2}} \int_{-1}^{+1} (1 - y^4)^{\frac{1}{2}} a dy = \left(n + \frac{1}{2}\right)\pi$; change the upper and lower limit of the integration.

Or, $\sqrt{\frac{2mE}{\hbar^2}}\, a \int_{-1}^{+1} (1 - y^4)^{\frac{1}{2}} dy = \left(n + \frac{1}{2}\right)\pi$

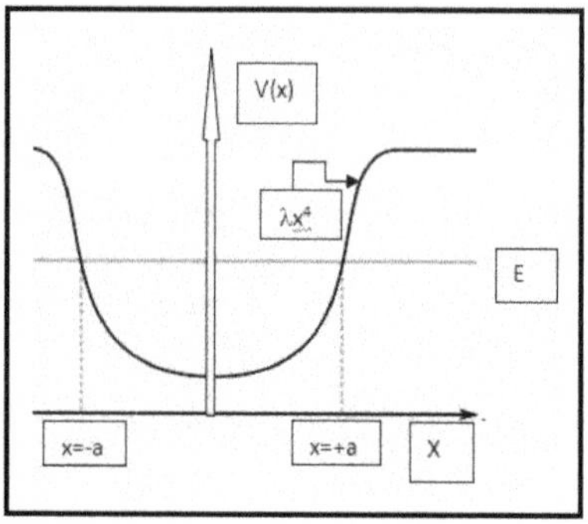

Fig.[17.9]: Shows the Energy levels for a given potential $V(x) = \lambda x^4$

Or, $\sqrt{\frac{2mE}{\hbar^2}} \left(\frac{E}{\lambda}\right)^{\frac{1}{4}} \int_{-1}^{+1} (1 - y^4)^{\frac{1}{2}} dy = \left(n + \frac{1}{2}\right)\pi$

Or, $\sqrt{\frac{2mE}{\hbar^2}} \left(\frac{E}{\lambda}\right)^{\frac{1}{4}} 2 \int_{0}^{+1} (1 - y^4)^{\frac{1}{2}} dy = \left(n + \frac{1}{2}\right)\pi$ [5]

Now consider the integral

$$I = \int_0^{+1} (1 - y^4)^{\frac{1}{2}} dy \qquad [6]$$

Let $y^4 = Z$, $y = z^{\frac{1}{4}}$ or, $y^3 = (Z)^{\frac{3}{4}}$

$4y^3 dy = dZ$, $dy = \frac{dZ}{4y^3} = \frac{dZ}{4(Z)^{\frac{3}{4}}}$

$\therefore$ $I = \int_0^{+1} (1 - Z)^{\frac{1}{2}} \frac{dZ}{4(Z)^{\frac{3}{4}}}$

Or, $I = \frac{1}{4} \int_0^{+1} (Z)^{-\frac{3}{4}} (1 - Z)^{\frac{1}{2}} dZ$

Or, $I = \frac{1}{4} \int_0^{+1} (Z)^{\frac{1}{4}-1} (1 - Z)^{\frac{3}{2}-1}\, dZ$ [7]

From standard integral [beta function]

$\int_0^{+1} (x)^{\alpha-1} (1 - x)^{\beta-1}\, dx = B(\alpha, \beta)$ $or,$ $B(\alpha, \beta) = \frac{\Gamma\alpha\Gamma\beta}{\Gamma(\alpha+\beta)}$ [8]

Comparing, equation [7] and equation [8] we can write

$$I = \frac{1}{4} B\left(\frac{1}{4}, \frac{3}{2}\right) \quad [10]$$

Therefore, equation [5] becomes

$$\sqrt{\frac{2mE}{\hbar^2}} \left(\frac{E}{\lambda}\right)^{\frac{1}{4}} 2 \; \frac{1}{4} \; B\left(\frac{1}{4}, \frac{3}{2}\right) = \left(n + \frac{1}{2}\right)\pi$$

Or, $\left(\frac{2mE}{\hbar^2}\right)^2 \frac{E}{\lambda} \frac{1}{16} B^4\left(\frac{1}{4}, \frac{3}{2}\right) = \left(n + \frac{1}{2}\right)^4 \pi^4 \quad or, \quad \frac{m^2 E^3}{4\hbar^4 \lambda} B^4\left(\frac{1}{4}, \frac{3}{2}\right) = \left(n + \frac{1}{2}\right)^4 \pi^4$

Or, $E^3 = \frac{4\lambda\hbar^4\pi^4}{m^2 B^4\left(\frac{1}{4}, \frac{3}{2}\right)} \left(n + \frac{1}{2}\right)^4$

Or, $$E = \left[\frac{4\lambda\hbar^4\pi^4}{m^2 . B^4\left(\frac{1}{4}, \frac{3}{2}\right)}\right]^{\frac{1}{3}} \left(n + \frac{1}{2}\right)^{\frac{4}{3}} \quad [11]$$

Equation [11] gives the energy levels for a given potential of the type $V(x) = \lambda x^4$.

Note: $\Gamma(\alpha + 1) = \alpha\Gamma\alpha$; $\Gamma n = (n - 1)!$ if n is an integer $\Gamma 1 = \Gamma 2 = 1$; $\Gamma\frac{1}{2} = \sqrt{\pi}$; $\Gamma\frac{3}{2} = \frac{\sqrt{\pi}}{2}$

17.10 Energy levels of a ball bouncing off a perfectly reflecting plane in a uniform gravitational field

The energy levels of the potential $V(x) = Fx = mgx \; ; \; x > 0 \; and \; V(x) = \infty \; ; x \leq 0$

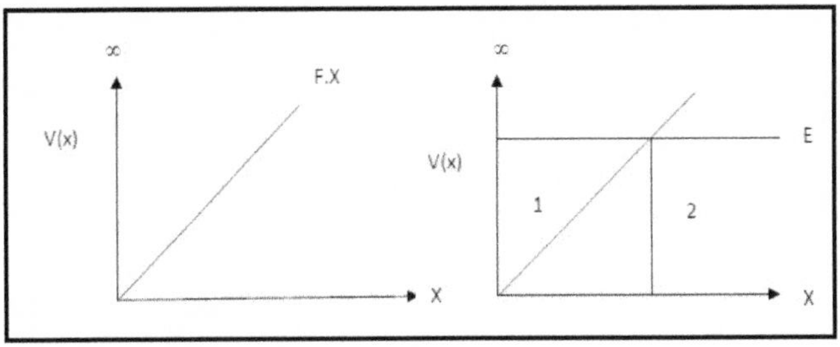

Fig.[17.10]: Shows Energy levels of a ball bouncing off a perfectly reflecting plane in a uniform gravitational field.

There is only one turning point at $x = b$.Where $E = F.b \; or, b = \frac{E}{F}$ [1]

Now to the far right of b, the solution must be exponentially decaying. Therefore we can write

$$\Psi_2(x) = \frac{A}{\sqrt{|k(x)|}} \exp\left(-\int_b^x |k(x)| \, dx\right) \quad [2]$$

By using the connection formulas, the WKB wave function in region 1 is

$$\Psi_1(x) = \frac{2A}{\sqrt{k(x)}} \text{Sin}\left(\int_x^b k(x)\,dx + \frac{\pi}{4}\right) \qquad [3]$$

Now, since the potential goes to infinity at x = 0; the wave function must vanish there. I.e.,

$$\Psi_1(x=0) = 0 \quad or, \ \text{Sin}\left(\int_0^b k(x)\,dx + \frac{\pi}{4}\right) = 0$$

Or, $$\text{Sin}\left(\int_0^b k(x)\,dx + \frac{\pi}{4}\right) = m\pi\,; \quad m = 1, 2, 3, \dots\dots\dots$$

We note that the phase of Sin I.e., $\text{Sin}\left(\int_0^b k(x)\,dx + \frac{\pi}{4}\right)$ is positive.

$$\int_0^b k(x)\,dx = \left(m - \frac{1}{4}\right)\pi \qquad [4]$$

We can write the above equation in the following form

$$\int_0^b k(x)\,dx = \left(n + \frac{3}{4}\right)\pi\,; \qquad n = 0,1,2,3, \dots\dots\dots \qquad [5]$$

Where $k(x) = \sqrt{\frac{2m}{\hbar^2}(E - V(x))}$; So equation [5] becomes

$$\int_0^b \sqrt{\frac{2m}{\hbar^2}(E - Fx)}\,dx = \left(n + \frac{3}{4}\right)\pi$$

Or, $$\frac{\sqrt{2m}}{\hbar}\int_0^b \sqrt{(E - Fx)}\,dx = \left(n + \frac{3}{4}\right)\pi$$

Or, $$\sqrt{2mF}\int_0^b \sqrt{\left(\frac{E}{F} - x\right)}\,dx = \left(n + \frac{3}{4}\right)\pi\hbar$$

Or, $$\sqrt{2mF}\int_0^b (b - x)^{\frac{1}{2}}\,dx = \left(n + \frac{3}{4}\right)\pi\hbar$$

Or, $$\sqrt{2mF}\left[\frac{-(b-x)^{\frac{3}{2}}}{\frac{3}{2}}\right]_0^b = \left(n + \frac{3}{4}\right)\pi\hbar$$

Or, $$\sqrt{2mF}\,\frac{2}{3}\left[-(b - x)^{\frac{3}{2}}\right]_0^b = \left(n + \frac{3}{4}\right)\pi\hbar$$

Or, $$\sqrt{2mF}\,\frac{2}{3}b^{\frac{3}{2}} = \left(n + \frac{3}{4}\right)\pi\hbar$$

Or, $$\sqrt{2mF}\,\frac{2}{3}\left(\frac{E}{F}\right)^{\frac{3}{2}} = \left(n + \frac{3}{4}\right)\pi\hbar$$

Or, $$\frac{E^{\frac{3}{2}}\,2\sqrt{2m}}{3F} = \left(n + \frac{3}{4}\right)\pi\hbar$$

Or, $$E^{\frac{3}{2}} = \frac{3F}{2\sqrt{2m}}\left(n + \frac{3}{4}\right)\pi\hbar$$

Or, $$E^3 = \frac{9\,\pi^2\hbar^2F^2}{8m}\left(n + \frac{3}{2}\right)^2$$

Or, $$E = \left(\frac{9}{8}mg^2\hbar^2\pi^2\right)^{\frac{1}{3}}\left(n+\frac{3}{2}\right)^{\frac{2}{3}} \qquad [6]$$

Equation [6] represents the bound state Eigen values for the potential

$$V(x) = Fx = mgx\,;\ x > 0\ and\ V(x) = \infty\ ;\ x \leq 0$$

17.11 Transmission and reflection coefficient of single potential barrier of general shape using WKB method

We have a single potential barrier of general shape. To calculate transmission and reflection coefficient of single step potential barrier of such general shape in which $V(x)$ is slowly varying function of x. In figure, $x = a$ and $x = b$ are classical turning points. For a particle having total energy, I.e., kinetic energy plus potential energy E. We take three regions as shown in figure away from $x = a$ and $x = b$. Hence WKB solutions of one dimensional Schrödinger equation are valid in all the three regions.

$$\Psi(x) = exp\left(\pm\frac{i}{\hbar}\int p(x)\,dx\right)\frac{C}{\sqrt{|p(x)|}} = \frac{C}{\sqrt{|p(x)|}}exp\left(\pm\frac{i}{\hbar}\int p(x)\,dx\right) \qquad [A]$$

In region *III*: $E > V(x)$ and WKB solution taken to be as

$$\Psi_{III} = \frac{C}{\sqrt{|p(x)|}}exp\left(\pm\frac{i}{\hbar}\int p(x)\,dx\right) = \frac{C}{\sqrt{k(x)}}e^{\pm i\int_b^x k(x)\,dx} \qquad [1a]$$

If $V(x)$ were constant or zero in region III, then Ψ_{III} would have been e^{+ikx} ; which is transmitted plane wave travelling along Positive x -axis. When the potential energy function $V(x)$ is not a simple form and when it is slowly varying function of position then the solution of the Schrödinger equation even in one dimension is usually a complicated problem, which requires the use of approximate methods. A particular method however is found is known as WKB approximate method, which can be solved the problems exactly.

Since $V(x)$ is not constant of motion, but it is slowly varying function of position, e^{+ikx} modifies to WKB solution given by equation [1]. Equation [1] gives WKB version of transmitted plane wave. Only we expect no reflection in region III, we have excluded WKB version of e^{-ikx} which is $\frac{C}{\sqrt{k(x)}}e^{-i\int_b^x k(x)\,dx}$ from equation [1]. Ψ_{III} of equation [1] can be called transmitted wave Ψ_t. I.e.,

$$\Psi_{trans} = \Psi_{III} = \frac{C}{\sqrt{k(x)}}e^{+i\int_b^x k(x)\,dx} \qquad [1b]$$

Now we are ready for using WKB connection formula between region II and III; the connection formula are[17]

$$\frac{1}{\sqrt{k'(x)}}e^{-\int_x^b k'(x)\,dx} \overset{II-III}{\longleftrightarrow} \frac{2}{\sqrt{k(x)}}\sin\left(\int_b^x k(x)\,dx+\frac{\pi}{4}\right) \quad [2]$$

$$\frac{1}{\sqrt{k'(x)}}e^{+\int_x^b k'(x)\,dx} \overset{II-III}{\longleftrightarrow} \frac{1}{\sqrt{k(x)}}\cos\left(\int_b^x k(x)\,dx+\frac{\pi}{4}\right) \quad [3]$$

Equation [1b] gives

$$\Psi_{III}=\frac{C}{\sqrt{k(x)}}e^{+i\int_b^x k(x)\,dx}=\frac{Ce^{-i\frac{\pi}{4}}}{\sqrt{k(x)}}e^{+i\left(\int_b^x k(x)\,dx+\frac{\pi}{4}\right)}=\frac{C'}{\sqrt{k(x)}}e^{+i\left(\int_b^x k(x)\,dx+\frac{\pi}{4}\right)}$$

Or, $$\Psi_{III}=\frac{C'}{\sqrt{k(x)}}\cos\left(\int_b^x k(x)\,dx+\frac{\pi}{4}\right)+\frac{iC'}{2\sqrt{k(x)}}2\sin\left(\int_b^x k(x)\,dx+\frac{\pi}{4}\right) \quad [4]$$

Using equations [2] and [3] in equation [4], we get

$$\Psi_{II}=\frac{C'}{\sqrt{k'(x)}}e^{+\int_x^b k'(x)\,dx}+\frac{i}{2}\frac{C'}{\sqrt{k'(x)}}e^{-\int_x^b k'(x)\,dx} \quad [5]$$

To get expression of Ψ_I we now need to use WKB connection formula between region I and II in equation [5] . The two connection formulae are

$$\frac{1}{\sqrt{k'(x)}}e^{-\int_a^x k'(x)\,dx} \overset{II-I}{\longleftrightarrow} \frac{2}{\sqrt{k(x)}}\sin\left(\int_x^a k(x)\,dx+\frac{\pi}{4}\right) \quad [6]$$

$$\frac{1}{\sqrt{k'(x)}}e^{+\int_a^x k'(x)\,dx} \overset{II-I}{\longleftrightarrow} \frac{1}{\sqrt{k(x)}}\cos\left(\int_x^a k(x)\,dx+\frac{\pi}{4}\right) \quad [7]$$

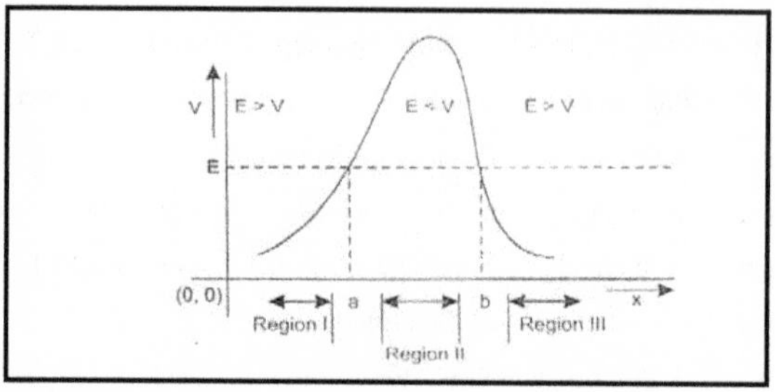

Fig.[17.11]: Shows a single potential barrier of general shape

We now adjust limits of integrals of equation [5] to use the WKB connection formulas [6] and [7] in equation [5]. Equation [5] gives

$$\Psi_{II}=\frac{C'}{\sqrt{k'(x)}}e^{\int_a^b k'(x)\,dx-\int_a^x k'(x)\,dx}+\frac{i}{2}\frac{C'}{\sqrt{k'(x)}}e^{-\int_a^b k'(x)\,dx+\int_a^x k'(x)\,dx} \quad [8]$$

Let $$e^{-\int_a^b k'(x)}=\theta \quad [9]$$

Equation [8] becomes

$$\Psi_{II}=\frac{C'}{\theta}\frac{1}{\sqrt{k'(x)}}e^{-\int_a^x k'(x)\,dx}+\frac{iC'\theta}{2}\frac{1}{\sqrt{k'(x)}}e^{+\int_a^x k'(x)\,dx} \quad [10]$$

Use of equations [6], [7] in equation [10]

$$\Psi_I = \frac{C'}{\theta}\frac{2}{\sqrt{k(x)}}\sin\left(\int_x^a k(x)\,dx + \frac{\pi}{4}\right) + \frac{iC'\theta}{2}\frac{1}{\sqrt{k(x)}}\cos\left(\int_x^a k(x)\,dx + \frac{\pi}{4}\right) \qquad [11]$$

Or, $$\Psi_I = \frac{A'}{\sqrt{k(x)}}e^{-i\left(\int_x^a k(x)\,dx + \frac{\pi}{4}\right)} + \frac{B'}{\sqrt{k(x)}}e^{+i\left(\int_x^a k(x)\,dx + \frac{\pi}{4}\right)} \qquad [12]$$

Or, $\Psi_I = \frac{A}{\sqrt{k(x)}}e^{-i\int_x^a k(x)\,dx} + \frac{B}{\sqrt{k(x)}}e^{+i\left(\int_x^a k(x)\,dx\right)}$ [Here $A = A'e^{-i\frac{\pi}{4}}$ and B= $B'e^{+i\frac{\pi}{4}}$.]

Or, $$\Psi_I = \frac{A}{\sqrt{k(x)}}e^{+i\int_a^x k(x)\,dx} + \frac{B}{\sqrt{k(x)}}e^{-i\int_a^x k(x)\,dx} \qquad [13]$$

Or, $$\Psi_I = \Psi_{incident} + \Psi_{reflected} \qquad [13]$$

From [13], we note that:

The first term $\Psi_{incident}$ is WKB version of e^{+ikx} which is the incident wave on the barrier from the left.

And the second term $\Psi_{reflected}$ is WKB version of e^{-ikx} which is the reflected wave from the barrier in region I.

Equations [11] and [12] are same. Thus

$$A' + B' = \frac{iC'\theta}{2}$$

Or, $$Ae^{+i\frac{\pi}{4}} + Be^{-i\frac{\pi}{4}} = \frac{iC'\theta}{2} \quad [\because A = A'e^{-i\frac{\pi}{4}} \text{ and } B = B'e^{+i\frac{\pi}{4}}.] \qquad [14]$$

And $$-iA' + iB' = \frac{2C'}{\theta}$$

Or, $$-Ae^{+i\frac{\pi}{4}} + Be^{-i\frac{\pi}{4}} = \frac{2C'}{i\theta} = -\frac{i2C'}{\theta} \qquad [15]$$

Subtracting [15] from [14]

$$2Ae^{i\frac{\pi}{4}} = iC'\left(\frac{\theta}{2} + \frac{2}{\theta}\right)$$

Or, $$A = iC'\left(\frac{\theta}{4} + \frac{1}{\theta}\right)e^{-i\frac{\pi}{4}} \qquad [16]$$

Adding [14] and [15]

$$2Be^{-i\frac{\pi}{4}} = iC'\left(\frac{\theta}{2} - \frac{2}{\theta}\right)$$

Or, $$B = iC'\left(\frac{\theta}{4} - \frac{1}{\theta}\right)e^{i\frac{\pi}{4}} \qquad [17]$$

Now the Probability current density can be defined as

$$J_t = \frac{\hbar}{2mi}\left(\Psi_t^{\times}\frac{\partial\Psi_t}{\partial x} - \Psi_t\frac{\partial\Psi_t^{\times}}{\partial x}\right)$$

Probability current density associated with transmitted WKB wave $\Psi_{trans} = \frac{C}{\sqrt{k(x)}}e^{+i\int_b^x k(x)\,dx}$ given by equation [1b] is

$$J_t = \frac{\hbar}{2mi}\left(\Psi_t^{\times}\frac{\partial\Psi_t}{\partial x} - \Psi_t\frac{\partial\Psi_t^{\times}}{\partial x}\right)$$

$$= \frac{\hbar}{2mi}\left[\frac{C^\times}{\sqrt{k(x)}}e^{-i\int_b^x k(x)\,dx}\frac{\partial}{\partial x}\left(\frac{C}{\sqrt{k(x)}}e^{+i\int_b^x k(x)\,dx}\right) - \frac{C}{\sqrt{k(x)}}e^{+i\int_b^x k(x)\,dx}\frac{\partial}{\partial x}\left(\frac{C^\times}{\sqrt{k(x)}}e^{-i\int_b^x k(x)\,dx}\right)\right]$$

$$= \frac{\hbar}{2mi}\left[\frac{C^\times}{\sqrt{k(x)}}e^{-i\int_b^x k(x)\,dx}\left(\frac{C}{\sqrt{k(x)}}e^{+i\int_b^x k(x)\,dx}ik - \frac{1}{2}\frac{C}{k^{3/2}}\frac{dk}{dx}e^{+i\int_b^x k(x)\,dx}\right)\right]$$

$$-\frac{\hbar}{2mi}\left[\frac{C}{\sqrt{k(x)}}e^{+i\int_b^x k(x)\,dx}\left(\frac{C^\times}{\sqrt{k(x)}}e^{-i\int_b^x k(x)\,dx}(-ik) - \frac{1}{2}\frac{C^\times}{k^{3/2}}\frac{dk}{dx}e^{-i\int_b^x k(x)\,dx}\right)\right]$$

$$J_t = \frac{\hbar}{2mi}\left[i|C|^2 - \frac{|C|^2}{2k^2}\frac{dk}{dx} + i|C|^2 + \frac{|C|^2}{2k^2}\frac{dk}{dx}\right]$$

$$= \frac{\hbar}{2mi}2i|C|^2 = \frac{\hbar}{m}|C|^2$$

$$J_t = \frac{\hbar}{m}|C|^2$$

Again from equation [13]

$$\Psi_I = \frac{A}{\sqrt{k(x)}}e^{+i\int_a^x k(x)\,dx} + \frac{B}{\sqrt{k(x)}}e^{-i\int_a^x k(x)\,dx} = \Psi_{incident} + \Psi_{reflected}$$

From the above equation it is clear that the probability current density associated with incident WKB wave

$$\Psi_i = \frac{A}{\sqrt{k(x)}}e^{+i\int_a^x k(x)\,dx} \text{ is } J_i = \frac{\hbar}{m}|A|^2.$$

And probability current density associated with reflected WKB wave

$$\Psi_r = \frac{A}{\sqrt{k(x)}}e^{-i\int_a^x k(x)\,dx} \text{ is } J_r = \frac{\hbar}{m}|B|^2.$$

Therefore, the Transmission coefficient [Using equation (16)]

$$T = \frac{J_{transmitted}}{J_{incident}} = \frac{\frac{\hbar}{m}|C|^2}{\frac{\hbar}{m}|A|^2} = \frac{|C|^2}{|A|^2} = \frac{\left|C'e^{i\frac{\pi}{4}}\right|^2}{|A|^2} = \frac{|C'|^2}{|A|^2} = \left|\frac{C'}{A}\right|^2 = \frac{1}{\left|i\left(\frac{\theta}{4}+\frac{1}{\theta}\right)e^{-i\frac{\pi}{4}}\right|^2} = \frac{1}{\left(\frac{\theta}{4}+\frac{1}{\theta}\right)^2} \quad [18]$$

In most cases, $\theta \ll 1$ and hence [18] becomes $T \approx \theta^2$

Again from equation [9] $\theta = e^{-\int_a^b k'(x)\,dx}$;

$$T \approx \theta^2 \approx e^{-2\int_a^b k'(x)\,dx} \quad [19]$$

Reflection coefficient is [Using equations (16) and (17)]

$$R = \frac{J_r}{J_i}$$

$$= \frac{\frac{\hbar}{m}|B|^2}{\frac{\hbar}{m}|A|^2} = \frac{|B|^2}{|A|^2} = \frac{\left|iC'\left(\frac{\theta}{4}-\frac{1}{\theta}\right)e^{i\frac{\pi}{4}}\right|^2}{\left|iC'\left(\frac{\theta}{4}+\frac{1}{\theta}\right)e^{-i\frac{\pi}{4}}\right|^2}$$

Thus $$R = \frac{\left(\frac{\theta}{4}-\frac{1}{\theta}\right)^2}{\left(\frac{\theta}{4}+\frac{1}{\theta}\right)^2} \quad [20]$$

The Sum of the transmission and the reflection coefficient is [Using equations (18) and (20)]

$$T+R=\frac{1}{\left(\frac{\theta}{4}+\frac{1}{\theta}\right)^2}+\frac{\left(\frac{\theta}{4}-\frac{1}{\theta}\right)^2}{\left(\frac{\theta}{4}+\frac{1}{\theta}\right)^2}=\frac{1+\left(\frac{\theta}{4}\right)^2+\left(\frac{1}{\theta}\right)^2-2\frac{\theta}{4}\frac{1}{\theta}}{\left(\frac{\theta}{4}+\frac{1}{\theta}\right)^2}$$

Or, $$T+R=\frac{\left(\frac{\theta}{4}\right)^2+\left(\frac{1}{\theta}\right)^2+2\frac{\theta}{4}\frac{1}{\theta}}{\left(\frac{\theta}{4}+\frac{1}{\theta}\right)^2}=\frac{\left(\frac{\theta}{4}+\frac{1}{\theta}\right)^2}{\left(\frac{\theta}{4}+\frac{1}{\theta}\right)^2}=1$$

Thus $T+R=1$

17.12 Validity of the WKB approximation method

Wentzel-Kramers-Brillouin [WKB] approximation gives direct and approximate solution of Schrödinger equation. WKB approximation method is applied to those Quantum mechanical problems where the potential function is a slowly varying function of the position. *A slowly changing potential means the variation of V(r) is slightly over several wave lengths (de-Broglie waves) of the particle.* Thus when the potential energy function V(x) is not a simple form and when it is slowly varying function of position then the solution of the Schrödinger equation even in one dimension is usually a complicated problem, which requires the use of approximate methods. A particular method however is found is known as WKB approximate method, which can be solved the problems exactly. Problem in one dimension and also of three dimensions reducible in one dimensional problem are solved by this method[5]. As such that

$$k(x)=\sqrt{\frac{2m}{\hbar^2}\big(E-V(x)\big)}=\frac{p(x)}{\hbar} \qquad [1]$$

To obtain a criterion for the validity of the WKB approximation, we note that

$$\left(\frac{\partial S(x)}{\partial x}\right)^2-p^2(x)-i\hbar\,\frac{\partial^2 S(x)}{\partial^2 x}=0 \qquad [2]$$

In the limit $\hbar\to 0$

$$\left(\frac{\partial S(x)}{\partial x}\right)^2-p^2(x)=0 \qquad [3]$$

Or, $$\left(\frac{\partial S(x)}{\partial x}\right)^2=p^2(x) \qquad [4]$$

From Classical relation $\vec{p}=\vec{\nabla}S$; clearly this is possible only when

$$\left(\frac{\partial S(x)}{\partial x}\right)^2\geq\geq\hbar\left|\frac{\partial^2 S}{\partial^2 x}\right| \quad \text{Or, } \left(\vec{\nabla}S\right)^2\geq\geq\hbar\left|\vec{\nabla}^2 S\right| \quad \text{Or, } (p)^2\geq\geq\hbar\left|\vec{\nabla}.\vec{\nabla}S\right|$$

Or, $$(p)^2\geq\geq\hbar\left|\vec{\nabla}.\vec{p}\right| \text{ Or, } (p)^2\geq\geq\hbar\left|\frac{\partial p}{\partial x}\right|$$

Or, $$\hbar\left|\frac{\partial p}{\partial x}\right|\leq\leq(p)^2 \text{ Or, } \hbar\left|\frac{\partial(\hbar k)}{\partial x}\right|\leq\leq(\hbar k)^2 \qquad [5]$$

Or, $$\left|\frac{\partial k}{\partial x}\right|\leq\leq k^2 \text{ Or, } \frac{1}{k^2}\left|\frac{\partial k}{\partial x}\right|\leq\leq 1 \text{ Or, } \left|\frac{1}{k}\frac{\partial k}{\partial x}\right|\leq\leq k \qquad [6]$$

I.e., Fractional change of k is much smaller than k. Equation [6] is called condition of validity of WKB approximation. Again from equation [5]

$$\hbar^2 \left|\frac{\partial k}{\partial x}\right| \leq\leq \hbar^2 k^2 \text{ ; if we use } k(x) = \frac{2\pi}{\lambda} \quad [7]$$

Or, $\left|\frac{\partial\left(\frac{2\pi}{\lambda}\right)}{\partial x}\right| \leq\leq \frac{(2\pi)^2}{\lambda^2}$ Or, $\left|\frac{\partial}{\partial x}\left(\frac{1}{\lambda}\right)\right| \leq\leq \frac{2\pi}{\lambda^2}$ Or, $\left|-\frac{1}{\lambda^2}\frac{\partial \lambda}{\partial x}\right| \leq\leq \frac{2\pi}{\lambda^2}$

$$\therefore \quad \left|\frac{\partial \lambda}{\partial x}\right| \leq\leq 2\pi \text{ Or, } \left|\frac{\partial \lambda}{\partial x}\lambda\right| \leq\leq 2\pi\lambda \text{ Or, } \left|\frac{\partial \lambda}{\partial x}\lambda\right| \leq\leq \lambda \quad [8]$$

I.e., Change of de-Broglie wavelength over a wavelength is much smaller than de-Broglie wavelength. Equation [8] is called condition of validity of WKB approximation.

17.13 Transmission through a barrier

We assume that the particle is incident on the barrier from the left. In region 3, $I.e., x > b$, we have only the transmitted wave, so the wave function is of the form

$$\Psi_3(x) = A\frac{1}{\sqrt{k(x)}}\exp\left[i\int_b^x k(x)dx + i\,\frac{\pi}{4}\right] \quad [1]$$

Where A is a constant & the phase factor $\frac{\pi}{4}$ has been included to facilate the application of the connection formulas. In terms of trigonometric functions, $\Psi_3(x)$ can be written as

$$\Psi_3(x) = A\frac{1}{\sqrt{k(x)}}\left[\cos\int_b^x k(x)dx + \frac{\pi}{4}\right] + i\,\sin\left(\int_b^x k(x)dx + \frac{\pi}{4}\right) \quad [2]$$

Using the connection formulas, the WKB wave function in region 2 is

$$\Psi_2(x) = \frac{A}{\sqrt{|k(x)}}\left[\exp\left(\int_x^b |k(x)dx + \frac{i}{2}\exp\left(-\int_x^b |k(x)dx\right)\right] \quad [3]$$

In order to find the appropriate wave function in region 1, we rewrite the integrals

$$\int_x^b |k(x)dx) = \int_a^b |k(x)|dx - \int_a^x |k(x)|dx \quad [4]$$

Thus

$$\Psi_2(x) = \frac{A}{\sqrt{|k(x)|}}\left[\exp\left(\int_a^b |k(x)|dx\right)\exp\left(-\int_a^x |k(x)|dx\right)\right]$$
$$+\frac{A}{\sqrt{|k(x)|}}\left[\frac{i}{2}\exp\left(-\int_a^b |k(x)|dx\right)\exp\left(\int_a^x |k(x)|dx\right)\right] \quad [5]$$

Using the connection formulas across the turning point a, We can now write the WKB wave function in region 1.We have

$$\Psi_1(x) = \frac{A}{\sqrt{k(x)|}}\left[\exp\left(\int_a^b |k(x)|dx\right) 2\text{Sin}\left(\int_x^a k(x)dx + \frac{\pi}{4}\right]\right.$$
$$+\frac{A}{\sqrt{k(x)|}}\left[\frac{i}{2}\exp\left(-\int_a^b |k(x)|dx\right).\text{Cos}\left(\int_x^a k(x)dx + \frac{\pi}{4}\right)\right] \quad [6]$$

Let $\theta = \exp\left(-\int_a^b |k(x)|dx\right)$ [7]

Therefore,

$$\Psi_1(x) = \frac{A}{\sqrt{k(x)|}}\left[2\theta^{-1}\mathrm{Sin}(\int_x^a k(x)dx + \frac{\pi}{4}) + \frac{i}{2}\theta\mathrm{Cos}\,((\int_x^a k(x)dx + \frac{\pi}{4})\right] \quad [8]$$

We now express $\Psi_1(x)$ in terms of exponential functions by using the identities

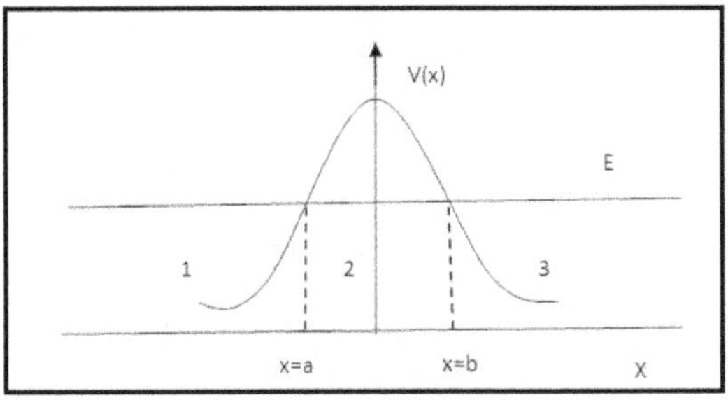

Fig.[17.13]: Shows a potential barrier

We now express $\Psi_1(x)$ in terms of exponential functions by using the identities

$$Sin\alpha = \frac{e^{i\alpha} - e^{-i\alpha}}{2i} \text{ and } Cos\alpha = \frac{e^{i\alpha} + e^{-i\alpha}}{2i}$$; Then we get

$$\Psi_1(x) = \frac{A}{\sqrt{k(x)|}}\left[2\theta^{-1}\frac{1}{2i}\{\exp\left[\mathrm{i}\left(\int_x^a k(x)dx + \frac{\pi}{4}\right)\right] - \exp\left[-i(\int_x^a k(x)dx + \frac{\pi}{4})\right]\}\right.$$

$$+\frac{i}{2}\theta\frac{1}{2i}\{ exp\left[\mathrm{i}\left(\int_x^a k(x)dx + \frac{\pi}{4}\right)\right] + \exp\left[-i\left(\int_x^a k(x)dx + \frac{\pi}{4}\right)\right]\}$$

$$\psi_1((x) = \frac{A}{i\sqrt{k(x)}}[(\theta^{-1} - \frac{\theta}{4})\exp\{i(\int_x^a k(x)dx + \frac{\pi}{4})\} - (\theta^{-1} + \frac{\theta}{4})\exp\{-i(\int_x^a k(x)dx + \frac{\pi}{4})\}] \quad [9]$$

The first term of this equation is identified with the reflected wave & the second term is identified with incident wave. Now the incident flux is

$$Si = \frac{|A|^2}{k(x)}\left(\theta^{-1} + \frac{\theta}{4}\right)^2 \frac{\hbar k(x)}{m} \qquad [\frac{p(x)}{m} = \frac{\sqrt{2m/\hbar^2}(E-V)}{m}]$$

$$Si = |A|^2\left(\theta^{-1} + \frac{\theta}{4}\right)^2 \frac{\hbar}{m} \quad [10]$$

The transmitted flux is obtained form $\psi_3((x)$ given in equation [1], we have

$$S_t = \frac{|A|^2}{k(x)}\frac{\hbar k(x)}{m} = |A|^2\frac{\hbar}{m} \quad [11]$$

Thus, the transmission probability T is

$$T = \frac{transmission\ flux}{Incident\ Flux} = \frac{S_t}{S_i}$$

$$T = \frac{1}{\left(\theta^{-1} + \frac{\theta^2}{4}\right)^2} = \frac{1}{\left(\theta^{-1}\right)^2 \left(1 + \frac{\theta^2}{4\theta^{-1}}\right)^2} = \frac{\theta^2}{\left(1 + \frac{\theta^2}{4}\right)^2} \quad [12]$$

WKB approximation to be valid if |k(x)| must be large & hence θ is small. We can thus neglect all powers of θ higher than the first. So that transmission probability $T = \theta^2$

Now from equation [7] $\theta = \exp\{-(\int_a^b | k(x) | dx$

The transmission probability T is

$$\mathrm{T} = [\exp\{-(\int_a^b | k(x) | dx]^2$$

$$\mathrm{T} = \exp[-2(\int_a^b k(x) dx] \quad [14]$$

17.14 The energy Eigen value of the hydrogen atom by using Somerfield-Wilson quantization rule

From the Somerfield-Wilson quantization rule

$$\int_{r_{\min}}^{r_{\max}} \mathrm{P}_r \, dr = (n + \frac{1}{2})\pi\hbar \quad [1]$$

Here $\mathrm{P}_r = \sqrt{2m(E - V(r)}$ *and for* $H_2 - atom$ $V(r) = \frac{e^2}{r}$

$$\therefore \int_{r\min}^{r\max} \sqrt{2m(E - V(r)} \, dr = (n + \frac{1}{2})\pi\hbar \quad \text{Or,} \quad 2\int_0^{r\max} \sqrt{2mE\left(1 + \frac{e^2}{rE}\right)^{\frac{1}{2}}} \, dr = (n + \frac{1}{2})\pi\hbar$$

$$2\sqrt{2mE} \int_0^{r\max} \left(1 + \frac{e^2}{rE}\right)^{\frac{1}{2}} dr = (n + \frac{1}{2})\pi\hbar$$

At the turning point $V(r) = E$

$$Or,\ E - V(r) = 0 \quad Or,\ E + \frac{e^2}{r} = 0 \quad Or,\ r = -\frac{e^2}{E}$$

$$Let\ \frac{e^2}{Er} = -\frac{1}{u} \quad Or,\ r = -\frac{e^2}{E}u \quad Or, u = -\frac{E}{e^2}r \quad \&,\ dr = -\frac{e^2}{E}du$$

$$If\ r = 0,\ u = 0\ and\ if\ r = -\frac{e^2}{E},\ u = 1$$

$$Therefore, \quad 2\sqrt{2mE}\int_0^1 \left(1 - \frac{1}{u}\right)\left(-\frac{e^2}{E}du\right) = \left(n + \frac{1}{2}\right)\pi\hbar$$

$$Or, \quad -\frac{2e^2}{E}\sqrt{2mE}\int_0^1 \sqrt{\left(\frac{-(1-u)}{u}\right)} du = (\mathrm{n} + \frac{1}{2})\pi\hbar$$

$$Or, \quad -i\,2e^2\sqrt{\frac{2m}{E}}\int_0^1 (1-u)^{\frac{1}{2}}\, u^{-\frac{1}{2}} du \; = (\text{n}+\frac{1}{2})\pi\hbar \qquad [\because \text{i}^2=-1,\;\; i=\sqrt{-1}]$$

$$Or, \quad -2ie^2\sqrt{\frac{2m}{E}}(1-u)^{\frac{3}{2}-1}\int_0^1 u^{\frac{1}{2}-1} du \; = (\text{n}+\frac{1}{2})\pi\hbar$$

$$Or, \quad -2ie^2\sqrt{\frac{2m}{E}}\,\beta\left(\frac{1}{2},\frac{3}{2}\right) = (\text{n}+\frac{1}{2})\pi\hbar$$

$$Or, \quad -2ie^2\sqrt{\frac{2m}{E}}\,\frac{\Gamma\left(\frac{1}{2}\right)\Gamma\left(\frac{3}{2}\right)}{\Gamma\left(\frac{1}{2}+\frac{3}{2}\right)} = (\text{n}+\frac{1}{2})\pi\hbar$$

$$Or, \quad -2ie^2\sqrt{\frac{2m}{E}}\,\frac{\sqrt{\pi}\,.\,\frac{\sqrt{\pi}}{2}}{\Gamma 2} = (\text{n}+\frac{1}{2})\pi\hbar$$

$$Or, \quad -2ie^2\sqrt{\frac{2m}{E}}\,.\,\frac{\pi}{2} = (\text{n}+\frac{1}{2})\pi\hbar \qquad [\because \Gamma 2 = 1! = 1]$$

$$Or, \quad -ie^2\sqrt{\frac{2m}{E}} = (\text{n}+\frac{1}{2})\hbar$$

$$Or, \quad -e^4\,\frac{2m}{E} = \left(\text{n}+\frac{1}{2}\right)^2 \hbar^2 \;\; [\textit{squaring both sides}]$$

$$Or, \qquad E = \frac{-2me^4}{(\text{n}+\frac{1}{2})^2\hbar^2} \quad Or, \quad E_n = -\frac{2me^4}{(\text{n}+\frac{1}{2})^2\hbar^2}$$

Exercise

1. What is WKB approximation [Wentzel–Kramers–Brillouin]? Discuss the importance of WKB [Wentzel, Kramer's and Brillouin] approximation method.
2. Obtain WKB approximation is a linear combination of two solutions.
3. Show that the WKB approximation is linear combination of two solutions

$$\Psi_+(x) = \frac{C_+}{\sqrt{|p(x)|}}\, exp\left(+\frac{i}{\hbar}\int p(x)\, dx\right)$$

And

$$\Psi_-(x) = \frac{C_-}{\sqrt{|p(x)|}}\, exp\left(-\frac{i}{\hbar}\int p(x)\, dx\right)$$

4. Define Classical turning points.
5. Derive the WKB connection formula for rising potential.
6. Derive the WKB connection formula for falling potential.
7. Establish the well-known Somerfield-Wilson quantization condition.
8. Show that the well-known Somerfield-Wilson quantization condition

$$\int_a^b k(x)\, dx = \left(n+\frac{1}{2}\right)\pi \text{ Or, } J = \oint p(x)dx = \pi\left(n+\frac{1}{2}\right)\hbar$$

9. Show that the Energy levels of the potential $V(x) = \frac{1}{2}kx^2$ for one dimensional harmonic oscillator by using WKB approximation method is

$$E = \left(n+\frac{1}{2}\right)\hbar\,\omega \text{ ; Where } n = 0, 1, 2, 3 \ldots\ldots\ldots\ldots\ldots$$

10. Determine the Energy levels of the potential $V(x) = \lambda x^4$ by using WKB approximation method.
11. Determine the Energy levels of the potential $V(x) = \frac{1}{2}kx^2$ for one dimensional harmonic oscillator by using WKB approximation method.
12. Calculate the energy levels of a ball bouncing off a perfectly reflecting plane in a uniform gravitational field. Given the energy levels of the potential

$$V(x) = Fx = mgx \ ; \ x > 0 \ and \ V(x) = \infty \ ; x \leq 0$$

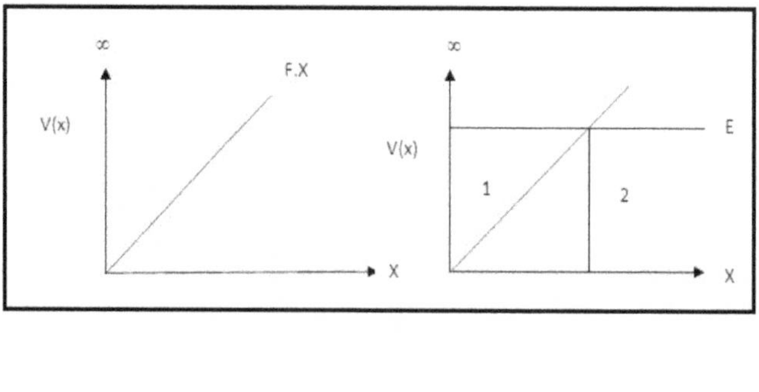

13. Determine the transmission and reflection coefficient of single potential barrier of general shape using WKB method.

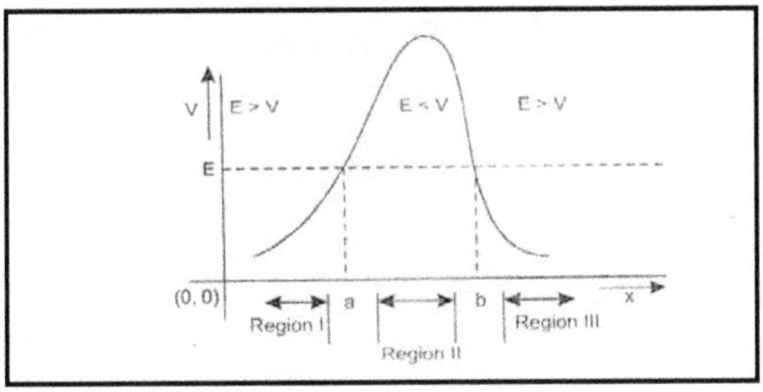

14. Explain the Validity of the WKB approximation method

15. Calculate the transmission coefficients through a barrier as shown in figure:

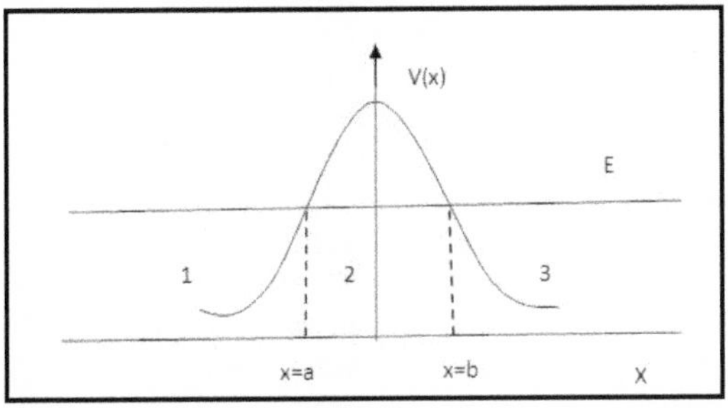

16. Calculate the energy Eigen value of the hydrogen atom by using Somerfield-Wilson quantization rule.
17. Calculate the energy Eigen values of the hydrogen atom using Sommerfeld –Wilson quantization condition.
18. Calculate transmission probability of a particle through a potential well with the help of WKB approximation method.
19. Use the Wilson- Sommerfeld quantization conditions to calculate the allowed energy level of a ball, bouncing elastically in a vertical direction.
20. Apply Wilson- Sommerfeld quantization conditions to a harmonic oscillator and show that its energy levels are discrete.
21. What is the Wilson- Sommerfeld quantization condition? Apply it to a particle in one dimensional box of length $e^{-19}cm$ and calculate first four energy levels in eV.

Citations

[1] Advance Quantum Theory — Paul Roman
[2] Quantum Mechanics — Eugen Merzbacher
[3] Quantum Mechanics — Leonard Schiff
[4] Quantum Mechanics — Messiah
[5] Quantum Mechanics — AM Harun ar Rashid
[6] Lectures on Physics — Feynmann
[7] Introduction to Quantum Mechanics — David J. Griffiths
[8] Quantum Mechanics — B.H. Bransden & C.J Joachain
[9] Quantum Mechanics — John L. Powell & Bernd Creasemann
[10] Quantum Mechanics — Ajoy Ghatak & S. Lokanathan
[11] A Text Book of Quantum Mechanics — P.M Mathews & K.Venkatesun
[12] Introduction to Quantum Mechanics — Linus Pauling & E. Right Wilson
[13] Quantum Mechanics — B.N. Srivastava
[14] Fundamentals of Quantum Mechanics — Y.R. Waghmare
[15] Quantum Mechanics — V.K. Thankapan
[16] Quantum Mechanics — Fraz Schwbal
[17] Quantum Mechanics — Gupta Kumar Sharma
[18] Classical Mechanics — Goldstein
[19] Solid State Physics — Charles Kittle
[20] Mathematical Physics — BD Gupta
[21] Mathematical Physics — Rajput
[22] Perspective of Modern Physics — Arthur Beiser
[23] Concept Modern Physics — Arthur Beiser
[24] Quantum Mechanics — Sujaul Chowdhury
[25] Advanced Quantum Mechanics [I&II] — S. Chowdhury
[26] Quantum Mechanics — S.P. Singh M.K. Bagde
[27] Quantum Mechanics — Sazedur Rahman
[28] Buttkus — Spectral Analysis and Filter theory in Applied Geophysics-@ Springer-Verlag Berlin, Heidelberg- 2000
[29] Quantum Mechanics — A. I. Lvovsky.
[30] Physics I — Robert Resnick and David Halliday
[31] Physics II — Robert Resnick and David Halliday

Deep Link attached at the end of my document.

Chapter 1 Development of Quantum Mechanics

Fig.[1.3]: The left side shows a potential well in which a particle can be trapped. The right side shows a particle attached to a spring. The particle is subject to the force due to the spring, but it can also be described by the force due to a potential well. **[Author's own work]**

Fig.[1.5]: Shows Max Planck's black body radiation
[http://www.canadaconnects.ca/_quantumphysics/photos/bb_vs_temp.jpg]

Fig.[1.6(a)]: Shows Quantum mechanics versus Classical mechanics [https://encrypted-
tbn0.gstatic.com/images?q=tbn:ANd9GcSDv1mbTABe3kQSqP1C1VjrnZOkQ1rliVQ-
oAl4pTjyiMWFShe_]

Fig.[1.6(b)]: Graphical presentation of Quantum mechanics versus Classical Mechanics
[https://i.stack.imgur.com/l2LIC.png]
[https://slideplayer.com/slide/16899934/97/images/6/Basics+of+Quantum+Mechanics+-
+The+Correspondence+Principle+-.jpg]

Fig.[1.9(a)]: Shows Energy distribution of black body radiation
[https://cnx.org/resources/59b25759762bc1bf6c57838ee7e7fff722a5092c/CNX_UPhysics_39_0
1_quartz.jpg] &
[https://www.google.com/search?q=energy+distribution+in+black+body+radiation&source=lnm
s&tbm=isch&sa=X&ved=0ahUKEwjpl5fey9rjAhUGqY8KHc4fBsIQ_AUIESgB&biw=1094&bi
h=506#imgrc=4fhO10T6gBJqxM:]

Fig.[1.9(b)]: Shows Temperature Dependence Specific heat of Solids
[https://www.google.com/search?q=temperature+dependence+specific+heat+of+solids&source=
lnms&tbm=isch&sa=X&ved=0ahUKEwjK8pH5ydrjAhVHLY8KHVtrAEAQ_AUIESgB&biw=
1094&bih=506#imgrc=ls65ybJrqCM0TM:] & [http://hyperphysics.phy-
astr.gsu.edu/hbase/Solids/imgsol/debyecurves.gif]

Fig.[1.18(a)]: Shows the Photoelectric Effect
[https://www.google.com/search?tbm=isch&q=photoelectric+effect&chips=q:photoelectric+effe
ct,g_1:diagram:cSjqDvLZE9U%3D&usg=AI4_-
kRTkryKGm0zdtrbAnCgekG2wWiskw&sa=X&ved=0ahUKEwjohsHStNrjAhVEvo8KHVmHB
r8Q4lYIKSgA&biw=1094&bih=506&dpr=1.25#imgrc=WiQXrndIx678bM:]

Fig.[1.18(b)]: Shows the Compton Scattering
[https://www.google.com/search?biw=1094&bih=506&tbm=isch&sa=1&ei=bg4_XezsM8m8vA
TTgZjwAQ&q=compton+effect&oq=ompton+effect&gs_l=img.1.0.0i10i24.738151.745807..75
0534...0.0..0.1274.3931.2j11j7-2......0....1..gws-wiz-img.......0i7i30j0i67j0.lgoNIJ-
UpH8#imgrc=lCwgkfSbGt5b-M:]

Fig.[1.19]: Shows Photo-Electric Effect and Einstein Photoelectron Energy [https://physicsfoundationsblog.files.wordpress.com/2017/01/download.png?w=1040] & [https://www.google.com/search?biw=1094&bih=506&tbm=isch&sa=1&ei=hCo_Xfy2JsP3vATOlqPYAg&q=photoelectric+effect&oq=+photoelectric+effect&gs_l=img.1.0.0i67j0l8j0i67.50924.50924..53524...0.0..0.129.129.0j1......0....1..gws-wiz-img.LrA1wsXYYOA#imgrc=R_ysNo1F9ZKSHM:]

Fig.[1.20]: Shows the Photo Electric Effect and the Compton Effect [https://encrypted-tbn0.gstatic.com/images?q=tbn:ANd9GcS_Jm8gwGzyCGCUK-aRP2YuRxkm1nkvxhhLzPg_LF_q0DDkdjTZ]

Fig.[1.24]: Shows Phase velocity and Group velocity [https://www.google.com/search?q=phase+velocity+and+group+velocity&source=lnms&tbm=isch&sa=X&ved=0ahUKEwiTnKDK0trjAhVdk3AKHcvLAUUQ_AUIEigC&biw=1094&bih=506#imgrc=jeZ-5nh8tKk-1M:] & [https://www.google.com/search?q=phase+velocity+and+group+velocity&source=lnms&tbm=isch&sa=X&ved=0ahUKEwiTnKDK0trjAhVdk3AKHcvLAUUQ_AUIEigC&biw=1094&bih=506#imgrc=rN3SQy9Ie2grvM:]

Chapter 2 Basic Concepts of Quantum Mechanics

Fig.[2.1]: Shows Particle is propagating along the positive X-direction [https://s3-us-west-2.amazonaws.com/courses-images/wp-content/uploads/sites/752/2016/09/26194428/wavelength.png]

Fig.[2.3]: Shows $|\psi(x,t)|^2 d^3x$ expresses the probability density in the volume element d^3x at time t. [https://ecee.colorado.edu/~bart/book/book/chapter1/gif/fig1_2_11.gif]

Fig.[2.6]: Two body system: Hydrogen atom [https://quantumanthropology2.files.wordpress.com/2012/03/picture-hydrogen-2.png]

Fig.[2.12]: Shows Bohr's quantization condition [https://www.google.com/search?q=de+broglie+and+wave+model&source=lnms&tbm=isch&sa=X&ved=0ahUKEwjJ36KF6drjAhVEpo8KHW-6B88Q_AUIESgB&biw=1094&bih=506#imgrc=JDJ1WZjgKrniMM:]

Fig.[2.13(a)]: Shows at the bottom $\Delta\, x\; small\; and\;\; \Delta\, p_x\; large$ [https://encrypted-tbn0.gstatic.com/images?q=tbn:ANd9GcRcS-yvgeSo81I3HMSQvjFYDunNLxc-duPreLtFTkSQ2QzLnAdyaQ]

Fig.[2.13(b)]: Shows on the top $\Delta\, x\; large\;\; and\;\; \Delta\, p_x\; small$ [https://encrypted-tbn0.gstatic.com/images?q=tbn:ANd9GcRcS-yvgeSo81I3HMSQvjFYDunNLxc-duPreLtFTkSQ2QzLnAdyaQ]

Fig.[2.14] Shows of Heisenberg Uncertainty relation [https://encrypted-tbn0.gstatic.com/images?q=tbn:ANd9GcRqsfPNOGE7SLHRi9PRxlTimjq0EZKA7l831JJ1nTNEuRVLEhTwuw]

Fig.[2.15]: Shows the correspondence principle by Niels Bohr [https://slideplayer.com/slide/16899934/97/images/6/Basics+of+Quantum+Mechanics+-+The+Correspondence+Principle+-.jpg]

Fig.[2.17]: Shows Heisenberg uncertainty principle (between energy and time) **[Author's own work]**

Fig.[2.18]: Shows Heisenberg uncertainty principle [https://encrypted-tbn0.gstatic.com/images?q=tbn:ANd9GcQ4DuTwx1-tP16Scsq_Y-Qy5FR5-STX7RLDA0QnKnoAgGKBUyWUCg]

Chapter 3 Mathematical Structure of Quantum mechanics

Fig. [3.37]: Plots of a typical wave packet $\psi(x)$ and its Fourier transform [https://readingpenrose.files.wordpress.com/2014/04/example-of-wave-packet.jpg] & [https://www.google.com/search?q=fourier+transform&source=lnms&tbm=isch&sa=X&ved=0ahUKEwiJwvGk-trjAhVHOI8KHbTEDiEQ_AUIESgB&biw=1094&bih=506#imgrc=nta04UY_4W2MZM:]

Fig. [3.41]: Illustration of the sifting property of the Dirac delta function [https://www.probabilitycourse.com/images/chapter4/Delta-Pulse_b.png] & [https://encrypted-tbn0.gstatic.com/images?q=tbn:ANd9GcRHNEWVa9CuED7IpQoHgdJ-bkenbH_lpfQI_KuISIest__PYXt4]

Chapter 4 Matrix formulation of Quantum Mechanics

Figure Nil

Chapter 5 Some application of Quantum mechanics: Solution of simple one dimensional problems

Fig.[5.1]: Shows a one dimensional single step Potential barrier **[Author's own work]**

Fig.[5.2]: Shows Quantum theory for one dimensional single step Potential barrier **[Author's own work]**

Fig.[5.3]: Shows A rectangular Potential barrier **[Author's own work]**

Fig.[5.4]: Shows Energy levels for one dimensional square-well potential of finite depth **[Author's own work]**

Fig.[5.5]: Shows a square-well with infinitely high sides **[Author's own work]**

Fig.[5.6]: Shows one dimensional potential well **[Author's own work]**

Fig.[5.7]: Shows a particle of mass m is confined within one dimensional potential well **[Author's own work]**

Fig.[5.8]: Shows an Attractive Square well Potential **[Author's own work]**

Chapter 6 Quantum Mechanics of Linear Harmonic Oscillator

Fig.[6.1]: Shows a Linear Harmonic Oscillator [L.H.O] [http://hyperphysics.phy-astr.gsu.edu/hbase/images/oscda.gif]

Fig.[6.4]: Sows Quantum Mechanically Harmonic Oscillator [http://hyperphysics.phy-astr.gsu.edu/hbase/quantum/imgqua/qhar.gif]

Fig.[6.6]: Shows Energy Eigen value and wave functions of one dimensional Linear Harmonic Oscillator[L.H.O.] [https://encrypted-tbn0.gstatic.com/images?q=tbn:ANd9GcQ_MTpmcro3C7yEwIQyGn4TEhDVCGqqE_x5MTtZ8Jd1UJT-rP2euw] & [http://hyperphysics.phy-astr.gsu.edu/hbase/quantum/imgqua/hoscom2.gif]

Fig.[6.7]: Shows non-degeneracy of Linear Harmonic Oscillator [L.H.O.] [http://hyperphysics.phy-astr.gsu.edu/hbase/quantum/imgqua/hoscwf3.gif]

Chapter 7 Atomic Orbitals of Hydrogen Atom

Fig.[7.1] Shows visible spectrum of atomic hydrogen [https://encrypted-tbn0.gstatic.com/images?q=tbn:ANd9GcSIDeoyIfut2oHDqIOOSXcy2iVFaO2GUMu4_8oeOFxENLFw1w0QQ]

Fig.[7.2]: Shows Neils Bohr Atomic Model [https://quantumanthropology2.files.wordpress.com/2012/03/picture-hydrogen-2.png] & [https://media.licdn.com/dms/image/C4E12AQHSt7sLh9aeVQ/article-inline_image-shrink_1500_2232/0?e=1569456000&v=beta&t=evWGJ9pnpM_MAgSBuWIKBtbtFRuBXngXcF33kKQfU5w]

Fig.[7.3]: Bohr-Sommerfeld orbits for n= 1; 2; 3 [https://cdn.britannica.com/s:500x350/33/6033-004-162F0D99.jpg]

Fig.[7.5(a)]: Types of atomic orbitals of Hydrogen atom [https://upload.wikimedia.org/wikipedia/commons/5/5b/D_orbital.gif]

Fig.[7.5(b)]: Types of atomic orbitals of Hydrogen atom [https://upload.wikimedia.org/wikipedia/commons/5/5b/D_orbital.gif]

Fig.[7.6.1]: Shows the wave functions for 1s and 2s orbital for atomic hydrogen. The 2s-function [scaled by a factor of 2] has a node at r = 2 Bohr

[https://files.mtstatic.com/site_4334/18830/0?Expires=1564461736&Signature=iE0763KmQqGaTJPvAVCras3hx9vc7pp8qPrJfroacHm10uH6FJNAK0DPoDwxby50PTvoGkFQjkPd6IqSu7PR7djdgScj8jLqEtQfsui1ZhAnG~UGE2N1gQuLbaSY~vdINA82Hsel9RFdt6PIUhgaBuvA66DT-sxQKYKM0vUIst4_&Key-Pair-Id=APKAJ5Y6AV4GI7A555NA]

Fig.[7.6.2]: Shows the Contour map of 1s orbital in the x, y plane **[Author's own work]**

Fig.[7.6.3]: Shows the Scatter plot of electron position measurements in hydrogen 1s orbital **[Author's own work]**

Fig.[7.6.4]: Shows Density $\rho(r)$ and RDF $D(r)$ for hydrogen 1s orbital [https://encrypted-tbn0.gstatic.com/images?q=tbn:ANd9GcR83_tESHohArc3V9su6-4VwzGO6joSnNyyGirkbQzoMA7-4gzULQ]

Fig.[7.7.1]: Shows Contour plot of $2p_Z$ orbital. Negative values are shown in red in the lower portion. Scale units in Bohrs [https://encrypted-tbn0.gstatic.com/images?q=tbn:ANd9GcS_j2elgjfZnV9ZHyo3_LYeptOSOnLgI6QoJ-1Vq150RS63OvCI]

Fig.[7.7.2]: Shows Contour plots of 3d orbitals [https://encrypted-tbn0.gstatic.com/images?q=tbn:ANd9GcTkBP7D5mKpFIJsGm5TN-uwb2FQ6wPZ7KruXPvGgPVRxtutq0C0]

Fig.[7.8]: Some radial distribution function[RDF] [https://encrypted-tbn0.gstatic.com/images?q=tbn:ANd9GcRTkjJABPks7nPWsrAgl3w6utEITYzUiI4Pgy2gj_ajkyC5AfsBwA] & [https://encrypted-tbn0.gstatic.com/images?q=tbn:ANd9GcQFLA6C-yH2VhbbED0IfQxt_o6vMFng6Mv9bnHI_2fax92QN2WfhQ]

Fig.[7. 13]The angular momentum quantum number refers to the shapes of orbitals [https://encrypted-tbn0.gstatic.com/images?q=tbn:ANd9GcRa7MaGvkPE8Q8mqQk60Y1bRNyPD4_CBS_cAtZj5jLAHzSQJIW73w]

Fig.[7. 14]The Spin angular momentum quantum number refers to the shapes of orbitals [http://web.fscj.edu/Milczanowski/psc/lect/Ch8/SLIDE96.jpg]

Chapter 8 Quantum mechanics of Hydrogen like atoms

Fig.[8.2]: Shows a comparison of Laboratory and centre of mass coordinates. [http://images.books24x7.com/bookimages/id_11656/fig37_01.jpg] & [https://d2vlcm61l7u1fs.cloudfront.net/media%2Ff63%2Ff63ae6e6-96cc-49fb-89ba-69791f106412%2FphpMEOx8J.png]

Fig.[8.3]: Energy levels of atomic hydrogen
[http://images.tutorvista.com/cms/images/101/hydrogen-spectrum.png] &
[http://spiff.rit.edu/classes/phys301/lectures/spec_lines/hyd_jump_down.gif]

Fig.[8.7.1]: Shows the wave functions for 1s and 2s orbital for atomic hydrogen. The 2s-function [scaled by a factor of 2] has a node at r = 2 Bohr
[https://files.mtstatic.com/site_4334/18830/0?Expires=1564461736&Signature=iE0763KmQqGaTJPvAVCras3hx9vc7pp8qPrJfroacHm10uH6FJNAK0DPoDwxby50PTvoGkFQjkPd6IqSu7PR7djdgScj8jLqEtQfsui1ZhAnG~UGE2N1gQuLbaSY~vdINA82Hsel9RFdt6PIUhgaBuvA66DT-sxQKYKM0vUIst4_&Key-Pair-Id=APKAJ5Y6AV4GI7A555NA]

Fig.[8.7.2]: Shows the degeneracy of Hydrogen atom
[http://www.sliderbase.com/images/referats/1106/image013.png]

Fig.[8.10]: Shows Volume element in terms of spherical polar coordinates [https://encrypted-tbn0.gstatic.com/images?q=tbn:ANd9GcQiov5wVXjzoiypCU0jwLfi63YrXZ88tZ3t4cj5vMZCp3rHne-fjA]

Fig.[8. 11]: Shows Average value $\langle r \rangle$ [http://hyperphysics.phy-astr.gsu.edu/hbase/quantum/imgqua/hravg.gif]

Chapter 9 Orbital angular momentum in Quantum mechanics

Fig.[9.1]: Shows the orbital angular momentum **[Author's own work]**
Fig.[9.7.1]: Shows the orbital angular momentum **[Author's own work]**
Fig.[9.7.2] Shows the Spherical polar coordinates **[Author's own work]**
Fig.[9.8.1]: Shows different orbitals of magnetic quantum number [?]
Fig.[9.8.2]: Shows there are 3 possible orientations of $\vec{L}$ **[Author's own work]**
Fig.[9.8.3]: Shows there are 5 possible orientations of $\vec{L}$ **[Author's own work]**

Chapter 10 Generalized Angular momentum in Quantum Mechanics

Fig.[10.7]: Shows addition of two angular momentum **[Author's own work]**

Chapter 11 Identical Particles and Spin

Fig. [11.17]: Shows scattering of identical particles in the centre of mass system **[Author's own work]**

Chapter 12 Quantum Dynamics: Schrödinger picture, Heisenberg picture and Dirac or Interaction picture

Figure Nil

Chapter 13 Quantum Theory of Scattering

Fig.[13.1]: Shows, the scattering of an incident beam. The particles scattered into the solid angle $d\Omega$ are received by the detector. **[Author's own work]**

Fig.[13.4]: Shows, Laboratory system and Centre of mass system [https://che.gg/2Kblebn]

Chapter 14 Relativistic Quantum Mechanics

Fig.[14.19(a)]: Shows Dirac's hole theory [https://upload.wikimedia.org/wikipedia/commons/thumb/9/9b/Dirac_sea.svg/1200px-Dirac_sea.svg.png] & [https://link.springer.com/chapter/10.1007/978-3-642-23728-7_1]

Fig.[14.19(b)]: Illustrating Principle of Dirac Current Positron Generator [https://tesla3.com/paul-dirac/]

Chapter 15 The Variational method: An approximation method

Fig.[15.3]: Two body system: Hydrogen atom [https://quantumanthropology2.files.wordpress.com/2012/03/picture-hydrogen-2.png]

Fig[15.4]: Shows many electron atom (*In atoms with more than one electron)*[https://quatr.us/chemistry/helium-atoms-elements-chemistry.htm]

Fig.[15.4a]: Shows many electrons atom: Helium atom [http://scientificsentence.net/Quantum_Mechanics/Helium.jpg]

Fig.[15.4b]: Shows many electrons atom: Helium atom in coordinate system [https://chem.libretexts.org/@api/deki/files/51421/HeAtomReduced.gif?revision=1&size=bestfit&width=499&height=247]

Fig.[15.5]: Shows the energy level of Helium atom [http://hyperphysics.phy-astr.gsu.edu/hbase/quantum/imgqua/qnenergy4.gif]

Chapter 16 Approximation Method: Perturbation Theory

Fig.[16.5]: Shows Fermi's Golden rule and states that the transition probability per unit. **[Author's own work]**
Fig.[16.9]: Shows the splitting of energy level in stark effect. **[Author's own work]**
[Fig.16.10]: illustrates the Zeeman Effect. [https://encrypted-tbn0.gstatic.com/images?q=tbn:ANd9GcQClfd76VzPN-23mNCwkNpYKkVUNt5KdSesv2hPd6b7N3wYCePEDA]

[Fig.16.11]: Shows the possible transitions between two levels. [https://encrypted-tbn0.gstatic.com/images?q=tbn:ANd9GcSYFvISFDN7DfCgxQKLM6qTVNIdjOS7LaIP0e7eVun8K2T0kU5r] And [https://casper.ssl.berkeley.edu/astrobaki/index.php/Zeeman_splitting]

Fig.[16.14a]: Shows two body systems like hydrogen atom. **[Author's own work]**
Fig.[16.14b]: Shows selection rules $\Delta\ell = \pm 1, \Delta m_\ell = 0, \pm 1$. **[Author's own work]**

Chapter 17 WKB approximation method

Fig. [17.4a]: Shows the classical turning point if $V$is a increasing function of x. **[Author's own work]**
Fig.[17.4b]: Shows the classical turning point if $V$is a decreasing function of x. **[Author's own work]**
Fig.[17.4c]: Shows classical turning points ($a$and b) for a particle of total energy E encountering a single potential barrier of general shape. **[Author's own work]**
Fig.[17.4d]: Shows classical turning points ($a$and b) for a particle of total energy E residing in the quantum well of general shape. **[Author's own work]**
Fig. [17.5a]: Shows the classical turning point if $V$is a increasing function of x. **[Author's own work]**
Fig.[17.5b]: Shows WKB connection formula for falling potential I.e., the classical turning point if $V$is a decreasing function of x. **[Author's own work]**
Fig. [17.6]: Shows WKB connection formula for falling potential I.e., the classical turning point if $V$is a decreasing function of x. **[Author's own work]**
Fig.[17.7a]: Shows a simple well-shaped potential with two classical turning points. **[Author's own work]**
Fig. [17.7b]Shows: Phase space representation of the periodic motion of a particle confined between the classical turning points $x = a$ and $x = b$. **[Author's own work]**
Fig.[17.8]: Shows is a parabolic potential well. **[Author's own work]**
Fig.[17.9]: Shows the Energy levels for a given potential $V(x) = \lambda x^4$. **[Author's own work]**
Fig.[17.10]: Shows Energy levels of a ball bouncing off a perfectly reflecting plane in a uniform gravitational field. **[Author's own work]**
Fig.[17.11]: Shows a single potential barrier of general shape. **[Author's own work]**
Fig.[17.13]: Shows a potential barrier. **[Author's own work]**